PLANT HORMONES IN CROP IMPROVEMENT

PLANT HORMONES IN CROP IMPROVEMENT

Edited by

M. IQBAL R. KHAN
Department of Botany, Jamia Hamdard, New Delhi, India

AMARJEET SINGH
National Institute of Plant Genome Research, New Delhi, India

PÉTER POÓR
*Department of Plant Biology, Faculty of Science and Informatics,
University of Szeged, Szeged, Hungary*

ACADEMIC PRESS
An imprint of Elsevier

ELSEVIER

Academic Press is an imprint of Elsevier
125 London Wall, London EC2Y 5AS, United Kingdom
525 B Street, Suite 1650, San Diego, CA 92101, United States
50 Hampshire Street, 5th Floor, Cambridge, MA 02139, United States
The Boulevard, Langford Lane, Kidlington, Oxford OX5 1GB, United Kingdom

Notices

Knowledge and best practice in this field are constantly changing. As new research and experience broaden our
understanding, changes in research methods, professional practices, or medical treatment may become necessary.

Practitioners and researchers must always rely on their own experience and knowledge in evaluating and using any
information, methods, compounds, or experiments described herein. In using such information or methods they should be
mindful of their own safety and the safety of others, including parties for whom they have a professional responsibility.

To the fullest extent of the law, neither the Publisher nor the authors, contributors, or editors, assume any liability for any
injury and/or damage to persons or property as a matter of products liability, negligence or otherwise, or from any use or
operation of any methods, products, instructions, or ideas contained in the material herein.

ISBN: 978-0-323-91886-2

For Information on all Academic Press publications
visit our website at https://www.elsevier.com/books-and-journals

Publisher: Nikki P. Levy
Acquisitions Editor: Nancy J. Maragioglio
Editorial Project Manager: Kyle Gravel
Production Project Manager: Sajana Devasi P. K.
Cover Designer: Christian J. Bilbow

Typeset by MPS Limited, Chennai, India

Working together
to grow libraries in
developing countries

www.elsevier.com • www.bookaid.org

Contents

4. Role of phytohormones in regulating agronomically important seed traits in crop plants   65

Rubi Jain, Namrata Dhaka, Pinky Yadav and Rita Sharma

5. Phytohormone signaling in osmotic stress response   89

Riddhi Datta, Ananya Roy and Soumitra Paul

6. Role of phytohormones in plant response to drought and salinity stresses   109

Tanushree Agarwal and Sudipta Ray

7. Regulation of plants nutrient deficiency responses by phytohormones   129

Deepika Deepika, Kamankshi Sonkar and Amarjeet Singh

Bhuvnesh Kapoor, Pankaj Kumar, Rajnish Sharma and Mohammad Irfan

Ágnes Szepesi and Péter Poór

Harmanjit Kaur, Tashima, Sandeep Singh and Sofi J. Hussain

List of contributors

Tanushree Agarwal Department of Botany, University of Calcutta, Kolkata, West Bengal, India

Iffat Zareen Ahmad Natural Product Laboratory, Department of Bioengineering, Integral University, Lucknow, Uttar Pradesh, India

Ankit Ankit National Institute of Plant Genome Research, New Delhi, India

Mohammad Israil Ansari Department of Botany, University of Lucknow, Lucknow, Uttar Pradesh, India

Shamim Akhtar Ansari Institute of Forest Research and Productivity, Ranchi, Jharkhand, India

Paramita Bera National Institute of Plant Genome Research (NIPGR), New Delhi, Delhi, India

Halidev Krishna Botta Lab 203, National Institute of Plant Genome Research, New Delhi, Delhi, India

Zalán Czékus Department of Plant Biology, Faculty of Science and Informatics, University of Szeged, Szeged, Hungary

Riddhi Datta Department of Botany, Dr. A.P.J. Abdul Kalam Government College, Kolkata, West Bengal, India

Deepika Deepika National Institute of Plant Genome Research, New Delhi, India

Loitongbam Lorinda Devi National Institute of Plant Genome Research, New Delhi, India

Namrata Dhaka Department of Biotechnology, Central University of Haryana, Mahendergarh, Haryana, India

Sandhya Hora Department of Molecular and Cellular Medicine, Institute of Liver and Biliary Sciences, New Delhi, India

Sofi J. Hussain Department of Botany, Central University of Kashmir, Ganderbal, Jammu and Kashmir, India

Nadeem Iqbal Department of Plant Biology, Faculty of Science and Informatics, University of Szeged, Szeged, Hungary

Mohammad Irfan Plant Biology Section, School of Integrative Plant Science, Cornell University, Ithaca, NY, United States

Rubi Jain School of Computational and Integrative Sciences, Jawaharlal Nehru University, New Delhi, India

Gábor Jakab Department of Plant Biology, Faculty of Sciences, University of Pécs, Pécs, Hungary

Syed Uzma Jalil Amity Institute of Biotechnology, Amity University Uttar Pradesh, Lucknow Campus, Lucknow, Uttar Pradesh, India

Saravanappriyan Kamali National Institute of Plant Genome Research, New Delhi, India

Bhuvnesh Kapoor University Institute of Biotechnology (UIBT), Chandigarh University, Mohali, Punjab, India

Harmanjit Kaur Post Graduate Department of Botany, Government College for Girls, Ludhiana, Punjab, India

Pankaj Kumar Department of Biotechnology, Dr. Yashwant Singh Parmar University of Horticulture and Forestry, Solan, Himachal Pradesh, India

Pritha Kundu National Institute of Plant Genome Research (NIPGR), New Delhi, Delhi, India

Ashverya Laxmi Lab 203, National Institute of Plant Genome Research, New Delhi, Delhi, India

Gyöngyi Major Department of Plant Biology, Faculty of Sciences, University of Pécs, Pécs, Hungary

Shruti Mishra National Institute of Plant Genome Research (NIPGR), New Delhi, Delhi, India

Attila Ördög Department of Plant Biology, Faculty of Science and Informatics, University of Szeged, Szeged, Hungary

Priti Pal Shri Ramswaroop Memorial Group of Professional Colleges, Lucknow, Uttar Pradesh, India

Soumitra Paul Department of Botany, University of Calcutta, Kolkata, West Bengal, India

Péter Poór Department of Plant Biology, Faculty of Science and Informatics, University of Szeged, Szeged, Hungary

Priyanka Prajapati Centre for Advanced Studies in Botany, Department of Botany, Institute of Science, Banaras Hindu University, Varanasi, Uttar Pradesh, India

Sudipta Ray Department of Botany, University of Calcutta, Kolkata, West Bengal, India

Ananya Roy Department of Botany, University of Calcutta, Kolkata, West Bengal, India

Harshita B. Saksena Lab 203, National Institute of Plant Genome Research, New Delhi, Delhi, India

Manvi Sharma Umeå Plant Science Centre, Department of Forest Genetics and Plant Physiology, Swedish University of Agricultural Sciences UMEÅ, Sweden

Mohan Sharma Lab 203, National Institute of Plant Genome Research, New Delhi, Delhi, India; Institute of Biology III, Albert-Ludwigs-University Freiburg, Schänzlestraße, Freiburg im Breisgau, Germany

Rajnish Sharma Department of Biotechnology, Dr. Yashwant Singh Parmar University of Horticulture and Forestry, Solan, Himachal Pradesh, India

Rita Sharma Department of Biological Sciences, Birla Institute of Technology & Science, Pilani, Rajasthan, India

Amar Pal Singh National Institute of Plant Genome Research, New Delhi, India

Amarjeet Singh National Institute of Plant Genome Research, New Delhi, India

Devendra Singh Department of Biotechnology, Motilal Nehru National Institute of Technology Allahabad, Allahabad, Uttar Pradesh, India

Sandeep Singh Department of Botany, Kanya Maha Vidyalaya, Jalandhar, Punjab, India

Sneha Singh Department of Phytochemistry, Council of Scientific and Industrial Research (CSIR)—Central Institute of Medicinal and Aromatic Plants, Lucknow, Uttar Pradesh, India

Kamankshi Sonkar National Institute of Plant Genome Research, New Delhi, India

Ágnes Szepesi Department of Plant Biology, Faculty of Science and Informatics, University University of Szeged, Szeged, Hungary

Tashima Department of Botany, Akal University, Bathinda, Punjab, India

Jyothilakshmi Vadassery National Institute of Plant Genome Research (NIPGR), New Delhi, Delhi, India

Kanchan Vishwakarma Department of Microbial Technology, Amity Institute of Microbial Technology, Amity University, Noida, Uttar Pradesh, India

Gaurav Yadav Natural Product Laboratory, Department of Bioengineering, Integral University, Lucknow, Uttar Pradesh, India

Pinky Yadav Department of Biotechnology, Central University of Haryana, Mahendergarh, Haryana, India

Ritesh Kumar Yadav National Institute of Plant Genome Research, New Delhi, India

Regulatory role of phytohormones in plant growth and development

Priti Pal[1], Shamim Akhtar Ansari[2], Syed Uzma Jalil[3] and Mohammad Israil Ansari[4]

[1]Shri Ramswaroop Memorial Group of Professional Colleges, Lucknow, Uttar Pradesh, India [2]Institute of Forest Research and Productivity, Ranchi, Jharkhand, India [3]Amity Institute of Biotechnology, Amity University Uttar Pradesh, Lucknow Campus, Lucknow, Uttar Pradesh, India [4]Department of Botany, University of Lucknow, Lucknow, Uttar Pradesh, India

1.1 Introduction

Phytohormones are plant-produced chemical substances that perform diverse regulatory effects on plant growth and improvement. In reaction to their environment, all plants produce hormones that regulate metabolism, growth, and development (i.e., plant shape, size, and function) (Salehin et al., 2019). Various plant parts, for example, auxin in shoot apices and kinetin in root apices, synthesize phytohormones transported to their action site and behave as chemical messengers to bind to receptors and trigger a specific response in particular cells/organs (Mitchell, 1942; Rademacher, 2015). Phytohormones control cell shape and function, cell division (cell number), cell expansion (cell size), organ shedding (cell differentiation), sex determination, roots, blooming, fruiting, dormancy, and germination as well.

Phytohormones modulate plant responses, such as target cells/tissues, nutrition and water availability, plant developmental stage, absorption, and storage, for better adaptation and survival under stress conditions that enhanced crop productivity (Ferguson and Grafton-Cardwell, 2014). Many phytohormones work together to control a specific developmental phase, while a single kind of phytohormone can be engaged in multiple developmental processes. Nevertheless, phytohormones elude precise quantification in tissues/organs due to their low endogenous concentrations (Rademacher, 2015).

Consequently, researchers resort to exogenous application of their synthetic analogs called plant growth regulators (PGRs) to investigate roots, fruits, shoots, buds, and flowers (Flasinski and Hac-Wydro, 2014). PGRs like strigolactones (SLs) and peptide phytohormones have recently gained a lot of interest. For decades, phytohormone research has focused on auxins, cytokinin (CK), gibberellin (GA), abscisic acid (ABA), and ethylene (ET). Several compounds of low molecular weights like brassinosteroid (BR), jasmonic acid (JA), salicylic acid (SA), SLs, and peptides have been identified to perform regulatory roles. BRs influence plant development and adaptation. Because they are active in plant responses to a range of stressors (Northey et al., 2016). JA and SA are usually referred to as resistance-related phytohormones. Different phytohormones regulate plant growth differently, yet in many cases, two or more phytohormones synergistically perform the same function and generate the same response. Phytohormones play a critical role in plants' metabolism, defense, and signaling activities and have additionally been utilized in agriculture, viticulture, and horticulture to increase crop yield and make cultivation simpler in less-than-ideal or stressful conditions (Rademacher, 2015). Many researchers strive to create effective techniques to reduce abiotic stresses and improve hormone action mechanisms to improve crop output in the future.

1.2 Phytohormones and their impact on plant growth and development

Darwin and Darwin (1881) were the first to suggest that certain chemicals, phytohormones, can help plants grow and develop more quickly. Phytohormones are endogenous chemical messengers playing critical roles in various physiological and biological systems in higher plants (Fig. 1.1). A single phytohormone can govern a vast range of cellular and developmental processes, even though numerous hormones may be involved in controlling a specific activity simultaneously (Gray, 2004). Phytohormones like indole-3-acetic acid (IAA), ET, GA, ABA, CKs, SA, BR, JA, and the recently discovered SLs help plant growth and development, signaling and cross talk (Bücker-Neto et al., 2017). Phytohormones take cues from endogenous signals and environmental changes to balance various metabolic pathways that ultimately benefit plants' survival and proper growth and development (Borghi et al., 2015).

1.2.1 The significance of auxin in plant growth and development

Auxins are low-molecular-weight natural phytohormones that govern many plant growth and development aspects, including morphogenesis, cell division, and elongation (Saini et al., 2013; George et al., 2008). Auxins are hydrophilic chemical compounds that may move long distances via the vascular tissue (George et al., 2008). The research synthesized auxins in meristematic and flourishing organs (i.e., developing seeds, leaves, and buds). Expanding leaves and buds have the largest auxin concentrations, whereas the roots have the lowest concentrations (George et al., 2008). As an outcome, auxins are more abundant in younger plants (such as seedlings and juveniles) and show a more significant role in plant development throughout the early stages. Auxins are responsible for phototropism, apical dominance, lateral root development, vascular formation, and gravitropism (Rademacher, 2015). Auxins regulate plant development during abiotic stress adaptation in concurrence with ABA and SA. It is thought that foliar application of natural and artificial auxins to vitally growing plants improves physiological mechanisms that drive plant

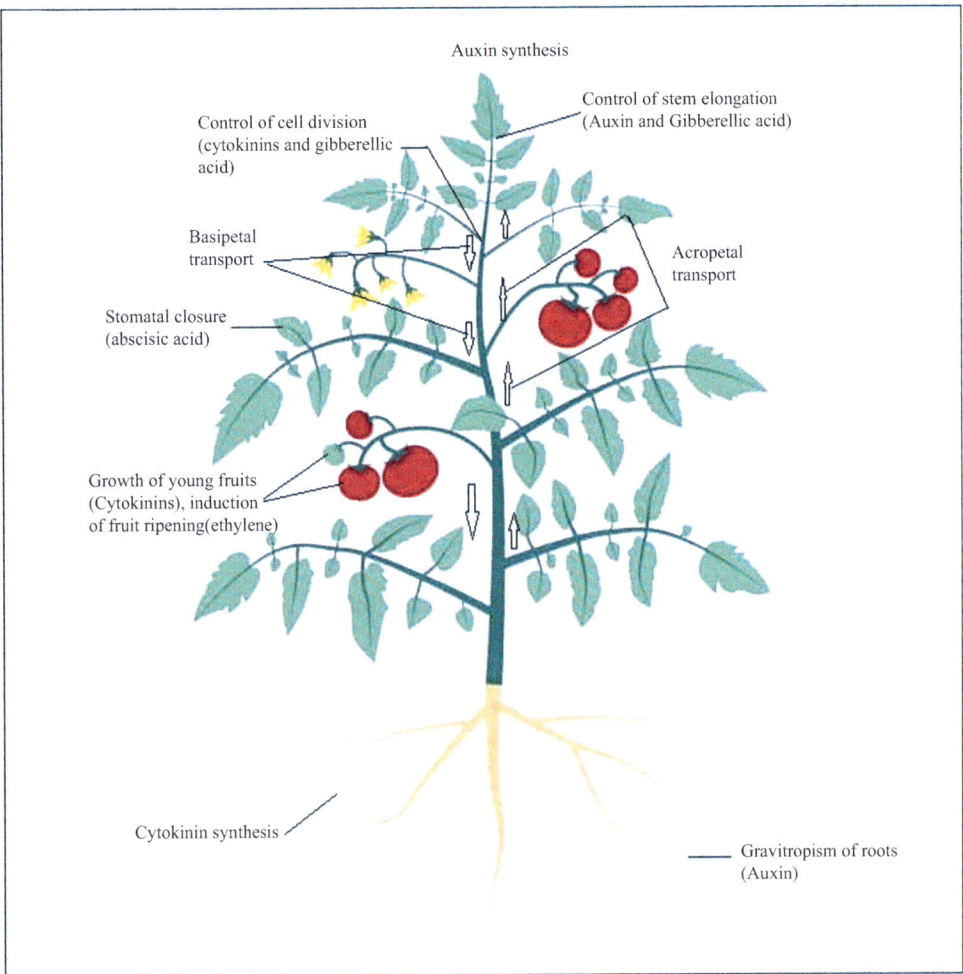

FIGURE 1.1 Phytohormones' role in plant growth and development.

development. Exogenous applications of auxins to plant cuttings help induction and grow adventitious roots (Harms and Oplinger, 1988).

Further, stimulatory response to auxin application depends upon plant species, absorption, and transport rates to target tissue, endogenous auxin levels, auxin sensitivity, metabolic rate, and hormone interactions (George et al., 2008). Even though auxins are essential for plant viability, excessive levels can result in oxidative stress and cell death (Flasinski and Hac-Wydro, 2014). Auxins are produced by bacteria or derived from plants; natural auxin decays quickly, restricting their use (Reich, 2013). The least stable form is indole-3-acetic acid (IAA), which degrades fast in sunlight and is destroyed by the enzyme IAA-oxidase in plants (George et al., 2008). Because artificial auxins do not get metabolized within plant tissues, they are thought to be more efficient than native auxin (Reich, 2013).

1.2.2 The significance of cytokinins in plant growth and development

CKs are frequently regarded as master controllers. The term "CKs" refers to a category of nonuniform chemicals that are adenine derivatives with an N^6 substitution (Wang and Irving, 2011). The highest concentrations have been found in meristematic tissues, fresh leaves, seeds, and fruits in plants (Sakakibara, 2010). CKs stimulate lateral plant growth and the formation of lateral shoots and roots and counteract auxins in apical dominance (Rademacher, 2015). CKs have been shown to stimulate cell division, protein synthesis, cell elaboration, senescence, and amino acid transport in plants, as well as to boost shoot generation from internodes, chloroplast development, the release of buds from dormancy, the formation of callus, and the spreading of heavy roots (George et al., 2008). CKs promote root growth and increase the rate of root development (Carrow and Duncan, 2012). Plants' responses to light, nutrition, water availability, abiotic stresses, and biotic challenges are all controlled by CKs. They prevent cellular disintegration, boost protective enzymes like antioxidants, and signal protein production (Carrow and Duncan, 2012). Adding organic or artificial CKs to a plant helps mobilize nutrients to a particular application site (Werner et al., 2001). When CKs are applied directly to plant leaves, they retard the rate of chlorophyll and protein breakdown. Exogenously administered CKs enhance plant growth via boosting nitrogen consumption.

The quantities of CKs in plants have been recommended as a stress indicator. Auxins (cell cycle initiators that inhibit DNA replication) and CKs (cell division simulators that inhibit mitosis) are thought to be essential for plant cell division.

1.2.3 Gibberellic acid and its significance in plant growth and development

The phytohormone GA belongs to the tetracyclic diterpenoid acid family. GA was discovered for the first time in *Gibberella fujikuroi* (disease-causing stupid seedling). A large amount of GA produced in plants causes the plant to become long and slender (often damaged due to its inability to support its weight). They also become chlorotic and essentially unproductive (Binenbaum et al., 2018). Although over 130 GAs have been found in plants, bacteria, and fungi, only four (GA_1, GA_3, GA_4, and GA_7) exhibit bioactive properties as phytohormones. In addition, GAs has been recognized as a harmful protein (Gupta and Chakrabarty, 2013). GA is mainly synthesized near the place of activity and transported long distance (Hedden and Sponsel, 2015). GA is a hormone that has been linked to many plant activities, for example, seed maturation, organ extension, cell enlargement and growth, and flower, fruit, and seed improvement (Binenbaum et al., 2018). GAs facilitates the shift throughout developmental stages and is also necessary for plants' adaptive response to abiotic stress (Colebrook et al., 2014).

Generally, three GA transporters are involved in importing bioactive GA molecules. One is NPF3.1, which has been discovered to cause plasma membrane influx and operates as a putative transporter in the roots (endodermal cells) (Tal et al., 2016). In addition, the endodermis in roots is the site of GA response. When the function was lost, GA phenotypes were not observed in NPF3.1 plants, attributable to functional continuance (Tal et al., 2016). Another is NPF2.10/GTR1, which has been found to be expressed in elongating stamen filaments (Jackson, 1993). This is because mutant GA transporters (gtr) are badly affected during filament elongation and dehiscence, but the phenotype is restored after treatment with this phytohormone. NPF2.10/GTR1 is important in transferring glucosinolates and defense chemicals (mainly belonging to the Brassicaceae family). Moreover, it can only import GA_3 and not traffic GAs in their bioactive forms. According to Saito et al. (2015), failure of function in this transporter produces a problem in stamen formation and a fall in fertility rate due to a defect in GAs transportation. Gtr1 is also involved in the transport of jasmonyl isoleucine into oocytes and the formation of stamens (Reganult et al., 2015). NPF4.1/AIT3, the third transporter, assists in the import of bioactive GAs. NPF4.1/AIT3 is involved in the transfer of GAs GA_1, GA_3, and GA_4. It is vital to note that GA, like auxin, has a similar transport mechanism that restricts the capability to exit cells through the ion trap procedure (Kramer, 2006). A total of 45 out of 53 NPF family members may import GAs through the yeast membrane (Chiba et al., 2015). GA transfer from a photosynthetic resource to a sink involves a range of processes, including floral transition, internode elongation, and secondary growths in stems (Eriksson et al., 2006).

1.2.4 The significance of abscisic acid in plant growth and development

The significance of ABA in regulating and integrating stress signals and responses in changing environmental and physiological situations is well documented (Tuteja, 2007). When plants are exposed to inorganic environmental factors, such as drought, higher and lower temperatures, salt, and flooding, ABA synthesis enhances plant acclimatization and stress tolerance. Two phytopathogenic fungi that produce ABA are *Cercospora rosicola* and *Botrytis cinerea*. ABA synthesized by *B. cinerea* is currently commercially isolated for exogenous application to plants (Rademacher, 2015). ABA production in plants is usually connected to decrease plant development in general. Changes in cellular membrane permeability and water/nutrient accessibility intake are also conceivable. The reduction in leaf stomatal opening during drought is hypothesized to be caused by ABA, resulting in reduced intercellular dehydration and transpiration (Rademacher, 2015; Tuteja, 2007). To accomplish this, ABA causes guard cells to activate, which seal the stomata (Tuteja, 2007). ABA helps seedlings overcome stress and germinate when growth conditions are favorable (Tuteja, 2007). In the absence of ABA, plants subjected to continuous environmental stress droop grow slowly and finally die (Tuteja, 2007). Roots control water, ion absorption, leaf abscission, leaf senescence, and seedling growth; biosynthesis of storage proteins and lipids; pathogen defense; morphogenesis and tissue growth; and modulation of embryo development and maturation gene expression are all activities of ABA (Flasinski and Hac-Wydro, 2014).

1.2.5 The significance of ethylene in plant growth and development

ET is a well-known phytohormone involved in ripening fruits, flower inflorescence, abscission of leaf and petal, and many other plant growth and development processes. Because of its involvement in plant maturation, ET is termed an "aging" phytohormone. In numerous plants, ET synthesis has been connected to cell size control and is a byproduct of rapidly proliferating cells (George et al., 2008). ET is a familiar phytohormone that plays a role in fruit ripening, flower inflorescence, leaf and petal abscission, and other plant growth and development processes. S-Adenosyl-L-methionine (AdoMet) and 1-aminocyclopropane-1-carboxylic acid (ACC) are two essential precursors used for the synthesis of ET. AdoMet is transformed to ACC by ACC synthase and converted to ET by ACC oxidase. This phytohormone influences physiological phenomena, including growth and aging. The first study of ET's influence on plants was observed in pea (etiolated sprouts), which resulted in growth retardation. ET acts as a growth inhibitor (Han et al., 2018). ET is required for seed germination, cell expansion, flowering, fruiting, senescence, abscission, and so on. ET aids leaf development and elongation, according to Dubois et al. (2018). Fiorani et al. (2002) showed that, in *Poa alpina* and *Poa compressa*, even a low concentration of ET causes an increase in leaf elongation rates. Being a gas by nature, ET can be easily transported long distances via aerenchyma or intercellular voids by cell-to-cell diffusion. ACC can move from the lower end of plants (roots) to the upper ends (shoots) with the help of xylem when plants are exposed to a hypoxic condition, which is predominant at the time of flooding (Van de Poel and Van der Straiten, 2014).

1.2.6 The significance of brassinosteroids in plant growth and development

BRs are steroid hormones present across the plant kingdom that possess the capability to promote growth and development in plants (Nazir et al., 2019, 2021, 2022). Dwarf rice mutant d2 with a typical phenotype of leaf inclination reduces upright leaf angle or left inclination, followed by restoration to its natural shape on receiving exogenous BRs (Hu et al., 2019). As a kind of BR, EBL (2,4-epibrassinolide) can protect against various biotic and abiotic stressors. After EBL treatment, the rate of superoxide anion formation in apricot fruit is drastically reduced. It also slows the formation of apricot fruit plaques and lowers malondialdehyde (MDA) levels and cell membrane permeability (Shi et al., 2019). According to transmission electron microscopy, EBL treatment preserves organelles' structural efficiency in apricot fruit, like mitochondria and chloroplasts throughout the deposit. EBL likewise mediates the reaction to heavy metal stress. Lead poisoning immediately impacts cell metabolism, resulting in damage to the antioxidant enzyme defense mechanism and free radical toxicity. The application of BRs according to Soares et al. (2020) restores the impacts of lead stress on seed germination and seedling development in *Brassica juncea* (L.). By increasing antioxidant enzymes such as superoxide dismutase (SOD), catalase (CAT), peroxidase (POD), and others, scientists have recently made fresh discoveries about BRs to promote plant growth and improvement (Zhang et al., 2020). Li et al. (2020) have discovered that starch in wild-type plants decomposes swiftly in light. In contrast, starch concentration improves in light in BRs-deficient and insensitive mutants, preventing stomata from opening usually.

1.2.7 The significance of jasmonates in plant growth and development

The jasmonates (JAs) regulate seed maturation, callus growth, primary root growth, blooming, gum and bulb establishment, and senescence, among other plant developmental processes. JA synthesis occurs in the leaves, and proof of a similar pathway in the roots has been identified. Furthermore, it is assumed that cellular organelles, including chloroplasts and peroxisomes, are the significant sources of JA production (Cheong and Choi, 2003). JA notably induces plant defense responses to different biotic and abiotic stressors. The liquid and vapor transition phases of JAs in plants impact a wide range of plant growth and development processes. In other studies, plants subjected to insect attack and disease enhance the JA concentrations (Hamayun et al., 2010). Plants produce JAs under abiotic and biotic stress as well as physical injury; instantaneous JA production in reaction to wounds during biotic stress benefits plants to help protect themselves from insects and diseases by producing systemic signaling molecules that interact with plasma membrane receptors (Creelman et al., 1997). The synthesis or exogenous administration of JAs results in the production of antifungal proteins, which improves fungus resistance. JAs have both stimulatory and suppressive effects on plants. Either exogenous or endogenous JAs enhance senescence, accelerate, or restrict plant development, thereby assisting stress management.

1.2.8 The significance of salicylic acid in plant growth and development

SA regulates plant growth and development, relationships with other organisms, and defense responses to external defense (Bastam et al., 2013). It plays an influential role in seed maturation, glycolysis, flowering, fruit production, ion absorption and transport, photosynthetic rate, stomatal conductance, transpiration, thermo-tolerance, senescence, and nodulation. SA also regulates biotic stress responses. SA has been shown to play a role in pathogenesis, systemic acquired resistance (SAR), and hypersensitive response gene expression (Shah, 2003). According to Hao et al. (2011), SA-induced proteins have been linked to a wide range of plant processes, such as protein folding, synthesis, antioxidative responses, and photosynthesis. The phytohormone regulates crop development, productivity, and yield by increasing leaf area and dry matter in crops, such as soybean, maize, and others (He et al., 2014). Seeds that have been pretreated with phytohormones have boosted germination and growth (Shakirova, 2007). A low-level dose of this phytohormone increases plant dry matter, whereas a high level causes a decrease in dry matter (Muthulakshmi and Lingakumar, 2016).

1.2.9 Strigolactones and their involvement in plant development

SLs are synthesized in roots and branches from a carotenoid seed germination stimulator. They are squeezed out from roots to induce hyphal branching, and they help regulate shoot development. Xie et al. (2010) revealed that this phytohormone transporter might be a member of the ABCG transporter family as it also functions in abiotic stress conditions. According to Borghi et al. (2015), a G-type ABC transporter called pleiotropic drug resistance discovered in a *Petunia* plant plays a significant role in enhancing SL release from a variety of plant tissues, including root cells, vascular tissues, and nodal tissues. This resistance gene also accumulates in the hypodermal cells of roots, where it facilitates acropetal trafficking of SLs and mycorrhizal fungus, Transporters, and plant osmotic stress. Also, the pdr1 mutant gene helps in lateral branching of the plant but does not help in SLs exudation in soil. This causes a reduction in symbiotic association (Kretzschmar et al., 2012). SLs stimulate plant growth and development and strengthen the symbiotic relationship with arbuscular mycorrhizal fungi, particularly the proliferation of fungal hyphae (Akiyama et al., 2005).

Furthermore, SLs limit the expansion of axillary buds, thereby essentially inhibiting shoot branching. Various genes encoding axillary bud development promote SL signaling. Additional axillary growth genes (MAX) include MAX1/AT2G26170 cytochrome, MAX3/AT2G44990, MAX2/AT2G42620 cytochrome, P450 monooxygenases, and MAX4/AT4G32810 that work in different ways and are mainly depending on the plant (Alder et al., 2012).

1.2.10 The significance of triazoles in plant growth and development

This class of chemicals can be categorized as PGRs or fungicides, but they have various degrees of both abilities. Plants can defend themselves against different types of biotic and abiotic stressors. Uniconazole is the most effective TR found in decreasing stressors in plants; however, its nonstop action has practical relevance for managing salinity in agriculture. TRs involved in the morphological response in plants as they are more potent than most other growth retardants and relatively low rates are required to inhibit shoot growth. However, even at high application rates, they generally are not phytotoxic. The most pronounced effect of TRs on plants is a reduction in height, with the treated plants being greener and more compact. The TRs show physiological response in plants. The foliage of triazole (TR)-treated plants typically exhibits intense dark green color compared to untreated controls. In most cases, this is due to enhanced chlorophyll content and more well-packed chloroplasts within a smaller leaf area. TR enhanced the photosynthesis rate. The effects of different concentrations of TRs [triademefon (TDM) and paclobutrazol (PBZ)] on net photosynthetic rate, transpiration rate, and stomatal resistance were studied in *Setaria italica* plants grown under field conditions.

1.3 Phytohormones cross talk and regulation of biotic and abiotic stresses

The interaction of phytohormones is critical in defining the type of response (Fig. 1.2). By boosting Ca^{2+} concentration and acting as a secondary messenger intracellularly, the recently discovered phytohormones, such as JA and BRs, play a crucial role in managing biotic and abiotic stressors. Under abiotic conditions, such as drought, cold, salinity, and heat stress, inositol phosphate is generated along with reactive oxygen species (ROS), which increases endogenous ABA levels (Leshem et al., 2007). Other plant defensive phytohormones, SA, JA, and ET, are vital in the biotic stress response, and their concentration rises quickly when pathogens infect the plant (Bari and Jones, 2009).

Short- and long-term growth responses, such as stomata closure to maintain water balance and stress-responsive gene expression, are aided by ABA. The interaction of ABA and SA aids plants in maintaining their water by regulating stomatal conductance, osmotic adjustment, leaf senescence, and photo-assimilate allocation. ET interacts with ABA during drought by reducing its function via modulating stomatal aperture, root, and shoot growth (Gazzarrini and Mccourt, 2003). ABA counteracts ET's effects on hyponasty and adventitious root development (Valluru et al., 2016). Tanaka and Fukuda (2006) revealed that JA and GA interactions are synergistic and are involved in various metabolic activities, including trichome initiation and sesquiterpenoid biosynthesis, while auxin promotes the WRKY23 gene, which controls root

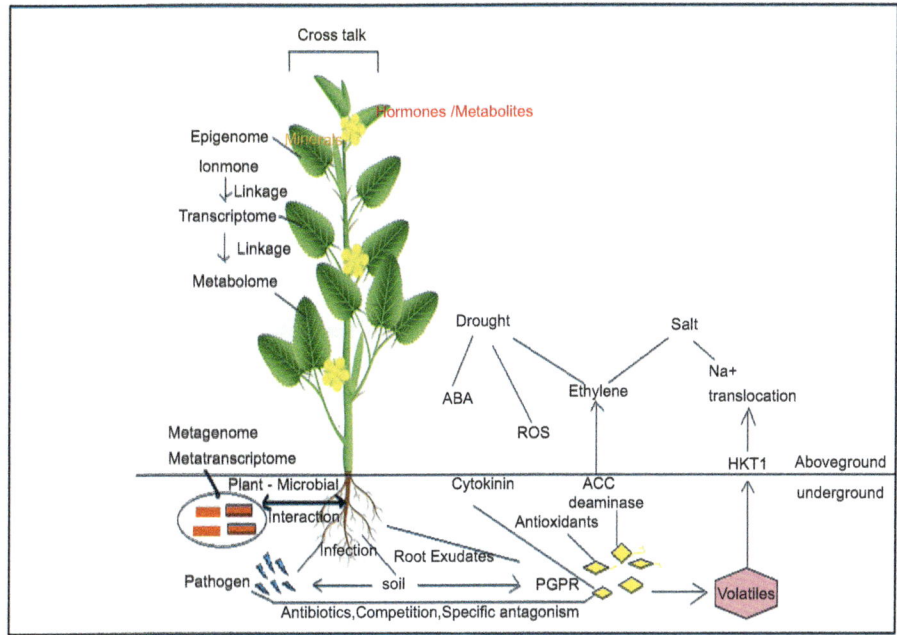

FIGURE 1.2 Plant Growth Regulators (PGRs) and hormonal cross talk under stress conditions.

growth and development. IAA and flavonoids compete to bind flavonoids to the glutathione 5-transferase phi-2 (STF2) genes, which regulate auxin transport and distribution and protect against various stresses. The phytohormonal connection between JA and GA regulates the growth-defense trade-offs. Through JAZs-DELLAs, JA also serves as an antidote to the GA-dependent growth response. In the absence of JA, *Arabidopsis* JAZ9 binds to the DELLA protein RGA, preventing it from inhibiting the growth-promoting TF PIF3 (phytochrome-interacting factor3). Furthermore, JA increases JAZ breakdown and delays GA-mediated DELLA degradation, reducing the GA-dependent growth response (Yang et al., 2012).

Furthermore, reducing enhancers like AUXs, JA and ET reduce leaf cell cycle processes and growth. AUXs have also been postulated as inhibitors of J.A. production and the J.A./Et-mediated nicotine response. However, following *Manduca sexta herbivore* JA and AUXs collaborate to keep *Nicotiana attenuata* regrowth to a minimum (Shi et al., 2012). MAPK cascades, which integrate many inputs and govern cross talk, are triggered by phytohormones. Mitogen-activated protein kinase (MAPK) cascades govern the transmission of environmental stimuli. Plants have several MAPK components activated in response to various types of stress. When a plant is exposed to abiotic stress, MAPK cascades such as MEKKI/MKK2/MPK4 and MPK6 are activated. In plants, MEKK1, MKKI/MKK2, and MPK4 work together to modulate innate immunity (Rodriguez et al., 2010). According to Ludwig et al. (2005), the CDPK and MAPK pathways are stimulated by abiotic and biotic stimuli. ET serves as a crucial cross talk site between the two pathways. The ABA and ET signaling pathways have been identified as a significant cross talk point and a key target for modifying plant responses to various stressors. Phytohormone-regulated specific gene expression protects plants from oxidative damage produced by abiotic and biotic stresses (Chinchilla et al., 2007). In plants under drought stress, *Arabidopsis* CML 42 (calmodulin-like protein 42) reduces ABA accumulation and JA-dependent insect resistance. The synergy of JA and ABA in various stress reactions also supports this conceptual framework (De Lucia et al., 2012). SA promotes plant tolerance to metal toxicity, salt, osmotic stress, drought, and heat stress, among other biotic and abiotic stressors. SA causes SAR, which raises the expression of pathogenesis-related (PR) proteins and reduces pathogen attack/increases defense responses against a broad range of infections (Ding et al., 2011; Horváth et al., 2007). The generation of antioxidant enzymes and their activity and the development of heat shock proteins are all increased by SA (Clarke et al., 2004). The SA, JA, and ET defensive signaling pathways communicate, allowing the plant to choose the optimum defense strategy. Temperature and humidity affect the expression of defense-related genes in response to pathogen infection (Fraire-Velázquez et al., 2011).

1.4 Conclusions and future perspectives

Phytohormones are critical for plant growth and development, primarily under abiotic stress conditions. Phytohormones are chemical messengers that help plants control and regulate various processes, such as germinating seeds, fruit ripening, and flowering. Phytohormones in plants help respond to stress in a variety of ways. Changes in phytohormonal levels have been observed in stress situations, enhancing the plant's survival adaptation. Some reactions include stomata activity, water level management, source-sink transportation, and antioxidant component restoration. There is a need to learn more about the unique method plants adjust for survival using phytohormones. This will promote the production of plant cultivars resistant to both abiotic and biotic stresses. Plants subjected to abiotic and biotic stresses may also gain tolerance power for these conditions through phytohormone engineering. Researchers have focused their efforts on understanding how plants respond to such frequent biotic and abiotic stress occurrences, but there are still some mysteries to be solved. It is generally understood that phytohormone participation, whether direct or indirect, aids the plant's ability as part of the plant's defense mechanism; it helps it manage abiotic stress.

Crop production is restricted by rising food demand, a diminishing natural resource base, and climatic uncertainty. Abiotic and biotic stressors are key constraints affecting agricultural output around the world. In recent years, the molecular processes directing phytohormonal synthesis, signaling, and action have been investigated, and phytohormones' responsibilities in regulating responses to adverse environmental conditions have long been known. Understanding phytohormone metabolism and transport channels will allow new approaches to control plant development by manipulating hormone levels. Many receptors for various hormone classes have been found, leading to novel ways to perceive hormones. A better knowledge of receptor function could aid the production of novel PGRs.

Meanwhile, many downstream signaling components have been found. Understanding how different signaling pathways interact during plant growth and development will be a significant challenge. For example, auxin, GA, and BR are all known to induce cell elongation in the plant stem. However, the relative contribution of each signaling pathway to growth regulation in response to environmental changes is unknown.

Similarly, it's unclear whether each hormone signal engages the same genes for cell elongation. Finally, our ability to create forecast plant growth models will increase as we learn more about hormone action mechanisms. Models for the shoot and root development regulation dependent on auxins have already been developed. We may expect these models to eventually contain a wide range of phytohormones and environmental signals, providing a more holistic view of plant development and growth.

References

Akiyama, K., Matsuzaki, K., Hayashi, H., 2005. Plant sesquiterpenes induce hyphal branching in arbuscular mycorrhizal fungi. Nature 435 (7043), 824−827.

Alder, A., Jamil, M., Marzorati, M., Bruno, M., Vermathen, M., Bigler, P., et al., 2012. The path from β-carotene to carlactone, a strigolactone-like plant hormone. Science 335 (6074), 1348−1351.

Bari, R., Jones, J.D., 2009. Role of plant hormones in plant defence responses. Plant Mol. Biol. 1, 473−488.

Bastam, N., Baninasab, B., Ghobadi, C., 2013. Improving salt tolerance by exogenous application of salicylic acid in seedlings of pistachio. Plant. Growth Regul. 69, 275−284.

Binenbaum, J., Weinstain, R., Shahi, E., 2018. Gibberellin localization and transport in plants. Trends Plant. Sci. 23 (5), 410−421.

Borghi, L., Kang, J., Ko, D., Lee, Y., Martinoia, E., 2015. The role of ABCG-type ABC transporters in phytohormone transport. Biochem. Soc. Trans. 43 (5), 924−930.

Bücker-Neto, L., Paiva, A.L.S., Machado, R.D., Arenhart, R.A., MargisPinheiro, M., 2017. Interactions between plant hormones and HMs responses. Genet. Mol. Biol. 40, 373−386.

Carrow, R.N., Duncan, R.R., 2012. Best Management Practices for Saline and Sodic Turfgrass Soils: Assessment and Reclamation. CRC Press, Boca Raton, FL, pp. 350−355.

Cheong, J.J., Choi, Y.D., 2003. Methyl jasmonate as a vital substance in plants. Trends Genet. 19, 409−413.

Chiba, Y., Shimizu, T., Miyakawa, S., Kanno, Y., Koshiba, T., Kamiya, Y., et al., 2015. Identification of *Arabidopsis thaliana* NRT1/PTR family (NPF) proteins capable of transporting plant hormones. J. Plant Res. 128, 679−686.

Chinchilla, D., Zipfel, C., Robatzek, S., Kemmerling, B., Nürnberger, T., Jones, J.D., et al., 2007. Aflagellin-induced complex of the receptor FLS2 and BAK1 initiates plant defence. Nature 448, 497.

Clarke, S.M., Mur, L.A., Wood, J.E., Scott, I.M., 2004. Salicylic acid dependent signaling promotes basal thermotolerance but is not essential for acquired thermotolerance in *Arabidopsis thaliana*. Plant J. 38, 432−447.

Colebrook, E.H., Thomas, S.G., Phillips, A.L., Hedden, P., 2014. The role of gibberellins signaling in plant response to abiotic stress. J. Exp. Biol. 217, 67−75.

Creelman, R.A., Bell, F., Mullet, McConn, M.J.E., Browse, J., 1997. Jasmonate is essential for insect defense in Arabidopsis. Proc. Natl. Acad. Sci. U.S.A. 94, 5473−5477.

Darwin, C., Darwin, F., 1881. The Power of Movement in Plants; D. 592Appleton and Company, New York, NY, USA.

De Lucia, E.H., Nabity, P.D., Zavala, J.A., Berenbaum, M.R., 2012. Climate change: resetting plant-insect interactions. Plant. Physiol. 160, 1677−1685.

Ding, L., Xu, H., Yi, H., Yang, L., Kong, Z., Zhang, L., et al., 2011. Resistance to hemi-biotrophic *F. graminearum* infection is associated with coordinated and ordered expression of diverse defense signaling pathways. PLoS One 6, e19008.

Dubois, M., Broeck, L.V.D., Inze, D., 2018. The pivotal role of ethylene in plant growth. Trends Plant. Sci. 23 (4), 311−323.

Eriksson, S., Bohlenius, H., Moritz, T., Nilsson, O., 2006. GA4 is the active gibberellins in the regulation of LEAFY transcription and Arabidopsis floral initiation. Plant Cell 18, 2172−2181.

Ferguson, L., Grafton-Cardwell, E.E., 2014. Citrus Production Manual. University of California, Agriculture and Natural Resources, Richmond, CA.

Fiorani, F., Bogemann, G.M., Visser, E.J.W., Lambers, H., Voesenek, L.A.C.J., 2002. Ethylene emission and responsiveness to applied ethylene vary among Poa species that inherently differ in leaf elongation rates. Plant Physiol. 129, 1382–1390.

Flasinski, M., Hac-Wydro, K., 2014. Natural vs synthetic auxin: studies on the interactions between plant hormones and biological membrane lipids. Environ. Res. 133, 123–134.

Fraire-Velázquez, S., Rodríguez-Guerra, R., Sánchez-Calderón, L., 2011. Abiotic and biotic stress response crosstalk in plants. In: Shanker, A.K., Venkateswarlu, B. (Eds.), Abiotic Stress Response in Plants-Physiological, Biochemical and Genetic Perspectives. InTech, Rijeka, pp. 3–26.

Gazzarrini, S., Mccourt, P., 2003. Crosstalk in plant hormone signalling: what Arabidopsis mutants are telling us. Ann. Bot. 91, 605–612.

George, E.F., Hall, M.A., De Klerk, G., 2008. Plant growth regulators I: introduction; auxins, their analogues and inhibitors, Plant Propagation by Tissue Culture, 3rd ed. Springer, The Netherlands, Dordrecht, pp. 175–204, Netherlands (Ed.).

Gray, W.M., 2004. Hormonal regulation of plant growth and development. PLoS Biol. 2 (9), e311.

Gupta, R., Chakrabarty, S.K., 2013. Gibberellic acid in plant: still a mystery unresolved. Plant Signal. Behav. 8 (9), e25504.

Hamayun, M., Yoon, J.Y., Lee, S.K., Lee, I.J., 2010. Methyl jasmonate alleviated salinity stress in soybean. J. Crop. Sci. Biotechnol. 12, 63–68.

Han, X., Zeng, H., Bartocci, P., Fantozzi, F., Yan, Y., 2018. Phytohormones and effects on growth and metabolites of microalgae: a review. Fermentation 4 (25), 1–15.

Hao, J.H., Donga, C.J., Zhanga, Z.G., Wanga, X.L., Shang, Q.M., 2011. Insights into salicylic acid responses in cucumber (Cucumis sativus L.) cotyledons based on a comparative proteomic analysis. Plant Sci. 187, 69–82.

Harms, C.L., Oplinger, E.S., 1988. Plant Growth Regulators: Their Use in Crop Production. North Central Region Extension Publication 303, Specialized Soil Amendments, Products and Growth Stimulants. U.S. Department of Agriculture and Cooperative Extension Services. Illinois, IA.

He, Q., Zhao, S., Ma, Q., et al., 2014. Endogenous salicylic acid levels and signaling positively regulate Arabidopsis response to polyethylene glycol-simulated drought stress. J. Plant Growth Regul. 33, 871–880. Available from: https://doi.org/10.1007/s00344-014-9438-9.

Hedden, P., Sponsel, V., 2015. A century of gibberellins research. J. Plant Growth Regul. 34, 740–760.

Horváth, E., Szalai, G., Janda, T., 2007. Induction of abiotic stress tolerance by salicylic acid signaling. J. Plant Growth Regul. 26, 290–300.

Hu, J., Lin, H., Xu, N., Jiao, R., Dai, Z.J., Lu, C.L., et al., 2019. Advances in molecular mechanism and breeding application of leaf inclination in rice. Chin. J. Rice Sci. 33, 391–400.

Jackson, M.B., 1993. Are plant hormones involved in root to shoot communication? Adv. Bot. Res. 19, 103–187.

Kramer, E.M., 2006. How far can a molecule of weak acid travel in the apoplast or xylem? Plant Physiol. 141, 1233–1236.

Kretzschmar, T., Kohlen, W., Sasse, J., Borghi, L., Schlegel, M., Bachelier, J.B., et al., 2012. A petunia ABC protein controls strigolactone-dependent symbiotic signalling and branching. Nature 483, 341–344.

Leshem, Y., Seri, L., Levine, A., 2007. Induction of phosphatidylinositol 3-kinase-mediated endocytosis by salt stress leads to intracellular production of reactive oxygen species and salt tolerance. Plant J. 51 (2), 185–197.

Li, J.G., Fan, M., Hua, W., Tian, Y., Chen, L.G., Sun, Y., et al., 2020. Brassinosteroid and hydrogen peroxide interdependently induce stomatal opening by promoting guard cell starch degradation. Plant Cell 32 (4), 984–999.

Ludwig, A.A., Saitoh, H., Felix, G., Freymark, G., Miersch, O., Wasternack, C., et al., 2005. Ethylene-mediated crosstalk between calcium-dependent protein kinase and MAPK signaling controls stress responses in plants. Proc. Nat. Acad. Sci. U.S.A. 102 (30), 10736–10741.

Mitchell, J.W., 1942. Plant-Growth Regulators. Department of Agriculture., Misc. Pub, U.S, pp. 495.

Muthulakshmi, S., Lingakumar, K., 2016. Salicyclic acid induced responses on growth and biochemical constitutents in Vigna mungo (L.) hepper. Eur. J. Exp. Biol. 6 (1), 9–14.

Nazir, F., Hussain, A., Fariduddin, Q., 2019. Interactive role of epibrassinolide and hydrogen peroxide in regulating stomatal physiology, root morphology, photosynthetic and growth traits in Solanum lycopersicum L. under nickel stress. Environ. Exp. Bot. 162, 479–495.

Nazir, F., Hussain, A., Fariduddin, Q., Tanveer, A.K., 2021. Brassinosteroid and hydrogen peroxide improve photosynthetic efficiency and maintain chloroplast ultrastructure, stomatal movement, root morphology, cell viability and reduce Cu-triggered oxidative burst in tomato. Ecotox. Environ. Saf 207, 111081.

Nazir, F., Fariduddin, Q., Tanveer, A.K., 2022. Interaction Between Brassinosteroids and Hydrogen Peroxide Networking Signal Molecules in Plants: Brassinosteroids Signalling, Intervention with Phytohormones and Their Relationship in Plant Adaptation to Abiotic Stresses (pp. 59–79).

Northey, J.G.B., Liang, S., Jamshed, M., et al., 2016. Farnesylation mediates brassinosteroid biosynthesis to regulate abscisic acid responses. Nat. Plants 2, 16114. Available from: https://doi.org/10.1038/nplants.2016.114.

Rademacher, W., 2015. Plant growth regulators: backgrounds and uses in plant production. J. Plant Growth Regul. 34, 845–872.

Reganult, T., Daviere, J.M., Wild, M., Sakvarelidze-Achard, L., Heintz, D., Carrera Bergua, E., et al., 2015. The gibberellins precursor GA 12 acts as long distance growth signal in Arabidopsis. Nat. Plants 1, 15073.

Reich, L., 2013. Plant Hormones, Natural or Synthetic, Can Help Cuttings Take Root. The Associated Press. Available [online]: <http://www.huffingtonpost.ca/2013/07/09/plant-hormones-natural-o_n_3567961.html> Accessed August 10, 2016.

Rodriguez, Cutler, S.R.P.L., Finkelstein, R.R., Abrams, S.R., 2010. Abscisic acid: emergence of a core signaling network. Annu. Rev. Plant Biol. 61, 651–679.

Saini, S., Sharma, I., Kaur, N., Pati, P.K., 2013. Auxin: a master regulator in plant root development. Plant. Cell Rep. 32, 741–757.

Saito, H., Oikawa, T., Hamamoto, S., Ishimaru, Y., Kanamori-Sato, M., Sasaki-Sekimoto, Y., et al., 2015. The jasmonate-responsive GTR1 transporter is required for gibberellin-mediated stamen development in Arabidopsis. Nat. Commun. 4 (6), 6095.

Sakakibara, H., 2010. C.K. biosynthesis and metabolism. In: Davies, P.J. (Ed.), The Plant Hormones: Biosynthesis, Signal Transduction, Action!, third ed. Springer Netherlands, Dordrecht, pp. 95–114.

Salehin, M., Li, B., Tang, M., et al., 2019. Auxin-sensitive Aux/IAA proteins mediate drought tolerance in Arabidopsis by regulating glucosinolate levels. Nat. Commun. Available from: https://doi.org/10.1038/s41467-019-12002-1.

Shah, J., 2003. The salicylic acid loop in plant defense. Curr. Opin. Plant Biol. 6, 365–371.

Shakirova, F.M., 2007. Role of hormonal system in the manifestation of growth promoting and anti-stress action of salicylic acid. In: Hayat, S., Ahmad, A. (Eds.), Salicylic Acid. Springer, Dordrecht, Netherlands (A plant hormone).

Shi, Y., Tian, S., Hou, L., Huang, X., Zhang, X., Guo, H., et al., 2012. Ethylene signaling negatively regulates freezing tolerance by repressing expression of CBF and type-a ARR genes in Arabidopsis. Plant Cell 24, 2578–2595.

Shi, L., Li, L.H., Zhang, R.J., Li, Y.L., Li, L., Zhang, Y., et al., 2019. 24-epbrassinolactone regulates active oxygen metabolism and enhances disease resistance of apricot fruits after harvest. Food Sci. 41, 126–132.

Soares, T., Dias, D., Oliveira, A.M.S., Ribeiro, D.M., Dias, L., 2020. Exogenous brassinosteroids increase lead stress tolerance in seed germination and seedling growth of *Brassica juncea* L. Ecotoxicol. Environ. Saf. 193, 110296.

Tal, I., Zhang, Y., Jørgensen, M.E., Pisanty, O., Barbosa, I.C.R., Zourelidou, M., et al., 2016. The Arabidopsis NPF3 protein is a GA transporter. Nat. Commun. 7, 11486.

Tanaka, Y., Fukuda, A., 2006. Effects of ABA, auxin and gibberellin on the expression of genes for vacuolar H+ inorganic pyrophosphatase, H+ ATPase subunit A, and Na+/H+ antiporter in barley. Plant Physiol. Biochem. 44, 351–358.

Tuteja, N., 2007. Abscisic acid and abiotic stress signaling. Plant Signal. Behav. 2, 135–138.

Valluru, R., Davies, W.J., Reynolds, M.P., Dodd, I.C., 2016. Foliarabscisic acid-to-ethylene accumulation and response regulate shoot growth sensitivity to mild drought in wheat. Front. Plant Sci. 7, 461. Available from: https://doi.org/10.3389/fpls.2016.00461.

Van de Poel, B., Van der Straiten, D., 2014. 1-Aminocyclopropane 1-carbolxylic acid (ACC) in plants: more than just the precursor of ethylene. Front. Plant Sci. 5, 640.

Wang, Y.H., Irving, H.R., 2011. Developing a model of plant hormone interactions. Plant Signal. Behav. 6 (4), 494–500.

Werner, T., Motyka, V., Strnad, M., Schmulling, T., 2001. Regulation of plant growth by Cytokinin. Proc. Natl. Acad. Sci. U.S.A. 98, 10487–10492.

Yang, D.L., Yao, J., Mei, C.S., Tong, X.H., Zeng, L.J., Li, Q., et al., 2012. Plant hormone jasmonate prioritizes defense over growth by interfering with gibberellin signaling cascade. Proc. Natl. Acad. Sci. U.S.A. 109, E1200.

Zhang, S., Li, C., Ren, H.H., Zhao, T., Li, Q., Wang, S.F., et al., 2020. BAK1 mediates light intensity to phosphorylate and activate catalases to regulate plant growth and development. Int. J. Mol. Sci. 21 (4), 1437–1454.

2

Deciphering the physiological and molecular functions of phytohormones

Manvi Sharma[1] and Ashverya Laxmi[2]

[1]Umeå Plant Science Centre, Department of Forest Genetics and Plant Physiology, Swedish University of Agricultural Sciences UMEÅ, Sweden [2]Lab 203, National Institute of Plant Genome Research, New Delhi, Delhi, India

2.1 Introduction

Plants exhibit an amazing variety of shapes and forms, rely on continuous growth, and are able to regenerate organs from undifferentiated meristematic cell population. Plant growth, differentiation, development, reproduction, and response to environmental stimuli are regulated by small endogenous compounds called phytohormones. Auxin was the first hormone to be discovered, followed by gibberellic acid (GA), cytokinin (CK), abscisic acid (ABA), and ethylene (ET) (Kende and Zeevaart, 1997). Additional ones, such as brassinosteroids (BRs), jasmonates (JAs), salicylic acid (SA), and strigolactones (SLs), were added to the list and new ones continue to be discovered.

The topic on hormone action brings into focus the molecular mechanism that drives hormones to bring a physiological effect on plants. Hormone action involves its perception, initiation of a specific response, and finally, a sustained response under the new steady-state growth conditions. Modern tools and genetic analyses provide superb opportunities for understanding the intricate details of the mechanism of hormone action. For instance, using yeast-one-hybrid (Y1H) screen, ChIP-qPCR, and a protoplast-based luciferase assay, a study by Šimášková et al. (2015a) has demonstrated that CYTOKININ RESPONSE FACTOR 2 (CRF2) and CRF6 are positive regulators of PIN7 and PIN1 expression, respectively. The arrival of whole plant genome sequencing has provided the scientific community with immense knowledge related to metabolism, receptors and mechanisms, and manipulation of the genome.

This chapter will focus on the complex and layered but ordered regulatory mechanisms of hormone homeostasis, transport, cross talk, epigenetic regulation, and their physiological effects on plant growth and development.

2.2 Shoot apical meristem maintenance and activity

Unlike animals that complete the majority of organogenesis during embryogenesis, plants establish shoot apical meristem (SAM) and root apical meristem (RAM) in the mature embryo. The meristems are responsible for the continuous generation of new cells and the formation of all postembryonic organs, such as the RAM forms the entire root, whereas SAM forms the above-ground part of the plant like leaves, stems, and flowers (Chang et al., 2020; Satbhai et al., 2015). The SAM can be divided into different zones: the peripheral zone (PZ) located at the meristem flanks is responsible for the initiation of the lateral organs; the central zone (CZ), at the peak of the SAM, contains slowly dividing cells that provide initials for the PZ (Ali et al., 2020; Bustamante et al., 2016). The latest emerging primordium is termed plastochron 1 (P_1), the next oldest primordium is termed P_2, and so on. The SAM shows autonomy in making lateral organs. However, changes in light, water, temperature, and nutrients to other parts of the plants are communicated to SAM, which then evokes morphological and physiological changes in organ growth (De Jonge et al., 2016; Sugiyama and Gotoh, 2010).

Several groups of transcription factors (TFs) participate in meristem maintenance and activity. Class I KNOTTED1-like homeobox (KNOXI) proteins participate in the maintenance of the indeterminate nature of the meristem and in the formation of organ boundaries (Hake et al., 2004). NAC group of TFs, such as CUP-SHAPED COTYLEDON1 (CUC1), CUC2, and CUC3, are essential for the formation of boundaries between organs and meristems and between adjacent organs (Aida and Tasaka, 2006). Receptor-like kinase CLAVATA1 (CLV1) and its ligand CLAVATA3 (CLV3) control stem cell proliferation and exit (Clark et al., 1997). CLV3 binds to CLV1 and initiates the signaling cascade, which restricts the expression of WUSCHEL (WUS) in the organizing center (Brand et al., 2000; Ogawa et al., 2008; Schoof et al., 2000). WUS protein moves from its site of synthesis through plasmodesmata and directly activates CLV3 in CZ and also represses genes involved in differentiation (Yadav et al., 2011, 2013).

Hormones play an essential role in controlling the SAM function and the role of auxin and CK is very well established (Veit, 2009). CK receptor triple-mutant *ahk2ahk3ahk4* showed intense reduction in meristem size and leaf-initiation rate, as well as impaired leaf development (Riefler et al., 2006). Misexpression of genes encoding CK oxidase resulted in reduced meristem size, leaf-initiation rate, and meristem abortion (Werner et al., 2003). Hence, suggesting that CK response is required for proper meristem function. In *Arabidopsis*, CK production and response are controlled by the activity of KNOX proteins thus, maintaining meristem indeterminacy (Hay et al., 2004). Yanai et al. (2005) reported that exogenous CK or the expression of a bacterial *IPT* gene through the STM promoter was able to partially rescue the meristem function of severe *stm* alleles. Several ARABIDOPSIS RESPONSE REGULATORS (ARRs), such as ARR5 and ARR7, have been shown to be rapidly repressed by WUS activation and positively regulated by CLV3

(Leibfried et al., 2005). WUS was shown to bind sequences in the *ARR7* promoter directly. Overexpression of a constitutively active ARR7 form caused variable phenotypic aberrations, the most severe of which was early meristem determination, similar to that of *wus* mutants (Leibfried et al., 2005). Mutant of maize *ABPH1* gene, a type A ARR, shows increased meristem size and initiates leaf primordia in pairs rather than in the alternating pattern seen in the wild-type, thus suggesting that one consequence of this negative regulation is to limit meristem growth (Jackson and Hake, 1999). Hence, meristem maintenance requires CK response, and this response is facilitated by the specific expression patterns of KNOXI and WUS TFs, type A ARRs, thus leading to differential CK effects.

Prior studies by Snow and Snow (1937) showed the potential role of auxin in perturbing patterning processes in the SAM. Direct application of auxin to dissected shoot tips modified positions at which leaf primordia formed. Auxin biosynthesis, perception, signaling, and transport are involved in regulating SAM function. Auxin micro-application experiments have suggested that a local auxin maximum is necessary and sufficient to trigger primordium initiation in the PZ of the SAM (Reinhardt et al., 2000). Moreover, polar auxin transport (PAT) establishes local auxin maxima that specify incipient primordia. This formation of local maxima results from the coordinated PIN1 polarity of each cell (Heisler et al., 2005). Several studies have shown that primordium initiation occurs when high relative auxin concentrations in the P_0 region downregulate the expression of *CUC* genes. In *Brassica juncea* embryos, the application of exogenous auxin resulted in cotyledon fusion and failure to produce further organs, similar to *Arabidopsis cuc1 cuc2* mutants (Hadfi et al., 1998). Mutants, such as pinoid (*pid*), enhancer of pinoid (*enp*), and *pidenp* showed defects in lateral organ formation and failure to produce cotyledons. This was accompanied by the expansion of the expression domains of *CUC* and *STM* genes into the P_0 region or the incipient cotyledon in the embryo (Furutani et al., 2004; Treml et al., 2005). Elimination of *CUC1* function in the *cucpinpid* restored cotyledon development in *pin1pid* double mutants. This indicates that the expansion of *CUC1* and *STM* expression domains mediates the failure of *pin1pid* mutants to develop cotyledons (Furutani et al., 2004). Hence, higher auxin concentrations in P_0 assist lateral organ initiation through the downregulation of CUC.

2.3 Flower development

Flowers are reproductive organs of angiosperms and display amazing morphological diversity. When to flower is an important decision for plants that marks changes in the SAM. The switch to flowering occurs when SAM in the vegetative state (V) receives appropriate signals and precedes bolting. After the floral transition, SAM enters the inflorescence stage (I), leading to the appearance of flowers instead of leaves, thus altering the above-ground plant architecture. Studies on *Arabidopsis* have been instrumental in advancing our knowledge on flower development and organ initiation. This section summarizes how different hormonal pathways control floral regulatory networks.

The role of GAs in transition to flowering is well established. Wilson et al. (1992) showed that mutation in GA1 locus in *Arabidopsis* leads to late flowering under noninductive short-day (SD) conditions. Additionally, lines deficient in GA signaling show late flowering phenotypes under noninductive SD conditions (Wilson et al., 1992; Moon et al., 2003). Genotypes showing enhanced GA signaling, such as overexpressors of *FLOWERING PROMOTIVE FACTOR1* (*FPF1*), *GA-insensitive* (*gai*), and *spindly* (*spy*). All these observations led to the fact that GA is an all-time floral timing promoter (Wilson et al., 1992; Jacobsen and Olszewski, 1993; Kania et al., 1997). Lang et al. demonstrated the ability of GAs to promote bolting and flower formation in long-day (LD) and biennial plants (Lang, 1957). Application of GA leads to flowering in plants with noninductive conditions (King et al., 2001). The process of bolting that signifies the transition to reproductive development in rosette plants involves GA-dependent cell division and elongation in crops, such as spinach, sugar beet, and field pennycress (Silverstone and Sun, 2000; Sachs and Lang, 1957). Several reports claim GAs as the mobile signals for flower induction. *Arabidopsis* growing under SD conditions showed the accumulation of GA_4 in the shoot apex before floral transition, and this accumulation did not correspond with changes in GA-biosynthetic gene expression in the apex, hence suggesting that GA originated from elsewhere (Eriksson et al., 2006). King et al. showed that the treatment of LD plants with GA-biosynthetic inhibitor paclobutrazol induced flowering, thus showing the requirement of GA for flower initiation (King et al., 2006). Following floral initiation, GA is required for the normal development of floral organs (Griffiths et al., 2006). Mild GA deficiency leads to impaired male fertility due to abnormal stamen development (Chhun et al., 2007; Hu et al., 2008), whereas extreme GA deficiency leads to female sterility (Nester and Zeevaart, 1988; Goto and Pharis, 1999). This finding is consistent with the reports of Goto and Pharis who demonstrated the requirement of high GA concentrations by *Arabidopsis* stamens than pistils, petals, and sepals (Goto and Pharis, 1999). Chromatin structural modifications for GA biosynthesis and signaling response play an important role to regulate flowering. H3K27me3 regulates many genes in the GA pathway as revealed by a genome-wide profiling (Zhang et al., 2012a). *PICKLE* (*PKL*), a chromatin remodeling enzyme in *Arabidopsis* promotes H3K27me3 (Zhang et al., 2007, 2008). The *pkl* mutant shows increased flowering time and delayed anther dehiscence (Ogas et al., 1999). It was found that the exogenous application of GA_3 to *pkl* mutants led to reduction in the flowering time; however, the endogenous levels of bioactive GA forms remained the same. Thus suggesting that PKL is involved in the perception but not in the biosynthesis of GA (Henderson et al., 2004). Recently, a connection between DELLA and PKL has been found to regulate GA-dependent flowering, presumably by increasing methylation at H3K27me3 (Zhang et al., 2012a; Park et al., 2017). DELLA proteins can also interact with FLC to repress the transition to flowering (Li et al., 2016), and the FLC chromatin can be silenced by PRC2-mediated silencing at H3K27me3 (Yuan et al., 2016; Buzas et al., 2011). Thus a study recently hypothesized a model for the regulation of flowering transition by GA in which DELLA interacts with PKL that further recruits PRC2. Thus PRC2 increases H3K27me3 levels in *FLC*, which regulate target genes, such as *FT* and *SOC1* (Campos-Rivero et al., 2017).

PAT plays an important role in floral organ initiation from the floral meristem (FM), which initiates as lateral organs at the inflorescence meristem (IM) periphery. The TF WUS acts as a rheostat and is responsible for the maintenance of IM by keeping a balance between the self-renewal of stem cells and differentiation (Ma et al., 2019). This balance is maintained by restricting auxin signaling in stem cells and at the same time allowing low auxin signaling output. However, the precise molecular mechanism is still unknown.

Another example of the role of auxin in floral organ initiation comes from the study of mutants, for instance, weak *pin1* mutant that lacks sepals and shows a variable number of abnormally shaped petals. The addition of auxin to *pin1−1* inflorescence restores flower formation (Okada et al., 1991). The *pid* mutants show supernumerary petals, few sepals, and stamens. Floral primordium initiation also requires the activity of AUXIN RESPONSE FACTORS (ARFs) (Przemeck et al., 1996; Yamaguchi et al., 2013). The mutants of ARF5/MONOPTEROS (MP) show naked inflorescence (Przemeck et al., 1996). Targets of MP have been identified, which promote flower initiation, such as *LEAFY* (*LFY*), *AINTEGUMENTA* (*ANT*), and *AINTEGUMENTA-LIKE 6* (*AIL6*) (Elliott et al., 1996; Krizek and Eaddy, 2012; Schultz and Haughn, 1991). A study by Wu et al. (2015) uncovers the MP-anchored chromatin state switch that underlies flower primordium initiation. In the presence of high auxin, MP recruits SWI/SNF chromatin remodeling ATPases to make the DNA more accessible to regulators of flower primordium initiation (Wu et al., 2015).

BRs play a positive role in floral induction since the mutants defective in BR show late flowering (Li et al., 2010). Molecular and physiological studies have shown that GA−BR cross talk provides additional clues about the action of BR in inducing flowering. DELLA has been shown to negatively regulate BR signaling through sequestering BRASSINAZOLE-RESISTANT 1 (BZR1) (activators of BR signaling). Once released from DELLA repression, BR-activated BZR1 causes BR-dependent responses; however, the precise mechanism is still unknown. One way of BZR1 activity is the transcriptional activation of several GA-biosynthetic genes (Unterholzner et al., 2015). Another finding suggests that BRI1-EMS-SUPPRESSOR 1 (BES1) recruits Jumonji N/C (JmjN/C) domain-containing proteins, such as EARLY FLOWERING 6 (ELF6) and RELATIVE OF EARLY FLOWERING 6 (REF6), to control flowering time (Yu et al., 2008). Also, ChIP experiments indicated that ELF6 and REF6 are histone demethylases. Thus suggesting that BES1 and its interacting partners control floral regulatory network and modulate gene expression through histone modifications (Yu et al., 2008).

In flowering plants, JA plays an important role in male reproductive development. The homeotic gene *AGAMOUS (AG)* controls stamen development by regulating the expression of the JA biosynthetic gene, *DELAYED IN ANTHER DEHISCENCE* (DAD1) (Ito et al., 2007). Also, JA application can rescue several male sterile JA biosynthesis mutants (Stintzi and Browse, 2000). The overexpression of truncated JAZ1 also causes early flowering, thus supporting the role of canonical JA signaling in flower development. Hou et al. presented a "relief of repression" model, wherein DELLA competes with MYC2 to bind with JAZ1, thus liberating MYC2 to promote JA response (Hou et al., 2010). Two MYB TFs, MYB21 and MYB24, are direct targets of *JAZs* and were shown to be specifically involved in JA-regulated anther development and filament elongation (Song et al., 2011). MYB108 works downstream to MYB21 to regulate stamen maturation (Mandaokar and Browse, 2009). Cross talk between JA and GA signalings also regulates stamen development. GA by suppressing DELLA proteins activates the expression of DAD1 leading to the activation of JA signaling that activates MYB21, 24, and 57 expressions to promote filament elongation (Cheng et al., 2009). Several other regulating factors like *DAD1-ACTIVATING FACTOR, AG* and *DAYU (DAU)* activate JA biosynthesis to influence anther dehiscence and pollen germination (Li et al., 2014), whereas ANTHER INDEHISCENCE FACTOR (AIF) and

INGUBANG (JGB) inhibit JA biosynthesis to repress stamen development (Shih et al., 2014; Ju et al., 2016).

2.4 Root growth and development

As the hidden part of the plant, roots play an important role in the absorption and translocation of water and nutrients. Primary root (PR) is formed embryonically, and its growth occurs via cell proliferation in the RAM and cell elongation in the elongation zone. Multiple phytohormones play important roles in the growth and development of the PR (Hu et al., 2017; Huang, 2018; Ni et al., 2010; Sharma et al., 2021; Wei and Li, 2016).

As an invincible regulator of root development, auxin, earlier called root-forming hormone, affects every major phase of PR growth. Low concentrations of auxin stimulate PR growth in *Arabidopsis* and maize, whereas higher concentrations inhibit via TRANSPORT INHIBITOR RESPONSE 1 (TIR1)-mediated signaling through an unknown nontranscriptional mechanism (Barbez et al., 2017; Fendrych et al., 2018). There are various enzymes and proteins that regulate root growth by affecting auxin homeostasis. For instance, a recent report highlights the role of a rice gene named *FPF1-like protein 4* of rice (*OsFPFL4*) in modulating root growth by affecting auxin levels. Plants overexpressing *OsFPFL4* displayed shorter roots that were due to higher auxin level (Guo et al., 2020). Another gene encoding for NITRILASE 1 (NIT1) enzyme regulates root growth by affecting auxin biosynthesis. *NIT1* overexpressing plants showed shorter PRs that were due to drastic changes of both free IAA and IAN levels (Lehmann et al., 2017). Nan and coworkers showed an auxin-dependent accumulation of cyclic GUANOSINE 3′,5′-MONOPHOSPHATE (cGMP) in *Arabidopsis* roots. Exogenous application of cGMP derivative 8-bromo-cGMP increases auxin-dependent PR growth through the degradation of AUX/IAA proteins (Nan et al., 2014) (Fig. 2.1A). Recently, auxin-triggered root tip phosphoproteome revealed novel regulators of *Arabidopsis* root growth such as the receptor-like kinases and MAP kinases. The study also suggested auxin, H + -ATPases, cell wall modifications, and cell wall sensing receptor-like kinases are tightly embedded in pathways underlying cell elongation. They also found MAP KINASE KINASE (MKK) as a potential novel regulator of root growth regulating auxin biosynthesis and signaling (Nikonorova et al., 2021).

Auxin regulated PR development by interacting with multiple phytohormones (Hu et al., 2017; Huang, 2018; Qin et al., 2019a; Street et al., 2015). The transcription of auxin genes is regulated by ET, and its treatment on root tips promotes auxin accumulation as revealed by Dr5 reporter and IAA measurements (Méndez-Bravo et al., 2019; Růžička et al., 2007a). ET-responsive HD-Zip gene HOMEOBOX PROTEIN52 (HB52) acts as an important node in mediating auxin-ET cross talk in controlling PR development. Biochemical and genetic analyses revealed HB52 to work downstream of *ETHYLENE-INSENSITIVE3* (*EIN3*). Overexpression of HB52 displayed shorter roots, and knockdown plants showed ET-insensitive phenotype during PR elongation (Fig. 2.1C). It also regulates the transcription of various auxin transport—related genes like *WAVY ROOT GROWTH1* (*WAG1*), *WAG2*, and *PIN2* by binding to their promoter regions. *WAG1* and *WAG2* then phosphorylate PIN2 and modulate auxin transport. JA signaling is also linked to auxin homeostasis leading to an inhibition of *Arabidopsis* root growth. Sun et al. (2009) revealed

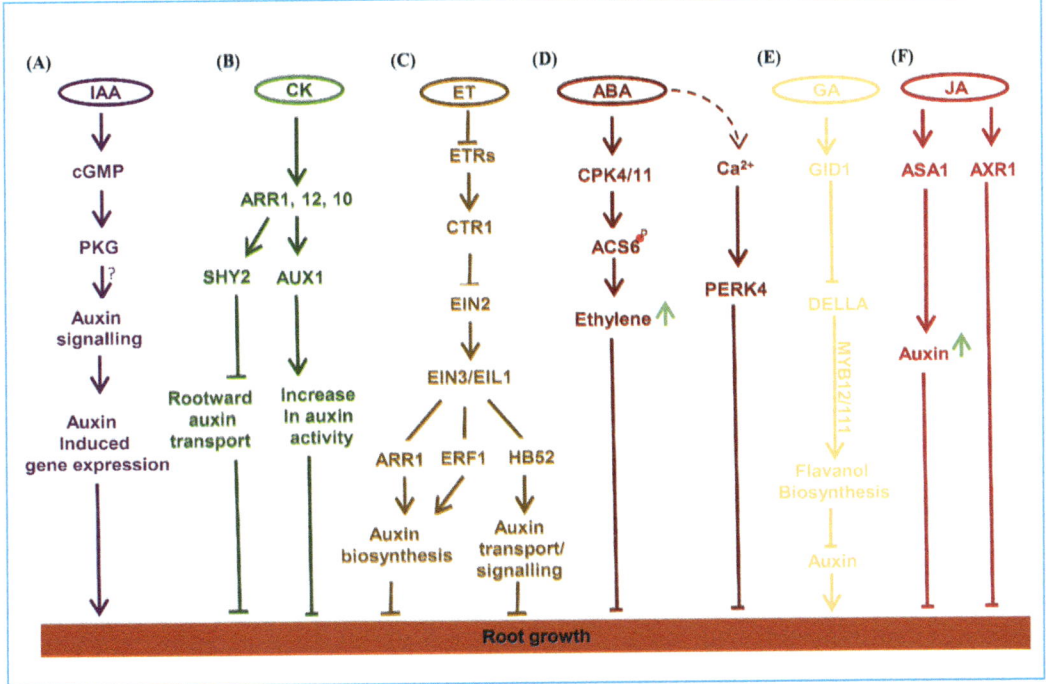

FIGURE 2.1 Regulation of PR growth by hormones and their cross talk: (A) Auxin induces cGMP accumulation, which then influences the auxin-dependent SCF^{TIR1} complex through PKG action, which results in Aux/IAA degradation, facilitates activation of gene expression, and finally, affects auxin-regulated root growth. (B) The auxin influx carrier AUX1 functions downstream of the cytokinin signaling pathway to mediate shootward auxin transport, leading to localized increases in auxin activity, and inhibition of cell elongation. The auxin signaling repressor SHY2 functions downstream of the cytokinin signaling pathway to inhibit rootward auxin transport, resulting in reduced cell proliferation of the RAM. (C) ET effects root growth by interacting with auxin signaling pathway to control PR elongation. (D) ABA activates CPK4 and 11 that phosphorylates ACS6 and then interacts with ET signaling to inhibit root growth. (E) GA effects root growth by modulating the auxin pathway. (F) JA regulates expression of *ASA1* and modulates auxin levels to cause root growth inhibition. JA and auxin signaling occurs via AXR1 control PR growth. Each color represents a distinct signaling pathway. Solid arrows indicate confirmed pathway. Dotted arrows and question marks indicate the possibility of additional routes and other factors. Light green arrow indicates high level of hormone. *ABA*, Abscisic acid; *ET*, ethylene; *GA*, gibberellic acid; *JAs*, jasmonates; *PR*, primary root; *RAM*, root apical meristem.

that JA induces the expression of auxin biosynthetic gene *ANTHRANILATE SYNTHASE A1 (ASA1)*, thereby leading to root growth inhibition (Fig. 2.1F). AXR1 also acts as a nodal point in JA—auxin cross talk. It shows JA-mediated root growth inhibition and pathogen susceptibility (Fig. 2.1F) (Tiryaki and Staswick, 2002).

Several reports have established the negative role of CK in root development (Werner et al., 2003, 2010; Li et al., 2006; Šimášková et al., 2015b; Gao et al., 2014; Ding et al., 2017; Márquez et al., 2019). Werner and group showed that the CK-deficient mutants had longer PR (Werner et al., 2003). Lowering of CK signaling through single and double mutants of CK receptors can increase the root meristem size and activity (Werner et al., 2003; Higuchi et al., 2004); however, CK receptor triple mutant is CK insensitive and shows decreased root meristem size and activity (Riefler et al., 2006; Higuchi et al., 2004; Nishimura et al., 2004). This suggests that a minimum of CK signaling is required for PR development and inhibitory levels of CK results in lowering of PR growth. Similarly, type B ARR higher order mutants of *Arabidopsis* displayed increased root length suggesting a negative role of CK in PR length (Mason et al., 2005; Dello Ioio et al., 2007; Hill et al., 2013). However, contrary to this general notion, mutants of type B ARR (*ARR21, ARR22, ARR23*) showed shorter roots due to less cell proliferation as compared to WT (Worthen et al., 2019). Recently, two *Arabidopsis* genes encoding the CK-degrading enzymes CYTOKININ OXIDASE/ DEHYDROGENASE 1 and 2 (CKX1/2) were expressed in *Hordeum vulgare* cv Golden Promise to reduce CK levels in roots. The transgenic lines showed increased root length from 24% to 70% and total root surface area from 12% to 50%. Moreover, the transgenic lines showed drought tolerance as there was less accumulation of ABA without any yield penalty. Hence, reducing CK levels in monocots can be an effective strategy to improve root biomass and combat abiotic stress (Ramireddy et al., 2018). A recent study uncovers regulatory circuit controlling auxin−CK interactions where auxin importer AUX1 functions as a positive regulator of CK responses in the root. CK regulates root cell elongation through ET-dependent and -independent mechanisms, both hormonal signals converging on AUX1 as a regulatory hub (Fig. 2.1B). AUX1 functions downstream to ARR1, ARR10, and ARR12 and mediates shootward auxin transport leading to the inhibition of cell elongation. SHY2, on the other hand, works downstream to CK signaling and inhibits rootward auxin transport and hence reduces the cell proliferation of SAM (Street et al., 2016) (Fig. 2.1B). *OsNAC*, a rice gene expressing in PR tips, lateral root primordia, and crown roots, integrates auxin and CK pathways to regulate root development (Mao et al., 2020). Yeast-one-hybrid assay, ChIP-seq assay, and ChIP-qPCR validated that OsNAC2 regulates IAA and CK functions by binding to the promoters of *GRETCHEN HAGEN* (*GH3.6* and *GH3.8*), *AUXIN RESPONSE FACTOR 25* (*OsARF25*), and *CYTOKININ OXIDASE 4* (*OsCKX4*). Furthermore, a genetic analysis of *ON11/osgh3.6* and *RNAi31/ osckx4* homozygote confirmed that OsCKX4 and OsGH3.6 functioned downstream of OsNAC2. The results also showed that OsNAC2 inhibits root initiation by inhibiting the expression of cell cycle marker genes *OsCDK* and *OsCRL* (Mao et al., 2020).

An increasing number of evidences have established that ET inhibits root elongation via two processes: cell proliferation and cell elongation (Street et al., 2015; Růžička et al., 2007a). Genetic and molecular analyses identified that the ET-induced suppression of cell elongation involves a subset of auxin perception and transport genes. In *Arabidopsis*, the chemical inhibition of auxin transport and some of the mutants of auxin transport render plants insensitive to ET responses (Růžička et al., 2007b). The cell type-specific site of action of ET in restricting cell elongation has been shown to be the epidermal layer in both root and shoot (Vaseva et al., 2018). An ET-induced inhibition of RAM employs canonical ET-response pathway involving ET receptor TFs EIN2, CTR1, and EIN3/EIL as determined by ET treatment and pathway mutant analyses (Street et al., 2015). Similar studies carried out in rice also suggest the role of ET in controlling root growth (Qin et al., 2019b). Also, in rice, ET mainly cross talks with auxin in mediating root growth responses (Qin et al., 2017). Some auxin biosynthesis and signaling mutants in rice, such as *maohuzi2* (*mhz2/sor1*) and *rice ethylene-insensitive7* (*rein7/yuc8*), display altered ET response in root growth (Qin et al., 2017; Chen et al., 2018).

In ET−auxin cross talk, EIN3/EIL1, Ethylene Response Factor 1 (ERF1) work as the nodal points (Fig. 2.1C). EIN3 enhances auxin biosynthesis in the root tip epidermis by enhancing the transcriptional activity of type B *Arabidopsis* Response Regulator (ARR1) via direct protein−protein interaction. The ARRs promote the expression of a key auxin biosynthetic gene *TRYPTOPHAN AMINOTRANSFERASE OF ARABIDOPSIS 1* (*TAA1*) resulting in enhanced auxin levels and reduced RAM (Yan et al., 2017). ERF1 also plays a major role in ET−auxin cross talk in the regulation of root elongation. ERF1 is the direct target of EIN3, and it upregulates *ASA1*, encoding *ANTHRANILATE SYNTHASE α1*, which is a rate-limiting enzyme in Trp biosynthesis from which auxin is derived consequently promoting auxin biosynthesis (Mao et al., 2016) (Fig. 2.1C). Also, ET controls auxin transport by regulating the members of auxin influx carrier AUX1 and efflux carrier PIN families. Mutations in the members of these carriers lead to ET-insensitive PR phenotype (Stepanova et al., 2007). ET increases auxin transport and transcript levels of *AUX1*, *PIN3*, and *PIN7* (Zemlyanskaya et al., 2018). The ET control over AUX1 is possibly mediated through EIN3 as suggested by the presence of EIN3 binding sites in the promoter region of AUX1 (Chang et al., 2013).

Studies by Zdarska and coworkers provide novel insights into basic principles of hormonal cross talk. They postulated that ETR1-mediated CK−ET cross talk controls cell differentiation in the root transition zone (TZ). They showed that both ETR1-controlled multistep phosphorelay (MSP) and ET signaling are essential for CK- and ET-mediated RAM shortening. MSP signaling is activated by CK in the root stele to increase ET production in roots. ET is recognized by ETR1, which then interacts with AHP's that then phosphorylate ARR-Bs, including ARR10. The canonical ET signal via EIN2 upregulates the level of phosphorylatable ARRs-B via transcriptional upregulation of *ARR10*. Phosphorylation of ARRs-B allows specific control of target genes, such as *ARR3−4*, *ARR6−8*, and *ARR16*, and activates cell differentiation in the root TZ. When exogenous ET is absent, histidine kinase (HK) activity of ETR1 maintains RAM size, possibly via upregulation of mitotic activity of meristematic cells and/or control over the balance between cell division/differentiation in the root TZ (Zdárská et al., 2013, 2019). Reports by Adams and Turner (2010) show COI1 to be involved in ET−JA cross talk in mediating root growth and development. The mutant *coi1−16* was unresponsive to ET-induced root growth inhibition only in light. Additionally, the inhibition of *Arabidopsis* root growth to ACC was found to be JA independent and occurs due to inhibition in cell elongation (Adams and Turner, 2010; Ellis and Turner, 2002).

BR is another player regulating root growth and development (Saini et al., 2015; Planas-Riverola et al., 2019). Studies have shown that both high and low doses of BR inhibit root length. BR signaling mutants as well as *bri1-ems-suppressor 1* (*bes1-D*) (gain-of-function) mutants have short roots (Chaiwanon and Wang, 2015; González-García et al., 2011; Hacham et al., 2011). On the contrary, a study showed that low concentrations of BRs promote root growth in WT plants as well as in BR-deficient mutants (Müssig et al., 2003); however, the changes were small (Müssig et al., 2003). BR modulates root meristem size by affecting both cell proliferation and cell elongation in a concentration-dependent manner (Wei and Li, 2016). Thus this indicates that an optimal level of BR is crucial for root growth and meristem homeostasis. There is a complex cross talk of BR signaling, BR catabolism, and auxin synthesis culminating in a pattern formation of BRASSINAZOLE-RESISTANT 1

(BZR1) in meristem and elongation zone. It has been found that low levels of BZR1 are required to maintain the QC, whereas high levels are required in the elongation zone. It has also been shown that a balance between auxin signaling and BR signaling is required to maintain normal root growth (Chaiwanon and Wang, 2015). Also, BR signaling behaves differently in the epidermal and stele region. BR signaling promotes stem cell proliferation in epidermal region; however, it induces the differentiation of stem cells in the stele region. BR signaling induces target genes in the epidermis particularly related to auxin signaling but mostly represses genes in the stele (Vragović et al., 2015).

Several reports suggest a dose-dependent role of ABA in regulating root growth in plants. In *Arabidopsis*, exogenous treatment of low concentrations of ABA promoted root growth, while higher concentrations inhibited it (Ghassemian et al., 2000; Li et al., 2017). The ABA-mediated inhibition of root growth involves the regulation at the level of cell division. A study found that ABA act at the quiescent center (QC) by promoting quiescence. Further, ABA negatively regulates stem cell differentiation. Through these mechanisms, ABA protects root stem cells in stress conditions (Zhang et al., 2010). ABA also regulates the expression of genes involved in stem cell regulation, such as cyclins, WOX5, MP, and PLTs (Zhang et al., 2010; Yang et al., 2014; Promchuea et al., 2017; Wang et al., 2011). ABA promotes ET signaling to suppress PR growth (Li et al., 2017; Thole et al., 2014). At the molecular level, ABA activates CALCIUM-DEPENDENT PROTEIN KINASE 4 and 11 (CPK4 and 11) that phosphorylate 1-AMINOCYCLOPROPANE-1-CARBOXYLIC ACID SYNTHASE 6 (ACS6) at C-terminus promoting its stability. ACSs are the rate-limiting enzymes in the ET biosynthesis (Luo et al., 2014). Thus ABA is directly involved in the enhancement of ET production which inhibits the PR growth (Fig. 2.1D). ABA negatively regulates the expression of PINs which alter the auxin gradient in the root tip leading to inhibition of root growth. ABA also regulates genes involved in auxin biosynthesis and metabolism to regulate root architecture (Li et al., 2017; Thole et al., 2014; Rowe et al., 2016). In *Arabidopsis*, forward genetic screening identified that ARF2 negatively regulates ABA-mediated regulation of root growth. ARF2 binds to the promoter of HOMEOBOX PROTEIN 33 (HB33), a positive regulator of ABA-mediated regulation of root growth, leading to its transcriptional repression (Wang et al., 2011). ABA also interacts with calcium signaling to regulate cell elongation in root. PROLINE-RICH EXTENSIN-LIKE RECEPTOR KINASE 4 (PERK4) is activated by ABA and Ca^{2+} to suppress root growth through inhibiting cell elongation (Bai et al., 2009) (Fig. 2.1D). PYL8 was found to be the major ABA receptor that coordinates with PP2C—SnRK2 components to regulate ABA-dependent inhibition of root growth (Antoni et al., 2013; Fujii et al., 2007).

Earlier studies overlooked the role of GA in root growth and development (Stowe and Yamaki, 1957); however, it is now evident that GA promotes the PR growth in a concentration-dependent manner and shows inhibitory effects only at higher concentrations (Tanimoto, 1991, 2012). GAs positively regulates the elongation of both dividing and postmitotic endodermal cells, thereby controlling the division and elongation of other types of root cells and the overall root meristem size that results in PR elongation (Ubeda-Tomás et al., 2009). The mechanism by which GAs affect root growth is by modulating the auxin pathway (Tan et al., 2019). TFs MYB12/111 along with DELLA bind to the promoter of *FLAVONOL SYNTHASE* (*FLS1*) and *CHALCONE SYNTHASE* (*CHS*), thus increasing the flavonol biosynthesis that results in the decrement of root length via inhibition of auxin pathway. Incompatible nature of GA toward DELLA promotes the root growth by displacing DELLA through GID1, thus making MYB12/111 inadequate to transcribe the flavanol biosynthesis genes (Tan et al., 2019) (Fig. 2.1E). Root emergence that is a result of the epidermal programed cell death (Steffens et al., 2012) has been shown to be increased by GA (Steffens and Sauter, 2005; Steffens et al., 2006). Along with cell elongation, GA also promotes cell proliferation by increasing the root meristem size through DELLA protein degradation (Ubeda-Tomás et al., 2008, 2009). In accordance with the previous phenomena, DELLA keeps a check on cell proliferation at root meristem by enhancing the cell cycle inhibitors (Achard et al., 2009).

2.5 Senescence

During their life cycle, plants undergo through various growth phases starting with germination, juvenile, and adult vegetative phase, after which they flower and reproduce followed by senescence and ultimately death. Apart from age-dependent, developmental senescence, hypoxia, darkness, and biotic/abiotic stressors are major factors that induce senescence (Pyung et al., 2007). The senescing organs relocate resources to the sink tissues to improve growth and development. For instance, upon floral initiation, leaves relocalize nutrients to reproductive organs to improve reproductive success (Pyung et al., 2007). Leaf senescence is an orderly, regulated degeneration process that is activated at the mature stage of leaf development. The first sign of senescence is the degradation of chlorophyll (Matile et al., 1996), followed by the degradation of macromolecules, such as proteins, membrane lipids, and RNA. Increased catabolism converts cellular materials accumulated during the growth phase of leaf into exportable nutrients that are supplied to developing seeds or to other growing organs. The hormonal pathways happen to play an intimate role at all stages of leaf senescence from the initiation phase of senescence, progression, and the terminal phases.

ET plays a major role in leaf senescence in *Arabidopsis*, crops, and trees. Mutations or downregulation of ET biosynthesis genes cause delay in leaf senescence, whereas exogenous application of ET promotes the process (Koyama, 2014). Thus suggesting the importance of endogenous signaling pathway as a positive regulator of leaf senescence. Preliminary reports have shown that ET biosynthesis is higher during the initial stages of leaf formation and declines upon leaf maturity. However, the levels increase again during the early stages of the senescence initiation (Hunter et al., 1999). ET induces a number of deleterious effects on plants undergoing senescence, such as abscission or induction of necrosis. Other signs include lack of growth, epinasty, and malformed or thickened leaves (Özgen et al., 2005). ET signaling mutant *ethylene-insensitive2* (*ein2*) displayed delayed developmental leaf senescence due to a low expression of senescence-associated genes (*SAGs*) that encode the polygalacturonases and pectinesterases enzymes responsible for cell wall decomposition (Buchanan-Wollaston et al., 2005). Also, EIN3, acting downstream to EIN2, has been shown to positively regulate leaf senescence by activating two senescence-promoting TFs, ORE1 and AtNAP (Kim et al., 2014) (Fig. 2.2). A feed-forward regulation of chlorophyll degradation during leaf senescence was reported where EIN3, ORE1, and chlorophyll catabolic genes (*CCGs*) are involved, advancing the understanding of the

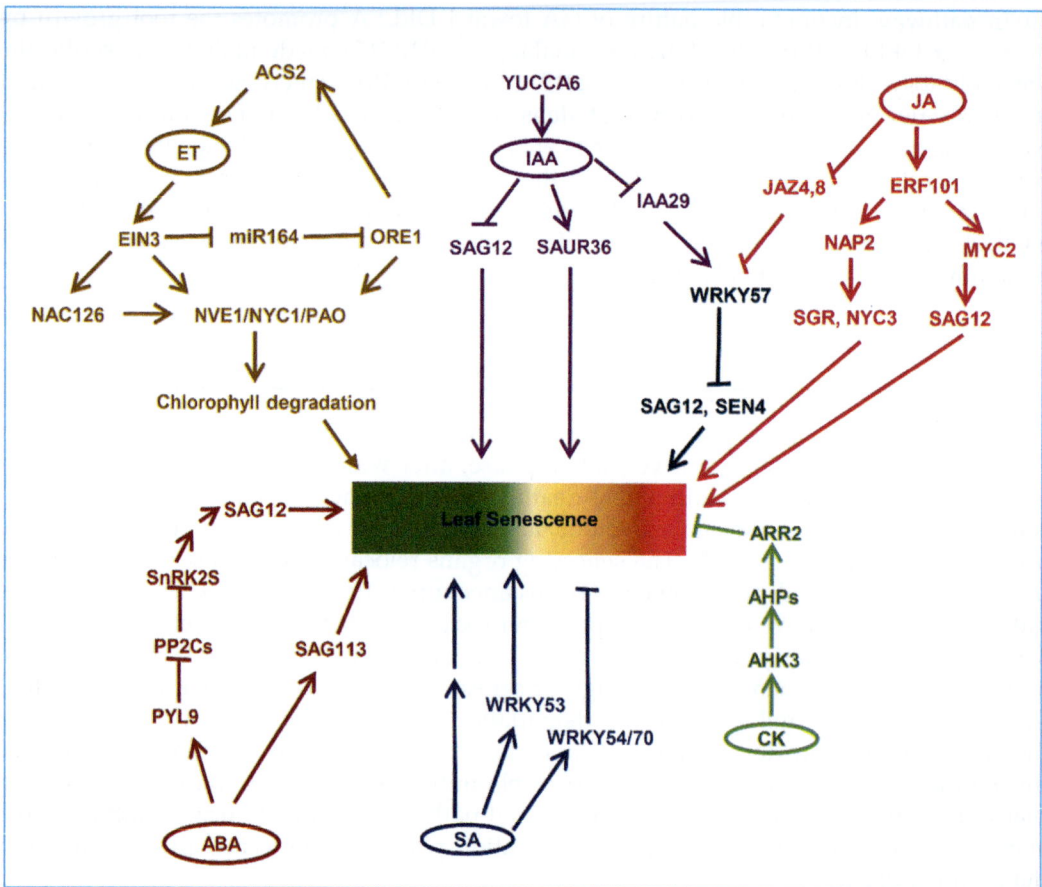

FIGURE 2.2 Regulation of leaf senescence by hormones and their cross talk. A feed-forward loop that involves EIN3 and ORE1 in regulating ethylene-mediated chlorophyll degradation. EIN3 directly represses the transcription of *miR164*, which negatively regulates *ORE1* at the posttranscriptional level. Meanwhile, EIN3 can directly bind to the *ORE1* promoter and induce *ORE1* transcription. Three *CCGs*, *NYE1*, *NYC1*, and *PAO*, are the direct targets of EIN3 and ORE1, which when expressed lead to chlorophyll degradation and leaf senescence. ORE1 directly activates the expression of *ACS2*, which presumably triggers a positive feedback regulation of ethylene synthesis. ERF101 promotes the onset and progression of leaf senescence through JA-mediated signaling pathway. It directly binds to the promoter regions of *OsNAP* and *OsMYC2*, which activate genes involved in chlorophyll degradation and leaf senescence. WRK57 acts as a nodal point between auxin and JA interaction in regulating JA-induced leaf senescence. JAZ4/8 and IAA29 act as negative regulators of JA and auxin pathway, respectively, and competitively interact with WRKY57 that acts as a negative regulator of JA-induced leaf senescence. SA controls the expression of WRKY53, 54, and 70 involved in senescence-related processes. In response to ABA, PYL9 inhibits PP2C activities, resulting in the activation of SnRK2s. Activated SnRK2s promote leaf senescence by elevating the expression of *SAGs*. SAG113 is also involved in leaf senescence. AHK3-mediated regulation of senescence is dependent on its phosphorylation activity of ARR2. Each color represents a distinct signaling pathway. Solid arrows indicate confirmed pathway. *JA*, Jasmonate.

molecular mechanism of leaf senescence (Qiu et al., 2015). EIN3 directly represses the transcription of miR164, which negatively regulates ORE1 at the posttranscriptional level. Meanwhile, EIN3 can directly bind to the *ORE1* promoter and induces *ORE1* transcription. In *ein3eil1* double mutant, the expression of three major *CCGs*, *NON-YELLOW 1* (*NYE1*), *NON-YELLOW COLORING 1* (*NYC1*), and *PHEOPHORBIDE A OXYGENASE* (*PAO*) by ET, was repressed. Additionally, EIN3 and ORE1 could directly bind to and activate the transcription of the previous chlorophyll catabolic genes. Moreover, ORE1 activates a positive feedback mechanism by activating the expression *ACS2* during senescence (Qiu et al., 2015) (Fig. 2.2). An NAC TF, ZmNAC126 was identified to be responsible for natural and ET-triggered leaf senescence in maize (Fig. 2.2). Ectopic overexpression of *ZmNAC126* in *Arabidopsis* and maize enhanced chlorophyll degradation and promoted leaf senescence and also could bind to major chlorophyll catabolic genes in maize. Additionally, ET and *ZmEIN3* were identified to induce *ZmNAC126* and bind to its promoter to transactivate its expression (Yang et al., 2020).

The first evidence of senescence promoting the role of JAs came when MeJA isolated from wormwood caused a rapid degradation of chlorophyll in *Avena sativa* (Ueda and Kato, 1980). It was also observed that *coi1* was defective in JA-dependent senescence, implying that the canonical JA signaling pathway is essential to promote leaf senescence (He et al., 2002). They also reported that exogenous JAs promote senescence in both attached and detached leaves of *Arabidopsis*. During the age-dependent senescence of leaves, an increased expression of JA synthesis genes (*LOX3*, *OPR3*, *AOC1*, and *AOC4*) and signaling genes (*MYC2*, *JAZ1*, *JAZ6*, and *JAZ8*) were reported (Ahmad and Guo, 2019; Van Der Graaff et al., 2006). Functional studies on a nuclear-localized CCCH-type zinc finger protein, OsDOS (*Oryza sativa* Delay of the Onset of Senescence), also support the involvement of MeJA in leaf senescence (Kong et al., 2006). Expression analyses revealed that *OsDOS* acts as a negative regulator for leaf senescence as *OsDOSRNAi* hastened the age-dependent process of leaf senescence, whereas its overexpression resulted in a marked delay of leaf senescence. A genome-wide expression analysis revealed the upregulation of several JA signaling—dependent genes in the RNAi line and downregulation in the overexpression line, thereby suggesting the involvement of OsDOS in JA signaling pathway (Kong et al., 2006). JA-induced leaf senescence is also induced by MYC2, MYC3, and MYC4 TFs. MYC2/3/4 binds to the promoter of *PAO* and increases its transcription leading to chlorophyll degradation. Also, the overexpression lines of MYC2/3/4 increased the transcript level of *PAO* promoter. Additionally, MYC2 could bind and activate the transcription of *SAG29*, which further activates JA-induced leaf senescing (Qi et al., 2015). The *ETHYLENE RESPONSE FACTOR 101* (ERF101) promotes the onset and progression of leaf senescence through JA-mediated signaling pathway. The *Oserf101* T-DNA knockout mutant showed delayed leaf senescence and higher chlorophyll content along with downregulation of several genes involved in chlorophyll degradation like *NYC1*, *NYC3*, *OsNAP2*, *STAY GREEN* (*SGR*) during DIS. Consistent with these observations, the expression of *MYC2* and *COI1* was downregulated in *Oserf101* leaves during DIS. Transient transactivation and chromatin immunoprecipitation assays revealed that OsERF101 directly binds to the promoter regions of *OsNAP* and *OsMYC2*, which activate genes involved in chlorophyll degradation and JA signaling—mediated leaf senescence (Lim et al., 2020) (Fig. 2.2).

As a positive regulator of leaf senescence, SA is involved in both onset and progression of leaf senescence (Ahmad and Guo, 2019; Van Der Graaff et al., 2006). It was reported that SA levels are four times higher in senescing leaves of *Arabidopsis*, which appear to be involved in the upregulation of several SAGs (Morris et al., 2000). SA signaling mutants like *npr1* and *pad4* mutants and *NahG* transgenic plants showed a reduced expression of SAGs, such as *PR1a*, *chitinase*, and *SAG12*. The changes in the expression pattern of genes were accompanied by delayed yellowing and reduced necrosis, suggesting a role for SA in the cell death that occurs at the final stage of senescence (Morris et al., 2000). SA also controls the expression of WRKY TFs *WRKY53, 54*, and *70*, involved in senescence-related processes (Pyung et al., 2007; Besseau et al., 2012) (Fig. 2.2).

Yu and coworkers suggested the involvement of EIN3 and EIL1 in SA-induced leaf senescence in *Arabidopsis*. Furthermore, it was shown that ET enhanced the effect of SA in promoting senescence. Moreover, the transcriptional activity of EIN3 was enhanced by interacting with *NON-EXPRESSER OF PR GENES 1* (NPR1), the master regulator of SA signaling. Thus their study suggests that SA and ET function coordinately in senescence (Yu et al., 2021). Another report of synergism between SA and ET signals revealed that EIN3 and NPR1 synergistically promoted the expression of the SAGs like *ORE1* and *SAG29*. The senescence-related phenotype was delayed in the *ein3eil1npr1* triple mutant as compared with *ein3eil1* or *npr1* upon ET and/or SA treatment (Wang et al., 2021). A recent study by Lee and group showed that RSE1, a glucosyltransferease like protein, acts as a negative regulator of *Arabidopsis* leaf senescence in an SA-dependent manner. Transcriptome analysis showed that SA signaling genes were upregulated in *rse1* mutant. Consistent with the previous observation, the early leaf senescence phenotype of *rse1* was further enhanced upon the addition of SA. A null mutation in the SA biosynthesis gene *SID2*, but not in *PAD4*, restored the early leaf senescence of *rse1*, implying that RSE1 negatively regulates leaf senescence through the *SID2*-dependent SA biosynthesis pathway (Lee et al., 2020). SA and JA signaling pathways also show antagonism in regulating leaf senescence. In the study, it was seen that an application of low concentrations of SA ($1-50\ \mu M$) delayed the process of MeJA-promoted leaf senescence by upregulating nitric oxide synthase activity. Additionally, low concentrations of SA enhanced antioxidant defenses and limited ROS accumulation in MeJA-treated leaves (Ji et al., 2016).

Several studies clearly indicate the positive role of ABA in regulating leaf senescence. Comparative transcriptome analysis revealed the induction of ABA biosynthesis and signaling genes during DIS in *Arabidopsis*. There was also an increase in endogenous ABA levels and signaling in different parts of the plants, such as tobacco, oat, rice, maize, and *Arabidopsis* (Zhang et al., 2012b). SAG113, a PP2C protein phosphatase that acts as a negative regulator of ABA signaling, is also involved in leaf senescence. Its expression is downregulated in biosynthesis and signaling mutants *aba2−1* and *abi4−1*. Also, since *sag113* shows delayed senescence and its overexpression line promotes senescence, SAG113 is a positive regulator of senescence (Zhang et al., 2012b). Another mechanistic regulation on the role of ABA in senescence comes from the study of RPK1, a membrane-bound leucine-rich repeat receptor-like kinase that acts upstream to ABA signaling. Its expression was increased in an ABA-dependent manner throughout the progression of leaf senescence, and ABA-induced leaf senescence was delayed in *rpk1* plants and hastened in the overexpression line, indicative of its promontory role in the pathway

(Lee et al., 2011). Another study has revealed the role of PYL9 in activating SnRK2s to promote drought resistance and leaf senescence. Activated SnRK2s regulate *SAG12* expression, and PYL cannot activate SAG12-LUC expression in *snrk2.2/3/6* triple-mutant protoplasts. Additionally, ABA treatment in *snrk2.2/3/6* mutant did not show enhanced expression of SAG12-LUC expression, suggesting *PYL9* as a promoter of ABA-induced leaf senescence (Zhao et al., 2016) (Fig. 2.2).

BRs are another class of hormones that act as positive regulators of leaf senescence as the application of at epibrassinolide (eBL) on detached wheat leaves accelerated and delayed senescence in high and low concentrations, respectively (Sağlam-Çağ, 2007). BRs also regulate leaf senescence by altering cellular lipid composition as observed in the leaves of *Pisum sativum* (Fedina et al., 2017). The leaves of early senescence mutant *O. sativa premature leaf senescence 1* (*ospls1*) showed suppression of genes encoding BR signaling proteins BRI1 kinase inhibitor 1 (BKI1) and BZR1, indicating the role of BR in leaf senescence. ATBS1-INTERACTING FACTOR 2 (AIF2) is a BIN2-interacting and non-DNA-binding bHLH protein that interacts with ICE1 and leads to retardation of dark triggered and BR-induced leaf senescence in *Arabidopsis*. Upon light illumination, PIF4 is degraded and BIN2-activated AIF2 binds to ICE1 to downregulate *PIF4* expression and antagonistically upregulate *CBF*, which promote senescence-retarding genes. However, upon darkness, PIF4 promotes BR biosynthesis and BZR1 activation leading to degradation of AIF2. As darkness progresses, an increasing expression of PIF4 suppresses senescence-retarding genes such as *CBF* and facilitates an expression of senescence-promoting gene, such as those involved in ET/JA biosynthesis and their signaling pathways (Kim et al., 2020).

The involvement of auxin in leaf senescence has been seen in studies utilizing the external application of auxin on leaves. Studies suggest that auxin level increases during age-dependent leaf senescence due to the increase in the expression of biosynthetic genes encoding tryptophan synthase (*TSA1*), aldehyde oxidase (*AO1*), and nitrilases (*NIT1−3*) are upregulated (Van Der Graaff et al., 2006). Also, an exogenous application of auxin represses the expression of SAG12, a marker gene for developmental senescence (Noh and Amasino, 1999). *Arabidopsis* plants overexpressing the auxin biosynthetic gene *YUCCA6* showed increased accumulation of endogenous auxin and hence delayed leaf senescence and decreased expression of *SAGs* (Kim et al., 2011). Microarray analysis showed the upregulation of *AUXIN RESPONSE FACTOR 2* (*ARF2*) in the senescing leaves (Ellis et al., 2005). In addition, T-DNA insertion mutant of *ARF2* showed a delay in leaf senescing leaves, suggesting ARF2 as a positive regulator of auxin signaling pathway. *NPH4/ARF7* and *ARF19* were also induced upon senescence, and the mutations in these genes enhanced *arf2* phenotypes (Ellis et al., 2005). A report shows the involvement of WRK57 as a nodal point between auxin and JA interaction in regulating JA-induced leaf senescence. JAZ4/8 and IAA29 act as negative regulators of JA and auxin pathway, respectively, and competitively interact with WRKY57, which acts as a negative regulator of JA-induced leaf senescence. It binds to the promoters of *SEN4* and *SAG12* and represses their transcription. At the protein level, JA mediates the degradation of WRKY57, whereas auxin stabilizes WRKY57. Thus WRKY57 works as a common component of auxin and JA signaling in regulating JA-induced leaf senescence (Jiang et al., 2014) (Fig. 2.2).

Richmond and Lang (1957) first revealed the negative role played by CK in leaf senescence with greater chlorophyll retention. Transcriptomic studies have previously reported a decrease in the accumulation in the transcripts of the CK synthesis genes like *cytokinin synthase* and *isopentyl phosphotransferase* (*IPT*) and increased level of CK degrading enzymes like CK oxidase and CK inactivating N- and O-glycosylases (Buchanan-Wollaston et al., 2005). Autoregulation of CK biosynthesis during senescence using an isopentenyl transferase (*IPT*) gene under the regulation of *SAG12* promoter further demonstrated the negative effect of CKs on leaf senescence (Gan and Amasino, 1995). When IPT gene is expressed under the control of SAG12 promoter, delay in leaf senescence was observed in many crops, suggesting that CK-induced leaf senescence is evolutionary conserved across plant species (Bustamante et al., 2016; Iqbal et al., 2017). In addition, senescence reduced ET biosynthesis in the transformed plants and also caused increased tolerance to various stresses and delayed senescences induced by stress (Guo and Gan, 2014). One of the CK receptors AHK3 plays an important role in CK-mediated leaf longevity (Kim et al., 2006) (Fig. 2.2). A gain-of-function mutation in AHK3 delays leaf senescence, whereas the loss-of-function mutation of only *AHK3* conferred reduced sensitivity to CK in CK-mediated delay of leaf senescence. The study also demonstrated that the AHK3-mediated regulation of senescence is dependent on its phosphorylation activity of ARR2; however, the exact mechanism is not known (Kim et al., 2006). A study has shown an interesting link between CK-mediated delay in senescence and extracellular invertase activity, the enzyme responsible for apoplastic phloem unloading. Upon inhibition of invertase, senescence was also repressed in the presence of CK, supporting the notion that CKs may act by modulating the activity of this enzyme (Lara et al., 2004).

2.6 Conclusion

Plant growth and development require the integration of exogenous and endogenous cues that work in tandem with the intrinsic genetic program to govern plant form. In recent years, there have been significant advances in our understanding of the role of phytohormones in regulating various aspects of plant growth and development. It has long been obvious that hormones do not function in simple discrete or linear pathways, rather they show extensive cross talk with each other that can either be synergistic or antagonistic. Not only this, phytohormones also integrate the developmental and environmental signaling pathways, the molecular bases of which have been unclear at many places. In this chapter, we have provided insight into the roles of different hormones and their point of convergence in deciphering molecular mechanisms governing various plant functions. Moreover, multiple reports in this chapter have made clear as to how almost all aspects of growth and development converge at auxin homeostasis (Tiryaki and Staswick, 2002; Sun et al., 2011), hence making it clear the essential role played by auxin transport and signaling in governing plant function. However, there are still lots of gaps in understanding the underlying mechanisms of various aspects of plant growth and development. Hopefully, the advancement of newer technologies and system biology—aided approach will enable us to integrate information related to the overall organization of how plants sense various endogenous and environmental signals, and the mechanisms by which they turn them into cellular responses.

Acknowledgments

The research in AL laboratory is supported by a project grant from Department of Biotechnology, Government of India (Grant no. BT/HRD/NWBA/37/01/2015) and a Core Grant from the National Institute of Plant Genome Research. Ms, duly acknowledge research fellowships from National Institute of Plant Genome Research, Govt of India. The authors acknowledge DBT-eLibrary Consortium (DeLCON) for providing access to e-resources.

Conflict of interest statement

The authors declare that the research was conducted in the absence of any commercial or financial relationships that could be construed as a potential conflict of interest.

Author contributions

A.L. and M.S. conceptualized the chapter. M.S. wrote the original draft and prepared the figures. A.L. supervised the chapter. A.L. and M.S. reviewed and finalized the chapter.

References

Achard, P., Gusti, A., Cheminant, S., Alioua, M., Dhondt, S., Coppens, F., et al., 2009. Gibberellin signaling controls cell proliferation rate in *Arabidopsis*. Curr. Biol.: CB 19 (14), 1188−1193.

Adams, E., Turner, J., 2010. COI1, a jasmonate receptor, is involved in ethylene-induced inhibition of *Arabidopsis* root growth in the light. J. Exp. Bot. 61, 4373−4386.

Ahmad, S., Guo, Y., 2019. Signal transduction in leaf senescence: progress and perspective [Internet]. Plants 8, 405. MDPI AG [cited 2021 May 27]. Available from: <http://www.mdpi.com/journal/plants>.

Aida, M., Tasaka, M., 2006. Genetic control of shoot organ boundaries. Curr. Opin. Plant. Biol. 9 (1), 72−77. Available from: https://doi.org/10.1016/j.pbi.2005.11.011.

Ali, S., Khan, N., Xie, L., 2020. Molecular and hormonal regulation of leaf morphogenesis in *Arabidopsis*. Int. J. Mol. Sci. [Internet]. 21 (14), 1−31. Jul 2 [cited 2022 Jan 31]. Available from: </pmc/articles/PMC7404056/>.

Antoni, R., Gonzalez-Guzman, M., Rodriguez, L., Peirats-Llobet, M., Pizzio, G.A., Fernandez, M.A., et al., 2013. PYRABACTIN RESISTANCE1-LIKE8 plays an important role for the regulation of abscisic acid signaling in root. Plant. Physiol. 161 (2), 931−941.

Bai, L., Zhang, G., Zhou, Y., Zhang, Z., Wang, W., Du, Y., et al., 2009. Plasma membrane-associated proline-rich extensin-like receptor kinase 4, a novel regulator of Ca^{2+} signalling, is required for abscisic acid responses in *Arabidopsis thaliana*. Plant. J. 60 (2), 314−327.

Barbez, E., Dünser, K., Gaidora, A., Lendl, T., Busch, W., 2017. Auxin steers root cell expansion via apoplastic pH regulation in *Arabidopsis thaliana*. Proc. Natl. Acad. Sci. U.S.A. 114 (24). Available from: https://doi.org/10.1073/pnas.1613499114.

Besseau, S., Li, J., Palva, E.T., 2012. WRKY54 and WRKY70 co-operate as negative regulators of leaf senescence in *Arabidopsis thaliana*. J. Exp. Bot. [Internet] 63 (7), 2667−2679. Available from: https://pubmed.ncbi.nlm.nih.gov/22268143/.

Brand, U., Fletcher, J.C., Hobe, M., Meyerowitz, E.M., Simon, R., 2000. Dependence of stem cell fate in *Arabidopsis* on a feedback loop regulated by CLV3 activity. Science. 289 (5479), 617−619. Available from: https://doi.org/10.1126/science.289.5479.617.

Buchanan-Wollaston, V., Page, T., Harrison, E., Breeze, E., Pyung, O.L., Hong, G.N., et al., 2005. Comparative transcriptome analysis reveals significant differences in gene expression and signalling pathways between developmental and dark/starvation-induced senescence in *Arabidopsis*. Plant. J. [Internet] 42 (4), 567−585. Available from: https://pubmed.ncbi.nlm.nih.gov/15860015/.

Bustamante, M., Tomás Matus, J., Riechmann, J.L., 2016. Genome-Wide Analyses for Dissecting Gene Regulatory Networks in the Shoot Apical Meristem. Available from: <https://academic.oup.com/jxb/article-abstract/67/6/1639/2885182/>.

Buzas, D.M., Robertson, M., Finnegan, E.J., Helliwell, C.A., 2011. Transcription-dependence of histone H3 lysine 27 trimethylation at the *Arabidopsis* polycomb target gene FLC. Plant. J. 65 (6), 872–881. Available from: https://doi.org/10.1111/j.1365-313X.2010.04471.x.

Campos-Rivero, G., Osorio-Montalvo, P., Sánchez-Borges, R., Us-Camas, R., Duarte-Aké, F., De-la-Peña, C., 2017. Plant hormone signaling in flowering: an epigenetic point of view. J. Plant. Physiol. 214, 16–27. Available from: https://doi.org/10.1016/j.jplph.2017.03.018.

Chaiwanon, J., Wang, Z.Y., 2015. Spatiotemporal brassinosteroid signaling and antagonism with auxin pattern stem cell dynamics in *Arabidopsis* roots. Curr. Biol. 25 (8), 1031–1042.

Chang, W., Guo, Y., Zhang, H., Liu, X., Guo, L., 2020. Same actor in different stages: genes in shoot apical meristem maintenance and floral meristem determinacy in *Arabidopsis*. Front. Ecol. Evol. 8, 89.

Chang, K.N., Zhong, S., Weirauch, M.T., Hon, G., Pelizzola, M., Li, H., et al., 2013. Temporal transcriptional response to ethylene gas drives growth hormone cross-regulation in *Arabidopsis*. eLife 2, 675.

Cheng, H., Song, S., Xiao, L., Soo, H.M., Cheng, Z., Xie, D., et al., 2009. Gibberellin acts through jasmonate to control the expression of MYB21, MYB24, and MYB57 to promote stamen filament growth in *Arabidopsis*. PLoS Genet. 5 (3). Available from: https://doi.org/10.1371/journal.pgen.1000440.

Chen, H., Ma, B., Zhou, Y., He, S.J., Tang, S.Y., Lu, X., et al., 2018. E3 ubiquitin ligase SOR1 regulates ethylene response in rice root by modulating stability of Aux/IAA protein. Proc. Natl. Acad. Sci. U.S.A. 115 (17), 4513–4518.

Chhun, T., Aya, K., Asano, K., Yamamoto, E., Morinaka, Y., Watanabe, M., et al., 2007. Gibberellin regulates pollen viability and pollen tube growth in rice. The Plant Cell 9 (12), 3876–3888. Available from: https://doi.org/10.1105/tpc.107.054759.

Clark, S.E., Williams, R.W., Meyerowitz, E.M., 1997. The CLAVATA1 gene encodes a putative receptor kinase that controls shoot and floral meristem size in *Arabidopsis*. Cell. 89 (4), 575–585. Available from: https://doi.org/10.1016/s0092-8674(00)80239-1.

Dello Ioio, R., Linhares, F.S., Scacchi, E., Casamitjana-Martinez, E., Heidstra, R., Costantino, P., et al., 2007. Cytokinins determine *Arabidopsis* root-meristem size by controlling cell differentiation. Curr. Biol. 17 (8), 678–682.

Ding, W., Tong, H., Zheng, W., Ye, J., Pan, Z., Zhang, B., et al., 2017. Isolation, characterization and transcriptome analysis of a cytokinin receptor mutant osckt1 in rice. Front. Plant. Sci. 8 (JANUARY), 1–13.

Elliott, R.C., Betzner, A.S., Huttner, E., Oakes, M.P., Tucker, W.Q.J., Gerentes, D., et al., 1996. AINTEGUMENTA, an APETALA2-like gene of *Arabidopsis* with pleiotropic roles in ovule development and floral organ growth. Plant. Cell 8 (8), 155–168. Available from: https://doi.org/10.1105/tpc.8.2.155.

Ellis, C.M., Nagpal, P., Young, J.C., Hagen, G., Guilfoyle, T.J., Reed, J.W., 2005. AUXIN RESPONSE FACTOR1 and AUXIN RESPONSE FACTOR2 regulate senescence and floral organ abscission in *Arabidopsis thaliana*. Development. 132 (20), 4563–4574.

Ellis, C., Turner, J.G., 2002. A conditionally fertile coi1 allele indicates cross-talk between plant hormone signalling pathways in *Arabidopsis thaliana* seeds and young seedlings. Planta. 215 (4), 549–556. Available from: https://doi.org/10.1007/s00425-002-0787-4.

Eriksson, S., Böhlenius, H., Moritz, T., Nilsson, O., 2006. GA4 is the active gibberellin in the regulation of LEAFY transcription and *Arabidopsis* floral initiation. Plant. Cell 18 (9), 2172–2181. Available from: https://doi.org/10.1105/tpc.106.042317.

Fedina, E., Yarin, A., Mukhitova, F., Blufard, A., Chechetkin, I., 2017. Brassinosteroid-induced changes of lipid composition in leaves of *Pisum sativum* L. during senescence. Steroids [Internet] 117, 25–28. Available from: https://pubmed.ncbi.nlm.nih.gov/27815033/.

Fendrych, M., Akhmanova, M., Merrin, J., Glanc, M., Hagihara, S., Takahashi, K., et al., 2018. Rapid and reversible root growth inhibition by TIR1 auxin signalling. Nat. Plants. 4 (7), 453–459. Available from: https://doi.org/10.1038/s41477-018-0190-1.

Fujii, H., Verslues, P.E., Zhu, J.-K., 2007. Identification of two protein kinases required for abscisic acid regulation of seed germination, root growth, and gene expression in *Arabidopsis*. Plant. Cell 19 (2), 485–494.

Furutani, M., Vernoux, T., Traas, J., Kato, T., Tasaka, M., Aida, M., 2004. PIN-FORMED1 and PINOID regulate boundary formation and cotyledon development in *Arabidopsis* embryogenesis. Development. 131 (20), 5021–5030. Available from: https://doi.org/10.1242/dev.01388.

Gan, S., Amasino, R.M., 1995. Inhibition of leaf senescence by autoregulated production of cytokinin [Internet]. Science 270, 1986–1988. American Association for the Advancement of Science [cited 2021 Jun 1]. Available from: <https://pubmed.ncbi.nlm.nih.gov/8592746/>.

Gao, S., Fang, J., Xu, F., Wang, W., Sun, X., Chu, J., et al., 2014. CYTOKININ OXIDASE/DEHYDROGENASE4 integrates cytokinin and auxin signaling to control rice crown root formation. Plant. Physiol. 165 (3), 1035–1046.

Ghassemian, M., Nambara, E., Cutler, S., Kawaide, H., Kamiya, Y., McCourt, P., 2000. Regulation of abscisic acid signaling by the ethylene response pathway in *Arabidopsis*. Plant. Cell 12 (7), 1117–1126.

González-García, M.P., Vilarrasa-Blasi, J., Zhiponova, M., Divol, F., Mora-García, S., Russinova, E., et al., 2011. Brassinosteroids control meristem size by promoting cell cycle progression in *Arabidopsis* roots. Development. 138 (5), 849–859.

Goto, N., Pharis, R.P., 1999. Role of gibberellins in the development of floral organs of the gibberellin-deficient mutant, ga1-1, of *Arabidopsis thaliana*. Can. J. Bot. 77 (7), 944–954. Available from: https://doi.org/10.1139/b99-090.

Griffiths, J., Murase, K., Rieu, I., Zentella, R., Zhang, Z.L., Powers, S.J., et al., 2006. Genetic characterization and functional analysis of the GID1 gibberellin receptors in *Arabidopsis*. Plant. Cell 18 (12), 3399–3414. Available from: https://doi.org/10.1105/tpc.106.047415.

Guo, Y., Gan, S.S., 2014. Translational researches on leaf senescence for enhancing plant productivity and quality [Internet]. J. Exp. Botany 65, 3901–3913. Oxford University Press [cited 2021 Jun 1]. Available from: <https://academic.oup.com/jxb/article/65/14/3901/2877519>.

Guo, Y., Wu, Q., Xie, Z., Yu, B., Zeng, R., Min, Q., et al., 2020. OsFPFL4 is involved in the root and flower development by affecting auxin levels and ROS accumulation in rice (*Oryza sativa*). Rice. 13 (1). Available from: https://doi.org/10.1186/s12284-019-0364-0.

Hacham, Y., Holland, N., Butterfield, C., Ubeda-Tomas, S., Bennett, M.J., Chory, J., et al., 2011. Brassinosteroid perception in the epidermis controls root meristem size. Development. 138 (5), 839–848.

Hadfi, K., Speth, V., Neuhaus, G., 1998. Auxin-induced developmental patterns in *Brassica juncea* embryos. Development. 125 (5), 879–887. Available from: https://doi.org/10.1242/dev.125.5.879.

Hake, S., Smith, H.M.S., Holtan, H., Magnani, E., Mele, G., Ramirez, J., 2004. The role of Knox genes in plant development. Annu. Rev. Cell Dev. Biol. 20, 125–151. Available from: https://doi.org/10.1146/annurev.cellbio.20.031803.093824.

Hay, A., Craft, J., Tsiantis, M., 2004. Plant hormones and homeoboxes: Bridging the gap? BioEssays. 26 (4), 395–404. Available from: https://doi.org/10.1002/bies.20016.

Heisler, M.G., Ohno, C., Das, P., Sieber, P., Reddy, G.V., Long, J.A., et al., 2005. Patterns of auxin transport and gene expression during primordium development revealed by live imaging of the *Arabidopsis* inflorescence meristem. Curr. Biol. 15 (21), 1899–1911. Available from: https://doi.org/10.1016/j.cub.2005.09.052.

Henderson, J.T., Li, H.C., Rider, S.D., Mordhorst, A.P., Romero-Severson, J., Cheng, J.C., et al., 2004. Pickle acts throughout the plant to repress expression of embryonic traits and may play a role in gibberellin-dependent responses. Plant. Physiol. 134 (3), 995–1005. Available from: https://doi.org/10.1104/pp.103.030148.

He, Y., Fukushige, H., Hildebrand, D.F., Gan, S., 2002. Evidence supporting a role of jasmonic acid in *Arabidopsis* leaf senescence. Plant. Physiol. [Internet] 128 (3), 876–884. Available from: http://www.plantphysiol.org/cgi/doi/10.1104/pp.010843.

Higuchi, M., Pischke, M.S., Mähönen, A.P., Miyawaki, K., Hashimoto, Y., Seki, M., et al., 2004. In planta functions of the *Arabidopsis* cytokinin receptor family. Proc. Natl. Acad. Sci. U.S.A. 101 (23), 8821–8826.

Hill, K., Mathews, D.E., Kim, H.J., Street, I.H., Wildes, S.L., Chiang, Y.H., et al., 2013. Functional characterization of type-B response regulators in the *Arabidopsis* cytokinin response. Plant. Physiol. 162 (1), 212–224.

Hou, X., Lee, L.Y.C., Xia, K., Yan, Y., Yu, H., 2010. DELLAs modulate jasmonate signaling via competitive binding to JAZs. Dev. Cell 19 (6), 884–894. Available from: https://doi.org/10.1016/j.devcel.2010.10.024.

Huang, R., 2018. Auxin controlled by ethylene steers root development. Int. J. Mol. Sci. 19 (11), 3656. Available from: https://doi.org/10.3390/ijms19113656.

Hunter, D.A., Yoo, S.D., Butcher, S.M., McManus, M.T., 1999. Expression of 1-aminocyclopropane-1-carboxylate oxidase during leaf ontogeny in white clover. Plant Physiol. [Internet] 120 (1), 131–141. Available from: https://doi.org/10.1104/pp.120.1.131. PMC59245.

Hu, J., Mitchum, M.G., Barnaby, N., Ayele, B.T., Ogawa, M., Nam, E., et al., 2008. Potential sites of bioactive gibberellin production during reproductive growth in *Arabidopsis*. Plant. Cell 20 (2), 320–336. Available from: https://doi.org/10.1105/tpc.107.057752.

Hu, Y., Vandenbussche, F., Van Der Straeten, D., 2017. Regulation of seedling growth by ethylene and the ethylene–auxin crosstalk. Planta.

Iqbal, N., Khan, N.A., Ferrante, A., Trivellini, A., Francini, A., Khan, M.I.R., 2017. Ethylene role in plant growth, development and senescence: interaction with other phytohormones. Front. Plant. Sci. [Internet] 8, 475. Available from: http://www.frontiersin.org.

Ito, T., Ng, K.H., Lim, T.S., Yu, H., Meyerowitz, E.M., 2007. The homeotic protein AGAMOUS controls late stamen development by regulating a jasmonate biosynthetic gene in *Arabidopsis*. Plant. Cell 19 (11), 3516–3529. Available from: https://doi.org/10.1105/tpc.107.055467.

Jackson, D., Hake, S., 1999. Control of phyllotaxy in maize by the abphyl1 gene. Development. 126 (2), 315–323. Available from: https://doi.org/10.1242/dev.126.2.315.

Jacobsen, S.E., Olszewski, N.E., 1993. Mutations at the SPINDLY locus of *Arabidopsis* alter gibberellin signal transduction. Plant. Cell 5 (8), 887–896. Available from: https://doi.org/10.1105/tpc.5.8.887.

Jiang, Y., Liang, G., Yang, S., Yu, D., 2014. *Arabidopsis* WRKY57 functions as a node of convergence for jasmonic acid- and auxin-mediated signaling in jasmonic acid-induced leaf senescence. Plant. Cell 26 (1), 230–245. Available from: https://doi.org/10.1105/tpc.113.117838.

Ji, Y., Liu, J., Xing, D., 2016. Low concentrations of salicylic acid delay methyl jasmonate-induced leaf senescence by up-regulating nitric oxide synthase activity. J. Exp. Bot. [Internet] 67 (17), 5233–5245. Available from: https://academic.oup.com/jxb/article/67/17/5233/2197712.

De Jonge, J., Kodde, J., Severing, E.I., Bonnema, G., Angenent, G.C., Immink, R.G.H., et al., 2016. Low temperature affects stem cell maintenance in *Brassica oleracea* seedlings. Front. Plant. Sci. 7 (June2016), 800.

Ju, Y., Guo, L., Cai, Q., Ma, F., Zhu, Q.Y., Zhang, Q., et al., 2016. *Arabidopsis* JINGUBANG is a negative regulator of pollen germination that prevents pollination in moist environments. Plant. Cell 28 (9), 2131–2146. Available from: https://doi.org/10.1105/tpc.16.00401.

Kania, T., Russenberger, D., Peng, S., Apel, K., Melzer, S., 1997. FPF1 promotes flowering in *Arabidopsis*. Plant. Cell 9 (8), 1327–1338. Available from: https://doi.org/10.1105/tpc.9.8.1327.

Kende, H., Zeevaart, J.A.D., 1997. The five "Classical" plant hormones. Plant. Cell 9, 1197–1210.

Kim, H.J., Hong, S.H., Kim, Y.W., Lee, I.H., Jun, J.H., Phee, B.K., et al., 2014. Gene regulatory cascade of senescence-associated NAC transcription factors activated by ETHYLENE-INSENSITIVE2-mediated leaf senescence signalling in *Arabidopsis*. J. Exp. Bot. [Internet] 65 (14), 4023–4036. Available from: https://pubmed.ncbi.nlm.nih.gov/24659488/.

Kim, J.I., Murphy, A.S., Baek, D., Lee, S.W., Yun, D.J., Bressan, R.A., et al., 2011. YUCCA6 over-expression demonstrates auxin function in delaying leaf senescence in *Arabidopsis thaliana*. J. Exp. Bot. [Internet] 62 (11), 3981–3992. Available from: https://pubmed.ncbi.nlm.nih.gov/21511905/.

Kim, Y., Park, S.U., Shin, D.M., Pham, G., Jeong, Y.S., Kim, S.H., 2020. ATBS1-INTERACTING FACTOR 2 negatively regulates dark-and brassinosteroid-induced leaf senescence through interactions with INDUCER of CBF EXPRESSION 1. J. Exp. Bot. [Internet] 71 (4), 1475–1490. Available from: https://www.ncbi.nlm.nih.gov/gds.

Kim, H.J., Ryu, H., Hong, S.H., Woo, H.R., Lim, P.O., Lee, I.C., et al., 2006. Cytokinin-mediated control of leaf longevity by AHK3 through phosphorylation of ARR2 in *Arabidopsis*. Proc. Natl. Acad. Sci. U.S.A. [Internet] 103 (3), 814–819. Available from: http://www.pnas.orgcgidoi10.1073pnas.0505150103.

King, R.W., Moritz, T., Evans, L.T., Junttila, O., Herlt, A.J., 2001. Long-day induction of flowering in Lolium temulentum involves sequential increases in specific gibberellins at the shoot apex. Plant. Physiol. 127 (2), 624–632.

King, R.W., Moritz, T., Evans, L.T., Martin, J., Andersen, C.H., Blundell, C., et al., 2006. Regulation of flowering in the long-day grass *Lolium temulentum* by gibberellins and the FLOWERING LOCUS T gene. Plant. Physiol. 141 (2), 498–507. Available from: https://doi.org/10.1104/pp.106.076760.

Kong, Z., Li, M., Yang, W., Xu, W., Xue, Y., 2006. A novel nuclear-localized CCCH-type zinc finger protein, OsDOS, is involved in delaying leaf senescence in rice. Plant. Physiol. [Internet] 141 (4), 1376–1388. Available from: https://doi.org/10.1104/pp.106.082941. PMC1533915.

Koyama, T., 2014. The roles of ethylene and transcription factors in the regulation of onset of leaf senescence [Internet]. Front. Plant Sci. 5. Frontiers Media S.A. [cited 2021 May 27]. Available from: </pmc/articles/PMC4243489/>.

Krizek, B.A., Eaddy, M., 2012. AINTEGUMENTA-LIKE6 regulates cellular differentiation in flowers. Plant. Mol. Biol. 78 (3), 199–209. Available from: https://doi.org/10.1007/s11103-011-9844-3.

Lang, A., 1957. The effect of gibberellin upon flower formation. Proc. Natl. Acad. Sci. U.S.A. 43 (8), 709−717.

Lara, M.E.B., Garcia, M.C.G., Fatima, T., Ehneß, R., Lee, T.K., Proels, R., et al., 2004. Extracellular invertase is an essential component of cytokinin-mediated delay of senescence. Plant. Cell [Internet] 16 (5), 1276−1287. Available from: https://doi.org/10.1105/tpc.018929. PMC423215.

Lee, I.C., Hong, S.W., Whang, S.S., Lim, P.O., Nam, H.G., Koo, J.C., 2011. Age-dependent action of an ABA-inducible receptor kinase, RPK1, as a positive regulator of senescence in *Arabidopsis* leaves. Plant. Cell Physiol. [Internet] 52 (4), 651−662. Available from: https://academic.oup.com/pcp/article/52/4/651/1853992.

Lee, S., Kim, M.H., Lee, J.H., Jeon, J., Kwak, J.M., Kim, Y.J., 2020. Glycosyltransferase-like RSE1 negatively regulates leaf senescence through salicylic acid signaling in *Arabidopsis*. Front. Plant. Sci. [Internet] 11, 551. Available from: http://www.frontiersin.org.

Lehmann, T., Janowitz, T., Sánchez-Parra, B., Alonso, M.M.P., Trompetter, I., Piotrowski, M., et al., 2017. *Arabidopsis* NITRILASE 1 contributes to the regulation of root growth and development through modulation of auxin biosynthesis in seedlings. Front. Plant. Sci. 8 (36). Available from: https://doi.org/10.3389/fpls.2017.00036.

Leibfried, A., To, J.P.C., Busch, W., Stehling, S., Kehle, A., Demar, M., et al., 2005. WUSCHEL controls meristem function by direct regulation of cytokinin-inducible response regulators. Nature 438 (7071), 1172−1175. Available from: https://doi.org/10.1038/nature04270.

Lim, C., Kang, K., Shim, Y., Sakuraba, Y., An, G., Paek, N.C., 2020. Rice ETHYLENE RESPONSE FACTOR 101 promotes leaf senescence through jasmonic acid-mediated regulation of OsNAP and OsMYC2. Front. Plant. Sci. [Internet] 11, 1096. Available from: http://www.frontiersin.org.

Li, M., An, F., Li, W., Ma, M., Feng, Y., Zhang, X., et al., 2016. DELLA proteins interact with FLC to repress flowering transition. J. Integr. Plant. Biol. 58 (7), 642−655. Available from: https://doi.org/10.1111/jipb.12451.

Li, X., Chen, L., Forde, B.G., Davies, W.J., 2017. The biphasic root growth response to abscisic acid in *Arabidopsis* involves interaction with ethylene and auxin signalling pathways. Front. plant. Sci. 8, 1493.

Li, J., Li, Y., Chen, S., An, L., 2010. Involvement of brassinosteroid signals in the floral-induction network of *Arabidopsis*. J. Exp. Bot. 61 (15), 4221−4230. Available from: https://doi.org/10.1093/jxb/erq241.

Li, X.R., Li, H.J., Yuan, L., Liu, M., Shi, D.Q., Liu, J., et al., 2014. *Arabidopsis* DAYU/ABERRANT PEROXISOME MORPHOLOGY9 is a key regulator of peroxisome biogenesis and plays critical roles during pollen maturation and germination in planta. Plant. Cell 26 (2), 619−635. Available from: https://doi.org/10.1105/tpc.113.121087.

Li, X., Mo, X., Shou, H., Wu, P., 2006. Cytokinin-mediated cell cycling arrest of pericycle founder cells in lateral root initiation of *Arabidopsis*. Plant. Cell Physiol. 47 (8), 1112−1123.

Luo, X., Chen, Z., Gao, J., Gong, Z., 2014. Abscisic acid inhibits root growth in *Arabidopsis* through ethylene biosynthesis. Plant. J. 79 (1), 44−55.

Mandaokar, A., Browse, J., 2009. MYB108 acts together with MYB24 to regulate jasmonate-mediated stamen maturation in *Arabidopsis*. Plant. Physiol. 149 (2), 851−862. Available from: https://doi.org/10.1104/pp.108.132597.

Mao, C., He, J., Liu, L., Deng, Q., Yao, X., Liu, C., et al., 2020. OsNAC2 integrates auxin and cytokinin pathways to modulate rice root development. Plant. Biotechnol. J. 18 (2), 429−442. Available from: https://doi.org/10.1111/pbi.13209.

Mao, J.L., Miao, Z.Q., Wang, Z., Yu, L.H., Cai, X.T., Xiang, C.B., 2016. *Arabidopsis* ERF1 mediates cross-talk between ethylene and auxin biosynthesis during primary root elongation by regulating ASA1 expression. PLoS Genet.

Mason, M.G., Mathews, D.E., Argyros, D.A., Maxwell, B.B., Kieber, J.J., Alonso, J.M., et al., 2005. Multiple type-B response regulators mediate cytokinin signal transduction in *Arabidopsis*. Plant. Cell 17 (11), 3007−3018.

Matile, P., Hörtensteiner, S., Thomas, H., Kräutler, B., 1996. Chlorophyll breakdown in senescent leaves [Internet]. Plant Physiol. American Society of Plant Biologists [cited 2021 May 27] 112, 1403−1409. Available from: <https://plantphysiol.org>.

Ma, Y., Miotk, A., Šutiković, Z., Ermakova, O., Wenzl, C., Medzihradszky, A., et al., 2019. WUSCHEL acts as an auxin response rheostat to maintain apical stem cells in *Arabidopsis*. Nat. Commun. 10 (1), 5093. Available from: https://doi.org/10.1038/s41467-019-13074-9.

Moon, J., Suh, S.S., Lee, H., Choi, K.R., Hong, C.B., Paek, N.C., et al., 2003. The SOC1 MADS-box gene integrates vernalization and gibberellin signals for flowering in *Arabidopsis*. Plant. J. 35 (5), 613−623. Available from: https://doi.org/10.1046/j.1365-313x.2003.01833.x.

Morris, K., Mackerness, S.A.H., Page, T., Fred John, C., Murphy, A.M., Carr, J.P., et al., 2000. Salicylic acid has a role in regulating gene expression during leaf senescence. Plant J. [Internet] 23 (5), 677−685. Available from: https://onlinelibrary.wiley.com/doi/full/10.1046/j.1365-313x.2000.00836.x.

Márquez, G., Alarcón, M.V., Salguero, J., 2019. Cytokinin inhibits lateral root development at the earliest stages of lateral root primordium initiation in maize primary root. J. Plant. Growth Regul. 38 (1), 83–92.

Méndez-Bravo, A., Ruiz-Herrera, L.F., Cruz-Ramírez, A., Guzman, P., Martínez-Trujillo, M., Ortiz-Castro, R., et al., 2019. CONSTITUTIVE TRIPLE RESPONSE1 and PIN2 act in a coordinate manner to support the indeterminate root growth and meristem cell proliferating activity in *Arabidopsis* seedlings. Plant. Sci. 280, 175–186.

Müssig, C., Shin, G.H., Altmann, T., 2003. Brassinosteroids promote root growth in *Arabidopsis*. Plant. Physiol. 133 (3), 1261–1271.

Nan, W., Wang, X., Yang, L., Hu, Y., Wei, Y., Liang, X., et al., 2014. Cyclic GMP is involved in auxin signalling during *Arabidopsis* root growth and development. J. Exp. Bot. 65 (6), 1571–1583. Available from: https://doi.org/10.1093/jxb/eru019.

Nester, J.E., Zeevaart, J.A.D., 1988. Flower development in normal tomato and a gibberellin-deficient (ga-2) mutant. Am. J. Bot. 75 (1), 45–55.

Nikonorova, N., Murphy, E., Fonseca de Lima, C.F., Zhu, S., van de Cotte, B., Vu, L.D., et al., 2021. Auxin-triggered changes in the *Arabidopsis* root tip (phospho)proteome reveal novel root growth regulators. bioRxiv. Available from: https://doi.org/10.1101/2021.03.04.433936.

Nishimura, C., Ohashi, Y., Sato, S., Kato, T., Tabata, S., Ueguchi, C., 2004. Histidine kinase homologs that act as cytokinin receptors possess overlapping functions in the regulation of shoot and root growth in *Arabidopsis*. Plant. Cell 16 (6), 1365–1377.

Ni, W., Dai, S., Karger, B.L., Zhou, Z.S., 2010. Analysis of isoaspartic acid by selective proteolysis with Asp-N and electron transfer dissociation mass spectrometry. Anal. Chem. [Internet] 82 (17), 7485–7491. Available from: https://pubs.acs.org/doi/10.1021/ac101806e.

Noh, Y.S., Amasino, R.M., 1999. Identification of a promoter region responsible for the senescence-specific expression of SAG12. Plant Mol. Biol. [Internet] 41 (2), 181–194. Available from: https://link.springer.com/article/10.1023/A:1006342412688.

Ogas, J., Kaufmann, S., Henderson, J., Somerville, C., 1999. PICKLE is a CHD3 chromatin-remodeling factor that regulates the transition from embryonic to vegetative development in *Arabidopsis*. Proc. Natl. Acad. Sci. U.S.A 96 (24), 13839–13844. Available from: https://doi.org/10.1073/pnas.96.24.13839.

Ogawa, M., Shinohara, H., Sakagami, Y., Matsubayash, Y., 2008. *Arabidopsis* CLV3 peptide directly binds CLV1 ectodomain. Science. 319 (5861). Available from: https://doi.org/10.1126/science.1150083.

Okada, K., Ueda, J., Komaki, M.K., Bell, C.J., Shimura, Y., 1991. Requirement of the auxin polar transport system in early stages of *Arabidopsis* floral bud formation. Plant. Cell 3 (7), 677–684. Available from: https://doi.org/10.1105/tpc.3.7.677.

Özgen, M., Park, S., Palta, J.P., 2005. Mitigation of ethylene-promoted leaf senescence by a natural lipid, lysophosphatidylethanolamine. HortScience. 40 (5), 1166–1167.

Park, J., Oh, D.H., Dassanayake, M., Nguyen, K.T., Ogas, J., Choi, G., et al., 2017. Gibberellin signaling requires chromatin remodeler PICKLE to promote vegetative growth and phase transitions. Plant. Physiol. 173 (2), 1463–1474. Available from: https://doi.org/10.1104/pp.16.01471.

Planas-Riverola, A., Gupta, A., Betegoń-Putze, I., Bosch, N., Ibaneş, M., Cano-Delgado, A.I., 2019. Brassinosteroid signaling in plant development and adaptation to stressVol Dev. (Cambridge). Co. Biologists Ltd.

Promchuea, S., Zhu, Y., Chen, Z., Zhang, J., Gong, Z., 2017. ARF2 coordinates with PLETHORAs and PINs to orchestrate ABA-mediated root meristem activity in *Arabidopsis*. J. Integr. Plant. Biol. 59 (1), 30–43.

Przemeck, G.K.H., Mattsson, J., Hardtke, C.S., Sung, Z.R., Berleth, T., 1996. Studies on the role of the *Arabidopsis* gene MONOPTEROS in vascular development and plant cell axialization. Planta. 22 (2), 229–237. Available from: https://doi.org/10.1007/BF00208313.

Pyung, O.L., Hyo, J.K., Hong, G.N., 2007. Leaf senescence [Internet]. Annu. Rev. Plant. Biol. [cited 2021 May 27]. 58, 115–36. Available from: <https://pubmed.ncbi.nlm.nih.gov/17177638/>.

Qin, H., He, L., Huang, R., 2019a. The coordination of ethylene and other hormones in primary root development. Front. Plant. Sci. 10, 874. Published 2019 Jul 10. Available from: https://doi.org/10.3389/fpls.2019.00874.

Qin, H., Wang, J., Chen, X., Wang, F., Peng, P., Zhou, Y., et al., 2019b. Rice OsDOF15 contributes to ethylene-inhibited primary root elongation under salt stress. N. Phytol 223 (2), 798–813.

Qin, H., Zhang, Z., Wang, J., Chen, X., Wei, P., Huang, R., 2017. The activation of OsEIL1 on YUC8M transcription and auxin biosynthesis is required for ethylene-inhibited root elongation in rice early seedling development. PLoS Genet. 13 (8).

Qiu, K., Li, Z., Yang, Z., Chen, J., Wu, S., Zhu, X., et al., 2015. EIN3 and ORE1 accelerate degreening during ethylene-mediated leaf senescence by directly activating chlorophyll catabolic genes in *Arabidopsis*. PLoS Genet. [Internet] 11 (7), e1005399. Available from: http://www.stcsm.gov.cn/http://www.nsfc.gov.cn/publish/portal0/default.htm.

Qi, T., Wang, J., Huang, H., Liu, B., Gao, H., Liu, Y., et al., 2015. Regulation of jasmonate-induced leaf senescence by antagonism between bHLH subgroup IIIe and IIId factors in *Arabidopsis*. Plant. Cell 27 (6), 1634–1649. Available from: https://doi.org/10.1105/tpc.15.00110.

Ramireddy, E., Hosseini, S.A., Eggert, K., Gillandt, S., Gnad, H., von Wirén, N., et al., 2018. Root engineering in barley: increasing cytokinin degradation produces a larger root system, mineral enrichment in the shoot and improved drought tolerance. Plant. Physiol. [Internet] 177 (3), 1078–1095. Available from: https://www.plant-physiol.org/cgi/doi/10.1104/pp.18.00199.

Reinhardt, D., Mandel, T., Kuhlemeier, C., 2000. Auxin regulates the initiation and radial position of plant lateral organs. Plant. Cell 12 (4), 507–518. Available from: https://doi.org/10.1105/tpc.12.4.507.

Richmond, A.E., Lang, A. 1957. Effect of kinetin on protein content and survival of detached xanthium leaves. Sci. [Internet]. Apr 5 [cited 2021 Jun 1]; 125 (3249), 650–651. Available from: <https://science.sciencemag.org/content/125/3249/650.2>.

Riefler, M., Novak, O., Strnad, M., Schmülling, T., 2006. *Arabidopsis* cytokinin receptor mutants reveal functions in shoot growth, leaf senescence, seed size, germination, root development, and cytokinin metabolism. Plant. Cell 18 (1), 40–54.

Rowe, J.H., Topping, J.F., Liu, J., Lindsey, K., 2016. Abscisic acid regulates root growth under osmotic stress conditions via an interacting hormonal network with cytokinin, ethylene and auxin. N. Phytol. 211 (1), 225–239.

Růžička, K., Ljung, K., Vanneste, S., Podhorská, R., Beeckman, T., Friml, J., et al., 2007a. Ethylene regulates root growth through effects on auxin biosynthesis and transport-dependent auxin distribution. Plant Cell [Internet] 19 (7), 2197–2212. Available from: http://www.plantcell.org/cgi/doi/10.1105/tpc.107.052126.

Růžička, K., Ljung, K., Vanneste, S., Podhorská, R., Beeckman, T., Friml, J., et al., 2007b. Ethylene regulates root growth through effects on auxin biosynthesis and transport-dependent auxin distribution. Plant. Cell 19 (7), 2197–2212.

Sachs, R.M., Lang, A., 1957. Effect of gibberellin on cell division in Hyoscyamus. Science. 125 (3258), 1144–1145. Available from: https://doi.org/10.1126/science.125.3258.1144.

Saini, S., Sharma, I., Pati, P.K., 2015. Versatile roles of brassinosteroid in plants in the context of its homoeostasis, signaling and crosstalksVol Front. Plant. Science. Front. Res. Found.

Satbhai, S.B., Ristova, D., Busch, W., 2015. Underground tuning: quantitative regulation of root growth. J. Exp. Bot. [Internet] 66 (4), 1099–1112. Available from: https://academic.oup.com/jxb/article/66/4/1099/593973.

Sağlam-Çağ, S., 2007. The effect of epibrassinolide on senescence in wheat leaves. Biotechnol. Biotechnol. Equip. [Internet] 21 (1), 63–65. Available from: https://doi.org/10.1080/13102818.2007.10817415.

Schoof, H., Lenhard, M., Haecker, A., Mayer, K.F.X., Jürgens, G., Laux, T., 2000. The stem cell population of *Arabidopsis* shoot meristems is maintained by a regulatory loop between the CLAVATA and WUSCHEL genes. Cell 100 (6), 635–644. Available from: https://doi.org/10.1016/s0092-8674(00)80700-x.

Schultz, E.A., Haughn, G.W., 1991. LEAFY, a homeotic gene that regulates inflorescence development in *Arabidopsis*. Plant. Cell 3 (8), 771–781. Available from: https://doi.org/10.1105/tpc.3.8.771.

Sharma, M., Singh, D., Saksena, H.B., Sharma, M., Tiwari, A., Awasthi, P., et al., 2021. Understanding the intricate web of phytohormone signalling in modulating root system architecture. Int. J. Mol. Sci. 22 (11), 5508. Available from: https://doi.org/10.3390/ijms22115508.

Shih, C.F., Hsu, W.H., Peng, Y.J., Yang, C.H., 2014. The NAC-like gene ANTHER INDEHISCENCE FACTOR acts as a repressor that controls anther dehiscence by regulating genes in the jasmonate biosynthesis pathway in *Arabidopsis*. J. Exp. Bot. 65 (2), 621–639. Available from: https://doi.org/10.1093/jxb/ert412.

Silverstone, A.L., Sun, T., 2000. Gibberellins and the green revolution. Trends Plant. Sci. 5 (1), 1–2. Available from: https://doi.org/10.1016/s1360-1385(99)01516-2.

Šimášková, M., O'Brien, J.A., Khan, M., Van Noorden, G., Ötvös, K., Vieten, A., et al., 2015a. Cytokinin response factors regulate PIN-FORMED auxin transporters. Nat. Commun. [Internet] 6. Available from: https://pubmed.ncbi.nlm.nih.gov/26541513/.

Šimášková, M., O'Brien, J.A., Khan, M., Van Noorden, G., Ötvös, K., Vieten, A., et al., 2015b. Cytokinin response factors regulate PIN-FORMED auxin transporters. Nat. Commun. 6.

Snow, M., Snow, R., 1937. Auxin and leaf formation. N. Phytol 36 (1), 1−18. Available from: https://doi.org/10.1111/j.1469-8137.1937.tb06899.x.

Song, S., Qi, T., Huang, H., Ren, Q., Wu, D., Chang, C., et al., 2011. The jasmonate-ZIM domain proteins interact with the R2R3-MYB transcription factors MYB21 and MYB24 to affect jasmonate-regulated stamen development in Arabidopsis. Plant. Cell 23 (3), 1000−1013. Available from: https://doi.org/10.1105/tpc.111.083089.

Steffens, B., Kovalev, A., Gorb, S.N., Sauter, M., 2012. Emerging roots alter epidermal cell fate through mechanical and reactive oxygen species signaling. Plant. Cell 24 (8), 3296−3306.

Steffens, B., Sauter, M., 2005. Epidermal cell death in rice is regulated by ethylene, gibberellin, and abscisic acid. Plant. Physiol. 139 (2), 713−721.

Steffens, B., Wang, J., Sauter, M., 2006. Interactions between ethylene, gibberellin and abscisic acid regulate emergence and growth rate of adventitious roots in deepwater rice. Planta. 223 (3), 604−612.

Stepanova, A.N., Yun, J., Likhacheva, A.V., Alonso, J.M., 2007. Multilevel interactions between ethylene and auxin in Arabidopsis roots. Plant. Cell 19 (7), 2169−2185.

Stintzi, A., Browse, J., 2000. The Arabidopsis male-sterile mutant, opr3, lacks the 12-oxophytodienoic acid reductase required for jasmonate synthesis. Proc. Natl. Acad. Sci. U.S.A. 97 (19), 10625−10630. Available from: https://doi.org/10.1073/pnas.190264497.

Stowe, B.B., Yamaki, T., 1957. The history and physiological action of the gibberellins. Annu. Rev. Plant. Physiol. 8 (1), 181−216.

Street, I.H., Aman, S., Zubo, Y., Ramzan, A., Wang, X., Shakeel, S.N., et al., 2015. Ethylene inhibits cell proliferation of the Arabidopsis root meristem. Plant. Physiol. 169 (1), 338−350.

Street, I.H., Mathews, D.E., Yamburkenko, M.V., Sorooshzadeh, A., John, R.T., Swarup, R., et al., 2016. Cytokinin acts through the auxin influx carrier AUX1 to regulate cell elongation in the root. Dev. (Camb.) 143 (21), 3982−3993.

Sugiyama, S.I., Gotoh, M., 2010. How meristem plasticity in response to soil nutrients and light affects plant growth in four Festuca grass species. N. Phytol. [Internet] 185 (3), 747−758. Available from: https://onlinelibrary.wiley.com/doi/full/10.1111/j.1469-8137.2009.03090.x.

Sun, J., Chen, Q., Qi, L., Jiang, H., Li, S., Xu, Y., et al., 2011. Jasmonate modulates endocytosis and plasma membrane accumulation of the Arabidopsis pin2 protein. N. Phytol. 191 (2), 360−375.

Sun, J., Xu, Y., Ye, S., Jiang, H., Chen, Q., Liu, F., et al., 2009. Arabidopsis ASA1 is important for jasmonate-mediated regulation of auxin biosynthesis and transport during lateral root formation. Plant. Cell 21 (5), 1495−1511.

Tanimoto, E., 1991. Gibberellin requirement for the normal growth of roots. Gibberellins 229−240. Available from: https://doi.org/10.1007/978-1-4612-3002-1_22.

Tanimoto, E., 2012. Tall or short? Slender or thick? A plant strategy for regulating elongation growth of roots by low concentrations of gibberellin. Ann. Bot. 110 (2), 373−381.

Tan, H., Man, C., Xie, Y., Yan, J., Chu, J., Huang, J., 2019. A crucial role of GA-regulated flavonol biosynthesis in root growth of Arabidopsis. Mol. Plant. 12 (4), 521−537.

Thole, J.M., Beisner, E.R., Liu, J., Venkova, S.V., Strader, L.C., 2014. Abscisic acid regulates root elongation through the activities of auxin and ethylene in Arabidopsis thaliana. G3 (Bethesda, Md) 4 (7), 1259−1274.

Tiryaki, I., Staswick, P.E., 2002. An Arabidopsis mutant defective in jasmonate response is allelic to the auxin-signaling mutant axr1. Plant. Physiol. 130 (2), 887−894. Available from: https://doi.org/10.1104/pp.005272.

Treml, B.S., Winderl, S., Radykewicz, R., Herz, M., Schweizer, G., Hutzler, P., et al., 2005. The gene ENHANCER OF PINOID controls cotyledon development in the Arabidopsis embryo. Development. 132 (18), 4063−4074. Available from: https://doi.org/10.1242/dev.01969.

Ubeda-Tomás, S., Federici, F., Casimiro, I., Beemster, G.T.S., Bhalerao, R., Swarup, R., et al., 2009. Gibberellin signaling in the endodermis controls Arabidopsis root meristem size. Curr. Biol. 19 (14), 1194−1199. Available from: https://doi.org/10.1016/j.cub.2009.06.023.

Ubeda-Tomás, S., Swarup, R., Coates, J., Swarup, K., Laplaze, L., Beemster, G.T.S., et al., 2008. Root growth in Arabidopsis requires gibberellin/DELLA signalling in the endodermis. Nat. Cell Biol. 10 (5), 625−628. Available from: https://doi.org/10.1038/ncb1726.

Ueda, J., Kato, J., 1980. Isolation and identification of a senescence-promoting substance from wormwood (Artemisia absinthium L.). Plant. Physiol.

Unterholzner, S.J., Rozhon, W., Papacek, M., Ciomas, J., Lange, T., Kugler, K.G., et al., 2015. Brassinosteroids are master regulators of gibberellin biosynthesis in Arabidopsis. Plant. Cell 27 (2261), 2272. Available from: https://doi.org/10.1105/tpc.15.00433.

Van Der Graaff, E., Schwacke, R., Schneider, A., Desimone, M., Flügge, U.I., Kunze, R., 2006. Transcription analysis of *Arabidopsis* membrane transporters and hormone pathways during developmental and induced leaf senescence. Plant. Physiol. [Internet] 141 (2), 776–792. Available from: http://www.plantphysiol.org/cgi/doi/10.1104/pp.106.079293.

Vaseva, I., Qudeimat, E., Potuschak, T., Du, Y., Genschik, P., Vandenbussche, F., et al., 2018. The plant hormone ethylene restricts *Arabidopsis* growth via the epidermis. Proc. Natl. Acad. Sci. U.S.A. 115 (17), E4130–E4139.

Veit, B., 2009. Hormone mediated regulation of the shoot apical meristem. Plant. Mol. Biol. 69 (4), 397–408. Available from: https://doi.org/10.1007/s11103-008-9396-3.

Vragović, K., Selaa, A., Friedlander-Shani, L., Fridman, Y., Hacham, Y., Holland, N., et al., 2015. Translatome analyses capture of opposing tissue-specific brassinosteroid signals orchestrating root meristem differentiation. Proc. Natl. Acad. Sci. U.S.A. 112 (3), 923–928.

Wang, C., Dai, S., Zhang, Z.L., Lao, W., Wang, R., Meng, X., et al., 2021. Ethylene and salicylic acid synergistically accelerate leaf senescence in Arabidopsis [Internet]. J. Integr. Plant Biol. 63, 828–33. Blackwell Publishing Ltd; [cited 2021 May 28]. Available from: <https://doi.org/10.1111/jipb.13075>.

Wang, L., Hua, D., He, J., Duan, Y., Chen, Z., Hong, X., et al., 2011. Auxin Response Factor2 (ARF2) and its regulated homeodomain gene HB33 mediate abscisic acid response in *Arabidopsis*. PLoS Genet. 7 (7), e1002172.

Wei, Z., Li, J., 2016. Brassinosteroids regulate root growth, development, and symbiosis. Mol. Plant. 86–100.

Werner, T., Motyka, V., Laucou, V., Smets, R., Van Onckelen, H., Schmülling, T., 2003. Cytokinin-deficient transgenic *Arabidopsis* plants show multiple developmental alterations indicating opposite functions of cytokinins in the regulation of shoot and root meristem activity. Plant. Cell 15 (11), 2532–2550.

Werner, T., Nehnevajova, E., Köllmer, I., Novák, O., Strnad, M., Krämer, U., et al., 2010. Root-specific reduction of cytokinin causes enhanced root growth, drought tolerance, and leaf mineral enrichment in *Arabidopsis* and tobacco. Plant. Cell 22 (12), 3905–3920.

Wilson, R.N., Heckman, J.W., Somerville, C.R., 1992. Gibberellin is required for flowering in *Arabidopsis thaliana* under short days. Plant. Physiol. 100 (1), 403–408. Available from: https://doi.org/10.1104/PP.100.1.403.

Worthen, J.M., Yamburenko, M.V., Lim, J., Nimchuk, Z.L., Kieber, J.J., Schaller, G.E., 2019. Type-B response regulators of rice play key roles in growth, development and cytokinin signaling. Dev. (Camb.) 146 (13), 1–11.

Wu, M.F., Yamaguchi, N., Xiao, J., Bargmann, B., Estelle, M., Sang, Y., et al., 2015. Auxin-regulated chromatin switch directs acquisition of flower primordium founder fate. eLife.

Yadav, R.K., Perales, M., Gruel, J., Girke, T., Jönsson, H., Venugopala Reddy, G., 2011. WUSCHEL protein movement mediates stem cell homeostasis in the *Arabidopsis* shoot apex. Genes. Dev. 25 (19), 2025–2030. Available from: https://doi.org/10.1101/gad.17258511.

Yadav, R.K., Perales, M., Gruel, J., Ohno, C., Heisler, M., Girke, T., et al., 2013. Plant stem cell maintenance involves direct transcriptional repression of differentiation program. Mol. Syst. Biol. 9 (654). Available from: https://doi.org/10.1038/msb.2013.8.

Yamaguchi, N., Wu, M.F., Winter, C.M., Berns, M.C., Nole-Wilson, S., Yamaguchi, A., et al., 2013. A molecular framework for auxin-mediated initiation of flower primordia. Dev. Cell 24 (3), 271–282. Available from: https://doi.org/10.1016/j.devcel.2012.12.017.

Yan Z, Liu X, Ljung K, Li S, Zhao W, Yang F, Wang M, Tao Y., 2017. Type B Response Regulators Act as central integrators in transcriptional control of the auxin biosynthesis enzyme TAA1. Plant Physiol. 175 (3), 1438–1454. Available from: https://doi.org/10.1104/pp.17.00878. PMID: 28931628; PMCID: PMC5664468.

Yanai, O., Shani, E., Dolezal, K., Tarkowski, P., Sablowski, R., Sandberg, G., et al., 2005. *Arabidopsis* KNOXI proteins activate cytokinin biosynthesis. Curr. Biol. 15 (17), 1566–1571. Available from: https://doi.org/10.1016/j.cub.2005.07.060.

Yang, Z., Wang, C., Qiu, K., Chen, H., Li, Z., Li, X., et al., 2020. The transcription factor ZmNAC126 accelerates leaf senescence downstream of the ethylene signalling pathway in maize. Plant Cell Environ. [Internet] 43 (9), 2287–2300. Available from: https://onlinelibrary.wiley.com/doi/full/10.1111/pce.13803.

Yang, L., Zhang, J., He, J., Qin, Y., Hua, D., Duan, Y., et al., 2014. ABA-mediated ROS in mitochondria regulate root meristem activity by controlling PLETHORA expression in *Arabidopsis*. PLoS Genet. 10 (12), e1004791.

Yuan, W., Luo, X., Li, Z., Yang, W., Wang, Y., Liu, R., et al., 2016. A cis cold memory element and a trans epigenome reader mediate Polycomb silencing of FLC by vernalization in *Arabidopsis*. Nat. Genet. 48 (12), 1527–1534. Available from: https://doi.org/10.1038/ng.3712.

Yu, X., Li, L., Li, L., Guo, M., Chory, J., Yin, Y., 2008. Modulation of brassinosteroid-regulated gene expression by jumonji domain-containing proteins ELF6 and REF6 in *Arabidopsis*. Proc. Natl. Acad. Sci. U.S.A. 105 (21), 7618–7623. Available from: https://doi.org/10.1073/pnas.0802254105.

Yu, X., Xu, Y., Yan, S., 2021. Salicylic acid and ethylene coordinately promote leaf senescence [Internet]. J. Integr. Plant Biol. 63, 823–827. Blackwell Publishing Ltd; [cited 2021 May 28]. Available from: <https://doi.org/10.1111/jipb.13074>.

Zdárská, M., Cuyacot, A.R., Tarr, P.T., Yamoune, A., Szmitkowska, A., Hrdinová, V., et al., 2019. ETR1 integrates response to ethylene and cytokinins into a single multistep phosphorelay pathway to control root growth. Mol. Plant. 12 (10), 1338–1352. Oct 7.

Zdárská, M., Zatloukalová, P., Benítez, M., Šedo, O., Potěšil, D., Novák, O., et al., 2013. Proteome analysis in *Arabidopsis* reveals shoot- and root-specific targets of cytokinin action and differential regulation of hormonal homeostasis. Plant. Physiol. [Internet] 161 (2), 918–930. Available from: https://pubmed.ncbi.nlm.nih.gov/23209126/.

Zemlyanskaya, E.V., Omelyanchuk, N.A., Ubogoeva, E.V., Mironova, V.V., 2018. Deciphering auxin-ethylene crosstalk at a systems levelVol Int. J. Mol. Sciences. MDPI AG.

Zhang, H., Bishop, B., Ringenberg, W., Muir, W.M., Ogas, J., 2012a. The CHD3 remodeler PICKLE associates with genes enriched for trimethylation of histone H3 lysine 27. Plant. Physiol. 159 (1), 418–432. Available from: https://doi.org/10.1104/pp.112.194878.

Zhang, X., Clarenz, O., Cokus, S., Bernatavichute, Y.V., Pellegrini, M., Goodrich, J., et al., 2007. Whole-genome analysis of histone H3 lysine 27 trimethylation in *Arabidopsis*. PLoS Biol 5 (5). Available from: https://doi.org/10.1371/journal.pbio.0050129.

Zhang, H., Han, W., De Smet, I., Talboys, P., Loya, R., Hassan, A., et al., 2010. ABA promotes quiescence of the quiescent centre and suppresses stem cell differentiation in the *Arabidopsis* primary root meristem. Plant. J. 64 (5), 764–774.

Zhang, H., Rider, S.D., Henderson, J.T., Fountain, M., Chuang, K., Kandachar, V., et al., 2008. The CHD3 remodeler PICKLE promotes trimethylation of histone H3 lysine 27. J. Biol. Chem. 159 (1), 418–432. Available from: https://doi.org/10.1104/pp.112.194878.

Zhang, K., Xia, X., Zhang, Y., Gan, S.S., 2012b. An ABA-regulated and Golgi-localized protein phosphatase controls water loss during leaf senescence in *Arabidopsis*. Plant. J. [Internet] 69 (4), 667–678. Available from: https://pubmed.ncbi.nlm.nih.gov/22007837/.

Zhao, Y., Chan, Z., Gao, J., Xing, L., Cao, M., Yu, C., et al., 2016. ABA receptor PYL9 promotes drought resistance and leaf senescence. Proc. Natl. Acad. Sci. U.S.A. [Internet] 113 (7), 1949–1954. Available from: http://www.pnas.org/cgi/doi/10.1073/pnas.1522840113.

Regulatory role of phytohormones in the interaction of plants with insect herbivores

Pritha Kundu, Paramita Bera, Shruti Mishra and Jyothilakshmi Vadassery

National Institute of Plant Genome Research (NIPGR), New Delhi, Delhi, India

3.1 Introduction

Herbivorous insects cause 20%−40% crop loss annually across the globe, posing a serious threat to agriculture (van der Meijden, 2015). The relationship of plants with insect herbivores has been in existence since 350 million years, which ranges from being beneficial as in the case of pollination to the most prevalent and harmful, predation of plants by herbivores (Gatehouse, 2002). This further resulted in coevolution with the gradual emergence of the ability in both plants and herbivores, to evade each other's defense system. The evolutionary arm's race led to the development of a highly sophisticated plant defense system with the unique ability to distinguish between self and nonself. Plants defend themselves against insects with a bewildering array of responses that may be categorized as direct or indirect responses and constitutive or induced defenses (Kessler and Baldwin, 2002). These responses can expand widely at the spatial scales ranging from well-studied cellular interactions to defensive actions at whole plant or community levels.

Phytohormones are defined as small molecule hormones that are present in very minute levels in the plants and play a pivotal role in plant growth, development, reproduction, survival, and fine-tuning the complex defense signaling network (Kundu and Sahu, 2021; Pieterse et al., 2009; Sahu et al., 2021b). Jasmonic acid (JA), salicylic acid (SA), ethylene (ET), abscisic acid (ABA), auxin, cytokinin (CK), gibberellins (GAs), and brassinosteroids (BRs) are the most prominent phytohormones reported in plants. However, the most important class of hormone in the plant–insect interaction against lepidopteran pests are the jasmonates (Erb et al., 2012). On the contrary, SA plays a key role in inducing defense against sucking pests, demonstrating the hormonal specificity in plant defense (Filgueiras et al., 2019). In plants, JA is a strong antagonist of the SA signaling, whereas JA and ET generally act synergistically. Deposition of caterpillar frass in maize resulted in the accumulation of SA, thereby lowering the JA level responsible for defense against herbivory (Ray et al., 2015). Thus in the plant system, SA, JA, and ET signaling pathways cross-communicate, forming an interconnecting network that fine-tune the defense signaling mechanism. A detailed description of the role of key phytohormones in plant–insect interaction and its crosstalk with others have been presented further in this chapter.

3.2 Insects of different feeding guilds

Herbivores can be broadly categorized into two types based on their host specificity. These are the generalists and the specialists. Generalists are the polyphagous herbivores that feed on a wide range of host species but may trigger different sets of responses in different host systems (Becerra et al., 2009). Common examples of generalists include cotton leafworm (*Spodoptera litura*), tobacco whitefly (*Bemisia tabaci*), Japanese beetle (*Popillia japonica*), and cotton bollworm (*Helicoverpa armigera*). Specialists are those herbivores that specifically feed on a selected set of plant species. Examples of specialists include cabbage butterfly (*Pieris rapae*), diamondback moth (*Plutella xylostella*), and cabbage looper (*Trichoplusia ni*).

Herbivorous insects employ diverse feeding strategies for driving nutrition from their host plants. The following feeding guilds/behaviors are common:

1. *Leaf chewer*: This group generally constitute the larvae of moths/butterflies that feed on the leaves by chewing, causing mechanical damage to the plant, leading to the release of damaged cell wall and intracellular components, thus triggering defense response in plants (Brutus et al., 2010; Carvalho et al., 2014). Leaf-chewing insects majorly belong to the order Lepidoptera and include cotton ballworm (*H. armigera*), *Spodoptera* sp. like beet armyworm (*Spodoptera exigua*), cotton leafworm (*S. litura*), fall armyworm (*Spodoptera frugiperda*).

2. *Phloem feeder*: Phloem-feeding insects represent some of the important crop pests belonging to the order Hemiptera. They generally feed on phloem sap via inserting stylets into the phloem sieve cells (Thompson and Goggin, 2006). Some of the most common examples include blue green aphid (*Acyrthosiphon kondoi*), silver leaf whitefly (*B. tabaci*), and brown planthopper (*Nilaparvata lugens*). Phloem feeders are also responsible for acting as vectors leading to the introduction of prokaryotic pathogens in the phloem sieve cells (Jiang et al., 2019).

3. *Root feeder*: Root-feeding insects generally belong to the order Coleoptera, which constitutes the beetles and are usually in the juvenile stages of insects living aboveground (Johnson and Rasmann, 2015). Examples include western corn rootworm (*Diabrotica virgifera*), rice water weevil (*Lissorhoptrus oryzophilus*), common cockchafer (*Melolontha melolontha*).

3.3 Plant defense responses against insect herbivory

Plants being immobile are subjected to a wide range of insects or herbivores in the natural ecosystem. Response to these large groups of arthropods is driven by an intricate and highly dynamic defense system in plants. This includes several defensive traits that can be broadly categorized as constitutive or induced. Both constitutive and induced modes of responses can be direct or indirect (Mithöfer and Boland, 2012). Constitutive responses are those that are always present in the plant, whereas induced defenses generally occur as a result of interaction with aggressors. Attack from herbivores is generally recognized by the perception of herbivore-associated molecular patterns (HAMPs) or damage-associated molecular patterns (DAMPs). Examples of HAMPs that are generally present in the insect regurgitate include caeliferins, volicitin, bruchins, and several other fatty acid (FA)–amino acid conjugates (Mithöfer and Boland, 2008; Stahl et al., 2018; Uemura and Arimura, 2019). Several examples of DAMPs produced from the wounded plant cells include cell wall–derived oligosaccharides, peptide signals, systemin, and eATP giving rise to wound-induced resistance (WIR) (Howe and Jander, 2008; Mithöfer et al., 2005). A brief description of direct and indirect defenses has been discussed in the following subsection.

3.3.1 Direct defenses

Direct defenses are mediated by specialized plant characteristics that modify the herbivore feeding behavior. Direct defenses include mechanical protection on the plants surface (e.g., hairs, trichomes, thorns, prickles, and higher levels of lignification) besides the production of defense-responsive toxic chemicals, such as terpenoids, alkaloids, anthocyanins, phenols, quinones, tannins, and latex (Aharoni et al., 2005). Morphological structures in the plant system constitute the first line of defense. Structural traits, such as spines, trichomes, and divaricated branching, play prime role in plant defense against herbivory (He et al., 2011). Trichomes that negatively affect the insect feeding behavior and oviposition can be straight, spiral, hooked, branched, unbranched, and glandular or nonglandular. Poisonous and toxic secondary metabolites that include flavonoids, terpenoids, and alkaloids are released by the glandular trichomes, thus employing conjugal chemical and structural defenses (Sharma et al., 2009). JA-dependent signaling cascade is most critical for the activation of induced direct defense against lepidopteran insects (Fig. 3.1). Activation of the jasmonate pathway in *Arabidopsis* results in defense responses that include the production of antinutritive proteins, such as trypsin inhibitors (TIs) and laccase-like multicopper oxidase, and secondary metabolites, such as glucosinolates (GSs) and benzoxazinoids (BXs) (Meena et al., 2019; Mithöfer and Boland, 2012).

3.3.2 Indirect defenses

Indirect mode of defenses in plants employs the release of a wide range of volatiles and other inorganic compounds. This further results in the attraction of predators from a different trophic level, essentially the enemies of the attacking herbivores. These volatiles can

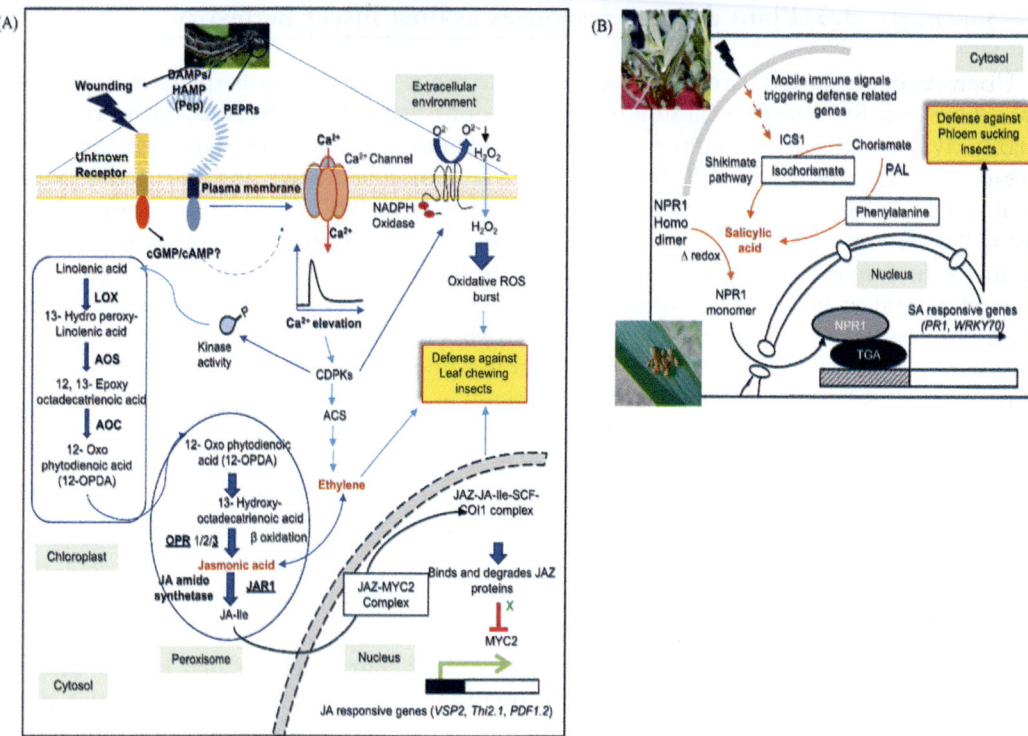

FIGURE 3.1   Schematic representation of the three key defense phytohormones (JA, ET, and SA) signaling pathways upon herbivory in plants. (A) Upon attack by chewing insects the external stimuli are perceived by several plasma membrane−bound receptors triggering membrane depolarization followed by cytosolic Ca^{2-} influx. This further leads to the synthesis of *cis*-OPDA in chloroplast that is converted to JA in peroxisomes. JA-lle, the bioactive form then enters the nucleus to switch "ON" the JA responsive genes. CDPKs also trigger the synthesis of ET that acts synergistically with JA. (B) Sucking insects rapidly induces SA that is synthesized either from phenylalanine or isochorismate by the action of PAL and ICS, respectively, in plant cells. Change in redox potential then triggers the transport of monomelic NPR1 in the nucleus further switching "ON" the SA-responsive genes. *ET*, Ethylene; *ICS*, isochorismate synthase; *JA*, jasmonic acid; *PAL*, phenylalanine ammonia lyase; *SA*, salicylic acid.

be triggered by (1) wound, (2) herbivore feeding, or (3) eggs/frass of insects and, thus, are categorized as herbivore-induced plant volatiles (HIPVs) and egg-induced plant volatiles (EIPVs) (Stahl et al., 2018). The HIPVs include terpenes, series of volatile organic compounds (VOCs), green leafy volatiles (GLVs), ET, and methyl salicylates (Dicke and Baldwin, 2010). Secondary metabolites from the glandular trichomes possess defensive characteristics. Interestingly, trichomes with dominant *O*-acyl sugars in *Nicotiana attenuata* has been reported to be the first choice of meal for the lepidopteran insects, thus modifying the headspace of the insect with the ingested sugar-derived long-chain aliphatic acids. This, in turn, acts as attractant for the ground hunting predator, the omnivorous ant, *Pogonomyrmex rugosus* (Weinhold and Baldwin, 2011).

3.4 Early signaling events prior to activation of phytohormones pathways

Upon successful recognition of the herbivore by the receptors, the plant triggers the first line of an induced defense system. This induces the concerted events of membrane depolarization leading to cytosolic calcium elevation that is decoded by a diverse range of calcium-binding proteins leading to a cascade of downstream signaling events (Sanders et al., 2002) like the production of reactive oxygen species (ROS) molecules, for example, H_2O_2. The signaling pathway culminates with the release of JA and other phytohormones, occurring within minutes (Kiep et al., 2015; Maffei et al., 2007a). Plants also activate local as well as systemic defenses upon herbivory (Kiep et al., 2015). Damage incurred in the plants by insects results in membrane depolarization leading to the development of slow wave potentials (SWPs) (Nguyen et al., 2018). Several plant genes that play a crucial role in the propagation of leaf-to-leaf SWPs in response to the leaf-chewing insects, *Pieris brassicae* and *Spodoptera littoralis*, have been discovered and include the ROS-producing NADPH oxidase, the channel TPC1, and the clade 3 glutamate receptor-like (GLR) proteins, which includes the phloem-based GLR3.3, and xylem-based GLR3.6 (Mousavi et al., 2013). Calcium nucleotide—gated channel, CNGC19, also plays a key role in *S. litura*—induced cytosolic $[Ca^{2+}]_{cyt}$ influx and subsequent upregulation of the JA pathway providing resistance in *Arabidopsis* (Meena et al., 2019). Hereafter, the role of different phytohormones in plant—insect interaction will be discussed.

3.5 Role of defense hormones in plant—insect interactions

3.5.1 Jasmonic acid

3.5.1.1 *Jasmonic acid biosynthesis and signaling*

JA constitutes the core signaling pathway activated in response to wounding and chewing insect herbivory in plants. JA and its derivatives, collectively called jasmonates (JAs), are lipid-derived signaling compounds and include methyl jasmonate (MeJA), 12-oxophytodienoate (OPDA), and iso-leucine conjugated form called (+)-7-*iso*-Jasmonoyl-L-isoleucine (JA-Ile) (Wasternack and Song, 2017). Among various forms of jasmonates, JA-Ile is the bioactive form of jasmonate, which is sensed by plants (Staswick and Tiryaki, 2004). JA is synthesized from α-linolenic acid (α-LeA) in chloroplast and peroxisome through a series of FA oxidation reaction called octadecanoid pathway (Erb et al., 2012; Staswick and Tiryaki, 2004; Wasternack and Hause, 2013). Oxygenation of α-LeA at C-13 position is the initial step for JA biosynthesis and requires the action of LIPOXYGENASE (LOX) enzyme. Among several branches of LOX pathway, the ALLENE OXIDE SYNTHASE (AOS) step leads to JA biosynthesis (Wasternack and Hause, 2013). *Arabidopsis* genome encodes six different *LOXs* of which four (*LOX2*, *LOX3*, *LOX4*, and *LOX6*) belong to 13-LOXs family, while *LOX1* and *LOX5* are 9-LOXs. All of 13-LOXs are involved in wound-inducible JA formation with different specificities (Wasternack and Hause, 2013). 13(*S*)-hydroperoxy-octadecatrienoic acid (13(*S*)-HPOT) formed by 13-LOXs is converted to epoxide by the action of AOS, and this is the first committed step to JA biosynthesis. *Arabidopsis* genome encodes a single copy of gene for *AOS*, whereas in other plants, *AOS* occurs as gene families (Wasternack and Song, 2017). Epoxide formed by AOS is catalyzed to form 12-oxo-phytodienoic acid (*cis*-(+)-OPDA) by the action of ALLENE OXIDE CYCLASE (AOC). *cis*-(+)-OPDA formed at this step has been proved to play a role as a signaling molecule, as it induces a set of distinct gene expression. All the enzymes involved in OPDA biosynthesis are located within the chloroplast; however, the reduction step of OPDA takes place in peroxisome. OPDA import into peroxisome seems to occur via COMATOSE exhibiting PXA1 activity (Theodoulou et al., 2005). The reduction reaction on OPDA takes place by OPDA reductase (OPR). *Arabidopsis* encodes six *OPRs*, and only *OPR3* is involved in JA biosynthesis in *Arabidopsis*. The ultimate step in the formation of the most bioactive form, JA-Ile, is the conjugation of amino acid iso-leucine to JA, which is catalyzed by the cytosolic enzyme jasmonoyl-isoleucine synthetase or JAR1, which is a member of GH3 gene family (Staswick and Tiryaki, 2004) (Fig. 3.1).

JA-Ile binds to its receptor CORONATINE INSENSITIVE 1 (COI1), an F box protein that is a part of E3 ubiquitin ligase to form an SCF-E3 ubiquitin ligase complex or SCF^{COI} (Katsir et al., 2008). At the resting stage, the repressor proteins JASMONATE ZIM DOMAIN (JAZ) bind to and prevent the basic helix–loop–helix (bHLH) transcription factor (TF), MYC2 from inducing the transcription of JA-responsive genes. Upon wounding and herbivory the increased endogenous JA-Ile binds to COI; this recruits the JAZ1 repressor to the receptor–ligand complex and leads to its proteasomal degradation. Degradation of JAZ derepresses MYC2 and turns "ON" the transcription of JA-responsive genes, such as *THIONIN2.1* (*Thi2.1*) and *VEGETATIVE STORAGE PROTEIN 2* (*VSP2*) (Sheard et al., 2010). *Arabidopsis* genome has 13 JAZs with a conserved C-terminal Jas-domain for binding COI and TFs like MYCs, and a highly conserved ZIM domain that interacts with corepressor proteins Novel Interactor of JAZ (NINJA)/TOPLESS (TPL) (Pauwels and Goossens, 2011). MYC2 interacts with JAZ via its N-terminal JAZ-interacting domain (JID), and it has a bHLH domain in C-terminal, which binds to G-box (JA-responsive) element. MYC3 and MYC4, other bHLH TFs, act redundantly with MYC2 to activate JA responses and therefore positively regulates defense against *S. littoralis* (Schweizer et al., 2013). Accumulated JA-Ile within the cell is quickly catabolized to attenuate the defense response, as it is costly for plants. Enzymes responsible for ω-oxidation that belongs to cytochrome P450 family—CYP94B1, CYP94B3, and CYP94C1 convert active JA-Ile to inactive 12-OH-JA-Ile and 12-COOH-JA-Ile. A group of hydrolyzing enzymes ILL6 and IAR3 hydrolyzes JA-Ile and 12-OH-JA-Ile to form JA and 12-OH-JA, respectively (Koo et al., 2011). Recently, another enzyme that belongs to the family of oxygenases, called JASMONATE-INDUCED OXYGENASES (JOX), has been shown to catalyze the hydrolysis of JA to 12-OH-JA (Caarls et al., 2017).

3.5.1.2 *Jasmonic acid signaling genes and their role in herbivory response*

The plant hormone JA and its derivatives are key players in the regulation of induced plant defense against insect herbivory and 95% of protein repatterning near wound sites is JA dependent (Gfeller et al., 2011). In *Arabidopsis*, many independent studies observed the accumulation of JA in wounded (local) and unwounded (systemic) leaves within minutes of damage and feeding by *S. littoralis* (Chauvin et al., 2013; Glauser et al., 2008; Koo et al., 2009; Reymond et al., 2004). Wound-induced JA accumulation occurs within 30 seconds in local leaf and within 90 seconds in vascularly connected systemic leaves (Chauvin et al.,

2013). Mousavi et al. (2013) demonstrated the role of electrical signals in propagating the long-distance wound signaling and the genetic basis behind the signal propagation by the plasma membrane—localized ion channel GLUTAMATE-LIKE RECEPTORs (GLRs) (Mousavi et al., 2013). Loss-of-function mutation in *GLR3.3* and *GLR3.6* hampered JA biosynthesis and JA-mediated systemic response to wounding. The calmodulin-like proteins CML42 and CML37 negatively and positively regulate the defense response during herbivory by modulating the COI1-mediated JA sensitivity (Scholz et al., 2014; Vadassery et al., 2012). The JJW complex comprising JAV1, JAZ8, and WRKY51 suppresses the JA biosynthesis genes in plants under healthy conditions. Upon insect attack, cells produce Ca^{2+} elevation that results in an activation of Ca^{2+} sensor calmodulin (CaMs), which leads to the Ca^{2+}/CaM-mediated phosphorylation of JAV1 in JJW complex, resulting in its ubiquitination followed by the derepression of the JA biosynthesis pathway (Kundu and Vadassery, 2021; Yan et al., 2018).

The role of jasmonates in herbivory has been identified by using genetic manipulation of genes involved in JA biosynthetic and signaling pathways. LOX2 is responsible for the bulk biosynthesis of JA upon the first hour of wounding. LOX3 and LOX4 are wound responsive and localized in vascular tissue; LOX6 also has wound-response activity and is localized in xylem cells; however, its activity is restricted during early stages of leaf differentiation (Wasternack and Hause, 2013). *lox2—1* and *aos* biosynthetic mutants in *Arabidopsis* have reduced resistance to the caterpillar *S. littoralis*. However, to establish individual roles of all four 13-LOXs in JA pathway and wound-induced JA signaling, single mutants and various combinations of double and triple mutants of four LOXs were used. A quadruple mutant for all 13-LOXs, *lox2—1 lox3B lox4A lox6A*, is severely compromised in JA accumulation upon wounding and is highly susceptible to *S. littoralis* feeding (Chauvin et al., 2013; Glauser et al., 2008). Double mutant *lox3B lox4A* has a modest reduction in JA accumulation in response to wounding in local leaves. *lox2—1 lox6A* double mutant exhibits a highly reduced accumulation of JA and JA-Ile in response to wounding in the local leaf. Also, double or triple mutants employing the *lox6A* allele demonstrates delayed JA and JA-Ile accumulation after wounding in systemic leaves. This shows the crucial role of *LOX6* in the systemic accumulation of JA upon wounding (Chauvin et al., 2013). While loss of function of JA biosynthesis genes compromises the antiherbivore defense, Laudert et al. (2000) showed that overexpressing *Arabidopsis thaliana AOS* gene in Nicotiana and *Arabidopsis* enhanced wound-induced JA accumulation as compared to WT plants (Laudert et al., 2000). *Arabidopsis* mutant *fad3 fad7 fad8* that fails to synthesize the precursor molecules for JA biosynthesis is extremely susceptible to feeding by the dipteran, *Bradysia impatiens* (Stintzi et al., 2001).

Arabidopsis plants with a loss-of-function mutation in JA receptor *coi1* are impaired in all the JA responses, including defense against generalist herbivore *S. littoralis* (Bodenhausen and Reymond, 2007). Herbivore *H. armigera* fed more on knockout lines of *MYC2*, called *jin1—2*, and, therefore, had compromised defense response; however, feeding was significantly reduced in *MYC2* overexpressor lines showing the direct role of *MCY2* in regulating herbivore defense (Dombrecht et al., 2007). In addition to MYC2, other bHLH TFs, MYC3 and MYC4, act additively to regulate JA responses, a triple knockout mutant *myc2 myc3 myc4* completely fails to accumulate GSs and is highly susceptible to *S. littoralis* feeding (Schweizer et al., 2013). A high-order loss-of-function mutation in 10 out of 13 *JAZ* genes, that is (*JAZ1−7, -9, -10,* and *-13*), *jaz* decuple or *jazD* leads to robust resistance against *T. ni* larvae as compared to WT plants (Guo et al., 2018). Knockout mutant *jasmonate resistant 1(jar1)*, which is impaired in JA-Ile formation, fails to induce wound response in *Arabidopsis* (Suza and Staswick, 2008). A quadruple knockout mutant for all four JOX genes responsible for JA-Ile catabolism in *Arabidopsis* called *joxQ* was also observed to have increased resistance to the generalist caterpillar, *Mamestra brassicae* (Caarls et al., 2017). Similarly, the analysis of single and double mutants of *CYP94B3* and *CYP94C1* revealed their role in JA-Ile catabolism as *cyp94b3* lines were impaired in 12-OH-JA-Ile levels and *cyp94c1* lines were impaired in 12-COOH-JA-Ile levels and the double mutant, *cyp94b3 cyp94c1* hyperaccumulated JA-Ile. Conversely, the overexpressor line CYP94B3-OE exhibited the phenotype reminiscent of JA deficient of insensitive mutants; it was insensitive to root growth inhibition by MeJA and susceptible to attack by *S. exigua* larvae (Heitz et al., 2012; Koo et al., 2011).

In addition to having a direct role in herbivory defense, the JA pathway also has a relevance in circadian-regulated antiherbivore defense. In *Arabidopsis*, JA levels, wound-inducible gene expression, and insect defense response peak at the subjective day; this is coincident with the circadian feeding behavior of *T. ni* larvae. Therefore the clock-regulated JA accumulation helps in plant defense against the specific herbivore (Goodspeed et al., 2012). Additionally, many JA pathway genes are regulated by the circadian clock. JA biosynthetic genes *AOS* and *OPR3* and TFs MYC2, MYC3, and MYC4 belong to the group of genes regulated by the circadian clock. During day time, red and blue lights stabilize the three MYCs, MYC2/3/4, and promotes insect defense, whereas in the dark, they are destabilized to suppress defense and promote growth. Also, in dark, all JAZ repressors, except JAZ2, JAZ6, and JAZ8, are stabilized, and their degradation by JA is reduced (Chico et al., 2014).

3.5.1.3 *Jasmonic acid-regulated secondary metabolite production upon herbivory*

JA regulates and induces the accumulation of several class of secondary metabolites, such as terpenoids, flavonoids, and nitrogen containing secondary metabolites, namely, GSs and alkaloids, and antinutritive molecules such as protease inhibitors-II (PI-II) in plants (Howe and Jander, 2008). GSs are the major secondary metabolites found in the family Brassicaceae. In *Arabidopsis*, activation of the JA pathway by chewing herbivore *S. exigua* leads to enhanced accumulation of GS (Mewis et al., 2005). Genes involved in the GS biosynthetic pathway are JA-inducible via the bHLH TFs MYC2, MYC3, and MYC4 (Schweizer et al., 2013). Secondary metabolite, anthocyanin, is a flavonoid that acts as antioxidant, and studies in several plant species have shown that anthocyanin production in response to herbivory damage is regulated by JA signaling mechanism. For example, in *Arabidopsis*, JAZ proteins interact with bHLH TF MYB to regulate anthocyanin biosynthesis (Qi et al., 2011). Anthocyanins are synthesized by phenylpropanoid pathway, and the first step is catalyzed by the enzyme; phenylalanine ammonia lyase (PAL) and chalcone synthase (CHS) are regulated by the JA pathway (Chen, 2008). In *Solanum lycopersicum*, insect herbivory leads to the JA-mediated accumulation of antinutritive protein, such as proteinase inhibitor and polyphenol oxidase (Constabel and Barbehenn, 2008; Kundu et al., 2021; Kundu et al., 2018). Induction of JA-responsive TF MYC2 in *N. attenuata* leads to the increased accumulation of the antiherbivore compound, nicotine (Pradhan et al., 2017).

Most plants release gaseous volatiles such as GLVs in response to wounding and attack by herbivores (Scala et al., 2013). *S. lycopersicum*, attacked by the white fly *B. tabaci*, releases

GLVs-(Z)-3-hexenol, which is a derivative of linolenic acid and linoleic acid (Silva et al., 2017). Yang et al. showed that, exogenous application of (Z)-3-hexenol reduced *B. tabaci* feeding and oviposition on *S. lycopersicum* plants (Yang et al., 2020). GLVs are also known to attract the natural enemies of insect pest to the site of feeding. For example, in *Z. maize*, feeding by *S. exigua* releases GLVs-(Z)-3-hexenol and (3Z)-hexenyl acetate, these HIPVs attract parasitoid wasps *Cotesia marginiventris*. HIPV emission is mediated by *LOX* genes as silencing of *LOX10* impaired GLV emission as well as impaired plant resistance against *S. exigua* (Christensen et al., 2013).

3.5.1.4 *Jasmonic acid signaling in crop plants*

The role of JA in defense against herbivores has been established in crop plants at a rapid pace, especially by RNAi- and VIGS-mediated silencing of genes involved in JA pathway. Loss-of-function mutation in tomato homolog of *COI* called *jasmonic acid-insensitive 1* or *jai1* renders tomato plants susceptible to cell content feeder—spider mites. *jai1* mutant lines have either reduced (*LoxD*, *AOS2*) or no accumulation of JA-responsive genes with severely compromised defense responses (Li et al., 2004). An et al. in 2019 also identified maize COIs—*ZmCOI1a*, *ZmCOI1b*, *ZmCOI1c*, and *ZmCOI2*; all but *ZmCOI2* were observed to activate wound-inducible JA responses (An et al., 2019). *S. lycopersicum* encodes six *LOX* genes, that is, *TomLoxA-F*, of which *TomLoxC* and *TomLoxD* are 13-LOXs localized in the chloroplast. Characterization of *spr8* mutant that is a point mutation in *TomLoxD* gene showed that only *TomLoxD* is involved in JA accumulation in response to herbivore feeding (Yan et al., 2013). Also, in *N. attenuata* and *Oryza sativa*, *LOX* genes have been identified, and its role in herbivory is well defined. Silencing of *LOX* in *N. attenuata- NaLOX3* reduced JA accumulation and enhanced *Manduca sexta* larvae feeding on the tobacco plants (Li et al., 2004). Lipoxygenase gene present in rice- *OsHI-LOX* accumulated in response to mechanical wounding and feeding by herbivore, rice strip stem borer (SSB)—*Chilo suppressalis*. Additionally, silencing of OsHI-LOX reduced JA and trypsin protease inhibitor accumulation and consequently increased larval feeding of SSB on rice (Zhou et al., 2009). *S. lycopersicum*, with an impaired accumulation of *AOC*, failed to induce wound responsive, protease inhibitor. Important role of OPRs in JA biosynthesis and its role in defense were recently elucidated in maize. *Zea mays* OPR7 and OPR8 have a redundant role in JA biosynthesis; while single knockout mutants had WT levels of JA, double mutants *opr7opr8* had drastically reduced level of JA and was highly susceptible to beet armyworm *S. exigua* (Yan et al., 2012). In cotton, *Gossypium hirsutum*, feeding by the caterpillar *S. littoralis*, induced JA-Ile accumulation in the systemic leaves of fed plants (Eisenring et al., 2018). In rice, *OsOPR7* is involved in JA biosynthesis, its expression is correlated with JA accumulation postwounding, and the heterologous expression of *OsOPR7* in *Arabidopsis* mutant *opr3* restored JA biosynthesis (Tani et al., 2008). Rice genome also encodes two JAR genes, *OsJAR1* and *OsJAR2*, of which *OsJAR1* undertakes the most important function of conjugating iso-leucine to JA. *Arabidopsis* homolog of *JAR1* is also found in *N. attenuata* called *JAR4*, is responsible for formation of the active JA-Ile, and silencing the gene via VIGS impairs defense against the chewing herbivore *M. sexta* (Kang et al., 2006).

Other players that contribute to JA-mediated insect defense have been uncovered more recently. Li, et al. (2020) discovered long noncoding RNAs (lncRNA) that are upregulated in *N. attenuata* plants treated with *M. sexta* oral secretion (OS). Temporal accumulation of lncRNAs correlated with JA accumulation postherbivory, and silencing of two of these lncRNAs rendered tobacco plants susceptible to *M. sexta* herbivory (Li et al., 2020). These studies clearly show that defense against herbivores is regulated by jasmonate signaling pathway in dicots and monocots. The best-studied hormones that alter JA-mediated defense responses and herbivore resistance are SA and ET. SA antagonizes JA-induced resistance, whereas ET can have both positive and negative effects (Erb et al., 2012).

3.5.2 Salicylic acid

SA is an important defense hormone against sucking insect pests and biotrophic pathogens (Glazebrook, 2005; Sahu et al., 2016a; Sahu et al., 2021b; Sahu et al., 2016b). It is a monohydroxybenzoic acid synthesized either from phenylalanine or isochorismate by the action of PAL and isochorismate synthase (ICS), respectively, in plastids of plant cells (Dempsey et al., 2011) (Fig. 3.1). SA signaling in response to stress requires ENHANCED DISEASE SUSCEPTIBILITY 1 (EDS1) and PHYTOALEXIN DEFICIENT 4 (PAD4) for feedback regulation of the signaling pathway. NONEXPRESSOR OF *PR* GENES 1 (NPR1) is the key regulatory protein that mediates SA signaling pathways. It acts downstream of EDS1 and PAD4—two of the major SA responsive genes—and activates the expression of glutaredoxin GRX480 and WRKY TF, WRKY70. WRKY70 in turn activates the expression of PATHOGENESIS RELATED 1 protein (PR1) to induce defense response (Pieterse et al., 2009; Sethi et al., 2021).

Numerous studies have elucidated the direct role of SA in defending against phloem sucking insects. Aphid (*Myzus persicae*) infestation on *Arabidopsis* plants leads to a strong induction of SA marker gene, *PR1* (Moran and Thompson, 2001). Role of SA on mounting a defense response against aphids has also been studied in the economically important crop, wheat (Sahu et al., 2021a). Russian wheat aphid (RWA), *Diuraphis noxia*, infestation induced the accumulation of SA selectively on resistant variety of wheat—cv. Tugela DN and not on susceptible near isogenic cv. Tugela (Mohase and van der Westhuizen, 2002). In the modern legume, *Medicago truncatula*, SA levels are significantly increased in response to piercing—sucking herbivore, *Tetranychus urticae* (Leitner et al., 2005). Similarly, resistance against potato aphid *Macrosiphum euphorbiae* in *S. lycopersicum* plants by *Mi-1* gene is mediated by SA signaling. Aphids' survival increased in the tomato plants in *NahG* mutants, which eliminates SA from plants (Li et al., 2006).

In addition to sucking insects, there are evidences that implicate the indirect role of SA in defense against chewing insects, and most of these defense responses are mediated by the production of ROS at the feeding site, which binds to the enzymes in the insect gut and prevents proper digestion of ingested plant tissue. It was observed that egg deposition of a chewing specialist herbivore, *P. brassicae*, induces SA levels in *Arabidopsis*. Also, feeding performance of *P. brassicae* larvae decrease on plants deposited with the eggs as compared to the control plants without egg deposition (Lortzing et al., 2019).

Some insects use the SA—JA antagonism to bypass plant insect defense and to successfully feed on plants. Oviposition by the cabbage butterfly (*P. brassicae*) induces SA accumulation and reduces the induction of JA-responsive genes, leading to reduce plant resistance against *S. littoralis* (Bruessow et al., 2010). *WRKY70*, which is activated upon SA signaling, suppresses JA-responsive gene *PDF1.2* (Pieterse et al., 2009). *Arabidopsis* plants infested with silver leaf white fly (SLWF) accumulated SA responsive genes—*NPR1, EDS1, PAD4*—in the insect-fed local leaves as well as in systemic leaves, whereas JA-responsive genes were repressed. In the same study, mutant plants *npr1* and *nahG* with constitutively suppressed SA signaling slowed SLWF nymph development, whereas mutant with enhanced SA level— *constitutive immunity 10* (*cmi10*) had accelerated nymph development (Zarate et al., 2007).

3.5.3 Abscisic acid

ABA is a signaling molecule with involvement in several developmental processes and adaptive stress responses in plants. ABA plays a major role in plant responses to drought stress, pests as well as pathogens (Adie et al., 2007b; Mauch-Mani and Mauch, 2005). The role of ABA in response to plant defense is very complex and varies from plant to plant and also depends on type of herbivore species. Normal ABA biosynthesis is prerequisite for JA production (Adie et al., 2007b), and COI-dependent ABA-induced MYC2 expression (Lorenzo and Solano, 2005). Mechanistic details on such interaction, however, are in infancy. A mechanism similar to the suppression of JA-induced TFs by JAZ—NINJA—TPL was identified for the ABA-dependent TF ABI5 (ABA insensitive5) in *Arabidopsis*. ABI5 binding proteins (AFPs) are NINJA homologs and contain the EAR motif to interact with the corepressors, TPLs for ABI5 inactivation (Pauwels et al., 2010). Thus JA—ABA interaction may exist at this JAZ—NINJA connection, downstream of JA biosynthesis, dependent on the binding specificity of different JAZs to NINJA or ABPs (Nguyen et al., 2016).

ABA and JA synergistically induce MYC2-dependent gene expression during wound responses and ABA deficiency in general increases plant susceptibility to herbivory (Erb et al., 2012). ABA synthesis and signaling pathway affects JA-inducible defense responses in maize, the herbivore-induced transcript levels and JA biosynthesis in *Arabidopsis*, and resistance to herbivores in tomato (Tooker and De Moraes, 2011a). Several research in *Solanum* species demonstrated that ABA was essential for wound-induced JA accumulation and normal JA signal transduction, indicating a close relationship and coordination of these two hormonal pathways (Dinh et al., 2013). ABA contributes to plant resistance against biotic invasion by upregulating JA-dependent defenses and by downregulating SA-dependent defenses, as well as repressing the synthesis of some secondary metabolites, such as indole GSs in *Arabidopsis* (Hillwig et al., 2016). ABA level was increased in maize in response to attack by the specialist herbivore, western corn rootworm (*D. virgifera*), but not by mechanical wounding alone (Erb et al., 2011b), and in *Arabidopsis* after induction with wounding and the OS of desert locust (*Schistocerca gregaria*) (Schäfer et al., 2015). Root herbivory by western corn rootworm did not lead to a jasmonate-dependent response in the leaves but specifically triggered water loss and ABA accumulation. The induction of ABA by itself was partly responsible for the induction of leaf defenses and reduced feeding by leaf herbivore, *S. littoralis* (Erb et al., 2011a). Thus ABA assimilates and adjusts a fine balance both biotic and abiotic stress response signaling networks. However, molecular mechanism involved in ABA-mediated plant defense against insect herbivory is still unsolved.

3.5.4 Ethylene

The gaseous phytohormone, C_2H_4 or ET, plays an important role in defense against pathogens and insect herbivory (Adie et al., 2007a). ET acts synergistically with JA in response to plant development and defense against necrotrophic pathogens (Fig. 3.1). However, both positive and negative effects are known between JA and ET in response to wounding and herbivory (Lorenzo and Solano, 2005), with several conflicting reports (O'donnell et al., 1996). The critical "fuel" for the first committed step in production of ET is methionine that deploys the enzyme S-AdoMet synthase (SAM synthase). However, the second step is considered the rate-limiting step in ET synthesis that involves the conversion of S-AdoMet to 1-aminocyclopropane-1-carboxilic acid (ACC) by ACC synthase (ACS). Upon oxidation of ACC, by ACC oxidase (ACO), ET is generated in plants (Adie et al., 2007a). In *Arabidopsis*, 8 out of 12 members of the ACS gene family are involved in ET synthesis (Yamagami et al., 2003). Function of ACS in governing the rate limiting step is posttranscriptionally regulated by phosphorylation balance and homo/hetero-dimerization. Recently, in tomato, it was reported that CDPK and MAPK6 plays a major role in regulating ACS stability leading to the synthesis of ET upon wounding (Chae and Kieber, 2005). In *Arabidopsis*, ETO-1 LIKE-2 (EOL2) protein was induced upon attack by cabbage white butterfly (*P. rapae*) for target recognition-mediated 26S/proteasome degradation of type 2 ACS subfamily, thus inhibiting ET biosynthesis (Christians et al., 2009).

Promoters of ET-regulated genes possess an 11 bp *cis*-element called GCC box or ET responsive element (ERE) in several plants (Chakravarthy et al., 2003). Several independent groups further reported diverse families of proteins that were capable of binding to this GCC box and regulate the ET signaling pathway. One such family was the ET-responsive-element-binding-proteins (EREBPs or ERFs) that were identified in tobacco, thus comprising the AP2/ERF superfamily (Ohme-Takagi and Shinshi, 1995). Some of the examples include *At*ERF1, *At*ERF2, *At*ERF3, *At*ERF4, *At*ERF13, and *At*EBP in *Arabidopsis*, *Ca*ERFLP1 in hot pepper, Pti4 in tomato, and Tsi1 and OPBP1 in tobacco that are known to either positively or negatively regulate plant defense (Adie et al., 2007a). In rice, a nucleus-localized gene, *OsERF3*, was reported to be induced rapidly upon attack by the rice striped stem borer (SSB), *C. suppressalis*, but increased the susceptibility of plants to the piercing sucking insect, the rice brown planthopper (BPH), *N. lugens*. Overexpression of *OsERF3* was found to upregulate MAPKs and WRKY genes along with induced concentrations of jasmonate (JA), salicylate (SA), and the activity of trypsin protease inhibitors, upon attack by SSB.

In *Arabidopsis*, two distinct branches governed by the MYC2 TF and the AP2/ERF TFs, ORA59 and ERF1, control the JA signaling response that acts in an antagonistic way, coregulated by ABA and ET, respectively. Upon attack by the cabbage white butterfly (*P. rapae*), a quick activation of the MYC2 branch followed by a collateral suppression of the ERF branch was reported in insect-fed leaves. This was followed by the accumulation of JA and (+)-7-iso-jasmonoyl-L-isoleucine in the local and systemic leaves (Vos et al., 2013). ET and JA act antagonistically in response to wounding in *Arabidopsis* and other plant species. It is proposed that ERFs might be involved in this ET-mediated suppression of JA genes, as the upregulation of ERFs hinders the expression of wound-induced JA-responsive genes (Lorenzo and Solano, 2005). However, the role of both ET and JA was reported to be necessary for the expression of wound-induced *pin2* gene in tomato or potato upon attack by adult Colorado potato beetles, *Leptinotarsa decemlineata* or their larvae (O'donnell et al., 1996). Constitutive production of JA and ET in the *cev1* mutant in *Arabidopsis* enhanced defense against aphids (Turner et al., 2002). This can be further explained by the selective mode of response based on the feeding pattern of the herbivores. Previous report suggests that the generalist *S. littoralis* performed better in ET-insensitive mutants (*ein2* and *hookless1*), but no change was detected in the feeding behavior of the specialist, diamond back moth (*P. xylostella*). Also, the application of an ET releasing chemical in *Arabidopsis*, ethephon, rendered the plant susceptible to the Egyptian cotton worm (Stotz et al., 2000). ET-mediated decreased level of nicotine was observed in local leaves of *N. attenuata* upon attack by larvae of *M. sexta*, although levels of JA and other volatile terpenoids remain unchanged. Such "adaptive tailoring of defenses" was proposed to play important role in reduced uptake of nicotine by the herbivores, thus rendering them more susceptible to nicotine-sensitive parasitoids (Kahl et al., 2000). In spite of this huge base of information the understanding of the mechanistic role of JA—ET crosstalk in response to insect herbivory still demands further in-depth molecular investigations.

3.6 Growth phytohormone—mediated defense responses in plant—insect interaction

Plant growth and defense are regulated by complex metabolic pathways and several small signaling molecules. These signaling molecules govern plant's adaptation to the environment and their perception of the environmental cues. Phytohormones, which are principally involved in plant growth and development, such as auxins, GAs, CKs, and BRs, are known as growth hormones (Shigenaga and Argueso, 2016). Here, we describe the role of growth hormones in plant defense response upon insect herbivory.

3.6.1 Auxins

Auxins regulate an enormous range of plant processes, including plant growth and development, as well as responses against insect attack (Machado et al., 2016). Plant resistance to insect herbivore and pathogen can also be modulated by changes in auxin sensitivity (Erb et al., 2012). Several studies showed the elevation in plant auxin levels upon gall-forming insects attack (Tooker and De Moraes, 2011a). It was observed that the gall-forming insect, sawfly (*Pontania* sp.), contain higher amount of auxins, and thus, auxin pool and signaling are increased in its host plant, *Salix japonica*, upon sawfly infestation (Tanaka et al., 2013). However, the plant responses were found to be different in the case of chewing insects. For example, it was found that the inactivation of auxin-mediated signaling activates defense responses in *N. attenuata* to *M. sexta* OS elicitation (Onkokesung et al., 2010). A reduced auxin levels were observed in *H. zea*—infested maize leaves, and no significant alterations were observed in auxin levels in goldenrod (*Solidago altissima*) and wheat (*Triticum aestivum*) attacked by *Heliothis virescens* caterpillars (Schmelz et al., 2003; Tooker and De Moraes, 2011a, b).

Auxin signaling may also affect and regulate plant responses to insect herbivory by modulating other hormonal pathways and subsequent defense responses (Erb et al., 2012). For example, exogenous auxin application decreases the herbivore-induced accumulation of nicotine and JA, expression of jasmonate-dependent proteinase inhibitors genes, and the vegetative storage proteins (Liu et al., 2005). This observation suggests a general negative role of auxin in JA-mediated nicotine accumulation (Onkokesung et al., 2010). Studies also suggest that in *Arabidopsis* root, auxin biosynthesis is increased by JA-mediated induction of auxin biosynthesis—related genes (Lackman et al., 2011). However, in *N. attenuata* leaves, wounding decreases auxin content due to negative regulation by JA, indicating that JA affects auxin biosynthesis tissue specifically and is species specific (Erb et al., 2012). Exogenous application of synthetic auxin to plants reduces the SA-induced defense responses to phloem-feeding insects (Li et al., 2006). Insects those are negatively affected by SA-induced defense (sucking pests) can change the auxin sensitivity or signaling to reduce host defense (Erb et al., 2012). These findings suggested that auxin is a potential modulator of herbivore-related defense responses and signify that plant may reconfigure the auxin levels to mediate specificity.

3.6.2 Cytokinins

CKs are a group of plant growth hormones that promote cell division and play an important role in various biological processes associated with plant growth, metabolism, and development, including seed germination, leaf expansion, delayed senescence, the formation of shoots from callus, and nutrient transport (Werner and Schmülling, 2009). The type and mode of action of CKs vary remarkably between different plant species and tissues, at different developmental stages and under various environmental conditions. The most abundant form of CKs is *trans*-Zeatin (*t*Z) and the enzyme, which converts isoprenoid to different forms of CKs, is isoprenyl transferases (IPTs).

Several reports on the role of CK in plant—insect interactions are known. CKs prime the plant to herbivory by inducing the wound-response gene expression and upregulating the accumulation of defense compounds (Dervinis et al., 2010; Giron et al., 2013). It was observed that after *M. sexta* herbivore attack, CK concentration is induced in wild tobacco, which in turn reconfigures the plant primary and secondary metabolism associated with plant defense (Giron et al., 2013). This was reported in interaction with several species like tobacco hornworm (*M. sexta*), the green peach aphid (*M. persicae*), or the gypsy moth (*Lymantria dispar*), whereby these responses might deter insect feeding, delay larval development, or reduce weight gain by insect larvae (Dervinis et al., 2010; Giron et al., 2013). Galling or leaf-mining insects manipulate the host plant's physiology, modulating source/sink ratio by moving CKs, leading to nutritional translocation toward the insect's feeding site (Giron et al., 2013. An increased CK level and alterations in CK-related transcript levels were observed when the sap-sucking bug *Tupiocoris notatus* was feeding on solanaceous plants such as *N. attenuata* (Brütting et al., 2018). Leaf miners have been shown to use CKs to modify the tissue surrounding their mines, resulting in the phenomenon of "green islands" (Engelbrecht et al., 1969) or certain sawflies that can induce "leaf galls" (Elzen, 1983; Schäfer et al., 2015). CK is responsible for the formation of green islands upon leaf-mining insect attack based on its amount in infected plants (Engelbrecht et al., 1969), and a higher amount of various CKs was found on infected tissues of *Malus domestica*/*Phyllonorycter blancardella* leaf-mining system (Akhtar et al., 2020). These studies suggested that CKs produced by *P. blancardella* might be the causal factor to manipulate plant defense responses (Zhang et al., 2017). Localized delay of senescence and formation of *stay green* phenomena on attacked leaves might be due to the biosynthesis of CKs by insects (Kaiser et al., 2010). Presence of high levels of CKs was found in insect excretion, gastrointestinal tract, and minor salivary glands of leaf-feeding larvae of *Stigmella argentipedella* and *Stigmella argyropeza*, which suggests a role of insects in CKs biosynthesis (Engelbrecht et al., 1969). The modulation of CK pathway upon chewing insect herbivory has very recently been elucidated. Leaf wounding rapidly induced transcriptional changes in multiple genes involved in CK pathway and increased isopentenyladenosine and *cis*-zeatin riboside levels. CK pathway changes were also observed in systemic leaves in response to OS application indicating a role in mediating long-distance systemic processes in response to herbivory (Schäfer et al., 2015). The CK status of a given tissue might determine the intensity of the defense response in that particular tissue after perception of herbivory and, thus, contribute to tissue-specific responses in herbivory-induced signaling (Ballaré, 2011).

3.6.3 Gibberellins

GA is biosynthesized from geranylgeranyl diphosphate (GGDP), a common C-20 precursor for diterpenoids. So far, 130 GAs have been identified in plants, fungi, and bacteria. Among them, GA1, GA3, GA4, and GA7 function as bioactive hormones; others are either precursor molecules of the bioactive forms or deactivated metabolites. The bioactive GAs in plants are GA1 and GA4. Though GA7 and GA3 are biologically active, they are present at very low levels (Binenbaum et al., 2018). In general, GAs modulate plant development, including seed germination, reproduction, and vegetative growth. Recent investigations have revealed GA involvement in biotic stress responses (Qi et al., 2018). In *Arabidopsis*, direct interaction between GAs and the GA INSENSITIVE DWARF1 (GID1) receptor induces the interaction between GID1 and DELLAs and provokes the degradation of DELLAs through E3 ubiquitin ligase-mediated ubiquitinylation and 26S proteasome-mediated proteolysis (Jang et al., 2020). GA affects JA signaling pathway negatively, via interaction of repressors DELLA with JAZ. In GA-free conditions, DELLAs directly interact with JAZs and allow MYC2 to promote the JA response, while in the presence of GA, JAZs are released from the DELLA−JAZ complex by degradation of DELLAs, and the free JAZs attenuate the JA response through direct interaction with MYC2 (Erb et al., 2012; Hou et al., 2013). Depending upon the kind of plant−insect interaction, GA can either suppress (chewing pest) or stimulate (sucking pest) plant defense response (Moosavi, 2017).

GA-deficient (*sd-1*, semidwarf-1 mutants) rice plants showed reduced resistance to sap-sucking brown planthopper (BPH) adults (*N. lugens*), whereas GA−excessive (*eui*, Elongated Uppermost Internodes) rice mutants were found to have increased resistance to BPH (Qi et al., 2018). Restoration of BPH resistance by exogenous GA application in transgenic *OsWRKY70* rice plants suggests that GA positively regulates phloem-feeding herbivore-induced plant response (Erb and Reymond, 2019; Kundu and Vadassery, 2021; Li et al., 2015; Qi et al., 2018). In contrast, no drastic alteration in defense metabolite accumulation was observed in the *Arabidopsis quad DELLA* mutant upon *S. exigua* feeding (Lan et al., 2014). Upon insect herbivory, many plant species change the direction of metabolic flux from growth into defense. Two key pathways modulating these processes are the GA/DELLA pathway and the jasmonate pathway. In response to stress, JA-mediated defense responses take priority over GA-dependent growth processes (Heinrich et al., 2013; Hou et al., 2013; Wild et al., 2012; Yang et al., 2012). *S. exigua* caterpillar labial salivary secretions suppress induced plant defenses, and this involves cross talk between JA—GA signaling pathways. In response to caterpillar herbivory, DELLA proteins are also involved in the regulation antiherbivore secondary metabolites, GSs, and also suppress laccase-like multicopper oxidase (LMCO) activity, which may be related to their role in plant cell wall fortification (Cai et al., 2006; Constabel and Barbehenn, 2008; Lan et al., 2014). However, the mechanism underlying the GA-mediated plant defense response to herbivores needs further study.

3.6.4 Brassinosteroids

BRs are plant steroid hormones that play crucial roles in various physiological processes, such as cell elongation, cell division, photomorphogenesis, xylem differentiation, and reproduction as well as both abiotic and biotic stress responses (Nolan et al., 2020). BRs are perceived by BR insensitive 1 (BRI1), a leucine-rich repeat receptor-like kinase. BRI1-associated kinase 1 (BAK1) interacts with BRI1 and plays an essential role in both BR signaling and multiple MAMP-elicited responses (Li and Chory, 1997; Erb et al., 2012). In BR signaling, two TFs, namely, brassinosteroid-INSENSITIVE1-EMS-SUPPRESSOR 1 (BES1) and RASSINAZOLE-RESISTANT 1 (BZR1) function as master regulators. (Gampala et al., 2007; Lian et al., 2020; Wang et al., 2002; Yin et al., 2002). BRs antagonize JA-activated plant defense against necrotrophic pathogens and herbivore insects through BES1. The BR-activated BES1 interacts with *PDF1.2a* and *PDF1.2b*, suppressing their transcriptional activities and attenuating JA signaling (Liao et al., 2020). BES1 inhibits the biosynthesis of the JA-induced insect defense-related metabolites, like indole GSs by interacting with TFs MYB and suppressing expression of genes encoding CYTOCHROME P450 FAMILY 79 SUBFAMILY B POLYPEPTIDE 3 (CYP79B3) and UDP-GLUCOSYL TRANSFERASE 74B1 (UGT74B1), thereby diminishing the defense responses against the insect herbivore *S. exigua* (Liao et al., 2020; Ortiz-Morea et al., 2020). Upon herbivory, a peptide hormone systemin functions as a mobile signal in the wounded tomato tissues (McGurl et al., 1992). Tomato BR receptor BRI1 homolog, LRR-RK160 (SR160), was identified as the receptor of systemin (Scheer and Ryan, 2002; Yu et al., 2018a). A role of SR160/BRI1 as systemin receptor could not be corroborated by others (Wang et al., 2018). Recent data suggests that SR160 is only a systemin-binding protein and has no role in systemin perception or signaling (Holton et al., 2007; Malinowski et al., 2009; Yu et al., 2018b). BRs contribute to plant defense responses against insect herbivores by negatively regulating GS biosynthesis through BES1 and/or BZR1, to achieve a trade-off between growth and defense (Guo et al., 2013; Lian et al., 2020; Yu et al., 2018b). *NaBAK1*, a tobacco homolog of *Arabidopsis* BAK1 in *N. attenuata*, plays an important role in herbivore-induced defense responses by modulating the JA accumulation and defense-related metabolites (Yang et al., 2011). BPH infestation suppresses the BR pathway, while successively activating the SA and JA pathways. BR-overproducing mutants and plants treated with 24-epibrassinolide (BL) showed increased susceptibility to BPH, whereas BR-deficient mutants were more resistant than the wild-type (Pan et al., 2018). Diamondback moth (*P. xylostella*), a specialist herbivore on *Arabidopsis*, feeds more on *bri 1—5* plants over the wild-type, indicating a negative role of BR in herbivore-induced plant defenses (Lee et al., 2018). Taken together, it is necessary to analyze mechanistically whether BR regulates DAMP or HAMP perception to modulate JA signaling.

3.7 Conclusions

Use of genetic engineering to target the manipulation of phytohormone levels in plants is an excellent approach to produce crops with improved herbivore resistance. Nevertheless, knowledge on the evolutionary conservation of hormone signaling networks in plants is inadequately available. Therefore, a comprehensive study on hormone signaling networks is essential for the conversion of understanding from model plants to crops (Berens et al., 2017). Hormonal cross talk is a crucial component in the hormonal network, which is activated via the perception of environmental cues. In response to the dynamic environments in which plants live, they have evolved mechanisms, such as hormone cross talk, to optimize fitness. The perception and response of each phytohormone in plant—herbivore interactions is generally shared with others, and therefore, this information provides a vital point for tailoring plant phytohormones pathway (Erb et al., 2012). To harness this process for the improvement of crops, it is very crucial to identify the molecular targets responsible for the cross talk and resource reallocation that facilitates prioritization of growth or defense (Huot et al., 2014).

Acknowledgments

We acknowledge Department of Biotechnology (DBT), India for NIPGR core grant, PK is funded by DBT-NE grant (BT/NER/95/SP42533/2021), PB is funded by National Agricultural Science Fund, ICAR (NASF/ABAP (SM)-8001/2019—20) and SM are funded by UGC doctoral fellowship. We acknowledge DBT-eLibrary Consortium (DeLCON) for providing access to e-resources.

References

Adie, B., Chico, J.M., Rubio-Somoza, I., Solano, R., 2007a. Modulation of plant defenses by ethylene. J. Plant Growth Regul. 26, 160—177.

Adie, B.A., Pérez-Pérez, J., Pérez-Pérez, M.M., Godoy, M., Sánchez-Serrano, J.-J., Schmelz, E.A., et al., 2007b. ABA is an essential signal for plant resistance to pathogens affecting JA biosynthesis and the activation of defenses in Arabidopsis. Plant. Cell 19, 1665—1681.

Aharoni, A., Jongsma, M.A., Bouwmeester, H.J., 2005. Volatile science? Metabolic engineering of terpenoids in plants. Trends Plant. Sci. 10, 594—602.

Akhtar, S.S., Mekureyaw, M.F., Pandey, C., Roitsch, T., 2020. Role of cytokinins for interactions of plants with microbial pathogens and pest insects. Front. Plant. Sci. 10, 1777.

An, L., Ahmad, R.M., Ren, H., Qin, J., Yan, Y., 2019. Jasmonate signal receptor gene family ZmCOIs restore male fertility and defense response of Arabidopsis mutant coi1-1. J. Plant. Growth Regul. 38, 479—493.

Ballaré, C.L., 2011. Jasmonate-induced defenses: a tale of intelligence, collaborators and rascals. Trends Plant. Sci. 16, 249—257.

Becerra, J.X., Noge, K., Venable, D.L., 2009. Macroevolutionary chemical escalation in an ancient plant—herbivore arms race. Proc. Natl. Acad. Sci. U.S.A. 106, 18062—18066.

Berens, M.L., Berry, H.M., Mine, A., Argueso, C.T., Tsuda, K., 2017. Evolution of hormone signaling networks in plant defense. Annu. Rev. Phytopathol. 55, 401—425.

Binenbaum, J., Weinstain, R., Shani, E., 2018. Gibberellin localization and transport in plants. Trends Plant. Sci. 23, 410–421.

Bodenhausen, N., Reymond, P., 2007. Signaling pathways controlling induced resistance to insect herbivores in Arabidopsis. Mol. Plant-Microbe Interact. 20, 1406–1420.

Bruessow, F., Gouhier-Darimont, C., Buchala, A., Metraux, J.P., Reymond, P., 2010. Insect eggs suppress plant defence against chewing herbivores. Plant. J. 62, 876–885.

Brütting, C., Crava, C.M., Schäfer, M., Schuman, M.C., Meldau, S., Adam, N., et al., 2018. Cytokinin transfer by a free-living mirid to *Nicotiana attenuata* recapitulates a strategy of endophytic insects. Elife 7, e36268.

Brutus, A., Sicilia, F., Macone, A., Cervone, F., De, Lorenzo, G., 2010. A domain swap approach reveals a role of the plant wall-associated kinase 1 (WAK1) as a receptor of oligogalacturonides. Proc. Natl. Acad. Sci. U.S.A. 107, 9452–9457.

Caarls, L., Elberse, J., Awwanah, M., Ludwig, N.R., De Vries, M., Zeilmaker, T., et al., 2017. *Arabidopsis* JASMONATE-INDUCED OXYGENASES down-regulate plant immunity by hydroxylation and inactivation of the hormone jasmonic acid. Proc. Natl. Acad. Sci. U.S.A. 114, 6388–6393.

Cai, C., Xu, C., Li, X., Ferguson, I., Chen, K., 2006. Accumulation of lignin in relation to change in activities of lignification enzymes in loquat fruit flesh after harvest. Postharvest Biol. Technol. 40, 163–169.

Carvalho, M.R., Wilf, P., Barrios, H., Windsor, D.M., Currano, E.D., Labandeira, C.C., et al., 2014. Insect leaf-chewing damage tracks herbivore richness in modern and ancient forests. PLoS One 9, e94950.

Chae, H.S., Kieber, J.J., 2005. Eto Brute? Role of ACS turnover in regulating ethylene biosynthesis. Trends Plant. Sci. 10, 291–296.

Chakravarthy, S., Tuori, R.P., D'Ascenzo, M.D., Fobert, P.R., Després, C., Martin, G.B., 2003. The tomato transcription factor Pti4 regulates defense-related gene expression via GCC box and non-GCC box cis elements. Plant. Cell 15, 3033–3050.

Chauvin, A., Caldelari, D., Wolfender, J.L., Farmer, E.E., 2013. Four 13-lipoxygenases contribute to rapid jasmonate synthesis in wounded *Arabidopsis thaliana* leaves: a role for lipoxygenase 6 in responses to long-distance wound signals. N. Phytol. 197, 566–575.

Chen, M.S., 2008. Inducible direct plant defense against insect herbivores: a review. Insect Sci. 15, 101–114.

Chico, J.-M., Fernández-Barbero, G., Chini, A., Fernández-Calvo, P., Díez-Díaz, M., Solano, R., 2014. Repression of jasmonate-dependent defenses by shade involves differential regulation of protein stability of MYC transcription factors and their JAZ repressors in *Arabidopsis*. Plant. Cell 26, 1967–1980.

Christensen, S.A., Nemchenko, A., Borrego, E., Murray, I., Sobhy, I.S., Bosak, L., et al., 2013. The maize lipoxygenase, Zm LOX 10, mediates green leaf volatile, jasmonate and herbivore-induced plant volatile production for defense against insect attack. Plant. J. 74, 59–73.

Christians, M.J., Gingerich, D.J., Hansen, M., Binder, B.M., Kieber, J.J., Vierstra, R.D., 2009. The BTB ubiquitin ligases ETO1, EOL1 and EOL2 act collectively to regulate ethylene biosynthesis in Arabidopsis by controlling type-2 ACC synthase levels. Plant. J. 57, 332–345.

Constabel, C.P., Barbehenn, R., 2008. Defensive roles of polyphenol oxidase in plants. Induced Plant Resistance to Herbivory. Springer, pp. 253–270.

Dempsey, D.M.A., Vlot, A.C., Wildermuth, M.C., Klessig, D.F., 2011. Salicylic acid biosynthesis and metabolism. Arabidopsis book/American Soc. Plant. Biologists 9.

Dervinis, C., Frost, C.J., Lawrence, S.D., Novak, N.G., Davis, J.M., 2010. Cytokinin primes plant responses to wounding and reduces insect performance. J. Plant. Growth Regul. 29, 289–296.

Dicke, M., Baldwin, I.T., 2010. The evolutionary context for herbivore-induced plant volatiles: beyond the 'cry for help'. Trends Plant. Sci. 15, 167–175.

Dinh, S.T., Baldwin, I.T., Galis, I., 2013. The HERBIVORE ELICITOR-REGULATED1 gene enhances abscisic acid levels and defenses against herbivores in *Nicotiana attenuata* plants. Plant. Physiol. 162, 2106–2124.

Dombrecht, B., Xue, G.P., Sprague, S.J., Kirkegaard, J.A., Ross, J.J., Reid, J.B., et al., 2007. MYC2 differentially modulates diverse jasmonate-dependent functions in *Arabidopsis*. Plant. Cell 19, 2225–2245.

Eisenring, M., Glauser, G., Meissle, M., Romeis, J., 2018. Differential impact of herbivores from three feeding guilds on systemic secondary metabolite induction, phytohormone levels and plant-mediated herbivore interactions. J. Chem. Ecol. 44, 1178–1189.

Elzen, G., 1983. Cytokinins and insect galls. Comparative Biochemistry and Physiology Part A: Physiology 76, 17–19.

Engelbrecht, L., Orban, U., Heese, W., 1969. Leaf-miner caterpillars and cytokinins in the "green islands" of autumn leaves. Nature 223, 319–321.

Erb, M., Köllner, T.G., Degenhardt, J., Zwahlen, C., Hibbard, B.E., Turlings, T.C., 2011a. The role of abscisic acid and water stress in root herbivore-induced leaf resistance. N. Phytol. 189, 308–320.

Erb, M., Meldau, S., Howe, G.A., 2012. Role of phytohormones in insect-specific plant reactions. Trends Plant. Sci. 17, 250–259.

Erb, M., Reymond, P., 2019. Molecular interactions between plants and insect herbivores. Annu. Rev. Plant. Biol. 70, 527–557.

Erb, M., Robert, C.A., Hibbard, B.E., Turlings, T.C., 2011b. Sequence of arrival determines plant-mediated interactions between herbivores. J. Ecol. 99, 7–15.

Filgueiras, C.C., Martins, A.D., Pereira, R.V., Willett, D.S., 2019. The ecology of salicylic acid signaling: primary, secondary and tertiary effects with applications in agriculture. Int. J. Mol. Sci. 20, 5851.

Gampala, S.S., Kim, T.-W., He, J.-X., Tang, W., Deng, Z., Bai, M.-Y., et al., 2007. An essential role for 14-3-3 proteins in brassinosteroid signal transduction in *Arabidopsis*. Dev. Cell 13, 177–189.

Gatehouse, J.A., 2002. Plant resistance towards insect herbivores: a dynamic interaction. New Phytologist 156, 145–169.

Gfeller, A., Baerenfaller, K., Loscos, J., Chetelat, A., Baginsky, S., Farmer, E.E., 2011. Jasmonate controls polypeptide patterning in undamaged tissue in wounded *Arabidopsis* leaves. Plant. Physiol. 156, 1797–1807.

Giron, D., Frago, E., Glevarec, G., Pieterse, C.M., Dicke, M., 2013. Cytokinins as key regulators in plant–microbe–insect interactions: connecting plant growth and defence. Funct. Ecol. 27, 599–609.

Glauser, G., Grata, E., Dubugnon, L., Rudaz, S., Farmer, E.E., Wolfender, J.-L., 2008. Spatial and temporal dynamics of jasmonate synthesis and accumulation in *Arabidopsis* in response to wounding. J. Biol. Chem. 283, 16400–16407.

Glazebrook, J., 2005. Contrasting mechanisms of defense against biotrophic and necrotrophic pathogens. Annu. Rev. Phytopathol. 43, 205–227.

Goodspeed, D., Chehab, E.W., Min-Venditti, A., Braam, J., Covington, M.F., 2012. *Arabidopsis* synchronizes jasmonate-mediated defense with insect circadian behavior. Proc. Natl. Acad. Sci. U.S.A. 109, 4674–4677.

Guo, H., Li, L., Aluru, M., Aluru, S., Yin, Y., 2013. Mechanisms and networks for brassinosteroid regulated gene expression. Curr. Opin. Plant. Biol. 16, 545–553.

Guo, Q., Yoshida, Y., Major, I.T., Wang, K., Sugimoto, K., Kapali, G., et al., 2018. JAZ repressors of metabolic defense promote growth and reproductive fitness in *Arabidopsis*. Proc. Natl. Acad. Sci. U.S.A. 115, E10768–E10777.

He, J., Chen, F., Chen, S., Lv, G., Deng, Y., Fang, W., et al., 2011. Chrysanthemum leaf epidermal surface morphology and antioxidant and defense enzyme activity in response to aphid infestation. J. Plant. Physiol. 168, 687–693.

Heinrich, M., Hettenhausen, C., Lange, T., Wünsche, H., Fang, J., Baldwin, I.T., et al., 2013. High levels of jasmonic acid antagonize the biosynthesis of gibberellins and inhibit the growth of N icotiana attenuata stems. Plant. J. 73, 591–606.

Heitz, T., Widemann, E., Lugan, R., Miesch, L., Ullmann, P., Désaubry, L., et al., 2012. Cytochromes P450 CYP94C1 and CYP94B3 catalyze two successive oxidation steps of plant hormone jasmonoyl-isoleucine for catabolic turnover. J. Biol. Chem. 287, 6296–6306.

Hillwig, M.S., Chiozza, M., Casteel, C.L., Lau, S.T., Hohenstein, J., Hernández, E., et al., 2016. Abscisic acid deficiency increases defence responses against *Myzus persicae* in A rabidopsis. Mol. Plant. Pathol. 17, 225–235.

Holton, N., Caño-Delgado, A., Harrison, K., Montoya, T., Chory, J., Bishop, G.J., 2007. Tomato BRASSINOSTEROID INSENSITIVE1 is required for systemin-induced root elongation in *Solanum pimpinellifolium* but is not essential for wound signaling. Plant. Cell 19, 1709–1717.

Hou, X., Ding, L., Yu, H., 2013. Crosstalk between GA and JA signaling mediates plant growth and defense. Plant. Cell Rep. 32, 1067–1074.

Howe, G.A., Jander, G., 2008. Plant immunity to insect herbivores. Annu. Rev. Plant. Biol. 59, 41–66.

Huot, B., Yao, J., Montgomery, B.L., He, S.Y., 2014. Growth–defense tradeoffs in plants: a balancing act to optimize fitness. Mol. Plant. 7, 1267–1287.

Jang, G., Yoon, Y., Choi, Y.D., 2020. Crosstalk with jasmonic acid integrates multiple responses in plant development. Int. J. Mol. Sci. 21, 305.

Jiang, Y., Zhang, C.-X., Chen, R., He, S.Y., 2019. Challenging battles of plants with phloem-feeding insects and prokaryotic pathogens. Proc. Natl. Acad. Sci. U.S.A. 116, 23390–23397.

Johnson, S.N., Rasmann, S., 2015. Root-feeding insects and their interactions with organisms in the rhizosphere. Annual Review of Entomology 60, 517–535.

Kahl, J., Siemens, D.H., Aerts, R.J., Gäbler, R., KuÈhnemann, F., Preston, C.A., et al., 2000. Herbivore-induced ethylene suppresses a direct defense but not a putative indirect defense against an adapted herbivore. Planta 210, 336–342.

Kaiser, W., Huguet, E., Casas, Jm, Commin, C., Giron, D., 2010. Plant green-island phenotype induced by leaf-miners is mediated by bacterial symbionts. Proc. R. Soc. B: Biol. Sci. 277, 2311–2319.

Kang, J.-H., Wang, L., Giri, A., Baldwin, I.T., 2006. Silencing threonine deaminase and JAR4 in *Nicotiana attenuata* impairs jasmonic acid–isoleucine–mediated defenses against *Manduca sexta*. Plant. Cell 18, 3303–3320.

Katsir, L., Schilmiller, A.L., Staswick, P.E., He, S.Y., Howe, G.A., 2008. COI1 is a critical component of a receptor for jasmonate and the bacterial virulence factor coronatine. Proc. Natl. Acad. Sci. U.S.A. 105, 7100–7105.

Kessler, A., Baldwin, I.T., 2002. Plant responses to insect herbivory: the emerging molecular analysis. Annu. Rev. Plant. Biol. 53, 299–328.

Kiep, V., Vadassery, J., Lattke, J., Maaß, J.P., Boland, W., Peiter, E., et al., 2015. Systemic cytosolic Ca2 + elevation is activated upon wounding and herbivory in *Arabidopsis*. N. Phytol. 207, 996–1004.

Koo, A.J., Cooke, T.F., Howe, G.A., 2011. Cytochrome P450 CYP94B3 mediates catabolism and inactivation of the plant hormone jasmonoyl-L-isoleucine. Proc. Natl. Acad. Sci. U.S.A. 108, 9298–9303.

Koo, A.J., Gao, X., Daniel Jones, A., Howe, G.A., 2009. A rapid wound signal activates the systemic synthesis of bioactive jasmonates in *Arabidopsis*. Plant. J. 59, 974–986.

Kundu, A., Mishra, S., Kundu, P., Jogawat, A., Vadassery, J., 2021. Piriformospora indica recruits host-derived putrescine for growth promotion in plants. Plant. Physiol.

Kundu, A., Mishra, S., Vadassery, J., 2018. *Spodoptera litura*-mediated chemical defense is differentially modulated in older and younger systemic leaves of *Solanum lycopersicum*. Planta 248, 981–997.

Kundu, P., Sahu, R., 2021. GIGANTEA confers susceptibility to plants during spot blotch attack by regulating salicylic acid signalling pathway. Plant. Physiol. Biochem.

Kundu, P., Vadassery, J., 2021. Role of WRKY transcription factors in plant defense against lepidopteran insect herbivores: an overview. J. Plant. Biochem. Biotechnol. 1–10.

Lackman, P., González-Guzmán, M., Tilleman, S., Carqueijeiro, I., Pérez, A.C., Moses, T., et al., 2011. Jasmonate signaling involves the abscisic acid receptor PYL4 to regulate metabolic reprogramming in *Arabidopsis* and tobacco. Proc. Natl. Acad. Sci. U.S.A. 108, 5891–5896.

Lan, Z., Krosse, S., Achard, P., van Dam, N.M., Bede, J.C., 2014. DELLA proteins modulate *Arabidopsis* defences induced in response to caterpillar herbivory. J. Exp. Bot. 65, 571–583.

Laudert, D., Schaller, F., Weiler, E.W., 2000. Transgenic Nicotiana tabacum and *Arabidopsis thaliana* plants overexpressing allene oxide synthase. Planta 211, 163–165.

Lee, J.H., Lee, J., Kim, H., Chae, W.B., Kim, S.J., Lim, Y.P., et al., 2018. Brassinosteroids regulate glucosinolate biosynthesis in *Arabidopsis thaliana*. Physiol. Plant. 163, 450–458.

Leitner, M., Boland, W., Mithöfer, A., 2005. Direct and indirect defences induced by piercing-sucking and chewing herbivores in *Medicago truncatula*. N. Phytol. 167, 597–606.

Li, J.M., Chory, J., 1997. A putative leucine-rich repeat receptor kinase involved in brassinosteroid signal transduction. Cell 90, 929–938.

Li, L., Zhao, Y., McCaig, B.C., Wingerd, B.A., Wang, J., Whalon, M.E., et al., 2004. The tomato homolog of CORONATINE-INSENSITIVE1 is required for the maternal control of seed maturation, jasmonate-signaled defense responses, and glandular trichome development. Plant. Cell 16, 126–143.

Li, Q., Xie, Q.-G., Smith-Becker, J., Navarre, D.A., Kaloshian, I., 2006. Mi-1-mediated aphid resistance involves salicylic acid and mitogen-activated protein kinase signaling cascades. Mol. Plant-Microbe Interact. 19, 655–664.

Li, R., Jin, J., Xu, J., Wang, L., Li, J., Lou, Y., et al., 2020. Long non-coding RNAs associate with jasmonate-mediated plant defence against herbivores. Plant, Cell Environ.

Li, R., Zhang, J., Li, J., Zhou, G., Wang, Q., Bian, W., et al., 2015. Prioritizing plant defence over growth through WRKY regulation facilitates infestation by non-target herbivores. Elife 4, e04805.

Lian, X., Tan, B., Yan, L., Jiang, C., Cheng, J., Zheng, X., et al., 2020. Transcript profiling provides insights into molecular processes during shoot elongation in temperature-sensitive peach (*Prunus persica*). Sci. Rep. 10, 1–12.

Liao, K., Peng, Y.-J., Yuan, L.-B., Dai, Y.-S., Chen, Q.-F., Yu, L.-J., et al., 2020. Brassinosteroids antagonize jasmonate-activated plant defense responses through BRI1-EMS-SUPPRESSOR1 (BES1). Plant. Physiol. 182, 1066–1082.

Liu, Y., Ahn, J.-E., Datta, S., Salzman, R.A., Moon, J., Huyghues-Despointes, B., et al., 2005. *Arabidopsis* vegetative storage protein is an anti-insect acid phosphatase. Plant. Physiol. 139, 1545–1556.

Lorenzo, O., Solano, R., 2005. Molecular players regulating the jasmonate signalling network. Curr. Opin. Plant. Biol. 8, 532–540.

Lortzing, V., Oberländer, J., Lortzing, T., Tohge, T., Steppuhn, A., Kunze, R., et al., 2019. Insect egg deposition renders plant defence against hatching larvae more effective in a salicylic acid-dependent manner. Plant, Cell Environ. 42, 1019–1032.

Machado, R.A., Robert, C.A., Arce, C.C., Ferrieri, A.P., Xu, S., Jimenez-Aleman, G.H., et al., 2016. Auxin is rapidly induced by herbivore attack and regulates a subset of systemic, jasmonate-dependent defenses. Plant. Physiol. 172, 521–532.

Maffei, M.E., Mithöfer, A., Boland, W., 2007a. Before gene expression: early events in plant–insect interaction. Trends Plant. Sci. 12, 310–316.

Malinowski, R., Higgins, R., Luo, Y., Piper, L., Nazir, A., Bajwa, V.S., et al., 2009. The tomato brassinosteroid receptor BRI1 increases binding of systemin to tobacco plasma membranes, but is not involved in systemin signaling. Plant. Mol. Biol. 70, 603–616.

Mauch-Mani, B., Mauch, F., 2005. The role of abscisic acid in plant–pathogen interactions. Current Opinion in Plant Biology 8, 409–414.

McGurl, B., Pearce, G., Orozco-Cardenas, M., Ryan, C.A., 1992. Structure, expression, and antisense inhibition of the systemin precursor gene. Science 255, 1570–1573.

Meena, M.K., Prajapati, R., Krishna, D., Divakaran, K., Pandey, Y., Reichelt, M., et al., 2019. The Ca2 + channel CNGC19 regulates *Arabidopsis* defense against Spodoptera herbivory. Plant. Cell 31, 1539–1562.

Mewis, I., Appel, H.M., Hom, A., Raina, R., Schultz, J.C., 2005. Major signaling pathways modulate *Arabidopsis* glucosinolate accumulation and response to both phloem-feeding and chewing insects. Plant. Physiol. 138, 1149–1162.

Mithöfer, A., Boland, W., 2008. Recognition of herbivory-associated molecular patterns. Plant. Physiol. 146, 825–831.

Mithöfer, A., Boland, W., 2012. Plant defense against herbivores: chemical aspects. Annu. Rev. Plant. Biol. 63, 431–450.

Mithöfer, A., Wanner, G., Boland, W., 2005. Effects of feeding *Spodoptera littoralis* on lima bean leaves. II. Continuous mechanical wounding resembling insect feeding is sufficient to elicit herbivory-related volatile emission. Plant. Physiol. 137, 1160–1168.

Mohase, L., van der Westhuizen, A.J., 2002. Salicylic acid is involved in resistance responses in the Russian wheat aphid-wheat interaction. J. Plant. Physiol. 159, 585–590.

Moosavi, M.R., 2017. The effect of gibberellin and abscisic acid on plant defense responses and on disease severity caused by *Meloidogyne javanica* on tomato plants. J. Gen. Plant. Pathol. 83, 173–184.

Moran, P.J., Thompson, G.A., 2001. Molecular responses to aphid feeding in *Arabidopsis* in relation to plant defense pathways. Plant. Physiol. 125, 1074–1085.

Mousavi, S.A., Chauvin, A., Pascaud, F., Kellenberger, S., Farmer, E.E., 2013. GLUTAMATE RECEPTOR-LIKE genes mediate leaf-to-leaf wound signalling. Nature 500, 422.

Nguyen, C.T., Kurenda, A., Stolz, S., Chételat, A., Farmer, E.E., 2018. Identification of cell populations necessary for leaf-to-leaf electrical signaling in a wounded plant. Proc. Natl. Acad. Sci. U.S.A. 115, 10178–10183.

Nguyen, D., Rieu, I., Mariani, C., van Dam, N.M., 2016. How plants handle multiple stresses: hormonal interactions underlying responses to abiotic stress and insect herbivory. Plant. Mol. Biol. 91, 727–740.

Nolan, T.M., Vukašinović, N., Liu, D., Russinova, E., Yin, Y., 2020. Brassinosteroids: multidimensional regulators of plant growth, development, and stress responses. Plant. Cell 32, 295–318.

O'donnell, P., Calvert, C., Atzorn, R., Wasternack, C., Leyser, H., Bowles, D., 1996. Ethylene as a signal mediating the wound response of tomato plants. Science 274, 1914–1917.

Ohme-Takagi, M., Shinshi, H., 1995. Ethylene-inducible DNA binding proteins that interact with an ethylene-responsive element. Plant. Cell 7, 173–182.

Onkokesung, N., Gális, I., von Dahl, C.C., Matsuoka, K., Saluz, H.-P., Baldwin, I.T., 2010. Jasmonic acid and ethylene modulate local responses to wounding and simulated herbivory in *Nicotiana attenuata* leaves. Plant. Physiol. 153, 785–798.

Ortiz-Morea, F.A., He, P., Shan, L., Russinova, E., 2020. It takes two to tango—molecular links between plant immunity and brassinosteroid signalling. J. Cell Sci. 133.

Pan, G., Liu, Y., Ji, L., Zhang, X., He, J., Huang, J., et al., 2018. Brassinosteroids mediate susceptibility to brown planthopper by integrating with the salicylic acid and jasmonic acid pathways in rice. J. Exp. Bot. 69, 4433–4442.

Pauwels, L., Barbero, G.F., Geerinck, J., Tilleman, S., Grunewald, W., Pérez, A.C., et al., 2010. NINJA connects the co-repressor TOPLESS to jasmonate signalling. Nature 464, 788–791.

Pauwels, L., Goossens, A., 2011. The JAZ proteins: a crucial interface in the jasmonate signaling cascade. Plant. Cell 23, 3089–3100.

Pieterse, C.M., Leon-Reyes, A., Van der Ent, S., Van Wees, S.C., 2009. Networking by small-molecule hormones in plant immunity. Nat. Chem. Biol. 5, 308–316.

Pradhan, M., Pandey, P., Gase, K., Sharaff, M., Singh, R.K., Sethi, A., et al., 2017. Argonaute 8 (AGO8) mediates the elicitation of direct defenses against herbivory. Plant. Physiol. 175, 927–946.

Qi, J., ul Malook, S., Shen, G., Gao, L., Zhang, C., Li, J., et al., 2018. Current understanding of maize and rice defense against insect herbivores. Plant. Divers. 40, 189–195.

Qi, T., Song, S., Ren, Q., Wu, D., Huang, H., Chen, Y., et al., 2011. The Jasmonate-ZIM-domain proteins interact with the WD-Repeat/bHLH/MYB complexes to regulate Jasmonate-mediated anthocyanin accumulation and trichome initiation in *Arabidopsis thaliana*. Plant. Cell 23, 1795–1814.

Ray, S., Gaffor, I., Acevedo, F.E., Helms, A., Chuang, W.-P., Tooker, J., et al., 2015. Maize plants recognize herbivore-associated cues from caterpillar frass. J. Chem. Ecol. 41, 781–792.

Reymond, P., Bodenhausen, N., Van Poecke, R.M., Krishnamurthy, V., Dicke, M., Farmer, E.E., 2004. A conserved transcript pattern in response to a specialist and a generalist herbivore. Plant. Cell 16, 3132–3147.

Sahu, R., Kundu, P., Sethi, A., 2021a. In vitro antioxidant activity and enzyme inhibition properties of wheat whole grain, bran and flour defatted with hexane and supercritical fluid extraction. LWT 146, 111376.

Sahu, R., Kundu, P., Shraff, M., Pradhan, M., Mishra, V., Chand, R., 2016a. Understanding the defense-related mechanism during the wheat's interaction with fungal pathogens. Indian. Phytopathol. 69, 203–205.

Sahu, R., Prabhakaran, N., Kundu, P., Kumar, A., 2021b. Differential response of phytohormone signalling network determines nonhost resistance in rice during wheat stem rust (*Puccinia graminis* f. sp. tritici) colonization. Plant. Pathol.

Sahu, R., Sharaff, M., Pradhan, M., Sethi, A., Bandyopadhyay, T., Mishra, V.K., et al., 2016b. Elucidation of defense-related signaling responses to spot blotch infection in bread wheat (*Triticum aestivum* L.). Plant. J. 86, 35–49.

Sanders, D., Pelloux, J., Brownlee, C., Harper, J.F., 2002. Calcium at the crossroads of signaling. Plant. Cell 14, S401–S417.

Scala, A., Allmann, S., Mirabella, R., Haring, M.A., Schuurink, R.C., 2013. Green leaf volatiles: a plant's multifunctional weapon against herbivores and pathogens. Int. J. Mol. Sci. 14, 17781–17811.

Schäfer, M., Brütting, C., Meza-Canales, I.D., Großkinsky, D.K., Vankova, R., Baldwin, I.T., et al., 2015. The role of cis-zeatin-type cytokinins in plant growth regulation and mediating responses to environmental interactions. J. Exp. Bot. 66, 4873–4884.

Scheer, J.M., Ryan, C.A., 2002. The systemin receptor SR160 from *Lycopersicon peruvianum* is a member of the LRR receptor kinase family. Proc. Natl. Acad. Sci. U.S.A. 99, 9585–9590.

Schmelz, E.A., Alborn, H.T., Engelberth, J., Tumlinson, J.H., 2003. Nitrogen deficiency increases volicitin-induced volatile emission, jasmonic acid accumulation, and ethylene sensitivity in maize. Plant. Physiol. 133, 295–306.

Scholz, S.S., Vadassery, J., Heyer, M., Reichelt, M., Bender, K.W., Snedden, W.A., et al., 2014. Mutation of the Arabidopsis calmodulin-like protein CML37 deregulates the jasmonate pathway and enhances susceptibility to herbivory. Mol. Plant. 7, 1712–1726.

Schweizer, F., Bodenhausen, N., Lassueur, S., Masclaux, F.G., Reymond, P., 2013. Differential contribution of transcription factors to *Arabidopsis thaliana* defense against *Spodoptera littoralis*. Front. Plant. Sci. 4, 13.

Sethi, A., Sharaff, M., Sahu, R., 2021. Deciphering common temporal transcriptional response during powdery mildew disease in plants using meta-analysis. Plant. Gene 27, 100307.

Sharma, H., Sujana, G., Rao, D.M., 2009. Morphological and chemical components of resistance to pod borer, *Helicoverpa armigera* in wild relatives of pigeonpea. Arthropod-Plant Interact. 3, 151–161.

Sheard, L.L.B., Tan, X., Mao, H., Withers, J., Ben-Nissan, G., Hinds, T.R., et al., 2010. Jasmonate perception by inositol-phosphate-potentiated COI1-JAZ co-receptor. Nature 468, 400–407.

Shigenaga, A.M., Argueso, C.T., 2016. No hormone to rule them all: interactions of plant hormones during the responses of plants to pathogens, Seminars in Cell & Developmental Biology, Vol. 56. Elsevier, pp. 174–189.

Silva, D.B., Weldegergis, B.T., Van Loon, J.J., Bueno, V.H., 2017. Qualitative and quantitative differences in herbivore-induced plant volatile blends from tomato plants infested by either *Tuta absoluta* or *Bemisia tabaci*. J. Chem. Ecol. 43, 53–65.

Stahl, E., Hilfiker, O., Reymond, P., 2018. Plant–arthropod interactions: who is the winner? Plant. J. 93, 703–728.

Staswick, P.E., Tiryaki, I., 2004. The oxylipin signal jasmonic acid is activated by an enzyme that conjugates it to isoleucine in Arabidopsis. Plant. Cell 16, 2117–2127.

Stintzi, A., Weber, H., Reymond, P., Farmer, E.E., 2001. Plant defense in the absence of jasmonic acid: the role of cyclopentenones. Proc. Natl. Acad. Sci. U.S.A. 98, 12837–12842.

Stotz, H.U., Pittendrigh, B.R., Kroymann, J., Weniger, K., Fritsche, J., Bauke, A., et al., 2000. Induced plant defense responses against chewing insects. Ethylene signaling reduces resistance of Arabidopsis against Egyptian cotton worm but not diamondback moth. Plant. Physiol. 124, 1007–1018.

Suza, W.P., Staswick, P.E., 2008. The role of JAR1 in jasmonoyl-L-isoleucine production during Arabidopsis wound response. Planta 227, 1221–1232.

Tanaka, Y., Okada, K., Asami, T., Suzuki, Y., 2013. Phytohormones and willow gall induction by a gall-inducing sawfly. Biosci. Biotechnol. Biochem. 77, 1942–1948.

Tani, T., Sobajima, H., Okada, K., Chujo, T., Arimura, S.-i, Tsutsumi, N., et al., 2008. Identification of the OsOPR7 gene encoding 12-oxophytodienoate reductase involved in the biosynthesis of jasmonic acid in rice. Planta 227, 517–526.

Theodoulou, F.L., Job, K., Slocombe, S.P., Footitt, S., Holdsworth, M., Baker, A., et al., 2005. Jasmonic acid levels are reduced in COMATOSE ATP-binding cassette transporter mutants. Implications for transport of jasmonate precursors into peroxisomes. Plant. Physiol. 137, 835–840.

Thompson, G.A., Goggin, F.L., 2006. Transcriptomics and functional genomics of plant defence induction by phloem-feeding insects. J. Exp. Bot. 57, 755–766.

Tooker, J.F., De Moraes, C.M., 2011a. Feeding by a gall-inducing caterpillar species alters levels of indole-3-acetic and abscisic acid in *Solidago altissima* (Asteraceae) stems. Arthropod-Plant Interact. 5, 115–124.

Tooker, J.F., De Moraes, C.M., 2011b. Feeding by Hessian fly (*Mayetiola destructor* [Say]) larvae on wheat increases levels of fatty acids and indole-3-acetic acid but not hormones involved in plant-defense signaling. J. Plant. Growth Regul. 30, 158–165.

Turner, J.G., Ellis, C., Devoto, A., 2002. The jasmonate signal pathway. Plant. Cell 14, S153–S164.

Uemura, T., Arimura, G.-I., 2019. Current opinions about herbivore-associated molecular patterns and plant intracellular signaling. Plant. Signal. Behav. 14, e1633887.

Vadassery, J., Reichelt, M., Hause, B., Gershenzon, J., Boland, W., Mithöfer, A., 2012. CML42-mediated calcium signaling coordinates responses to Spodoptera herbivory and abiotic stresses in Arabidopsis. Plant. Physiol. 159, 1159–1175.

van der Meijden, E., 2015. Herbivorous insects—a threat for crop production. Principles of Plant–Microbe Interactions. Springer, pp. 103–114.

Vos, I.A., Verhage, A., Schuurink, R.C., Watt, L.G., Pieterse, C.M., Van, et al., 2013. Onset of herbivore-induced resistance in systemic tissue primed for jasmonate-dependent defenses is activated by abscisic acid. Front. plant. Sci. 4, 539.

Wang, L., Einig, E., Almeida-Trapp, M., Albert, M., Fliegmann, J., Mithöfer, A., et al., 2018. The systemin receptor SYR1 enhances resistance of tomato against herbivorous insects. Nat. Plants 4, 152–156.

Wang, Z.-Y., Nakano, T., Gendron, J., He, J., Chen, M., Vafeados, D., et al., 2002. Nuclear-localized BZR1 mediates brassinosteroid-induced growth and feedback suppression of brassinosteroid biosynthesis. Dev. Cell 2, 505–513.

Wasternack, C., Hause, B., 2013. Jasmonates: biosynthesis, perception, signal transduction and action in plant stress response, growth and development. An update to the 2007 review in Annals of Botany. Ann. Bot. 111, 1021–1058.

Wasternack, C., Song, S., 2017. Jasmonates: biosynthesis, metabolism, and signaling by proteins activating and repressing transcription. J. Exp. Bot. 68, 1303–1321.

Weinhold, A., Baldwin, I.T., 2011. Trichome-derived O-acyl sugars are a first meal for caterpillars that tags them for predation. Proc. Natl. Acad. Sci. U.S.A. 108, 7855–7859.

Werner, T., Schmülling, T., 2009. Cytokinin action in plant development. Curr. Opin. Plant. Biol. 12, 527–538.

Wild, M., Davière, J.-M., Cheminant, S., Regnault, T., Baumberger, N., Heintz, D., et al., 2012. The Arabidopsis DELLA RGA-LIKE3 is a direct target of MYC2 and modulates jasmonate signaling responses. Plant. Cell 24, 3307–3319.

Yamagami, T., Tsuchisaka, A., Yamada, K., Haddon, W.F., Harden, L.A., Theologis, A., 2003. Biochemical diversity among the 1-amino-cyclopropane-1-carboxylate synthase isozymes encoded by the Arabidopsis gene family. J. Biol. Chem. 278, 49102–49112.

Yan, C., Fan, M., Yang, M., Zhao, J., Zhang, W., Su, Y., et al., 2018. Injury activates Ca2 + /calmodulin-dependent phosphorylation of JAV1-JAZ8-WRKY51 complex for jasmonate biosynthesis. Mol. Cell 70, 136–149. e137.

Yan, L., Zhai, Q., Wei, J., Li, S., Wang, B., Huang, T., et al., 2013. Role of tomato lipoxygenase D in wound-induced jasmonate biosynthesis and plant immunity to insect herbivores. PLoS Genet. 9, e1003964.

Yan, Y., Christensen, S., Isakeit, T., Engelberth, J., Meeley, R., Hayward, A., et al., 2012. Disruption of OPR7 and OPR8 reveals the versatile functions of jasmonic acid in maize development and defense. Plant. Cell 24, 1420–1436.

Yang, D.-L., Yao, J., Mei, C.-S., Tong, X.-H., Zeng, L.-J., Li, Q., et al., 2012. Plant hormone jasmonate prioritizes defense over growth by interfering with gibberellin signaling cascade. Proc. Natl. Acad. Sci. U.S.A. 109, E1192–E1200.

Yang, D.H., Hettenhausen, C., Baldwin, I.T., Wu, J., 2011. BAK1 regulates the accumulation of jasmonic acid and the levels of trypsin proteinase inhibitors in *Nicotiana attenuata*'s responses to herbivory. J. Exp. Bot. 62, 641–652.

Yang, F., Zhang, Q., Yao, Q., Chen, G., Tong, H., Zhang, J., et al., 2020. Direct and indirect plant defenses induced by (Z)-3-hexenol in tomato against whitefly attack. J. Pest. Sci. 93, 1243–1254.

Yin, Y., Wang, Z.-Y., Mora-Garcia, S., Li, J., Yoshida, S., Asami, T., et al., 2002. BES1 accumulates in the nucleus in response to brassinosteroids to regulate gene expression and promote stem elongation. Cell 109, 181–191.

Yu, M.-H., Zhao, Z.-Z., He, J.-X., 2018a. Brassinosteroid signaling in plant–microbe interactions. Int. J. Mol. Sci. 19, 4091.

Yu, M.H., Zhao, Z.Z., He, J.X., 2018b. Brassinosteroid signaling in plant(-)microbe interactions. Int. J. Mol. Sci. 19.

Zarate, S.I., Kempema, L.A., Walling, L.L., 2007. Silverleaf whitefly induces salicylic acid defenses and suppresses effectual jasmonic acid defenses. Plant. Physiol. 143, 866–875.

Zhang, H., Guiguet, A., Dubreuil, G., Kisiala, A., Andreas, P., Emery, R.N., et al., 2017. Dynamics and origin of cytokinins involved in plant manipulation by a leaf-mining insect. Insect Sci. 24, 1065–1078.

Role of phytohormones in regulating agronomically important seed traits in crop plants

Rubi Jain[1], Namrata Dhaka[2], Pinky Yadav[2] and Rita Sharma[3]

[1]School of Computational and Integrative Sciences, Jawaharlal Nehru University, New Delhi, India [2]Department of Biotechnology, Central University of Haryana, Mahendergarh, Haryana, India [3]Department of Biological Sciences, Birla Institute of Technology & Science, Pilani, Rajasthan, India

4.1 Introduction

Seeds are not only the means of the sexual propagation of plants but are also the primary source of the world's food supply. Seed development in angiosperms initiates after double fertilization and requires coordinated growth of all the three significant components of the seed, namely, embryo, endosperm, and seed coat (Savadi, 2017). The fertilization of the egg cell with one of the sperm cells, followed by zygotic proliferation, differentiation, patterning, and establishment of polarity, gives rise to the embryo. Simultaneously, the fertilization of the diploid central cell with the second sperm cell results in the development of triploid endosperm. Endosperm comprises a significant part of the seed in monocots, while in dicots, developing embryo cotyledons consume most of the endosperm (Sundaresan, 2005). Seed development comprises three distinct phases: morphogenesis, maturation, and desiccation. Morphogenesis involves the differentiation of embryo and cotyledons, endosperm cellularization, and seed coat development, providing mechanical and chemical protection to both embryo and endosperm. Subsequently, in the maturation phase, the embryo undergoes further expansion along with the seed filling, that is, accumulation of starch, proteins, oils, and other nutrients (Locascio et al., 2014). Finally, the loss of water during the desiccation phase leads to a more resilient and dry state, culminating with the establishment of dormancy (Leprince et al., 2017).

Plants accumulate large amounts of storage compounds and nutrients in the seeds, such as proteins, vitamins, carbohydrates, oils, and minerals, making them a significant source of food, feed, and biofuels (Baud et al., 2008). Seed yield is pivotal in staple crops, rice, wheat, maize, sorghum, barley, and horticultural crops, such as soybean, tomato, canola, and pea. Seeds of these crops are either directly used for consumption or extraction of oils and other high-value-added products. However, continuous seed yield and quality improvements are required to meet the world's increasing food, feed, and fuel requirements.

Seed number, size, shape, weight, nutritional content, and longevity are critical agronomic traits that determine overall seed yield. The seed number is mainly determined by genes regulating panicle branching and spikelet development. In contrast, other aspects of seed yield are influenced by complex genetic networks associated with individual seed traits and interactions among them. The roles of transcription factors (TFs), plant hormones, epigenetic factors, peptides, and sugar signaling molecules in determining seed size have already been reviewed elsewhere (Li and Li, 2015; Savadi, 2017). Since the role of phytohormones in regulating grain number and seed dormancy has also been extensively reviewed recently, previously published studies can be consulted for a more exhaustive understanding of these aspects (Deveshwar et al., 2020; Shu et al., 2016). Here, we focus specifically on the role of phytohormones in regulating key agronomic traits relevant to seed yield.

4.2 Role of auxins

Auxins (AUXs) regulate almost all aspects of seed development, including embryo patterning, endosperm cellularization, and seed coat development (Table 4.1) (Figueiredo et al., 2016; Forestan and Varotto, 2012; Matilla, 2020). The maternal AUXs, released after fertilization, are hypothesized to induce early stages of embryo and endosperm development, while endosperm-derived AUXs regulate the further growth of endosperm and seed coat.

Mutants affecting genes involved in AUX perception (*tir1, transport inhibitor response 1; afb1, afb2, afb3, auxin signaling F box protein*), signaling (*mp, monopteros; bdl, bodenlos*), and transport (*pin1 pin3 pin4 pin7, pin-formed*) impact embryogenesis in *Arabidopsis* (Cheng et al., 2007). *YUC* (flavin monooxygenases) and *TAA/TAR* (*tryptophan amino-transferases*) genes encode enzymes involved in AUX biosynthesis (Kasahara, 2016). In *Arabidopsis*, the quadruple mutant *yuc1yuc4yuc10yuc11* exhibits defects in embryo morphology at the globular stage, while the double mutant *yuc1yuc4* shows no obvious phenotype (Cheng et al., 2007) (Table 4.1). On the contrary, in maize, even the single mutant *ZmYUC1*, also known as *defective endosperm18 (de18)*, leads to an altered seed phenotype (Bernardi et al., 2012).

Genes encoding PIN efflux transporters, AUX influx carriers (AUX/LAX1), and PGP (P-glycoprotein) proteins are essential for AUX transport. The activity of *PIN* genes determines the pattern of division and specification of zygote-derived cells, while *PIN1* and 7 mobilizations are important for establishing embryo polarity (Friml et al., 2003). In maize, an AUX gradient determines the differentiation of both embryo and endosperm, with *ZmPIN1* regulating polar AUX transport during embryogenesis (Forestan et al., 2010; Chen et al., 2014). The *PIN* mutants display defects in embryo polarity, with the severity of phenotype enhanced in higher order mutants. For instance, quadruple mutant *pin1 pin3 pin4 pin7* leads to embryo lethality or generates a plant with severe apical defects, depending

TABLE 4.1   List of key phytohormones-associated genes with experimentally demonstrated roles in regulating agronomically important seed traits.

S. no.	Gene	Species	Role in seed development	References
Auxins				
1.	*ABP1 (Auxin binding protein1)*	*Arabidopsis*, tobacco	Regulates embryogenesis	Chen et al. (2001)
2.	*afb1,2,3 (Auxin signaling F box protein 1)*	*Arabidopsis*	Regulate seed coat and embryo development	Cheng et al. (2007)
3.	*ANT (Aintegumenta)*	*Arabidopsis*	Regulates seed size	Mizukami and Fischer (2000)
4.	*ARF2* and *8 (Auxin Response Factor 2)*	*Arabidopsis*	Regulates seed size	Mizukami and Fischer (2000)
5.	*ARF19*	*Arabidopsis; Jatropha*	Regulates seed size	Sun et al. (2017)
6.	*ARF3/ETT (ETTIN)*	*Arabidopsis*	Regulates seed coat development and integument formation	Kelley et al. (2012)
7.	*ARF5/MP (Monopteros)*	*Arabidopsis*	Regulates differentiation of radicle and hypophysis	Berleth and Jurgens (1993)
8.	*ARF18*	*Brassica napus*	Regulates seed size	Liu et al. (2015b)
9.	*BDL/Aux/1AA12 (BODENLOS/ Auxin/INDOLACETICACID 12)*	*Arabidopsis*	Regulates early embryogenesis	Hamann et al. (2002)
10.	*BG1 (Big Grain 1)*	Rice	Regulates grain size	Liu et al. (2015c)
11.	*COR15A (cold-regulated 15A)*	*Arabidopsis*	Regulates seed size	Meng et al. (2015)
12.	*OsSK41/OsGSK5 (SHAGGY-like Family /Glycogen synthase kinase 3)*	Rice	Negatively regulates grain size	Zejun et al. (2018)
13.	*PGP 10 (P-glycoprotein 10)*	*Arabidopsis*	Governs transport of auxins from endosperm to integuments	Figueiredo et al. (2016)
14.	*PIN (Pin-formed)*	*Arabidopsis*, Maize	Determines division pattern, specification and embryo polarity	Cheng et al. (2007), Friml et al. (2003)
15.	*sem1 (semaphore 1)*	Maize	Role in auxin transport, endosperm and embryo patterning	Scanlon et al. (2002)
16.	*TAR2 (tryptophan aminotransferase-related protein)*	Pea	Regulates seed filling	Meitzel et al. (2021)
17.	*Transport inhibitor response 1 (tir1)*	*Arabidopsis*	Regulates embryogenesis	Cheng et al. (2007)
18.	*YUC (Flavin monooxygenases)*	*Arabidopsis*, Maize	Regulates embryo, endosperm, and seed coat development	Cheng et al. (2007)

(Continued)

Plant Hormones in Crop Improvement

TABLE 4.1   (Continued)

S. no.	Gene	Species	Role in seed development	References
Gibberellins				
19.	AtGA3ox2 (Gibberellinn 3 oxidase 2)	*Arabidopsis*	Regulates GA levels in developing embryos	Curaba et al. (2004)
20.	ATHB25 (Arabidopsis HOMEOBOX 25)	*Arabidopsis*	Regulates seed longevity	Bueso et al. (2014)
21.	GA2ox2 (GA 2-oxidase2)	*Arabidopsis, grapes, peas*	Ectopic expression causes abortion of seeds	Singh et al. (2010)
22.	GASA4 (Gibberellic acid stimulated Arabidopsis 4)	*Arabidopsis*	Regulates seed size and weight	Roxrud et al. (2007)
23.	GW6 (Grain Width 6)	Rice	Positively regulates grain width and weight	Shi et al. (2020)
24.	HvGAMYB (MYB transcription factor)	Barley	Crucial for seed maturation	Diaz et al. (2002)
25.	LH1 and LH2	Pea	Regulates seed weight	Swain et al. (1997)
26.	OsGA20ox2 (Gibberellin 20 oxidase 2)	Rice	Crucial for seed morphogenesis and maturation	Ye et al. (2015)
27.	OsGA2ox1 (GA 2-oxidase2)	Rice	Determines grain yield	Sakamoto et al. (2003)
28.	OsGA3ox2 (GA 3-oxidase2)	Rice	Regulates grain yield	Sakamoto et al. (2003)
29.	TaGW2–6A (Thousand Grain Weight)	Wheat	Regulates seed size	Li et al. (2017)
Cytokinins				
30.	AHPs (Histidine-containing phosphotransfer proteins)	*Arabidopsis*	Positively regulates embryo and seed size	Day et al. (2008)
31.	AHP1 and AHP2 (Histidine phosphotransfer proteins)	*Arabidopsis*, Rice	Regulate seed number and size	Day et al. (2008)
32.	CKI1 (Cytokinin Independent 1)	*Arabidopsis*	Regulate seed number and size	Deng et al. (2010)
33.	ckx1,ckx2,ckx3 (Cytokinin oxidase/ dehydrogenase)	*Arabidopsis*	Regulate seed size and weight	Locascio et al. (2014)
34.	IPT (Isopentyl transferase)	Tobacco	Regulates seed mass	Ma et al. (2008)
35.	LP/EP3 (Large Panicle/Erect Panicle 3)	Rice	Regulates grain number	Li et al. (2011)
36.	OsCKX2 (Cytokinin oxidase/ dehydrogenase)	Rice	Regulates grain number	Ashikari et al. (2005)
37.	OsSPL14 (Squamosa promoter binding protein-like gene)	Rice	Seed size determination	Miura et al. (2010)

(Continued)

TABLE 4.1 (Continued)

S. no.	Gene	Species	Role in seed development	References
38.	*OsVIL2 (VIN3-like 2)*	Rice	Regulates grain number	Yang et al. (2019)
39.	*SP3 (Short Panicle 3)*		Regulates the number of spikelets per panicle	Huang et al. (2019)
40.	*ZmHKs (Histidine kinases)*	Maize	Regulates seed size	Wang et al. (2014)
Brassinosteroids				
41.	*BRD2 (BR-deficient dwarf2)*	Rice	Regulates seed size and shape	Hong et al. (2005)
42.	*D11 (Dwarf11)*	Rice	Regulates seed length	Tanabe et al. (2005)
43.	*Det2 (de-etiolated2)*	*Arabidopsis*	Regulates embryo development	Jiang et al. (2013)
44.	*DWF 1 (Dwarf 1)*	Rice	Regulates seed size	Hong et al. (2005)
45.	*dwf5* and *shk1-D (dwarf 5, shrink1-D)*	Rice	Regulates seed shape	Takahashi et al. (2005)
46.	*GS5 (Grain Size 5)*	Rice	Regulates grain size	Xu et al. (2015)
47.	*GSK2 (Glycogen synthase kinase)*	Rice	Regulates grain shape	Zhao et al. (2018)
48.	*OsBZR (Brassinazole-resistant1)*	Rice	Regulates seed size and weight	Zhu et al. (2015)
49.	*OsPPKL1 (Protein phosphatase)*	Rice	Negative regulator of grain length	Zhang et al. (2012)
50.	*PP2C-1 (Phosphatase 2C-1)*	Soybean	Positively regulates seed size	Lu et al. (2017)
51.	*SHORT GRAIN1 (SG1)*	Rice	Regulates seed size	Nakagawa et al. (2012)
52.	*slg-D (slender grain Dominant)*	Rice	Regulates seed size and shape	Feng et al. (2016)
53.	*SMG11/DWF2 (Small Grain 11/DWARF2)*	Rice	Negatively regulates seed yield	Fang et al. (2016)
54.	*TBP1(Top Bending Panicle 1)*	Rice	Regulates seed number and seed size	Lin et al. (2017)
Ethylene				
55.	*AcdS (ACC deaminase)*	*B. napus*	Regulates seed size	Walton et al. (2012)
56.	*AfAP2−2 (APETALA2)*	*Aechmea fasciata*	Regulates seed size	Lei et al. (2019)
57.	*ctr1 (Constitutive triple response 1)*	*Arabidopsis*	Regulates seed shape and establishment of polarity in seeds	Robert, Noriega, et al. (2008)
58.	*EIN2, EIN3 (Ethylene Insensitive like)*	*Arabidopsis*	Involved in the synergid cell death during seed development	Völz et al. (2013)

(Continued)

TABLE 4.1   (Continued)

S. no.	Gene	Species	Role in seed development	References
59.	ETR2 (Ethylene Response 2)	Rice	Regulates seed-setting rate	Wuriyanghan et al. (2009)
60.	Sh2 (Shrunken 2)	Maize	Regulates PCD in the endosperm	Young et al. (1997)
Abscisic acid				
61.	ABI3 (Abscisic acid insensitive 3)	*Arabidopsis*	Desiccation tolerance in seeds	To et al. (2006)
62.	ABI3,4,5 (Abscisic acid insensitive3)	*Arabidopsis*	Regulates seed maturation and storage reserves accumulation	Ali et al. (2021)
63.	ABI5 (ABA Insensitive 5)	*Arabidopsis*	Regulates seed size	Cheng et al. (2014)
64.	ARF2	*Arabidopsis*	Regulates seed size	Meng et al. (2015)
65.	CIBG1 (ß-glucosidase encoding gene)	Watermelon	Regulates seed size	Wang et al. (2021)
66.	DREB2B (Dehydration-Responsive Element-Binding Protein 2B)	*Arabidopsis*, Cotton	Negatively regulates seed longevity	Ali et al. (2020)
67.	NCED6 and NCED9 (9-Cis-Epoxycarotenoid dioxygenase)	*Arabidopsis*	Induces seed dormancy	Liu et al. (2020)
68.	OsACOT (Acyl-CoA thioesterase)	Rice	Regulates grain filling	Zhao et al. (2019)
69.	OsGH3−2 (Gretchen hagen3−2)	Rice	Negative regulator of seed longevity	Yuan et al. (2021)
70.	Pyl (PRLR gene)	*Arabidopsis*	Regulates seed dormancy	Ma et al. (2009)
71.	SRK2D//I (SnRK2 proteins)	*Arabidopsis*	Regulates seed desiccation and maturation	Yamada et al. (2019)
72.	ZmSSI (Starch synthase I)	Maize	Regulates endosperm filling	Hu et al. (2012)
Jasmonic acid				
73.	EG1 (Extra Glume 1)	Rice	Negatively regulates grain yield	Cai et al. (2014)
74.	EG2 (Extra Glume 2)	Rice	Negatively regulates grain yield	Cai et al. (2014)
75.	OsOPR7 (OPDA reductase)	Rice	Regulates spikelet development	You et al. (2019)
76.	JAZ6 (Jasmonate-zim-domain protein 6)	*Arabidopsis*	Promotes seed size	Hu et al. (2021)
77.	JMT (Jasmonic acid carboxyl methyltransferase)	Rice	Negatively regulates seed number	Kim et al. (2009)
78.	KAT-2B (Lysine Acetyltransferase 2B)	Wheat	Regulates grain weight and yield	Chen et al. (2020)
79.	MSD1 (Mutliseeded1)	*Sorghum*	Regulates grain number per panicle	Jiao et al. (2018)
80.	NOG1(Number of Grains 1)	Rice	Regulates grain yield	Huo et al. (2017)

(Continued)

TABLE 4.1 (Continued)

S. no.	Gene	Species	Role in seed development	References
81.	*OsAOS/ PRE (Allene Oxide Synthase/Precocious)*	Rice	Role in seed setting	Hibara et al. (2016)
82.	*Spr2 (Suppressor of prosystemin-mediated responses2)*	Tomato	Regulate embryo development and PCD in seed coat and endosperm	Goetz et al. (2012)
Strigolactones				
83.	*d53 (DWARF 53)*	Rice	Regulates seed size	Yamada et al. (2019)
84.	*HTD2,4 (High-tillering Dwarf)*	Rice	Negatively regulates seed size and grain weight	Liu et al. (2009)

on the genotype (Friml et al., 2003). In addition, several *PIN* genes in *Arabidopsis* and maize regulate embryogenesis based on their specific expression patterns and localization (Forestan and Varotto, 2012; Vieten et al., 2005). Another mutant *sem1* (*semaphore 1*) of maize exhibits a reduction in polar AUX transport and defects in endosperm and embryo patterning (Scanlon et al., 2002).

AUX signaling is mediated by *ARFs* (auxin-response factors), which, in turn, regulate the expression of AUX-responsive genes by directly binding to the AUX-response elements in their promoter regions (Ulmasov et al., 1999). The loss of function of *ARF5/Monopteros* (*MP*) causes several malformations in the differentiation of radicle and hypophysis, leading to lethal embryo phenotype in *Arabidopsis* (Berleth and Jurgens, 1993). Another critical component of AUX signaling is AUX/IAA (AUX/Indole-3-Acetic Acid) proteins, which suppress AUX signaling by sequestering ARFs. AUX receptor, TIR1 (transport inhibitor response 1) interacts with the SCF (E3-Ubiquitin ligase Skp1/Cullin/F-box) complex, leading to the ubiquitination of Aux/IAA proteins and degradation. The T-DNA mutation in an F-box gene *AtCUL1* (*AtCULLINI1*), involved in assembling the SCF-type ubiquitin ligase complex, led to arrested embryogenesis in *Arabidopsis* (Shen et al., 2002). *BDL* (*Bodenlos*) encoded IAA12 interact with MPs preventing it from targeting the AUX-response genes. However, AUX-mediated BDL degradation releases MP, allowing basal body initiation during early embryogenesis (Hamann et al., 2002). AUXs also regulate the growth of integuments, with *PGP10* (*P-glycoprotein 10*) by governing the transport of AUXs from endosperm to integuments (Figueiredo et al., 2016). The lack of function of *ARF3/ETTIN (ETT)* leads to the fusion of inner and outer integuments resulting in rounded seeds with aberrant morphology (Kelley et al., 2012). Similarly, *Arabidopsis Auxin binding protein 1 (ABP1)* mutation leads to embryo lethality due to aberrant cell division and elongation (Chen et al., 2001).

AUXs also play a role in seed filling in pea. Due to a maternally derived increase in sucrose levels, the concentration of crucial metabolite, T6P (trehalose-6-phosphate), rises as the seed transitions from the early to maturation stage. This further activates *TAR2* (*TRYPTOPHAN ANINOTRANSFERASE RELATED2*), thereby positively regulating IAA (indole acetic acid) biosynthesis. IAA-mediated de-repression of AUX signaling genes in turn leads to the activation of starch biosynthesis (Meitzel et al., 2021). AUX levels may also alter with change in sugar concentration. The mutants of the *Mn1* (*Miniature 1*) of maize, encoding cell wall invertase, exhibit smaller seeds. The increased concentration of sucrose and decreased glucose and fructose in basal kernels of this mutant correlate with decrease in AUX concentration and expression of AUX biosynthesis gene, *ZmYUC* (Chourey et al., 2010; LeCLere et al., 2010).

Genetic manipulation of AUX biosynthesis alters seed size and weight. For instance, an overexpression of *ARF19* in *Arabidopsis* and *Jatropha* leads to increased seed size (Sun et al., 2017). However, *ARF2* and *8* negatively regulate seed size by repressing cell division. *ARF2/MNT* (*MEGAINTEGUMENTA*) mutant in *Arabidopsis* displays increased seed size and enlarged seed coat due to extra anticlinal cell divisions (Schruff et al., 2006). In conjunction with *COR15A* (*COLD-REGULATED15A*) and *ANT* (*ANTEGUMENTA*), *ARF2* also participates in AUX-ABA cross talk to modulate seed mass in response to drought stress (Meng et al., 2015). In *Brassica napus*, *ARF18* maternally controls seed size in the silique wall by restricting cell expansion. Its allelic mutant exhibits enhanced seed weight (Liu et al., 2015b).

The qTGW3, a QTL (quantitative trait locus) for thousand-grain weight, encodes a glycogen synthase kinase *OsGSK3/OsSK41* and controls grain weight and length in rice. OsSK41 interacts with OsARF4 to repress AUX-responsive genes during grain development (Zejun et al., 2018). Another rice gene, *TGW6* (*Thousand Grain Weight 6*), important for grain weight, encodes an IAA-glucose hydrolase and negatively regulates early endosperm development. Its loss-of-function enhances grain length and weight (Ishimaru et al., 2013). A dominant mutant of rice, *big grain1-D* (*Bg1-D*) that produces extra-large grains is a primary AUX response gene involved in AUX distribution and transport, demonstrating the impact of modulating AUX transport on grain size and productivity (Liu et al., 2015c).

4.3 Role of gibberellins

The tetracyclic diterpenoid hormone gibberellin (GA) regulates diverse aspects of plant growth and development. Several genes regulating GA biosynthesis or response have been used to improve crop productivity (Gao and Chu, 2020). Endogenous levels of GAs impact seed development and associated agronomic traits like seed size, weight, dormancy, and longevity. GAs are also used in the development of seedless varieties. For example, ectopic expression of pea GA-inactivating enzyme, *GA2 Oxidase 2* in *Arabidopsis* leads to seed abortion (Singh et al., 2002). In grapes, an exogenous application of GA3 leads to seed abortion due to impairment of redox homeostasis, thus resulting in seedless fruits (Cheng et al., 2013). Similarly, an ectopic expression of *PsGA2ox2* (*GA 2-oxidase 2*), encoding a GA inactivating enzyme under the control of *MEA* (*MEDEA*) promoter, causes the abortion of seeds (Singh et al., 2010).

Since GAs regulate both vegetative and reproductive development, manipulating GAs in a tissue-specific manner is desirable for engineering seed traits. For example, a constitutive expression of *OsGA2ox1* using *Actin* promoter in rice leads to dwarf plants with reduced grain yields. However, an expression of the same gene under the upper internode-specific *OsGA3ox2* promoter results in higher grain yield (Sakamoto et al., 2003). Spatiotemporal regulation of genes involved in GA biosynthesis and catabolism correlates with their multifaceted roles in regulating seed growth, development, and maturation. For example, the loss of function of the GA biosynthesis gene, *OsGA20ox2*, enhances seed dormancy, while an expression of this gene during early seed development is associated with seed morphogenesis and maturation (Ye et al., 2015).

Functional genomic studies in pea highlighted the role of GAs in regulating seed set. Loss of function of *LH* genes, *LH1* and *LH2*, required to synthesize GA biosynthetic enzyme ent-kaurene oxidase, reduce seed weight and survival (Swain et al., 1997). Tissue-specific transcriptional profiling of GA metabolism genes and quantification of GAs during seed filling in pea revealed their involvement in the acquisition and storage of photoassimilates and seed coat growth (Nadeau et al., 2011). In barley, *HvGAMYB* expresses in endosperm in response to GAs and interacts with the PB (Prolamine box)-binding factor BPBF and endosperm-specific DOF (DNA binding with One Finger). DOF, in turn, activates the transcription of *HOR2*, an endosperm-specific gene crucial for seed maturation, by binding to the PB (Diaz et al., 2002). The implication of GAs in seed maturation is also evident by the fact that *LEC2* (*LEAFY COTYLEDON* 2) and *FUS3* (*FUSCA3*), the master regulators for embryogenesis and maturation, regulate GA levels in developing embryos in *Arabidopsis* by repressing *AtGA3ox2* (Curaba et al., 2004). Besides regulating early embryogenesis and maturation, the cross talk of GAs with other hormones is also crucial for seed germination. For example, in developing maize seeds, GAs promote vivipary, while ABAs inhibit it. Thus, levels of ABAs and GAs determine whether seeds would undergo maturation or precocious germination (White et al., 2000).

Manipulating GAs by genetic alteration can also optimize seed yield or related traits in crop plants (Table 4.1). In wheat, a RING E3 ubiquitin-ligase encoding gene *TaGW2−6A* is involved in GA perception and signaling. Allelic variation in *TaGW2−6A* in Chinese spring and *NIl31* contribute to increased seed size by regulating the expression of GA biosynthesis genes, *GA3-oxidase*, and *GASA4* (*Gibberellic Acid−Stimulated Arabidopsis 4*). Overexpression of *GASA4* leads to enhanced cell division and elongation in endosperm during grain filling (Li et al., 2017). In *Arabidopsis*, an overexpression of *GASA4* causes an increase in seed size and weight. Although the loss of function of *GASA4* results in smaller seeds, the increase in number of branches results in overall increase in seed yield per plant (Roxrud et al., 2007). Another QTL *GW6* (*Grain Width 6*) that encodes a GA-regulated GAST protein positively regulates grain width and weight in rice. Induction of *GW6* in an *indica* rice variety leads to a 19.4% increase in grain yield (Shi et al., 2020). Apart from seed size, GAs are also involved in determining seed longevity. For example, *Arabidopsis HOMEOBOX 25 (ATHB25)* regulates seed longevity by regulating the expression of the GA biosynthetic gene, *GA3-Oxidase2*. Loss of function of *ATHB25* adversely impacts seed longevity and reinforcement of seed coat (Bueso et al., 2014).

4.4 Role of cytokinins

The adenine-derived phytohormone, cytokinin (CK), plays a vital role in seed development by controlling cell divisions in both embryo and endosperm (Werner et al., 2001). With the sharp increase in CKs immediately after anthesis, a positive correlation is established between cell divisions during early seed development and CK levels in several plant species. Although the CK requirement immediately after anthesis depends on maternal tissues, developing seeds can synthesize CKs a few days after pollination.

Several genes, involved in CK biosynthesis, activation, deactivation, reactivation, and degradation, regulate cytokinin levels in plants and seed yield (Table 4.1) (Jameson and Song, 2016). Histidine kinase receptors (HKs), histidine phosphotransfer proteins (HPs,) and response regulators (RRs) are essential components of CK perception and signaling. Loss of function of *CKI1* (*Cytokinin Independent 1*), encoding an HK, leads to fewer but larger seeds due to the lack of CK perception (Deng et al., 2010). Conversely, an ectopic expression of maize HKs (*ZmHKs*) in *Arabidopsis* causes a reduction in seed size (Wang et al., 2014). The triple mutants, *ahk1ahk2ahk3* and *ahp-1ahp-2ahp-3* with impaired HPs and RR proteins, exhibit enlarged embryos and seeds (Day et al., 2008). In rice, RNAi-mediated knockdown of two HPs, *OsAHP1* and *OsAHP2*, decreases seed yield, tiller number, and fertility (Sun et al., 2014).

IPTs (*Isopentyl transferase*) and *CKXs* (*Cytokinin oxidase/dehydrogenases*) are involved in CK biosynthesis and degradation, respectively. The maintenance of CK homeostasis by these enzymes regulates seed size. Enhanced expression of *IPT* in tobacco increases seed mass (Ma et al., 2008). The increase in panicle branching, in response to the application of nitrogen fertilizers before panicle initiation, is associated with an enhanced expression of *OsIPT* genes in rice (Ding et al., 2014). A fascinating example of the regulation of seed size by maintaining CK homeostasis comes from *IPA1* (*ideal plant architecture*)/*WFP* (*Wealthy Farmer's Panicle*) QTL in rice. *IPA1* encodes a squamosa promoter binding protein-like gene, *OsSPL14* (Miura et al., 2010). *OsSPL14* enhances CK levels by positively regulating the expression of *LOG*, a CK activating gene, and *DEP1* (*Dense and Erect Panicle 1*), which, in turn, downregulates the expression of *OsCKX2*, thereby downregulating CK degradation (Huang et al., 2009).

A major QTL *Gn1a* of rice encodes OsCKX2. Its mutant displays an increase in grain number (Ashikari et al., 2005). Rice mutant, *regulator of Gn1a* (*reg1*), a semidominant allele of a TF, *DST* (*Drought and Salt Tolerance*), also determines grain number by regulating *OsCKX2* expression (Li et al., 2013). Another mutant of an F-box protein-encoding gene, *Large Panicle* (*LP*)/*Erect Panicle 3* (*EP3*) of rice, exhibits a decreased expression of *OsCKX2*, leading to an increase in grain number (Li et al., 2011). The enhanced grain yield due to the loss of function of *OsNAC096* is due to decreased *OsCKX2* expression (Kang et al., 2019). *Short Panicle 3* (*SP3*), encoding a Dof TF, also regulates the number of spikelets per panicle by regulating the expression of both *CKX* and *IPT* genes (Huang et al., 2019).

Several genes associated with the IKU (HAIKU) pathway also regulate seed size (Li and Li, 2016). Among these, a WRKY TF, *MINI3* (*MINI-SEED 3*), governs seed size by activating *CKX2* expression, while an overexpression of *CKX2* in *iku2* mutant leads to seed enlargement. Authors further showed that *CKX2* expression is affected by H3K27m3 deposition in response to the demethylation of male gametes and maternal genome dosage imbalance (Li et al., 2013). *OsVIL2* (*VIN3-like 2*) is a chromatin-modifying factor that regulates *OsCKX2* expression by methylating its regulatory region, thereby enhancing CK levels. Loss of function of *OsVIL2* results in fewer grains per panicle, while *OsVIL2* overexpression causes an enhanced number of grains per panicle (Yang et al., 2019). Therefore, seed yield in crop plants can be optimized by maintaining a delicate balance of *IPT* and *CKX* expression. IPT genes can enhance cytokinin biosynthesis under source-limiting situations, while cytokinin degradation may be controlled by regulating *CKX* expression under sink-limited situations (Jameson and Song, 2016). However, an enhanced expression of CK-associated genes using constitutive or inducible promoters has not been a successful strategy for enhancing seed yield due to pleiotropic effects on the other aspects of plant growth and development. Higher CK levels prevent root growth while promoting shoot growth, leading to a bushy appearance. Therefore, the use of seed-specific promoters is recommended for enhancing seed-related traits.

4.5 Role of brassinosteroids

Brassinosteroids (BRs) are a group of plant-specific steroid hormones that regulate seed/grain size by promoting cell division and elongation (Jiang and Lin, 2013). The foliar spray of BRs positively influences seed yield and nutrition content in several crop plants. Therefore, spraying BRs on reproductive organs is traditionally used to enhance seed yield. However, due to the inability of BRs to move long distances, expenses involved in an external application, and stability issues, the large-scale application of BRs is not feasible. Also, since BRs negatively regulate panicle number, another critical factor in determining the overall yield of the plant, their use for agricultural production is not recommended.

Therefore, the genetic manipulation of BRs has been explored to combat this challenge (Table 4.1). Screening of BR-deficient or BR-insensitive mutants and overexpression of BR biosynthetic genes in several crop plants confirmed that BRs positively regulate seed growth and size (Hong et al., 2005; Jiang et al., 2013; Jiang and Lin, 2013; Morinaka et al., 2006; Nomura et al., 2007; Takahashi et al., 2005; Tanabe et al., 2005). For example, BR-deficient mutant *BRD2* of rice exhibits smaller and shorter seeds, while an overexpression of *Dwarf 1*, involved in BR biosynthesis, produces larger seeds (Hong et al., 2005). Similarly, BR-insensitive mutant *SG1* (*Short Grain 1*) produces longer grains (Nakagawa et al., 2012). *Arabidopsis's* BR-deficient mutant *det2* (*de-etiolated 2*) produces smaller and less elongated seeds due to reduced endosperm volume, integument cell length, and seed cavity. Further molecular and genetic studies with *det2* have confirmed that BRs produced by maternal tissues are required for seed elongation, while BRs produced by embryo and endosperm contribute to seed size (Jiang et al., 2013). Overexpression of BR biosynthetic gene *ZmD11*, an ortholog of rice *D11* (*DWARF11*), significantly enhances seed size and quality in maize (Sun et al., 2021).

The BR signaling TF *BZR1* (*Brassinazole Resistant 1*) also plays a significant role in seed development and differentiation by regulating the expression of key genes associated with integument and endosperm development (Jiang et al., 2013; Jiang and Lin, 2013; Li et al., 2019; Li and Li, 2015). An elite allele, *PP2C-1* (*Protein Phosphatase 2C-1*), in a soybean QTL, contributes to seed size and weight by enhancing the cell size of integuments and activating key genes associated with seed development. Molecular studies have revealed that PP2C-1 facilitates the accumulation of dephosphorylated *GmBZR1*, which also contributes to seed weight/size (Lu et al., 2017). Another study demonstrated that the allelic variation in a major QTL *GL2*, which encodes *OsGRF4*, determines grain size, weight, and filling rate in rice by regulating the brassinosteroid-induced genes (Che et al., 2015). The BR signaling gene *OsPPKL1* of rice, orthologous to *Arabidopsis BSU1*, in the qGL3/qGL3.1, increases grain length by an aspartate-to-glutamate change in the conserved domain in *OsPPKL1*. Conversely, an overexpression of *OsPPKL1* negatively impacts grain length (Zhang et al., 2012).

Another interesting rice QTL, *GRAIN WIDTH 5 (GW5)*, regulates grain weight and width by suppressing BR signaling component *OsGSK2*, thereby facilitating the accumulation of active forms of *OsBZR1* and *DLT (dwarf and low-tillering)* (Hwang et al., 2021; Liu et al., 2017). Overexpression of *OsGSK2* decreases grain size and weight (Morinaka et al., 2006). The gain-of-function mutant *slg-D (slender grain Dominant)* of rice exhibits longer and narrower grains by regulating BR homeostasis (Feng et al., 2016). A major QTL for grain size *GSE5* that interacts with *GSK2* acts as a negative regulator for grain width and weight. Another rice gene, *Grain Size 5 (GS5)*, encodes a serine carboxypeptidase that enhances BR signaling and positively regulates grain size (Xu et al., 2015). Another major QTL, *ZmGS5*, is a candidate for positively influencing maize kernel development as enhanced *ZmGS5* expression increases seed size (Li et al., 2019; Liu et al., 2015a). BRs also positively regulate grain number and weight by improving the efficiency of carbohydrate mobilization from source to sink in rice. Overexpression of *OsBZR1*, a BR signaling factor in rice, increases accumulated sugar levels in seeds, while knockdown leads to reduced seed size and weight (Zhu et al., 2015).

BRs also regulate seed shape. The *dwf5* and *shk1-D* mutants of BRs exhibit changes in seed shape (Takahashi et al., 2005). Similarly, BR-deficient mutant *ik* of pea exhibits the irregular shape of seeds (Nomura et al., 2007). *GSK2* regulates cell division to control grain shape by modulating the activity of GS9 (Grain Shape 9) with OsOFP8 and OsOFP14 proteins (Zhao et al., 2018). Although an overexpression of BR biosynthesis genes enhances seed yield and nutritional status, the pleiotropic effects, such as larger laminar joints, increased plant height, reduction in lodging resistance, and decreased planting density, are problematic. Tissue-specific modulation of specific BR signaling components, such as *BRI1*, *BIN2*, and *BZE1/BES1*, is established in the model system *Arabidopsis* (Zu et al., 2019).

Careful manipulation of BRs by enhancing the levels in reproductive organs while decreasing in vegetative organs may enhance crop yields (Lin, 2020). Also, the regulation of seed yield by *OsBAK1 (BRI1-Associated Receptor Kinase)/TBP1 (Top Bending Panicle 1)*, a somatic embryogenesis receptor kinase (SERK) domain protein, suggests that the level of overexpression needs to be controlled to minimize the penalty on total yield. Similarly, *TBP1* mutant shows an increase in seed number due to enhanced panicle branching but a reduction in seed size, thereby reducing the overall yield of the plant. Though the overexpression of *TBP1* increases the grain size but adversely affects seed yield (Lin et al., 2017). *SMG11 (Small Grain 11)*, a novel allele of *DWARF2* encoding cytochrome P450, is involved in BR biosynthesis in rice. High expression of *SMG11* reduces yield due to reduced panicle branching, while a slight increase in *SMG11* expression provides greater yields, indicating that the expression of BR-associated genes needs to be fine-tuned to optimize grain yields (Fang et al., 2016).

4.6 Role of ethylene

The gaseous hormone ethylene (ET) can act as a plant growth promoter or inhibitor depending on the concentration, developmental stage, and plant species. The role of ET in regulating seed traits is also contrary in different species. ACC (1-aminocyclopropane-1-carboxylate) is the precursor of ET, and therefore, ACC levels are directly associated with ethylene production. Overexpression of ACC deaminase that breaks down ACC reduces ET levels and seed size in *Brassica* (Walton et al., 2012). However, an ectopic overexpression of an ethylene response element-binding protein APETALA2 (*AfAP2–2*) of *Aechmea fasciata* in *Arabidopsis* negatively affects seed size and weight (Lei et al., 2019).

ET also influences grain filling in rice by affecting starch biosynthesis and carbohydrate mobilization (Table 4.1). Overexpression of ethylene receptor *ETR2* (*Ethylene Response 2*) causes a reduction in seed set due to the blocked translocation of sugars from stems to grains. In contrast, *etr2* mutants and RNAi lines show higher grain weight in rice (Wuriyanghan et al., 2009). Higher expression of *S*-adenosyl methionine synthase, involved in synthesizing *S*-adenosyl methionine precursor or ethylene, is associated with a higher rate of ethylene evolution in inferior spikelets of rice panicle (Yang et al., 2006). The lower ratio of ABA to ethylene in these spikelets correlates with the reduced grain filling rate and weight. External application of ethylene inhibitor 1-methylcyclopropene (1-MCP) at the early postanthesis stage enhances the expression of genes associated with cell cycle and starch biosynthesis resulting in significant improvement in grain filling of basal spikelets (Panda et al., 2016).

Some mutants affecting key genes associated with ethylene signaling have also impacted seed shape and polarity in *Arabidopsis*. For example, *ctr1–1*, *eto1–1*, *etr1–1*, *ein2–1*, *ers1–2*, and *etr1–7* exhibit altered seed shape in response to imbibition due to reduced curvature values on both the poles (Robert et al., 2008).

Ethylene is also associated with the onset of programed cell death (PCD) in the endosperm of cereals, with an external application of ethylene exhibiting enhanced PCD. The starch deficient mutant *shrunken 2* of maize exhibits enhanced PCD associated with increased ET levels (Young et al., 1997). Further studies have shown that the coordinated action of ET and JA is necessary for PCD in pericarp, while GAs and AUXs participate in PCD of maternal tissues (Domínguez and Cejudo, 2014).

Ethylene is also critical in regulating seed dormancy in several crop species (Bogatek and Gniazdowska, 2018). External application of ET can overcome seed dormancy in a dose-dependent manner. Ethylene-insensitive mutants, *etr1* (*Ethylene-Resistant 1*) and *ein2* (*Ethylene-Insensitive 2*) exhibit enhanced primary dormancy in *Arabidopsis* seeds associated with higher ABA sensitivity (Bleecker et al., 1988; Chiwocha et al., 2005). The impacts of antagonistic and synergistic interactions between ET and ABA and, GAs and ET, respectively, on seed dormancy have been reviewed elsewhere (Corbineau et al., 2014).

4.7 Role of abscisic acid

The abiotic stress hormone, abscisic acid (ABA), produced from both maternal and zygotic tissues, plays a crucial role in regulating the accumulation of storage compounds, maturation, desiccation, and dormancy in seeds (Table 4.1) (Kermode, 2005; Locascio et al., 2014; Kanno et al., 2010). Synthesis of ABA in maternal tissues is required for embryo development, while ABA synthesized in the zygote is responsible for seed desiccation and storage of food reserves in the aleurone layer (Kanno et al., 2010). The *LEC1/2* (*LEAFY COTYLEDON 1/2*),

FUS3 (FUSCA3), and *ABI3 (Abscisic acid insensitive 3)* genes of the LAFL network play an essential role in ABA-mediated seed development. Any defect in *ABI3*, *FUS3*, *LEC1*, and *LEC2* leads to abnormal seed development (Holdsworth et al., 2008). LAFL network controls embryogenesis by regulating *BBM (BABY BOOM)*, *AP2 (APETAL2)*, seed storage proteins, *At2S1 (ALBUMIN STORAGE PROTEIN 1)*, *CRC (CRUCFIERIN C)*, and *FLC (FLOWERING LOCUS C)*, whereas *ABI3/VP1*, *ABI4*, *ABI5*, *LEC1*, *LEC2*, and *FUS3* induce an expression of ABA-responsive seed-specific LEAs (Late Embryogenesis Abundant) and storage proteins (Finkelstein et al., 2002; Jia et al., 2013, 2014; Kwong et al., 2003; Parcy et al., 1997). *WRINKLED1 (WRI1)* is also involved in seed maturation *via* ABA-mediated regulation of sugars and oil content in seeds (To et al., 2012).

The miR1432-mediated regulation of *OsACOT (Acyl-CoA thioesterase)* plays a crucial role in grain size and filling by regulating both AUX transport and ABA signaling (Zhao et al., 2019). However, an overexpression of *OsACOT* leads to the perturbation of genes involved in starch biosynthesis and signaling, as well as fatty acid metabolism. In wheat, enhanced ABA levels significantly reduce the conversion of sucrose into starch, thus affecting the grain filling and yield (Ahmadi and Baker, 1999). ABA regulates the accumulation of storage compounds in aleurone cell layers by regulating the expression of bZIP and DOF TFs (Mönke et al., 2012). In maize, the ABA receptor *ABI4* regulates the expression of the starch biosynthesis gene, *ZmSSI*, thereby regulating endosperm filling (Hu et al., 2012). ABA and ethylene contents negatively correlate with starch biosynthesis (Zhu et al., 2011). Conforming to the role of ABAs in the negative regulation of seed development, the ABA-deficient mutants, *abi2−1* and *abi5*, exhibit bigger and heavier seeds with increased embryo cell number but delayed endosperm cellularization. The ABA signaling component, *ABI5 (ABSCISIC ACID-INSENSITIVES5)*, reduces seed growth by regulating the expression of the *SHB1* (Cheng et al., 2014). A signaling cascade comprising *ARF2*, *ANT (AINTEGUMENTA)*, and *COR15A (COLD-REGULATED15A)* regulate seed size in *Arabidopsis*, in which *ARF2*, in response to ABA, negatively regulates *ANT* expression, which further regulates *COR15A*, a negative regulator of seed mass (Meng et al., 2015). Targeted mutagenesis of *CIBG1*, a ß-glucosidase gene of *Citrullus lanatus*, involved in the hydrolysis of ABA-glucose ester and regulating free levels of ABA, reduced seed size and weight (Wang et al., 2021).

ABAs are also crucial for maintaining seed longevity (Zhou et al., 2020). Seed longevity is attained by accumulating protective proteins like heat shock, LEA proteins, and other molecules like nonreducing sugars during seed maturation. Accumulation of these proteins and molecules coincides with chlorophyll degradation during seed maturation. Further, the impact of AUXs on seed longevity is dose-dependent, which gets abolished by the loss of activity of *ABI3*, the master regulator of seed longevity (Pellizzaro et al., 2020). In *Arabidopsis*, the AUX signaling gene, *ABI3(ABA-INSENSITIVE3)*, is known to regulate the accumulation of seed-specific heat shock protein HSFA9 (HEAT SHOCK FACTOR 9) (Kotak et al., 2007). Ectopic overexpression of *HSFA9* from *Helianthus* promotes seed longevity in tobacco (Personat et al., 2014). Rice *GRETCHEN HAGEN3−2 (OsGH3−2)* mapped on the QTL for seed storability (qSS1) encodes an IAA-amido synthetase and acts as a negative regulator of seed longevity. It modulates ABA signaling and eventually inhibits the LEA genes to reduce longevity (Yuan et al., 2021). Another TF DREB2B (dehydration-responsive element-binding protein 2B) interacts with RCD1 (Radical-Induced Cell Death1) and acts as a negative regulator of seed longevity as well as seed vigor in both *Arabidopsis* and cotton. DREB2B acts downstream of ABA, but in a pathway parallel to *ABI3*, it interacts with HsfA9, which is partially controlled by *DREB2B*. Therefore, synergistic interactions between DREB2B and RCD1 through ABI3 regulate seed vigor and longevity (Ali et al., 2020). However, the loss of function of *ABI3* in *Arabidopsis* leads to the loss of desiccation tolerance in seeds (To et al., 2006).

The role of ABA in seed maturation and dormancy has been recently reviewed (Ali et al., 2021). The SnRK2 proteins, SRK2D/E/I, regulate seed desiccation and maturation *via* PP2C-dependent feedback regulation. The mutants deficient in ABA (*aba, nceds*) or insensitive to ABA (*pyls, snrks, abi3/4/5*) exhibit reduced seed dormancy (Nakashima et al., 2009). Some studies suggest that ABA induces seed dormancy by downregulating GA biosynthesis (Seo et al., 2006; Shu et al., 2013). The ABA receptor PYLs (*PYRABACTIN RESISTANCE*) are essential for ABA signaling. The quadruple mutant *pyr1/prl1/prl2/prl4* and *pyl* duodecuple mutant in *Arabidopsis* exhibit reduced seed dormancy (Ma et al., 2009). A recent study demonstrated the role of *ODR1* (for *reversal of RDO51* phenotype), a homolog of rice *Seed dormancy 4*, and its interacting partner bHLH57, in controlling ABA biosynthesis, thereby regulating seed dormancy (Liu et al., 2020). bHLH57 induces seed dormancy by activating ABA biosynthesis genes, *NCED6* (*9-Cis-Epoxycarotenoid dioxygenase*), and *NCED9*. ODR1, however, interacts with bHLH57 and inhibits it from activating ABA biosynthesis genes, *NCED6* and *9*. ABI3 directly represses *ODR1* itself. ABA controls dormancy with AUXs, GA, BR, and ethylene. For example, in rice, an overexpression of *GERMIN-LIKE PROTEIN 2−1* (*OsGLP2−1*) is induced by *ABI5* and suppressed by GAMYB, leading to enhanced seed dormancy, mediated by both ABA and GA signaling (Wang et al., 2020).

4.8 Role of jasmonic acid

The role of jasmonic acid (JA) in regulating seed development and related agronomic traits is well demonstrated (Table 4.1) (Wasternack et al., 2013). The impact of JAs on grain yield in rice is attributed to their role in regulating spikelet morphology, number of spikelets per panicle, and withering of lodicules postanthesis. The mutants impaired in JA biosynthesis gene, *EG1* (*Extra Glume1*), and JA signaling repressor, *EG2/OsJAZ1* (*Extra Glume 2*), exhibit compromised grain yield in rice due to defects in spikelet development (Cai et al., 2014). Another mutant, *dfo2* (*Deformed floral organ 2*), impaired in peroxisomal targeting sequence 1 receptor *OsPEX5*, also exhibits lower seed set due to defects in spikelet morphogenesis. *OsPEX5* is important for importing JA biosynthetic protein OsOPR7 into peroxisomes. A knockout mutant of *OsOPR7* exhibits abnormal spikelet morphology (You et al., 2019).

Overexpression of *Arabidopsis JMT* (*Jasmonic acid carboxyl methyltransferase*), which converts JA into MeJA (Methyl Jasmonate), significantly reduces seed production (Cipollini, 2007). Overexpression of *AtJMT* in rice reduces grain yield associated with increased MeJA and ABA levels (Kim et al., 2009). Conversely, a reduction in JA levels in rice plants overexpressing *NOG1* (*Number of Grains 1*) enhances grain yield (Huo et al., 2017).

JAs also regulate seed size and integument proliferation. The positive regulators of JA signaling, namely, *MYC2, MED25, JAR1*, and *CO11*, repress seed size, while JA signaling repressor JAZ6 promotes it (Hu et al., 2021). In wheat, JAs positively influence grain weight. The *tgw1* (*triticale grain weight 1*) mutant, due to the loss of activity of *KAT-2B* (*keto-acyl thiolase 2B*) involved in JA biosynthesis, exhibits reduced JA content and grain weight. *KAT-2B* not only complements the grain defect in the *tgw1* mutant, but an overexpression of *KAT-2B* also boosts seed yield and grain weight (Chen et al., 2020). On the contrary, *msd1* (*multi seeded 1*) mutant of sorghum exhibits enhanced grain yield with reduced JA levels (Jiao et al., 2018).

The negative impact of the decrease in JA biosynthesis on seed set in rice is associated with the withering of lodicules. Swelling of lodicules at the time of anthesis is essential for flower opening, while withering of lodicules facilitates its closing postanthesis. Loss of function of *OsAOS1* (*Allene Oxide Synthase*)/*PRE* (*Precocious*), involved in JA biosynthesis, led to no seed set due to nonwithering of lodicules (Hibara et al., 2016).

JA precursor OPDA (12-oxo-phytodienoic acid) also regulates seed development. The *spr2* mutant of tomato, deficient in OPDA and JA, exhibited delayed embryo development and enhanced PCD in seed coat and endosperm (Goetz et al., 2012). OPDA generated from seed coats in tomatoes regulates embryo development in a JA-independent manner (Wasternack et al., 2012). It downregulates cell wall invertase inhibitor *INVINH1*, thereby restoring cell wall invertase *LIN5* activity required for hexose generation during embryo development (Wasternack et al., 2013).

4.9 Role of strigolactones

The carotenoid-derived hormones, strigolactones (SLs), play critical roles in tillering and panicle branching in plants. Although their direct involvement in regulating seed development has not been established, several mutants impaired in SL biosynthesis/signaling, such as *HTD4*, and allelic mutants, *d14*, *d88*, *HTD2*, and *DHTA-34*, exhibit reduced seed set, size, or weight (Table 4.1) (Liang et al., 2019; Liu et al., 2009). Their role in inducing somatic embryogenesis in tomato mutants suggest possible involvement in zygotic embryogenesis as well (Leljak-Levanić et al., 2015; Wu et al., 2017). The recent screening of *dwarf* (*d*) mutants, *d3*, *d10*, *d14*, *d17*, *d27*, and *d53*, impaired in SL signaling in rice, revealed smaller seed size with reduced cell area in the endosperm (Yamada et al., 2019). Though the underlying mechanism is yet to be explored, cross talk of SLs with other hormones might be crucial for the observed phenotype.

4.10 Conclusions and future perspectives

Plant hormones play a major role in governing key agronomic seed traits. Although the role of AUXs, GAs, CKs, BRs, and ABA are now well established, the knowledge about the influence and underlying mechanism of ET, JA, and SL-mediated regulation of seed traits is still in infancy. Overall, most of the hormones discussed here influence seed size and weight, while seed shape is mainly affected by GAs, AUXs, CKs, and ET (Fig. 4.1). Similarly, seed longevity is mainly affected by AUXs, GAs, and ABA, while seed number

HORMONE \ TRAIT	Seed size/weight	Seed shape	Seed longevity	Seed number	Seed dormancy
Auxin	✓ At, Bn, Os, Ps	✓ Os	✓ At	–	–
Gibberellin	✓ At, Os, Ps, Ta	✓ At, Os	✓ At	✓ At, Os	✓ At
Cytokinin	✓ At, Os, Zm	✓ At	–	✓ At, Os	–
Brassinosteroid	✓ At, Gm, Os	✓ Os	–	✓ At, Os	–
Ethylene	✓ Bn	✓ At	–	✓ Os	✓ At
Abscicic Acid	✓ At, Os	–	✓ At, Gh	–	✓ At
Jasmonic Acid	✓ At, Ta	–	–	✓ Os, Ta, Sb	✓ At
Strigolactone	✓ Os	–	–	–	–

FIGURE 4.1 Role of phytohormones in regulating agronomically important seed traits. The "tick marks" represent the traits where regulatory roles of respective hormones are experimentally demonstrated, while "–" represents the traits where experimental data are lacking. The species in which the roles of respective phytohormones were demonstrated are also listed (At—*Arabidopsis thaliana*, Bn—*Brassica napus*, Gm—*Glycine max*, Gh—*Gossypium hirsutum*, Os—*Oryza sativa*, Ps—*Pisum sativum*, Ta—*Triticum aestivum*, Sb—*Sorghum bicolor*, Zm—*Zea mays*).

is influenced by GAs, CKs, BRs, and JAs (Fig. 4.1). Further, synergistic and antagonistic interactions between GAs, BRs, ABA, ET, and JA regulate seed dormancy.

In-line with their established roles in regulating seed yield, both the external application and genetic manipulation of hormones have been utilized to optimize seed yield in crop plants. However, pleiotropic effects on plant growth and development are significant bottlenecks. In some cases, this problem can be circumvented by using tissue-specific promoters or identifying candidates with seed-specific expression. However, the bigger challenge would be to establish the genotype to phenotype relationships for many phytohormone biosynthesis, transport, and signaling genes, as many are members of large gene families with significant functional redundancy. A high-throughput screening would be required to shortlist target genes for manipulation. Further, extensive cross talk between hormone pathways and shared signaling components can have severe implications in predicting the outcome of genetic engineering. Since many hormones, especially ABA, ET, and JAs, are primarily steered by environmental perturbations, the impact of engineering these hormones on biotic and abiotic stress responses must also be tested to avoid undesirable consequences.

Furthermore, due to an ease of experimentation and availability of resources, most of the studies have been done in the model systems, rice, and *Arabidopsis*. However, as the influence of hormones on seed traits varies with the stage of development and plant species, similar studies in other crop plants would also be required to understand the full spectrum of hormone functions and their utilization for improving seed traits.

Acknowledgments

ND and RS thank SERB for the Core Research grants, CRG/2019/001695 and CRG/2020/003466, respectively. ND also acknowledges DST for INSPIRE Research Grant (IFA17-LSPA90). RJ acknowledges financial support in the form of an ICMR-SRF fellowship.

References

Ahmadi, A., Baker, D.A., 1999. Effects of abscisic acid (ABA) on grain filling processes in wheat. Plant Growth Regul 28, 187–197.

Ali, F., Wei, Z., Li, Y., Gan, L., Yang, Z., Li, F., et al., 2020. An uncanonical transcription factor-DREB2B regulates seed vigor negatively through ABA pathway. bioRxiv.

Ali, F., Qanmber, G., Li, F., Wang, Z., 2021. Updated role of ABA in seed maturation, dormancy, and germination. J. Adv. Res 35, 199–214.

Ashikari, M., Sakakibara, H., Lin, S., Yamamoto, T., Takashi, T., Nishimura, A., et al., 2005. Cytokinin oxidase regulates rice grain production. Science 309 (5735), 741–745.

Baud, S., Dubreucq, B., Miquel, M., Rochat, C., Lepiniec, L., 2008. Storage reserve accumulation in arabidopsis: metabolic and developmental control of seed filling. Arabidopsis Book 6, e0113.

Berleth, T., Jurgens, G., 1993. The role of the monopteros gene in organising the basal body region of the arabidopsis embryo. Development 118 (2), 575–587.

Bernardi, J., Lanubile, A., Li, Q.-B., Kumar, D., Kladnik, A., Cook, S.D., et al., 2012. Impaired auxin biosynthesis in the defective endosperm18 mutant is due to mutational loss of expression in the zmyuc1 gene encoding endosperm-specific YUCCA1 protein in maize. Plant Physiol 160 (3), 1318–1328.

Bleecker, A.B., Estelle, M.A., Somerville, C., Kende, H., 1988. Insensitivity to ethylene conferred by a dominant mutation in *Arabidopsis thaliana*. Science 241 (4869), 1086–1089.

Bogatek, R., Gniazdowska, A., 2018. Ethylene in seed development, dormancy and germination. Ann. Plant Rev. 44.

Bueso, E., Muñoz-Bertomeu, J., Campos, F., Brunaud, V., Martínez, L., Sayas, E., et al., 2014. ARABIDOPSIS THALIANA HOMEOBOX25 uncovers a role for gibberellins in seed longevity. Plant Physiol 164 (2), 999–1010.

Cai, Q., Yuan, Z., Chen, M., Yin, C., Luo, Z., Zhao, X., et al., 2014. Jasmonic acid regulates spikelet development in rice. Nat. Commun. 5, 3476.

Che, R., Tong, H., Shi, B., Liu, Y., Fang, S., Liu, D., et al., 2015. Control of grain size and rice yield by GL2-mediated brassinosteroid responses. Nat. Plants 2, 15195.

Chen, J.-G., Ullah, H., Young, J.C., Sussman, M.R., Jones, A.M., 2001. ABP1 is required for organized cell elongation and division in arabidopsis embryogenesis. Genes Dev 15 (7), 902–911.

Chen, X., Grandont, L., Li, H., Hauschild, R., Paque, S., Abuzeineh, A., et al., 2014. Inhibition of cell expansion by rapid ABP1-mediated auxin effect on microtubules. Nature 516, 90–93.

Chen, Y., Yan, Y., Wu, T.T., Zhang, G.L., Yin, H., Chen, W., et al., 2020. Cloning of wheat keto-acyl thiolase 2B reveals a role of jasmonic acid in grain weight determination. Nat. Commun. 11, 6266.

Cheng, Y., Dai, X., Zhao, Y., 2007. Auxin synthesized by the YUCCA flavin monooxygenases is essential for embryogenesis and leaf formation in arabidopsis. Plant Cell 19 (8), 2430–2439.

Cheng, C., Xu, X., Singer, S.D., Li, J., Zhang, H., Gao, M., et al., 2013. Effect of GA3 treatment on seed development and seed-related gene expression in grape. PLoS One 8 (11), e80044.

Cheng, Z.J., Zhao, X.Y., Shao, X.X., Wang, F., Zhou, C., Liu, Y.G., et al., 2014. Abscisic acid regulates early seed development in arabidopsis by ABI5-mediated transcription of SHORTHYPOCOTYL UNDER BLUE1. Plant Cell 26 (3), 1053−1068.

Chiwocha, S.D.S., Cutler, A.J., Abrams, S.R., Ambrose, S.J., Yang, J., Ross, A.R.S., et al., 2005. The etr1−2 mutation in arabidopsis thaliana affects the abscisic acid, auxin, cytokinin and gibberellin metabolic pathways during maintenance of seed dormancy, moist-chilling and germination. Plant J. 42 (1), 35−48.

Chourey, P.S., Qin-Bao, L., Dibyendu, K., 2010. Sugar−Hormone cross-talk in seed development: two redundant pathways of IAS biosynthesis are regulated differentially in the invertase-deficient miniature1 (Mn1) seed mutant in maize. Mol. Plant 3 (6), 1026−1036.

Cipollini, D., 2007. Consequences of the overproduction of methyl jasmonate on seed production, tolerance to defoliation and competitive effect and response of arabidopsis thaliana. New Phytol. 173 (1), 146−153.

Corbineau, F., Xia, Q., Bailly, C., El-Maarouf-Bouteau, H., 2014. Ethylene, a key factor in the regulation of seed dormancy. Front. Plant Sci. 5, 539.

Curaba, J., Moritz, T., Blervaque, R., Parcy, F., Raz, V., Herzog, M., et al., 2004. AtGA3ox2, a key gene responsible for bioactive gibberellin biosynthesis, is regulated during embryogenesis by LEAFY COTYLEDON2 and FUSCA3 in arabidopsis. Plant Physiol 136 (3), 3660−3669.

Day, R.C., Herridge, R.P., Ambrose, B.A., Macknight, R.C., 2008. Transcriptome analysis of proliferating arabidopsis endosperm reveals biological implications for the control of syncytial division, cytokinin signaling, and gene expression regulation. Plant Physiol 148 (4), 1964−1984.

Deng, Y., Dong, H., Mu, J., Ren, B., Zheng, B., Ji, Z., et al., 2010. Arabidopsis histidine kinase cki1 acts upstream of HISTIDINE PHOSPHOTRANSFER PROTEINS to regulate female gametophyte development and vegetative growth. Plant Cell 22 (4), 1232−1248.

Deveshwar, P., Prusty, A., Sharma, S., Tyagi, A.K., 2020. Phytohormone-mediated molecular mechanisms involving multiple genes and QTL govern grain number in rice. Front. Genet. 11, 586462.

Diaz, I., Vicente-Carbajosa, J., Abraham, Z., Martínez, M., Moneda, I.I.L., Carbonero, P., 2002. The GAMYP protein from barley interacts with the DOF transcription factor BPBF and activates endosperm-specific genes during seed development. Plant J 29 (4), 453−464.

Ding, C., You, J., Chen, L., Wang, S., Ding, Y., 2014. Nitrogen fertilizer increases spikelet number per panicle by enhancing cytokinin synthesis in rice. Plant Cell Rep 33 (2), 363−371. Available from: https://doi.org/10.1007/s00299-013-1536-9. PMID: 24258242.

Domínguez, F., Cejudo, F.J., 2014. Programmed cell death (PCD): an essential process of cereal seed development and germination. Front. Plant Sci. 5, 366.

Fang, N., Xu, R., Huang, L., Zhang, B., Duan, P., Li, N., et al., 2016. SMALL GRAIN 11 controls grain size, grain number and grain yield in rice. Rice 9 (1), 64.

Feng, Z., Wu, C., Wang, C., Roh, J., Zhang, L., Chen, J., et al., 2016. SLG controls grain size and leaf angle by modulating brassinosteroid homeostasis in rice. J. Exp. Bot 67 (14), 4241−4253.

Figueiredo, D.D., Batista, R.A., Roszak, P.J., Hennig, L., Köhler, C., 2016. Auxin production in the endosperm drives seed coat development in arabidopsis. eLife 5, e20542.

Finkelstein, R.R., Gampala, S.S.L., Rock, C.D., 2002. Abscisic acid signaling in seeds and seedlings. Plant Cell 14, S15−S45.

Forestan, C., Varotto, S., 2012. The role of PIN auxin efflux carriers in polar auxin transport and accumulation and their effect on shaping maize development. Mol. Plant. 5 (4), 787−798.

Forestan, C., Meda, S., Varotto, S., 2010. ZmPIN1-mediated auxin transport is related to cellular differentiation during maize embryogenesis and endosperm development. Plant Physiol 152 (3), 1373−1390.

Friml, J., Vieten, A., Sauer, M., Weijers, D., Schwarz, H., Hamann, T., et al., 2003. Efflux-dependent auxin gradients establish the apical-basal axis of arabidopsis. Nature 426 (6963), 147−153.

Gao, S., Chu, C., 2020. Gibberellin metabolism and signaling: targets for improving agronomic performance of crops. Plant Cell Physiol 61 (11), 1902−1911.

Goetz, S., Hellwege, A., Stenzel, I., Kutter, C., Hauptmann, V., Forner, S., et al., 2012. Role of cis-12-Oxo-phytodienoic acid in tomato embryo development. Plant Physiol 158 (4), 1715−1727.

Hamann, T., Benkova, E., Bäurle, I., Kientz, M., Jürgens, G., 2002. The arabidopsis BODENLOS gene encodes an auxin response protein inhibiting MONOPTEROS-mediated embryo patterning. Genes Dev 16 (13), 1610–1615.

Hibara, K.-I., Isono, M., Mimura, M., Sentoku, N., Kojima, M., Sakakibara, H., et al., 2016. Jasmonate regulates juvenile-to-adult phase transition in rice. Development 143 (18), 3407–3446.

Holdsworth, M.J., Bentsink, L., Soppe, W.J.J., 2008. Molecular networks regulating arabidopsis seed maturation, after-ripening, dormancy and germination. New Phytol 179 (1), 33–54.

Hong, Z., Ueguchi-Tanaka, M., Fujioka, S., Takatsuto, S., Yoshida, S., Hasegawa, Y., et al., 2005. The rice brassinosteroid-deficient dwarf2 mutant, defective in the rice homolog of arabidopsis DIMINUTO/DWARF1, is rescued by the endogenously accumulated alternative bioactive brassinosteroid, dolichosterone. Plant Cell 17 (8), 2243–2254.

Hu, Y.-F., Li, Y.-P., Zhang, J., Liu, H., Tian, M., Huang, Y., 2012. Binding of ABI4 to a CACCG motif mediates the ABA-induced expression of the ZmSSI gene in maize (Zea mays L.) endosperm. J. Exp. Bot 63 (16), 5979–5989.

Hu, S., Yang, H., Gao, H., Yan, J., Xie, D., 2021. Control of seed size by jasmonate. Sci. China Life Sci 64 (8), 1215–1226.

Huang, X., Qian, Q., Liu, Z., Sun, H., He, S., Luo, D., et al., 2009. Natural variation at the DEP1 locus enhances grain yield in rice. Nat. Genet. 41, 494–497.

Huang, Y., Bai, X., Luo, M., Xing, Y., 2019. Short Panicle 3 controls panicle architecture by upregulating APO2/RFL and increasing cytokinin content in rice. J. Integr. Plant Biol. 61 (9), 987–999.

Huo, X., Wu, S., Zhu, Z., Liu, F., Fu, Y., Cai, H., et al., 2017. NOG1 increases grain production in rice. Nat. Commun. 8 (1), 1497.

Hwang, H., Ryu, H., Cho, H., 2021. Brassinosteroid signaling pathways interplaying with diverse signaling cues for crop enhancement. Agronomy 11 (3), 556.

Ishimaru, K., Hirotsu, N., Madoka, Y., Murakami, N., Hara, N., Onodera, H., et al., 2013. Loss of function of the IAA-glucose hydrolase gene TGW6 enhances rice grain weight and increases yield. Nat. Genet. 45 (6), 707–711.

Jameson, P.E., Song, J., 2016. Cytokinin: a key driver of seed yield. J. Exp. Bot 67 (3), 593–606.

Jia, H., McCarty, D.R., Suzuki, M., 2013. Distinct roles of LAFL network genes in promoting the embryonic seedling fate in the absence of VAL repression. Plant Physiol. 163 (3), 1293–1305.

Jia, H., Suzuki, M., McCarty, D.R., 2014. Regulation of the seed to seedling developmental phase transition by the LAFL and VAL transcription factor networks. Wiley Interdiscip. Rev. Dev. Biol. 3 (1), 135–145.

Jiang, W.-B., Lin, W.-H., 2013. Brassinosteroid functions in arabidopsis seed development. Plant Signal Behav. 8 (10), e25928.

Jiang, W.-B., Huang, H.-Y., Hu, Y.-W., Wang, Z.-Y., Lin, W.-H., 2013. Brassinosteroid regulates seed size and shape in arabidopsis. Plant Physiol. 162 (4), 1965–1977.

Jiao, Y., Lee, Y.K., Gladman, N., Chopra, R., Christensen, S.A., Regulski, M., et al., 2018. MSD1 regulates pedicellate spikelet fertility in sorghum through the jasmonic acid pathway. Nat. Commun. 9, 822.

Kang, K., Shim, Y., Gi, E., An, G., Paek, N.-C., 2019. Mutation of ONAC096 enhances grain yield by increasing panicle number and delaying leaf senescence during grain filling in rice. Int. J. Mol. Sci. 20 (20), 5241.

Kanno, Y., Jikumaru, Y., Hanada, A., Nambara, E., Abrams, S.R., Kamiya, Y., et al., 2010. Comprehensive hormone profiling in developing arabidopsis seeds: examination of the site of aba biosynthesis, ABA transport and hormone interactions. Plant Cell Physiol. 51 (12), 1988–2001.

Kasahara, H., 2016. Current aspects of auxin biosynthesis in plants. Biosci. Biotechnol. Biochem. 80 (1), 34–42.

Kelley, D.R., Arreola, A., Gallagher, T.L., Gasser, C.S., 2012. ETTIN (ARF3) physically interacts with KANADI proteins to form a functional complex essential for integument development and polarity determination in arabidopsis. Development 139 (6), 1105–1109.

Kermode, A.R., 2005. Role of abscisic acid in seed dormancy. J Plant Growth Regul 24, 319–344.

Kim, E.H., Kim, Y.S., Park, S.-H., Koo, Y.J., Choi, Y.D., Chung, Y.-Y., et al., 2009. Methyl jasmonate reduces grain yield by mediating stress signals to alter spikelet development in rice. Plant Physiol 149 (4), 1751–1760.

Kotak, S., Vierling, E., Bäumlein, H., Koskull-Döring, P.V., 2007. A novel transcriptional cascade regulating expression of heat stress proteins during seed development of arabidopsis. Plant Cell 19 (1), 182–195.

Kwong, R.W., Bui, A.Q., Lee, H., Kwong, L.W., Fischer, R.L., Goldberg, R.B., et al., 2003. LEAFY COTYLEDON1-LIKE defines a class of regulators essential for embryo development. Plant Cell 15 (1), 5–18.

LeCLere, S., Schmelz, E.A., Chourey, P.S., 2010. Sugar levels regulate tryptophan-dependent auxin biosynthesis in developing maize kernels. Plant Physiol 153 (1), 306–318.

Lei, M., Li, Z.-Y., Wang, J.-B., Fu, Y.-L., Xu, L., 2019. Ectopic expression of the *Aechmea fasciata* APETALA2 gene AfAP2-2 reduces seed size and delays flowering in arabidopsis. Plant Physiol. Biochem. 139, 642–650.

Leljak-Levanić, D., Mihaljević, S., Bauer, N., 2015. Somatic and zygotic embryos share common developmental features at the onset of plant embryogenesis. Acta Physiol. Plant. 37, 127.

Leprince, O., Pellizzaro, A., Berriri, S., Buitink, J., 2017. Late seed maturation: drying without dying. J. Exp. Bot. 68 (4), 827–841.

Li, N., Li, Y., 2015. Maternal control of seed size in plants. J. Exp. Bot 66 (4), 1087–1097.

Li, N., Li, Y., 2016. Signaling pathways of seed size control in plants. Curr. Opin. Plant Biol. 33, 23–32.

Li, M., Tang, D., Wang, K., Wu, X., Lu, L., Yu, H., et al., 2011. Mutations in the F-box gene LARGER PANICLE improve the panicle architecture and enhance the grain yield in rice. Plant Biotechnol. J. 9 (9), 1002–1013.

Li, S., Zhao, B., Yuan, D., Duan, M., Qian, Q., Tang, L., et al., 2013. Rice zinc finger protein DST enhances grain production through controlling Gn1a/OsCKX2 expression. Proc. Natl. Acad. Sci. U.S.A. 110 (8), 3167–3172.

Li, Q., Li, L., Liu, Y., Lv, Q., Zhang, H., Zhu, J., et al., 2017. Influence of TaGW2−6A on seed development in wheat by negatively regulating gibberellin synthesis. Plant Sci 263, 226–235.

Li, N., Xu, R., Li, Y., 2019. Molecular networks of seed size control in plants. Annu. Rev. Plant Biol. 70, 435–463.

Liang, R., Qin, R., Yang, C., Zeng, D., Jin, X., Shi, C., 2019. Identification and characterization of a novel strigolactone-insensitive mutant, dwarfism with High Tillering Ability 34 (dhta-34) in rice (*Oryza sativa* L.). Biochem. Genet. 57 (3), 403–420.

Lin, W.-H., 2020. Designed manipulation of the brassinosteroid signal to enhance crop yield. Front. Plant Sci. 11, 854.

Lin, Y., Zhao, Z., Zhou, S., Liu, L., Kong, W., Chen, H., et al., 2017. Top bending panicle1 is involved in brassinosteroid signaling and regulates the plant architecture in rice. Plant Physiol. Biochem. 121, 1–13.

Liu, W., Wu, C., Fu, Y., Hu, G., Si, H., Zhu, L., et al., 2009. Identification and characterization of HTD2: a novel gene negatively regulating tiller bud outgrowth in rice. Planta 230 (4), 649–658.

Liu, J., Deng, M., Guo, H., Raihan, S., Luo, J., Xu, Y., et al., 2015a. Maize orthologs of Rice GS5 and their transregulator are associated with kernel development. J. Integr. Plant Biol. 57 (11), 943–953.

Liu, J., Hua, W., Hu, Z., Yang, H., Zhang, L., Li, R., et al., 2015b. Natural variation in ARF18 gene simultaneously affects seed weight and silique length in polyploid rapeseed. Proc. Natl. Acad. Sci. U.S.A. 112 (37), E5123–E5132.

Liu, L., Tong, H., Xiao, Y., Che, R., Xu, F., Hu, B., et al., 2015c. Activation of Big Grain1 significantly improves grain size by regulating auxin transport in rice. Proc. Natl. Acad. Sci. U.S.A. 112 (35), 11102–11107.

Liu, J., Chen, J., Zheng, X., Wu, F., Lin, Q., Heng, Y., et al., 2017. GW5 acts in the brassinosteroid signalling pathway to regulate grain width and weight in rice. Nature Plants 3, 17043.

Liu, F., Zhang, H., Ding, L., Soppe, W.J.J., Xiang, Y., 2020. REVERSAL OF RDO5 1, a homolog of rice seed dormancy4, interacts with bHLH57 and controls ABA biosynthesis and seed dormancy in arabidopsis. Plant Cell 32 (6), 1933–1948.

Locascio, A., Roig-Villanova, I., Bernardi, J., Varotto, S., 2014. Current perspectives on the hormonal control of seed development in arabidopsis and maize: a focus on auxin. Front. Plant Sci. 5, 412.

Lu, X., Xiong, Q., Cheng, T., Li, Q.-T., Liu, X.-L., Bi, Y.-D., et al., 2017. A PP2C-1 allele underlying a quantitative trait locus enhances Soybean 100-seed weight. Mol. Plant. 10 (5), 670–684.

Ma, Q.-H., Wang, X.-M., Wang, Z.-M., 2008. Expression of isopentenyl transferase gene controlled by seedspecific lectin promoter in transgenic Tobacco influences seed development. J. Plant Growth Regul 27, 68–76.

Ma, Y., Szostkiewicz, I., Korte, A., Moes, D., Yang, Y., Christmann, A., et al., 2009. Regulators of PP2C phosphatase activity function as abscisic acid sensors. Science 324 (5930), 1064–1068.

Matilla, A.J., 2020. Auxin: hormonal signal required for seed development and dormancy. Plants 9 (6), 705.

Meitzel, T., Radchuk, R., McAdam, E.L., Thormählen, I., Feil, R., Munz, E., et al., 2021. Trehalose 6-phosphate promotes seed filling by activating auxin biosynthesis. New Phytol 229 (3), 1553–1565.

Meng, L.-S., Wang, Z.-B., Yao, S.-Q., Liu, A., 2015. The ARF2-ANT-COR15A gene cascade regulates ABAsignaling-mediated resistance of large seeds to drought in arabidopsis. J. Cell Sci 128 (21), 3922–3932.

Miura, K., Ikeda, M., Matsubara, A., Song, X.-J., Ito, M., Asano, K., et al., 2010. OsSPL14 promotes panicle branching and higher grain productivity in rice. Nat. Genet. 42 (6), 545–549.

Mizukami, Y., Fischer, R.L., 2000. Plant organ size control: AINTEGUMENTA regulates growth and cell numbers during organogenesis. Proc. Natl. Acad. Sci. U S A 97 (2), 942–947. 10.1073/pnas.97.2.942.

Mönke, G., Seifert, M., Keilwagen, J., Mohr, M., Grosse, I., Hähnel, U., et al., 2012. Toward the identification and regulation of the arabidopsis thaliana ABI3 regulon. Nucleic Acids Res 40 (17), 8240–8254.

Morinaka, Y., Sakamoto, T., Inukai, Y., Agetsuma, M., Kitano, H., Ashikari, M., et al., 2006. Morphological alteration caused by brassinosteroid insensitivity increases the biomass and grain production of rice. Plant Physiol 141 (3), 924–931.

Nadeau, C.D., Ozga, J.A., Kurepin, L.V., Jin, A., Pharis, R.P., Reinecke, D.M., 2011. Tissue-specific regulation of gibberellin biosynthesis in developing pea seeds. Plant Physiol. 156 (2), 897–912.

Nakagawa, H., Tanaka, A., Tanabata, T., Ohtake, M., Fujioka, S., Nakamura, H., et al., 2012. SHORT GRAIN1 decreases organ elongation and brassinosteroid response in rice. Plant Physiol. 158 (3), 1208–1219.

Nakashima, K., Ito, Y., Yamaguchi-Shinozaki, K., 2009. Transcriptional regulatory networks in response to abiotic stresses in arabidopsis and grasses. Plant Physiol. 149 (1), 88–95.

Nomura, T., Ueno, M., Yamada, Y., Takatsuto, S., Takeuchi, Y., Yokota, T., 2007. Roles of brassinosteroids and related mRNAs in pea seed growth and germination. Plant Physiol. 143 (4), 1680–1688.

Panda, B.B., Badoghar, A.K., Sekhar, S., Shaw, B.P., Mohapatra, P.K., 2016. 1-MCP treatment enhanced expression of genes controlling endosperm cell division and starch biosynthesis for improvement of grain filling in a dense-panicle rice cultivar. Plant Sci 246, 11–25.

Parcy, F., Valon, C., Kohara, A., Miséra, S., Giraudatag, J., 1997. The ABSCISIC ACID-INSENSITIVE3, FUSCA3, and LEAFY COTYLEDON1 loci act in concert to control multiple aspects of arabidopsis seed development. Plant Cell 9 (8), 1265–1277.

Pellizzaro, A., Neveu, M., Lalanne, D., Vu, B.L., Kanno, Y., Seo, M., et al., 2020. A role for auxin signaling in the acquisition of longevity during seed maturation. New Phytol 225 (1), 284–296.

Personat, J.-M., Tejedor-Cano, J., Prieto-Dapena, P., Almoguera, C., Jordano, J., 2014. Co-overexpression of two Heat Shock Factors results in enhanced seed longevity and in synergistic effects on seedling tolerance to severe dehydration and oxidative stress. BMC Plant Biol 14, 56.

Robert, C., Noriega, A., Tocino, Á., Cervantes, E., 2008. Morphological analysis of seed shape in arabidopsis thaliana reveals altered polarity in mutants of the ethylene signaling pathway. J. Plant Physiol 165 (9), 911–919.

Roxrud, I., Lid, S.E., Fletcher, J.C., Schmidt, E.D.L., Opsahl-Sorteberg, H.-G., 2007. GASA4, one of the 14-member arabidopsis GASA family of small polypeptides, regulates flowering and seed development. Plant and Cell Physiology 48 (3), 471–483.

Sakamoto, T., Morinaka, Y., Ishiyama, K., Kobayashi, M., Itoh, H., Kayano, T., Iwahori, S., Matsuoka, M., Tanaka, H., 2003. Genetic manipulation of gibberellin metabolism in transgenic rice. Nat. Biotechnol. 21 (8), 909–913.

Savadi, S., 2017. Molecular regulation of seed development and strategies for engineering seed size in crop plants. Plant Growth Regul 84, 401–422.

Scanlon, M.J., Henderson, D.C., Bernstein, B., 2002. SEMAPHORE1 functions during the regulation of ancestrally duplicated knox genes and polar auxin transport in maize. Development 129 (11), 2663–2673.

Schruff, M.C., Spielman, M., Tiwari, S., Adams, S., Fenby, N., Scott, R.J., 2006. The AUXIN RESPONSE FACTOR 2 gene of arabidopsis links auxin signalling, cell division, and the size of seeds and other organs. Development 133 (2), 251–261.

Seo, M., Hanada, A., Kuwahara, A., Endo, A., Okamoto, M., Yamauchi, Y., et al., 2006. Regulation of hormone metabolism in arabidopsis seeds: phytochrome regulation of abscisic acid metabolism and abscisic acid regulation of gibberellin metabolism. Plant J 48 (3), 354–366.

Shen, W.-H., Parmentier, Y., Hellmann, H., Lechner, E., Dong, A., Masson, J., et al., 2002. Null mutation of AtCUL1 causes arrest in early embryogenesis in arabidopsis. Mol. Biol. Cell 13 (6), 1916–1928.

Shi, C.-L., Dong, N.-Q., Guo, T., Ye, W.-W., Shan, J.-X., Lin, H.-X., 2020. A quantitative trait locus GW6 controls rice grain size and yield through the gibberellin pathway. Plant J. 103 (3), 1174–1188.

Shu, K., Zhang, H., Wang, S., Chen, M., Wu, Y., 2013. ABI4 regulates primary seed dormancy by regulating the biogenesis of abscisic acid and gibberellins in arabidopsis. PLoS Genet 9 (6), e1003577.

Shu, K., Liu, X.-D., Xie, Q., He, Z.-H., 2016. Two faces of one seed: hormonal regulation of dormancy and germination. Mol. Plant 9 (1), 34–45.

Singh, D.P., Filardo, F.F., Storey, R., Jermakow, A.M., Yamaguchi, S., Swain, S.M., 2010. Overexpression of a gibberellin inactivation gene alters seed development, KNOX gene expression, and plant development in arabidopsis. Physiol. Plant 138 (1), 74–90.

Sun, L., Zhang, Q., Wu, J., Zhang, L., Jiao, X., Zhang, S., et al., 2014. Two rice authentic histidine phosphotransfer proteins, OsAHP1 and OsAHP2, mediate cytokinin signaling and stress responses in rice. Plant Physiol. 165 (1), 335–345.

Singh, D.P., Jermakow, A.M., Swain, S.M., 2002. Gibberellins are required for seed development and pollen tube growth in *Arabidopsis*. The Plant Cell 14 (12), 3133–3147.

Sun, Y., Wang, C., Wang, N., Jiang, X., Mao, H., Zhu, C., et al., 2017. Manipulation of Auxin Response Factor 19 affects seed size in the woody perennial *Jatropha curcas*. Sci. Rep. 7, 40844.

Sun, H., Xu, H., Li, B., Shang, Y., Wei, M., Zhang, S., et al., 2021. The brassinosteroid biosynthesis gene, ZmD11, increases seed size and quality in rice and maize. Plant Physiol. Biochem. 160, 281–293.

Sundaresan, V., 2005. Control of seed size in plants. Proc. Natl. Acad. Sci. U.S.A. 102 (50), 17887–17888.

Swain, S.M., Reid, J.B., Kamiya, Y., 1997. Gibberellins are required for embryo growth and seed development in pea. Plant J 12 (6), 1329–1338.

Takahashi, N., Nakazawa, M., Shibata, K., Yokota, T., Ishikawa, A., Suzuki, K., et al., 2005. shk1-D, a dwarf arabidopsis mutant caused by activation of the CYP72C1 gene, has altered brassinosteroid levels. Plant J 42 (1), 13–22.

Tanabe, S., Ashikari, M., Fujioka, S., Takatsuto, S., Yoshida, S., Yano, M., et al., 2005. A novel cytochrome P450 is implicated in brassinosteroid biosynthesis via the characterization of a rice dwarf mutant, dwarf11, with reduced seed length. Plant Cell 17 (3), 776–790.

To, A., Valon, C., Savino, G., Guilleminot, J., Devic, M., Giraudat, J., et al., 2006. A network of local and redundant gene regulation governs arabidopsis seed maturation. Plant Cell 18 (7), 1642–1651.

To, A., Joubès, J., Barthole, G., Lécureuil, A., Scagnelli, A., Jasinski, S., et al., 2012. WRINKLED transcription factors orchestrate tissue-specific regulation of fatty acid biosynthesis in arabidopsis. Plant Cell 24 (12), 5007–5023.

Ulmasov, T., Hagen, G., Guilfoyle, T.J., 1999. Activation and repression of transcription by auxin-response factors. Proc. Natl. Acad. Sci. U.S.A. 96 (10), 5844–5849.

Vieten, A., Vanneste, S., Wisniewska, J., Benkova, E., Benjamins, R., Beeckman, T., et al., 2005. Functional redundancy of PIN proteins is accompanied by auxin-dependent cross-regulation of PIN expression. Development 132 (20), 4521–4531.

Völz, R., Heydlauff, J., Ripper, D., Lyncker, L.V., Groß-Hardt, R., 2013. Ethylene signaling is required for synergid degeneration and the establishment of a pollen tube block. Dev. Cell. 25 (3), 310–316.

Walton, L.J., Kurepin, L.V., Yeung, E.C., Shah, S., Emery, R.J.N., Reid, D.M., et al., 2012. Ethylene involvement in silique and seed development of canola, *Brassica napus* L. Plant Physiol. Biochem. 58, 142–150.

Wang, B., Chen, Y., Guo, B., Kabir, M.R., Yao, Y., Peng, H., et al., 2014. Expression and functional analysis of genes encoding cytokinin receptor-like histidine kinase in maize (*Zea mays* L.). Mol. Genet. Genomics 289 (4), 501–512.

Wang, H., Zhang, Y., Xiao, N., Zhang, G., Wang, F., Chen, X., et al., 2020. Rice GERMIN-LIKE PROTEIN 2-1 functions in seed dormancy under the control of abscisic acid and gibberellic acid signaling pathways. Plant Physiol 183 (3), 1157–1170.

Wang, Y., Wang, J., Guo, S., Tian, S., Zhang, J., Ren, Y., et al., 2021. CRISPR/Cas9-mediated mutagenesis of ClBG1 decreased seed size and promoted seed germination in watermelon. Hortic. Res. 8, 70.

Wasternack, C., Goetz, S., Hellwege, A., Forner, S., Strnad, M., Hause, B., 2012. Another JA/COI1-independent role of OPDA detected in tomato embryo development. Plant Signal. Behav. 7 (10), 1349–1353.

Wasternack, C., Forner, S., Strnad, M., Hause, B., 2013. Jasmonates in flower and seed development. Biochimie. 95 (1), 79–85.

Werner, T., Motyka, V., Strnad, M., Schmülling, T., 2001. Regulation of plant growth by cytokinin. Proc. Natl. Acad. Sci. U.S.A. 98 (18), 10487–10492.

White, C.N., Proebsting, W.M., Hedden, P., Rivin, C.J., 2000. Gibberellins and seed development in maize. I. Evidence that gibberellin/abscisic acid balance governs germination versus maturation pathways. Plant Physiol. 122 (4), 1081–1088.

Wu, Y., Dor, E., Hershenhorn, J., 2017. Strigolactones affect tomato hormone profile and somatic embryogenesis. Planta 245 (3), 583–594.

Wuriyanghan, H., Zhang, B., Cao, W.-H., Ma, B., Lei, G., Liu, Y.-F., Wei, W., Wu, H.-J., Chen, L.-J., Chen, H.-W., Cao, Y.-R., He, S.-J., Zhang, W.-K., Wang, X.-J., Chen, S.-Y., Zhang, J.-S., 2009. The ethylene receptor ETR2 delays floral transition and affects starch accumulation in rice. The Plant Cell 21 (5), 1473–1494.

Xu, C., Liu, Y., Li, Y., Xu, X., Xu, C., Li, X., et al., 2015. Differential expression of GS5 regulates grain size in rice. J. Exp. Bot. 66 (9), 2611–2623.

Yamada, Y., Otake, M., Furukawa, T., Shindo, M., Shimomura, K., Yamaguchi, S., et al., 2019. Effects of strigolactones on grain yield and seed development in rice. J. Plant Growth Regul. 38, 753–764.

Yang, J., Zhang, J., Wang, Z., Liu, K., Wang, P., 2006. Post-anthesis development of inferior and superior spikelets in rice in relation to abscisic acid and ethylene. J. Exp. Bot. 57 (1), 149–160.

Yang, J., Cho, L.-H., Yoon, J., Yoon, H., Wai, A.H., Hong, W.-J., et al., 2019. Chromatin interacting factor OsVIL2 increases biomass and rice grain yield. Plant Biotechnol. J. 17 (1), 178–187.

Ye, H., Feng, J., Zhang, L., Zhang, J., Mispan, M.S., Cao, Z., et al., 2015. Map-based cloning of seed dormancy1–2 identified a gibberellin synthesis gene regulating the development of endosperm-imposed dormancy in rice. Plant Physiol. 169 (3), 2152–2165.

You, X., Zhu, S., Zhang, W., Zhang, J., Wang, C., Jing, R., et al., 2019. OsPEX5 regulates rice spikelet development through modulating jasmonic acid biosynthesis. New Phytol 224 (2), 712–724.

Young, T.E., Gallie, D.R., DeMason, D.A., 1997. Ethylene-mediated programmed cell death during maize endosperm development of wild-type and shrunken2 genotypes. Plant Physiol 115 (2), 737–751.

Yuan, Z., Fan, K., Wang, Y., Tian, L., Zhang, C., Sun, W., et al., 2021. OsGRETCHENHAGEN3-2 modulates rice seed storability via accumulation of abscisic acid and protective substances. Plant Physiol. 186 (1), 469–482.

Zejun, H., Lu, S.-J., Wang, M.-J., He, H., Sun, L., Wang, H., et al., 2018. A novel QTL qTGW3 encodes the GSK3/SHAGGY-Like kinase OsGSK5/OsSK41 that interacts with OsARF4 to negatively regulate grain size and weight in rice. Mol. Plant 11 (5), 736–749.

Zhang, X., Wang, J., Huang, J., Lan, H., Wang, C., Yin, C., et al., 2012. Rare allele of OsPPKL1 associated with grain length causes extra-large grain and a significant yield increase in rice. Proc. Natl. Acad. Sci. U.S.A. 109 (52), 21534–21539.

Zhao, D.-S., Li, Q.-F., Zhang, C.-Q., Zhang, C., Yang, Q.-Q., Pan, L.-X., et al., 2018. GS9 acts as a transcriptional activator to regulate rice grain shape and appearance quality. Nat. Commun. 9, 1240.

Zhao, Y.-F., Peng, T., Sun, H.-Z., Teotia, S., Wen, H.-L., Du, Y.-X., et al., 2019. miR1432-OsACOT (Acyl-CoA thioesterase) module determines grain yield via enhancing grain filling rate in rice. Plant Biotechnol. J 17 (4), 712–723.

Zhou, W., Chen, F., Luo, X., Dai, Y., Yang, Y., Zheng, C., et al., 2020. A matter of life and death: molecular, physiological, and environmental regulation of seed longevity. Plant Cell Environ. 43 (2), 293–302.

Zhu, G., Ye, N., Yang, J., Peng, X., Zhang, J., 2011. Regulation of expression of starch synthesis genes by ethylene and ABA in relation to the development of rice inferior and superior spikelets. J. Exp. Bot. 62 (11), 3907–3916.

Zhu, X., Liang, W., Cui, X., Chen, M., Yin, C., Luo, Z., et al., 2015. Brassinosteroids promote development of rice pollen grains and seeds by triggering expression of carbon starved anther, a MYB domain protein. Plant J 82 (4), 570–581.

Zu, S.-H., Jiang, Y.-T., Hu, L.-Q., Zhang, Y.-J., Chang, J.-H., Xue, H.-W., et al., 2019. Effective modulating brassinosteroids signal to study their specific regulation of reproductive development and enhance yield. Front. Plant Sci. 10, 980.

Phytohormone signaling in osmotic stress response

Riddhi Datta[1], Ananya Roy[2] and Soumitra Paul[2]

[1]Department of Botany, Dr. A.P.J. Abdul Kalam Government College, Kolkata, West Bengal, India [2]Department of Botany, University of Calcutta, Kolkata, West Bengal, India

5.1 Introduction

Being sessile, the plants experience a wide array of environmental challenges in their life cycle. Fluctuations in abiotic factors like temperature, water, light, salt, oxygen, carbon dioxide, nutrients, and heavy metals impact broadly on their physicochemical properties. On the other hand, after perception, plants can respond to this change by altering their different signaling mechanisms and manipulating their physiological phenomena, including growth. Water deficit and hypersaline conditions cause osmotic stress, hindering plant growth, and the developmental process. Hypersalinity in soil lowers the soil water potential that leads to the reduced absorption of water by plant cells (Lambers et al., 1998). To overcome this, plants promote osmotic adjustment to lower the water potential by accumulating different solutes and by triggering appropriate cellular responses (Lambers et al., 1998). Ion toxicity and oxidative stress also accompany osmotic stress resulting in reactive oxygen species (ROS) generation causing severe damage to plant cells (Sharma et al., 2019). Plants respond to osmotic stress by altering various physiological mechanisms to switch from a "growth and development mode" to a "defense mode." Phytohormones not only act as plant growth regulators but also as essential stress managers. They coordinate with different cellular responses and signal transduction pathways to fine-tune the plant's response to osmotic stress (Datta et al., 2020). This chapter will highlight the role of different phytohormones in mitigating osmotic stress to improve plant growth and development.

5.2 Abscisic acid—mediated responses under osmotic stress

Abscisic acid (ABA), also known as the stress hormone, was first discovered in cotton while studying fruit abscission (Carns, 1954). Subsequently, its role was widely reported in regulating plant cellular processes in seed maturation, dormancy, stomatal closure, and initiation of osmotic stress responses (Chen et al., 2020). The ABA biosynthesis starts from a carotenoid intermediate with the conversion of five-carbon isopentenyl diphosphate (IPP) to zeaxanthin catalyzed by zeaxanthin epoxidase (ZEP) (Taylor et al., 2000). Zeaxanthin is then converted to violaxanthin which, in turn, is converted to 9′-cis-neoxanthin by 9-cis-epoxycarotenoid dioxygenase (NCED), which again cleaves neoxanthin to form xanthoxin. NCED is the rate-limiting enzyme in the ABA biosynthetic pathway (Qin and Zeevaart, 1999). Finally, xanthoxin is converted to ABA through a two-step reaction via ABA-aldehyde. These two steps are catalyzed by short-chain alcohol dehydrogenase/reductase (SDR) and ABA aldehyde oxidase (AAO), respectively (Xiong and Zhu, 2003). During osmotic stress, the transcript level of the *NCED* gene increases rapidly by a calcium (Ca^{2+})-dependent phosphorelay cascade leading to the accumulation of ABA (Xiong and Zhu, 2003). This transcriptional regulation also involves the activation of the other ABA biosynthetic enzymes like ZEP and AAO, but their induction is not as strong as NCED. Moreover, the vacuolar sorting receptor 1 (VSR1) has been demonstrated to be involved in the osmotic stress-responsive feedback regulation of ABA biosynthesis via changes in the pH gradients between cytoplasm and vacuoles (Wang et al., 2015b).

In ABA signaling pathway, protein phosphatase 2C (PP2C) acts as coreceptor and binds to SNF1-RELATED PROTEIN KINASE 2s (SnRK2s) to keep it repressed under control condition. Under osmotic stress, ABA is perceived by the PYR/PYL/RCAR receptor. The PYL receptor then interacts with PP2C thus relieving the repression from SnRK2. This activates SnRK2 by MAP kinase-mediated phosphorylation, which in turn phosphorylates the downstream bZIP transcription factors like ABRE-binding proteins (AREB/ABFs), enzymes, or ion channels and induce ABA-mediated stress responses as discussed in the following subsection (Fujii et al., 2009; Singh et al., 2016).

5.2.1 Regulation of stomatal closure and ion homeostasis

The regulation of stomatal opening and closure, primarily required for maintaining the cellular water balance and leaf temperature, is predominantly controlled by turgidity in guard cells. ABA accumulation mediates the generation of ROS, which acts as a secondary messenger in the guard cells that induces stomatal closure (Watkins et al., 2017). In response to osmotic stress, the elevated ABA level in the leaves activates the ABA receptor RCAR, thereby releasing a major kinase molecule open stomata 1 (OST1) from the inhibitory effect of the protein phosphatase type 2C (PP2C). This results in the phosphorylation of NADPH oxidase, RESPIRATORY BURST OF OXIDATIVE HOMOLOGUE F (RBOHF), and RBOHD. The activated RBOH now catalyzes the formation of ROS that in turn activates the Ca^{2+} ion channels in the guard cells (Yoshida et al., 2010; Fujita et al., 2013). The elevated cytosolic Ca^{2+} stimulates Ca^{2+}-dependent protein kinases (CPK) to induce phosphorylation and activation of the slow anion channel associated 1 (SLAC1) and KAT1 channels (Geiger et al., 2009). This, in turn, results in membrane depolarization by regulating the R-type (quick activating anionic channels) and S-type (slow activating anionic channels) anionic channels as well as the outward rectifying K^+ channels. This ionic change decreases the cellular turgidity and initiates stomatal closure (Geiger et al., 2009).

Further, vacuolar sorting receptor 3 (VSR3) has been reported to be essential in regulating ABA-mediated guard cell closure in response to osmotic stress (Fig. 5.1; Avila et al., 2008).

Plants undergo osmotic stress during high salt conditions that results in low osmotic potential due to the absorption of Na^+ and Cl^- and their retention in cells. To overcome

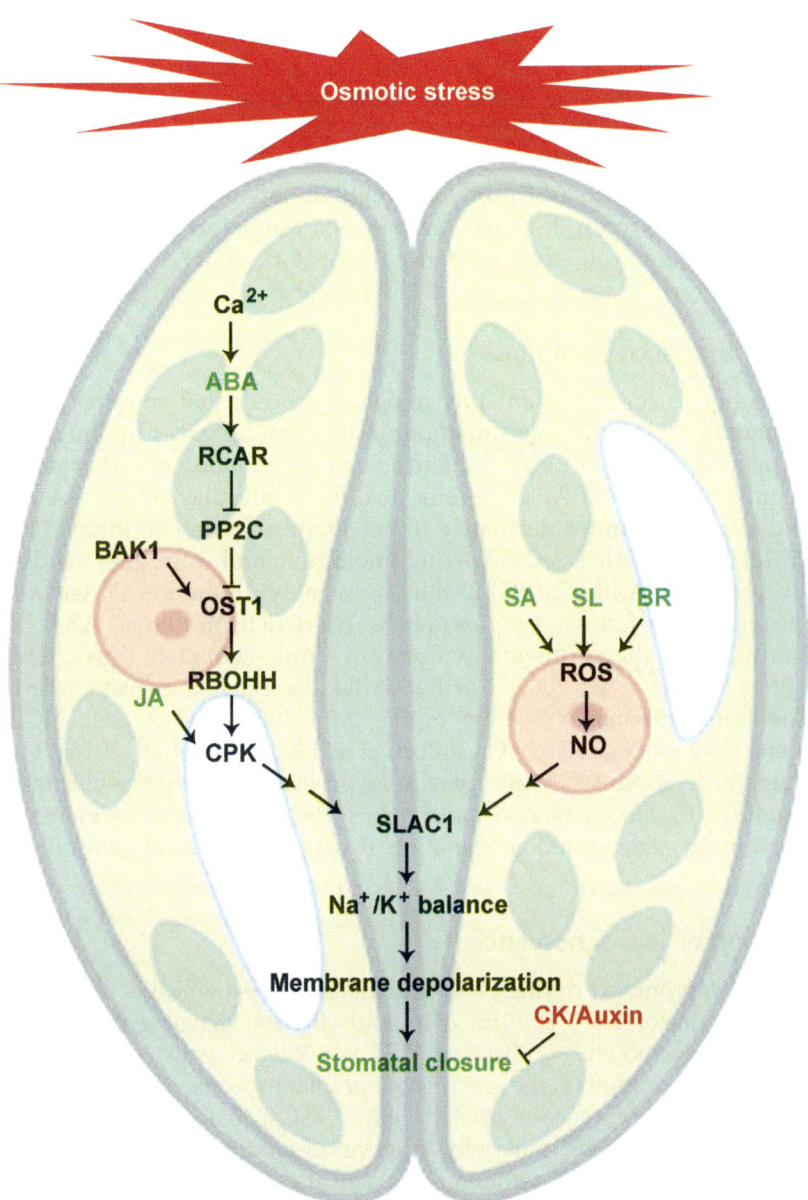

FIGURE 5.1   Phytohormone interplay in regulating stomatal closure in response to osmotic stress.

osmotic stress, the activation of ABA signaling is crucial for plants. During osmotic stress, ABA phosphorylates SNF1-related protein kinase 2 (SnRK2) and triggers the oscillation of Ca^{2+} in cells, which in turn activates the downstream salt overly sensitive (SOS) members like SOS2, SOS3, including calcineurin B like protein (CBL), and CBL-interacting protein kinase (CIPK) (Guo et al., 2002; Qiu et al., 2002). SOS3−SOS2 module further stimulates the SOS1, a Na^+/H^+ antiporter (NHX1) to regulate Na^+ ions transport in cells. This signaling mechanism helps plants to adapt against higher concentration of Na^+ ions during salt stress. In addition, PP2C family members like ABA Insensitive 1 (ABI1) and ABI2 act as a negative regulators in ABA signaling pathway, and the recessive mutants of *abi1* and *abi2* showed ABA hypersensitivity. However, the dominant mutants like *abi1−1* and *abi2−1* displayed reduced sensitivity to ABA (Merlot et al., 2001). Interestingly, SOS2 is found to interact with ABI protein to regulate this negative signaling. The disruption of the interaction between these two proteins led the plants sensitive to salt stress and justifies its positive role in salt stress tolerance (Ohta et al., 2003).

5.2.2 Inhibition of seed germination

Seed germination is an important physiological process that requires the presence of water. Consequently, restraining germination under osmotic stress by inducing dormancy constitutes a major adaptive strategy in plants. Seed germination and dormancy are antagonistically regulated by the ABA−gibberellic acid (GA) interplay where GA promotes germination, while ABA promotes dormancy under inappropriate conditions (Li et al., 2016; Shu et al., 2016). GA is synthesized following imbibition, and it triggers the degradation of the DELLA repressorRGA-like 2 (RGL2) thus inducing germination (Tyler et al., 2004). If, however, germination is followed by sudden water deficitconditions, ABA induces ABI3 and ABI5 that maintain the germinated embryo in a quiescent state thus conferring osmotolerance (Lopez-Molina et al., 2001). Further, ABI4 also plays a decisive role in this regulation via the transcriptional activation of the ABA biosynthesis gene NCED and GA catabolism gene GA2 oxidase (GA2ox; Cantoro et al., 2013; Shu et al., 2016). The regulation of seed germination via ABA-GA interplay is again integrated by the chromatin modifier protein PICKLE (PKL). Under mild osmotic stress, PKL limits the expression of ABI3 and ABI5 thus preventing exaggerated germination arrest (Belin and Lopez-Molina, 2008).

5.2.3 Promotion of leaf senescence

Leaf senescence is promoted during prolonged osmotic stress which is regulated by the phytochrome interacting factor 4 (PIF4), PIF5, ABA, and ethylene signaling (Sakuraba et al., 2014). The two light signaling molecules, PIF4 and PIF5, can interact with ABI5 and Enhanced EM level (EEL) proteins to promote the expression of plant-specific transcription factor NAC, Oresara 1 (ORE1), and senescence-associated genes (SAGs) to facilitate early leaf senescence during stress. Ethylene signaling molecule, Ethylene insensitive 3 (EIN3), also regulates this pathway (Sakuraba et al., 2014). In addition, ABA activates β amylase 1 (BAM1) and α amylase 3 (AMY3) enzymes in leaves through SnRK2-mediated phosphorylation for the breakdown of starch into sugars in response to osmotic stress. The activation of BAM1 and AMY3 helps in maltose synthesis from starch. The maltose is released from plastid and further metabolized into sucrose and glucose in cytosol. This sugar is then transported to roots to regulate osmolyte biosynthesis and helps root growth to cope with the adverse effect of osmotic stress (Thalmann et al., 2016). Together, ABA signaling plays a central role in regulating osmotic stress via modulating different physiologicalphenomena in plants; the interaction with other hormones further fine-tunes the osmotic stress responses to maintain plant growth and development.

5.3 Ethylene-mediated responses under osmotic stress

The gaseous hydrocarbon phytohormone, ethylene, was recognized as a plant growth regulator during the early 1930s. Ethylene is known to regulate plant growth, senescence, and fruit ripening depending on the endogenous levels. In addition, it has been widely reported to play a crucial role in biotic and abiotic stress responses.

Osmotic stress triggers ethylene synthesis and accumulation, which then binds to the receptor ETHYLENE RESPONSE1 (ETR1) and inactivates it. This, in turn, blocks the kinase activity of CONSTITUTIVE TRIPLE RESPONSE 1 (CTR1) and switches on the downstream ethylene signaling pathway via EIN2. Salt stress additionally downregulates the ETR1 receptor thus activating the ethylene response (Zhao and Schaller, 2004). Osmotic stress-mediated leaf growth restrain has also been reported to involve ethylene signaling. It has been demonstrated that salt or drought stress induces the accumulation of ethylene and its precursor, 1-aminocyclopropane-1-carboxylate (ACC) (Skirycz et al., 2011; Eun et al., 2019). In rice, the salt stress is sensed by the SALT INTOLERANCE 1 (SIT1) lectin receptor kinase in the root epidermis and induces ethylene synthesis via MAPK3/MAPK6-dependent pathway (Li et al., 2014). Ethylene accumulation triggers the redundant transcription factors ETHYLENE RESPONSE FACTOR 5 (ERF5) and ERF6 via MAPK3/MAPK6 pathway. EFR5/ERF6 performs two simultaneous functions. First, it induces various stress-responsive genes via the activation of WRKY33, MYB51, and STZ transcription factors. Second, it impedes leaf growth under osmotic stress by inhibiting cell proliferation. Early stress signal triggers ethylene-mediated reversible cell cycle arrest by inhibiting the cyclin-dependent kinase A activity in an EIN3-independent pathway (Skirycz et al., 2011). When the stress condition persists for longer duration, cells exit mitosis with an early onset of endoreduplication. This pathway involves the transcriptional induction of the GA catabolism-related gene, *GA2ox6*, and the stabilization of the DELLA repressor via the ERF5/ERF6 thus integrating the ethylene and GA signaling under osmotic stress (Claeys et al., 2012).

EIN3 transcription factor plays a central role in regulating ethylene response under osmotic stress. It has been reported that salt stress stabilizes EIN3 and EIN3-LIKE 1 (EIL1) proteins by promoting proteasomal degradation of the two F-box proteins, EIN3-binding F-box protein 1(EBF1) and EBF2, which degrades EIN3/EIL1 in the absence of ethylene (Peng et al., 2014). This enables EIN3/EIL1 to activate the transcription of downstream stress-responsive genes thus imparting salt tolerance in plants. The transcriptional induction of the ethylene-responsive downstream salinity-related genes proceeds via the EIN3–ESE1 transcriptional complex. Ethylene and salt-inducible ERF 1 (ESE1) is an domain-containing transcription factor that is induced in response to salinity and ACC treatment. In the presence of ethylene, EIN3 is activated, and it physically binds to the promoter region of the *ESE1* gene thus switching on its transcription. ESE1, in turn, binds to and activates the transcription of downstream salinity-related genes like *RD29A* and *COR15A* (Zhang et al., 2011b). Again, ethylene has been demonstrated to impart salinity tolerance by improving Na^+/K^+ homeostasis by regulating Na^+ accumulation in an RBOHF-dependent manner while modulating K^+ levels in an RBOH-independent pathway. The role of ethylene in regulating stomatal closure has also been documented during osmotic stress. Ethylene binding activates ETR1 that then stimulates H_2O_2 generation mediated by RBOHF, and this leads to stomatal closure via EIN2−ARR2 signaling. This ethylene-dependent stomatal closure occurs via ABA-independent pathway (Desikan et al., 2006). Further, ethylene and salt stress have been described to act antagonistically to regulate seed germination via CONSTITUTIVE PHOTOMORPHOGENESIS1 (COP1). In fact, salt stress retained the COP1 protein in the cytoplasm thus inhibiting seed germination. On the other hand, ethylene triggered the translocation of COP1 into the nucleus thus leading to germination (Yu et al., 2016).

Regulation of cortical microtubules in roots is essential for plant adaptation under osmotic stress (Wang et al., 2007). Salt stress induces a right-handed skewed phenotype in *Arabidopsis* roots along with the depolymerization of cortical microtubules. Stabilizing the microtubules using paclitaxel increased seedling death under salt stress, while disrupting them with oryzalin reversed the phenotype. When stress persisted longer, the depolymerized microtubules reorganized themselves indicating that microtubule depolymerization and reorganization are essential for adaptation under salt stress (Wang et al., 2007). The microtubule depolymerization was not induced under salt stress in the ethylene signaling mutant, *ein3*, suggesting a central role of EIN3 in microtubule regulation. Further, under prolonged salt stress, ethylene promotes microtubule reorganization by inducing the microtubule-stabilizing protein WAVE-DAMPENED2-LIKE5 (WDL5) enabling the plants to adapt to the saline condition (Dou et al., 2018). Under osmotic stress, a NAC2 transcription factor is also known to regulate lateral root growth that is mediated by ethylene response (He et al., 2005).

Surprisingly, several studies have suggested a negative role of ethylene in regulating salt stress in plants. For example, the ethylene biosynthesis genes, *ACC synthase 7* (*ACS7*) and *ACC oxidase 1* (*ACO1*), or the ethylene signaling genes, *EIL1* and *EIL2*, have been shown to negatively regulate salt stress in various plants (Dong et al., 2011; Chen et al., 2014; Yang et al., 2015). Again, the ACC deaminase enzyme is responsible for ethylene breakdown and plant growth−promoting bacteria that produce this enzyme has been shown to improve drought tolerance in maize and pepper (Mayak et al., 2004; Danish et al., 2020). These apparent contradictory reports can be explained by the fact that the ethylene-mediated osmotic stress response promotes growth restraints which is essential for stress adaptation.

5.4 Auxin-mediated response under osmotic stress

In 1928, F. W. Went isolated auxin from the tip of oat coleoptiles. It plays a pivotal role in plant development and growth. Besides, osmotic stress also regulates the auxin-responsive pathway at different levels through auxin biosynthesis, transport, or distribution. Indole-3-acetic acid (IAA) is the most common auxin found in plants, synthesized by a tryptophan-dependent or tryptophan-independent pathway (Bartel, 1997). The tryptophan (Trp)-dependent auxin biosynthesis pathway involves the removal of amino group on the tryptophan side chain to form indole-3-pyruvic acid (IPA), which is catalyzed by tryptophan aminotransferase (TAA). Finally, in an NADPH-dependent reaction, IPA is converted into IAA by YUC flavin-containing monooxygenases in the presence of oxygen (Zhao et al., 2001; Zhao, 2012).

5.4.1 Regulation of root growth

During osmotic stress, the YUC genes are highly regulated and play the predominant role in osmotic stress tolerance (Kilian et al., 2007; Kim et al., 2013). Interestingly, the expression of *YUC* genes is found to be shifted from columella region to epidermal and cortex region of roots in response to salinity stress. This phenomenon suggests the alteration of auxin transport and distribution, which is necessary to regulate lateral root development in plants. Further, under stress condition, IAA is synthesized through indole 3 acetaldoxime (IAOX) pathway where two key cytochrome P450 monooxygenase members like CYP79b2 and CYP79b3 help in the biosynthesis of IAOX from tryptophan (Zhao et al., 2002). The loss of function double mutants of these two genes showed lower IAA accumulation and reduced lateral root growth during salt stress (Zhao et al., 2002). On contrary, GRETCHEN HAGEN 3 (GH3) enzyme, encoded by *WES1* gene, forms an inactive amide conjugate of IAA (Park et al., 2007). Under drought stress, the upregulation of WES1 in rice reduces the activity of IAA and also interacts with salicylic acid (SA) and ABA signaling pathways. This enables plants to suppress growth during stress and helps to allocate the metabolites to counteract the stress-induced damage.

During salt stress, the auxin signaling pathway is also critically regulated to modify the plant growth. In auxin signaling pathway, AUX/IAA acts as a major repressor that blocks auxin-responsive factors (ARFs) to activate different auxin-responsive genes. During auxin-mediated responses, AUX/IAA repressor is degraded by the proteasomal degradation method and releases ARFs. During salt stress, NO accumulation decreases the accumulation of PIN protein and hence auxin accumulation. Further, the IAA17, one of the members of AUX/IAA family, is also stabilized by RGL protein that lowers both IAA and bioactive GA accumulation. This reduces plant growth and enables plants to exhibit salt tolerance phenotype (Shi et al., 2017). In addition, the miR390 is reported to regulate salt stress responses in poplar by regulating auxin signaling pathway. The miR390 helps to cleave AUXIN RESPONSE FACTOR (ARF) via trans-activating siRNA (tasiRNA) and enhances lateral root growth to exhibit salt tolerance (He et al., 2018). The stabilization of IAA17 reverses this miR390/tasARF4/ARF4-mediated signaling module and increased salt sensitivity.

5.4.2 Regulation of halotropism

Auxin-dependent root plasticity development such as lateral root growth or root gravitropism is an effective strategy to cope with osmotic stresses. Halotropism or bending of roots away from the zone of high salt concentration is an adaptive feature of plants mediated by PIN-FORMED 2 (PIN2) protein that helps to efflux auxin from one cell to another (Sun et al., 2008; Galvan-Ampudia et al., 2013). Salt stress triggers the degradation of amyloplasts in root columella cells, which leads to changes in ion distribution and SOS protein-mediated gravitropic responses. This can also reduce the accumulation of PIN2 protein and its allocation from root tips to the elongation zone to arrest cell growth. Further, it has been found that under salt stress, the activity of phospholipase D is enhanced in roots, which in turn facilitates the clathrin-mediated endocytosis of PIN2 protein to the side facing salt stress. This also accumulates auxin on the opposite side of the salinity in root tip and modulates gravitropism to adapt plants in salinity stress by salt escape mechanism (Galvan-Ampudia et al., 2013).

5.4.3 Cell wall modification

Plant growth is correlated with cell wall extension, which is essential for organ development. It has widely been reported that auxin positively regulates the cell wall extension by changing the apoplastic pH followed by cell wall loosening. Auxin activates a group of auxin-responsive genes small auxin up−RNA (*SAUR*) genes that inactivate PP2C (Spartz et al., 2014). The inactivation of PP2C generally induces plasma membrane H + ATPase activity causing H + efflux and cell wall acidification. This also facilitates membrane hyperpolarization and subsequent influx of K + into cell, which in turn drives water uptake (Hager, 2003; Hohm et al., 2014). The overexpression of the wheat *SAUR* gene, *SAUR75*, provides drought and salinity tolerance to *Arabidopsis* indicating a positive role of *SAUR* genes in response to osmotic stress (Guo et al., 2018). This increases the turgidity of the cell and expands cell wall. In this acid growth phenomenon, auxin also regulates cell wall−associated proteins like expansin and endotransglucosylase to activate cell wall loosening (Cosgrove, 2016). In contrast, under osmotic stress, two genes, *Arabidopsis* zinc finger protein 1 (*AZF1*) and *Arabidopsis* zinc finger protein 2 (*AZF2*), were found to be highly accumulated and repressed the activity of downstream *SAUR* genes (Kodaira et al., 2011). This may lead to restrict the cell wall acidification and extension resulting growth retardation (Ren and Gray, 2015). However, the growth retardation under osmotic stress is generally driven by an intricate signaling cross talk among auxin−ethylene−GA−ABA module that is an exciting area to explore more in future.

5.5 Gibberellic acid−mediated responses under osmotic stress

Gibberellic acids (GAs) belong to the large class of the terpenoid family. Gibberellin was first isolated from a fungus, *Gibberella fujikuroi*, in infected plants exhibiting high stem elongation. GA is known as growth hormone as it plays a central role in plant growth, such as seed germination and cell proliferation. GA-deficient mutants of rice or inhibition of GA synthesis in tef and finger millet significantly improved drought tolerance along with reduced plant height and lodging (Plaza-Wüthrich et al., 2016). Again, the expression of the *SPINDLY* (*SPY*) gene that encodes a negative regulator of GA signaling was found to be induced in response to osmotic stress in *Arabidopsis*. However, the loss of function mutation of the gene conferred tolerance against osmotic stress (Qin et al., 2011). GAs have been demonstrated to function antagonistically with ABA to regulate plant response under osmotic stress. A higher GA/ABA ratio promotes germination in brown-seeded *Suaeda salsa*, while a lower ratio in black-seeded plants inhibited germination under salt stress (Li et al., 2016). Similarly, salinity stress has been shown to inhibit seed germination in soybean by reducing the level of bioactive GAs like GA1, GA3, and GA4 while increasing ABA accumulation (Shu et al., 2016). Reversal of this stress-induced low GA/ABA ratio by inhibiting ABA biosynthesis could improve seed germination under salt stress indicating that the GA/ABA ratio is crucial for this physiological response. Supporting this observation, salt stress has been shown to lower the GA/ABA ratio in soybean seedlings with reduced biomass, shoot length, leaf area, chlorophyll content, and photosynthesis rate. Increasing the GA level using GA-producing endophytic fungus altered the GA/ABA ratio and increased biomass and leaf area under salt stress (Hamayun et al., 2017).

Intriguingly, the GA biosynthesis pathway genes like *ent-copalyl diphosphate synthase (CPS1)*, *ent-kaurene acid oxidase (KAO2)*, *2-oxoglutarate-dependent dioxygenases (GA20ox1 and GA20ox3)*, and *GA3ox1*were reported to be downregulated in response to dehydration, while *GA2ox1*, a gene involved in GA catabolism, was upregulated. This regulation was less pronounced in the ABA-deficient *nced3*−2 mutant line further supporting a GA-ABA cross talk. In line with this observation, the bioactive GA_4 level was undetectable in the late phase of dehydration stress (Urano et al., 2017). The downregulation of the *GA2ox* gene was further demonstrated to be mediated by the DWARF AND DELAYED FLOWERING 1 (DDF1) transcription factor (Magome et al., 2008). The GRAS family proteins, DELLA, function as important repressors in the GA signaling pathway and negatively regulate cell proliferation and expansion retarding growth under inappropriate conditions (Fleet and Sun, 2005). In presence of GA, DELLA repressors are degraded thus switching on the GA signaling pathway and promoting growth (Harberd, 2003). *Arabidopsis* has five DELLA genes, *GA INSENSITIVE (GAI)*, *REPRESSOR OF ga1*−3 *(RGA)*, *RGL1*, *RGL2*, and *RGL3*. The quadruple DELLA mutant of *Arabidopsis* lacking *GAI*, *RGA*, *RGL1*, and *RGL2* displayed less growth inhibition (leaf production rate, leaf expansion, and biomass) under salt stress than the wild type. These observations indicated that salt stress restrained plant growth in a DELLA-dependent manner, and this inhibition was also associated with the lowering of the bioactive GA levels. However, the induction of salt stress—responsive defense genes occurs in a DELLA-independent pathway. Salt stress has also been reported to delay flowering in *Arabidopsis* by altering both number of days to flowering and the number of leaves in rosette to bolting. This delay in flowering under salt stress conditions was not displayed by the quadruple DELLA mutant suggesting the involvement of DELLA in flowering. This DELLA-mediated delay in flowering was demonstrated to occur via regulation of the LEAFY (LFY) gene in an ABA-dependent fashion. The DELLA-mediated restrain in plant growth and flowering under osmotic stress, which is mediated by a low GA/ABA ratio, is actually advantageous for survival and enables plants to channelize its energy from a "growth phase" to the "defense phase." To top it all, smaller plants with less surface area may be less vulnerable to stress (Achard et al., 2006). In another study, a DELLA-mediated regulation of cell proliferation and differentiation was shown to occur via the anaphase-promoting complex/cyclosome activity in *Arabidopsis* leaves under stress (Claeys et al., 2012). Again, ethylene promotes salt tolerance in a DELLA-dependent manner. Thus, the two independent signaling pathways of ABA and ethylene are integrated by the DELLA proteins under osmotic stress (Achard et al., 2006). Further, during osmotic stress, the higher level of ethylene and DELLA in expanded

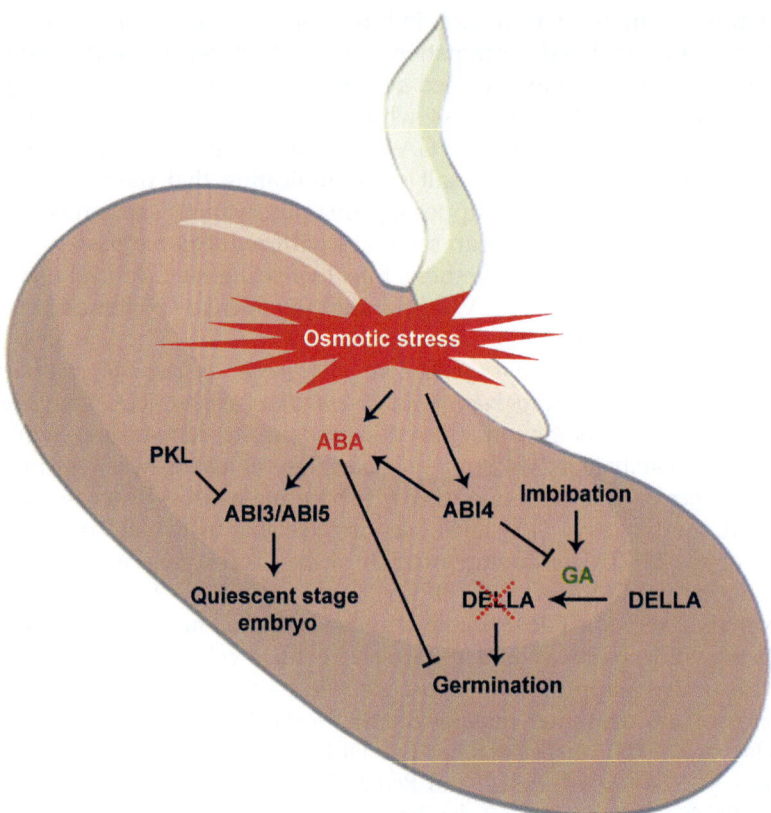

FIGURE 5.2 GA/ABA ratio determines seed germination during osmotic stress. *ABA*, Abscisic acid; *GA*, gibberellic acid.

leaves help to accumulate expansin proteins which facilitate cell wall loosening against low turgor pressure resulting in cell expansion. On the other hand, mature leaves showed increased sensitivity to ABA (Skirycz et al., 2010). Therefore, the salt-induced growth inhibition mediated by a lowering of bioactive GA, stabilization of DELLA proteins, and enhancement of ethylene and ABA represents an adaptive strategy essential for survival under stress (Fig. 5.2).

5.6 Cytokinin-mediated responses under osmotic stress

In the 1950s, Skoog and his team discovered a plant hormone that promotes cytokinesis or cell division and named it cytokinin (Skoog and Miller, 1957). The *adenosine phosphate−isopentenyl transferase* (IPT) gene encodes a key enzyme in the cytokinin biosynthetic pathway. Overexpression of the *IPT8* gene has been reported to impart osmotic stress sensitivity in *Arabidopsis* thus suggesting a negative role of cytokinin in osmotic stress response (Wang et al., 2015a). In plants, the cytokinin signaling comprises a two-component system. It binds to the CHASE domain of the CRE1 receptor and initiates its dimerization. This allows phosphorylation of the histidine kinase domain in each of two receptor molecules, AHK2 and AHK3. On activation, these kinase domains phosphorylate aspartate residues in their own receiver domain and later phosphorylate histidine residues in a separate AHP protein. The phosphate group is then transferred from AHP to either a B-type or A-type response regulator (RR), which activates the transcription of downstream genes (Nishimura et al., 2004; Ferreira and Kieber, 2005). Thus, the regulation of cytokinin signaling is essential for cytokinin-mediated responses.

Under osmotic stress conditions, the CRE1, AHK2, and AHK3 receptors have been demonstrated to function as negative regulators. Loss of function mutants of these genes displayed significantly improved stress tolerance by upregulating different stress and ABA-responsive genes indicating an antagonistic ABA-cytokinin cross talk (Tran et al., 2007). Similarly, mutations in AHP2, AHP3, and AHP5 genes also imparted osmotic stress tolerance via improved cell membrane integrity and increased ABA sensitivity (Nishiyama et al., 2013). Further downstream, the B-type RR proteins also act as negative regulators, and the *arr1arr12* double mutant exhibited enhance tolerance against salt stress. Supporting these observations, exogenous cytokinin treatment resulted in higher Na + accumulation in the shoots of *Arabidopsis*. This higher Na + accumulation was due to cytokinin-mediated repression of a high-affinity K + transporter 1.1 (HKT1.1) that removes excess Na + under salt stress (Mason et al., 2010). Furthermore, a subset of A-type RR proteins, ARR5, ARR15, and ARR22 were found to be induced under drought stress in a cytokinin-independent manner (Kang et al., 2012). Cytokinin has also been reported to be involved in a complex stress signaling network along with other phytohormones. Together with auxin, it inhibits ABA-induced stomatal closure under osmotic stress via an ethylene-dependent pathway (Tanaka et al., 2006).

5.7 Brassinosteroid-mediated response under osmotic stress

Brassinosteroids (BRs) are polyhydroxylated plant sterol hormones that regulate development and plant growth and have a role in adaptation to different stresses (Mitchell et al., 1970). In 1979, M.D. Grove and his team identified the active component brassinolide. BRs have been reported to reverse the harmful effect of osmotic stress on seed germination and seedling growth in various plants (Anuradha and Rao, 2001; Mahesh et al., 2013). It has also been reported to improve photosynthetic efficiency by activating Rubisco activase, chlorophyll accumulation, amylase activity, and stomatal conductance maintaining growth and yield under osmotic stress (Li et al., 2012; Zhao et al., 2017). Further, BR level has been reported to be induced in the young panicles under moderate water conditions that boosts spikelet differentiation and reduces spikelet degeneration. Under severe water deficit conditions, however, a reverse response is observed (Zhang et al., 2019). This BR-induced tolerance to osmotic stress has widely been reported to be modulated by the augmentation of the enzymatic and nonenzymatic antioxidant systems (Vardhini and Anjum, 2015).

Interestingly, a synergistic BR-ABA cross talk has been reported to regulate stomatal closure under osmotic stress. The BRI1-ASSOCIATED RECEPTOR KINASE 1 (BAK1) functions as an important signaling intermediate in the BR pathway. In the presence of BR, BRI1 recruits BAK1 to form a receptor complex in the cell membrane followed by their transphosphorylation. This activates the downstream BR signaling pathway (Nam and Li, 2002; Wang et al., 2008). Loss of function mutation of BAK1 gene perturbed the ABA-mediated stomatal closure suggesting a BR—ABA interplay. BAK1 was found to interact with and phosphorylate OST1 to induce ABA-dependent stomatal closure. This BAK1-OST1 interaction, and thus stomatal closure, was again negatively regulated by BR (Shang et al., 2016). In another study, BR was reported to improve drought tolerance by inducing ABA biosynthesis via NO signaling in maize (Zhang et al., 2011a). BR-induced stomatal closure was also documented in *Arabidopsis* via an activation of hydrogen peroxide and NO activation (Shi et al., 2015). These contrasting observations may be explained by the fact that the role of BR in regulating stomatal closure is dose-dependent. A low BR concentration promotes stomatal opening, while a higher concentration induces stomatal closure (Xia et al., 2014). However, the detailed mechanism needs to be explored in the future.

5.8 Salicylic acid -mediated responses under osmotic stress

SA is a phenolic compound that plays a crucial role in plant defense. The SA biosynthetic pathway gene, *benzoic acid 2-hydroxylase,* has been shown to be induced in response to salinity suggesting a role of SA in stress tolerance (Sawada et al., 2006). Endogenous SA level was also found to be strongly induced in response to osmotic stress (Munné-Bosch and Peñuelas, 2003). Besides, SA has been demonstrated to mitigate osmotic stress by increasing the proline, glycine betaine, and carbohydrate contents, photosynthesis rate, and antioxidant activity while lowering Na^+/K^+ ratio and membrane injury in different plant species (Khan et al., 2010; Marcińska et al., 2013; Khan et al., 2014). However, this beneficial effect of SA was exerted only at a low concentration, while higher SA concentration induced oxidative stress and growth retardation (Lee et al., 2010; Nazar et al., 2011). Supporting these observations, the pretreatment of wheat seedlings with SA could ameliorate the effects of osmotic stress by enhancing antioxidant activities, and this regulation involved nitric oxide (NO)-mediated signaling (Alavi et al., 2014). SA—NO cross talk was also found to be involved in regulating stomatal closure, an important adaptive strategy under osmotic stress. SA-induced stomatal closure via ROS generation and K + channel inactivation. This SA-mediated stomatal closure was perturbed in the absence of NO, while ROS accumulation in the guard cells was unaffected (Khokon et al., 2011). This regulation required SiZ1-mediated endogenous SA accumulation under osmotic stress. SiZ1, a small ubiquitin-like modifier, E3 ubiquitin ligase, negatively regulates SA accumulation and signaling responses in plants (Lee et al., 2007). Therefore, the *siz1* loss of function mutants showed enhanced SA-mediated ROS generation and reduced stomatal aperture to regulate drought stress tolerance in *Arabidopsis* (Miura et al., 2013). Moreover, the WRKY50 and WRKY74 transcription factors have been reported to function as a negative regulators of SA biosynthesis. The wrky50wrky74 double mutant accumulated higher SA level and exhibited enhanced tolerance to osmotic stress with improved water retention and stomatal closure (Li et al., 2013).

The *NahG* gene encodes salicylate hydroxylase enzyme that catalyzes the conversion of SA to catechol (Friedrich et al., 1995). SA-deficient transgenic *Arabidopsis* plants overexpressing *NahG* gene displayed tolerance to moderate salt stress (Borsani et al., 2001). The better stress adaptation of these plants involved a higher glutathione reductase and dehydroascorbate reductase activities, thereby maintaining a higher GSH/GSSG and ASA/DHA ratios under stress condition. However, the plants displayed sensitivity to severe salt stress when all the antioxidant activities and both the ratios diminished significantly (Cao et al., 2009). This indicates that SA potentiates the osmotic stress response by triggering ROS generation that could be somewhat neutralized by the higher antioxidant status in the *NahG* plants. SA-mediated regulation of GSH-ascorbate cycle was also documented in wheat under drought stress (Kang et al., 2013).

Besides, SA also helps to counter the salt-induced membrane depolarization by repressing the GORK channel-mediated $K+$ efflux and lowering the $Na+/K+$ ratio in shoots (Jayakannan et al., 2013). Furthermore, an interplay of SA with ABA and IAA under osmotic stress has also been reported (Shakirova et al., 2003; Szepesi et al., 2009).

5.9 Jasmonic acid–mediated responses under osmotic stress

In 1962, a lipid-derived plant hormone was isolated from jasmine oil and hence the name jasmonic acid (JA). In plants, JA is mainly involved in regulating mechanical stress, wound response, herbivory, and pathogen attack (Xie et al., 1998; Vijayan et al., 1998). In addition, it modulates several physiological phenomena like leaf senescence, seed germination, and pollen development (Stintzi and Browse, 2000; He et al., 2002). The role of JA in plant abiotic stress response has also been documented in different plant species (Wang et al., 2020).

JA has been reported to promote leaf senescence in apricot to check excessive water loss under drought stress (Ge et al., 2010). To counter the negative impact of drought stress, JA-mediated reduction in water loss also occurs through the regulation of stomatal closure. This regulation proceeds through the induction of SLAC1 via the Ca^{2+}-dependent protein kinase 6 (CPK6) in *Arabidopsis* (Munemasa et al., 2011). In fact, intracellular Ca^{2+} has been shown to mediate stomatal closure and is induced by ABA as well as MeJA indicating a probable convergence of both the pathways (Suhita et al., 2003). Conversely, in another study, drought stress has been shown to induce specifically the JA precursor, 12-oxo-phytodienoic acid (12-OPDA), rather than JA itself, that promotes ABA-mediated stomatal closure in *Arabidopsis* (Savchenko et al., 2014). Methyl jasmonate (MeJA) has also been shown to impart salt stress tolerance via the accumulation of the osmoprotectant glycine betaine in watermelon (Xu et al., 2018). Further, the JA-deficient mutant, *defenceless 1* (*def1*), exhibited increased sensitivity to salt stress due to ROS-associated injury and lower accumulation of enzymatic and non-enzymatic antioxidants (Abouelsaad and Renault, 2018). Again, exogenous JA treatment decreased root $Na+$ uptake in maize under salt stress (Shahzad et al., 2015). Pretreatment of barley plants with JA also drastically reduced $Na+$ accumulation in the shoot tissues under salt stress suggesting its role in maintaining ion balance. The salinity-induced inhibition of photosynthesis was also lessened in these JA pretreated plants via the regulation of ribulose 1.5-bisphosphate carboxylase/oxygenase (Rubisco) activase, apoplastic invertase, and arginine decarboxylase genes (Walia et al., 2007). Similarly, JA pretreatment reduced cell membrane injury and promoted leaf ABA accumulation under drought stress (Bandurska et al., 2003). Exogenous MeJA application was also demonstrated to mitigate salt stress in soybean seedlings along with an increase in the endogenous ABA content while the levels of bioactive GA4 decreased (Yoon et al., 2009).

In the JA signaling pathway, JASMONATE ZIM-domain (JAZ) proteins function as important negative regulators. In the presence of jasmonyl isoleucine or coronatine, JAZ interacts with the E3 ubiquitine ligase SCF^{COI1} complex and are degraded via 26S proteasome (Chini et al., 2007; Thines et al., 2007). The inhibitory role of JAZ proteins in drought stress tolerance has been documented. JAZ1 was found to interact with and inhibit the bHLH148 transcription factor. In the presence of JA or drought stress, JAZ1 was degraded by the SCF^{COI1} complex that released bHLH148. The activated bHLH148 then induced the transcription of its target genes like *DREB1* that led to drought stress tolerance in rice (Seo et al., 2011). Subsequently, the loss of function mutation in *JAZ1* gene was found to impart drought tolerance in rice along with hypersensitivity to MeJA and ABA treatments thus integrating the two phytohormone signaling pathways (Fu et al., 2017).

5.10 Strigolactone-mediated responses under osmotic stress

Strigolactones (SL) comprise a new class of plant hormones that are gaining importance for their crucial role in plant stress responses. SLs are carotenoid-derived lactones first discovered in cotton root exudates that stimulated seed germination in the parasitic weed, *Striga lutea* (Cook et al., 1966). Subsequently, their involvement in osmotic stress response has been reported in various plant species.

SLs have been demonstrated to play a positive role in drought and salinity tolerance in *Arabidopsis* primarily through shoot-related traits like stomatal closure and leaf senescence (Ha et al., 2014). Exogenous application of the synthetic SL (GR24) was able to improve photosynthetic efficiency along with other physiological traits in salt-stressed rice seedlings (Ling et al., 2020). In *Arabidopsis*, the loss of function mutation of the *more axillary growth* (*MAX*) genes that are involved in SL biosynthesis rendered the mutants hypersensitive toward drought and salt stress. The mutants further displayed slower ABA-induced stomatal closure indicating a possible SL-ABA cross talk in regulating osmotic stress in plants (Bu et al., 2014; Ha et al., 2014). Further studies with the SL-depleted mutants in tomato and lotus also supported the SL-ABA cross talk during drought stress (Liu et al., 2015; Visentin et al., 2016). The SL receptor mutant *atd14−5* also displayed hypersensitivity to drought stress similar to the *max2* mutants (Zhang et al., 2018). It was reported that osmotic stress induced a rapid decrease in SL levels via transcriptional regulation that was essential for the physiological increase in ABA concentration in lotus roots (Liu et al., 2015). Conversely, ABA has also been reported to be involved in regulating SL biosynthesis (López-Ráez et al., 2010). Moreover, SL was demonstrated to activate nitric oxide and hydrogen peroxide synthesis to regulate SLAC1-mediated stomatal closure via an ABA-independent pathway (Lv et al., 2018). However, this intricate SL-mediated regulation of stomatal closure needs further investigation in the future.

5.11 Conclusion

The intricate cross talk among the different phytohormone signaling pathways serves to fine-tune the plants response to stress. This knowledge has been exploited for crop improvement programs with foliar applications of various phytohormones in the initial years. Transgenic approach has also been undertaken to modulate the phytohormone signaling pathways that generated different stress-tolerant crops. With the advancement of molecular research, the CRISPR-Cas9-mediated genome-editing technology is also being applied for the generation of climate-resilient crops. However, an in-depth analysis of plant responses and signaling under various combination stress conditions will yield novel information in the future.

References

Abouelsaad, I., Renault, S., 2018. Enhanced oxidative stress in the jasmonic acid deficient tomato mutant *def-1* exposed to NaCl stress. J. Plant Physiol. 226, 136−144.

Achard, P., Cheng, H., De Grauwe, L., et al., 2006. Integration of plant responses to environmentally activated phytohormonal signals. Science 311 (5757), 91−94.

Alavi, S.M.N., Arvin, M.J., Kalantari, K.M., 2014. Salicylic acid and nitric oxide alleviate osmotic stress in wheat (*Triticum aestivum* L.) seedlings. J. Plant Interact. 9 (1), 683−688.

Anuradha, S., Rao, S., 2001. Effect of brassinosteroids on salinity stress induced inhibition of seed germination and seedling growth of rice (*Oryza sativa* L.). Plant Growth Regul. 33, 151−153.

Avila, E.L., Brown, M., Pan, S., et al., 2008. Expression analysis of Arabidopsis vacuolar sorting receptor 3 reveals a putative function in guard cells. J. Exp. Bot. 59 (6), 1149−1161.

Bandurska, H., Stroiński, A., Kubiś, J., 2003. The effect of jasmonic acid on the accumulation of ABA, proline and spermidine and its influence on membrane injury under water deficit in two barley genotypes. Acta Physiol. Plant 25, 279−285.

Bartel, B., 1997. Auxin biosynthesis. Annu. Rev. Plant Physiol. Plant Mol. Biol. 48, 51−66.

Belin, C., Lopez-Molina, L., 2008. Arabidopsis seed germination responses to osmotic stress involve the chromatin modifier PICKLE. Plant Signal Behav. 3 (7), 478−479.

Borsani, O., Valpuesta, V., Botella, M.A., 2001. Evidence for a role of salicylic acid in the oxidative damage generated by NaCl and osmotic stress in Arabidopsis seedlings. Plant Physiol. 126 (3), 1024−1030.

Bu, Q., Lv, T., Shen, H., et al., 2014. Regulation of drought tolerance by the F-box protein MAX2 in Arabidopsis. Plant Physiol. 164 (1), 424−439.

Cantoro, R., Crocco, C.D., Benech-Arnold, R.L., et al., 2013. In vitro binding of *Sorghum bicolor* transcription factors ABI4 and ABI5 to a conserved region of a GA 2-OXIDASE promoter: possible role of this interaction in the expression of seed dormancy. J. Exp. Bot. 64 (18), 5721−5735.

Cao, Y., Zhang, Z.W., Xue, L.W., et al., 2009. Lack of salicylic acid in Arabidopsis protects plants against moderate salt stress. J. Biosci. 64 (3−4), 231−238.

Carns, H.R., 1954. Present status of the abscission accelerator from young cotton bolls. In: 13th Cotton Defoliation Conference, National Cotton Council, Memphis, pp. 39.

Chen, D., Ma, X., Li, C., et al., 2014. A wheat aminocyclopropane-1-carboxylate oxidase gene, TaACO1, negatively regulates salinity stress in *Arabidopsis thaliana*. Plant Cell Rep. 33 (11), 1815−1827.

Chen, K., Li, G.-J., Bressan, R.A., et al., 2020. Abscisic acid dynamics, signaling, and function in plants. J. Integr. Plant Biol. 62 (1), 25−54.

Chini, A., Fonseca, S., Fernández, G., et al., 2007. The JAZ family of repressors is the missing link in jasmonate signalling. Nature 448 (7154), 666−671.

Claeys, H., Skirycz, A., Maleux, K., et al., 2012. DELLA signaling mediates stress-induced cell differentiation in Arabidopsis leaves through modulation of anaphase-promoting complex/cyclosome activity. Plant Physiol. 159 (2), 739−747.

Cook, C.E., Whichard, L.P., Turner, B., et al., 1966. Germination of witchweed (*Striga lutea* Lour.): isolation and properties of a potent stimulant. Science 154, 1189−1190.

Cosgrove, D.J., 2016. Catalysts of plant cell wall loosening. F1000Research 5:F1000 Faculty Rev-119.

Danish, S., Zafar-Ul-Hye, M., Mohsin, F., et al., 2020. ACC-deaminase producing plant growth promoting rhizobacteria and biochar mitigate adverse effects of drought stress on maize growth. PLoS One 15 (4), e0230615.

Datta, R., Sahid, S., Paul, S., 2020. Networking by small molecule hormones during drought stress in plants. In: Khan, M.I.R., Singh, A., Poór, P. (Eds.), Improving Abiotic Stress Tolerance in Plants. CRC Press, Boca Raton, FL, pp. 203–228.

Desikan, R., Last, K., Harrett-Williams, R., et al., 2006. Ethylene-induced stomatal closure in Arabidopsis occurs via AtrbohF-mediated hydrogen peroxide synthesis. Plant J. 47 (6), 907–916.

Dong, H., Zhen, Z., Peng, J., et al., 2011. Loss of ACS7 confers abiotic stress tolerance by modulating ABA sensitivity and accumulation in Arabidopsis. J. Exp. Bot. 62 (14), 4875–4887.

Dou, L., He, K., Higaki, T., et al., 2018. Ethylene signaling modulates cortical microtubule reassembly in response to salt stress. Plant Physiol. 176 (3), 2071–2081.

Eun, H.D., Ali, S., Jung, H., et al., 2019. Profiling of ACC synthase gene (ACS11) expression in Arabidopsis induced by abiotic stresses. Appl. Biol. Chem. 62, 42.

Ferreira, F.J., Kieber, J.J., 2005. Cytokinin signaling. Curr. Opin. Plant Biol. 8, 518–525.

Fleet, C.M., Sun, T.P., 2005. A DELLAcate balance: the role of gibberellin in plant morphogenesis. Curr. Opin. Plant Biol. 8 (1), 77–85.

Friedrich, L., Vernooij, B., Gaffney, T., et al., 1995. Characterization of tobacco plants expressing a bacterial salicylate hydroxylase gene. Plant Mol. Biol. 29 (5), 959–968.

Fu, J., Wu, H., Ma, S., et al., 2017. OsJAZ1 attenuates drought resistance by regulating JA and ABA signaling in rice. Front. Plant Sci. 8, 2108.

Fujii, H., Chinnusamy, V., Rodrigues, A., et al., 2009. In vitro reconstitution of an abscisic acid signalling pathway. Nature 462, 660–664.

Fujita, Y., Yoshida, T., Yamaguchi-Shinozaki, K., 2013. Pivotal role of the AREB/ABF-SnRK2 pathway in ABRE-mediated transcription in response to osmotic stress in plants. Physiol. Plant. 147, 15–27.

Galvan-Ampudia, C.S., Julkowska, M.M., Darwish, E., et al., 2013. Halotropism is a response of plant roots to avoid a saline environment. Curr. Biol. 23 (20), 2044–2050.

Ge, Y.X., Zhang, L.J., Li, F.H., et al., 2010. Relationship between jasmonic acid accumulation and senescence in drought-stress. Afr. J. Agric. Res. 5 (15), 1978–1983.

Geiger, D., Scherzer, S., Mumm, P., et al., 2009. Activity of guard cell anion channel SLAC1 is controlled by drought-stress signaling kinase-phosphatase pair. Proc. Natl. Acad. Sci. U.S.A. 106 (50), 21425–21430.

Guo, Y., Xiong, L., Song, C.P., et al., 2002. A calcium sensor and its interacting protein kinase are global regulators of abscisic acid signaling in Arabidopsis. Dev. Cell 3 (2), 233–244.

Guo, Y., Jiang, Q., Hu, Z., et al., 2018. Function of the auxin-responsive gene TaSAUR75 under salt and drought stress. Crop J. 6 (2), 181–190.

Ha, C.V., Leyva-González, M.A., Osakabe, Y., et al., 2014. Positive regulatory role of strigolactone in plant responses to drought and salt stress. Proc. Natl. Acad. Sci. U.S.A. 111 (2), 851–856.

Hager, A., 2003. Role of the plasma membrane H + -ATPase in auxin-induced elongation growth: historical and new aspects. J. Plant Res. 116 (6), 483–505.

Hamayun, M., Hussain, A., Khan, S.A., et al., 2017. Gibberellins producing endophytic fungus Porostereum spadiceum AGH786 rescues growth of salt affected soybean. Front. Microbiol. 8, 686.

Harberd, N.P., 2003. Botany, relieving DELLA restraint. Science 299 (5614), 1853–1854.

He, Y., Fukushige, H., Hildebrand, D.F., et al., 2002. Evidence supporting a role of jasmonic acid in Arabidopsis leaf senescence. Plant Physiol. 128 (3), 876–884.

He, X.J., Mu, R.L., Cao, W.H., et al., 2005. AtNAC2, a transcription factor downstream of ethylene and auxin signaling pathways, is involved in salt stress response and lateral root development. Plant J. 44 (6), 903–916.

He, F., Xu, C., Fu, X., et al., 2018. The MicroRNA390/TRANS-ACTING SHORT INTERFERING RNA3 module mediates lateral root growth under salt stress via the auxin pathway. Plant Physiol. 177 (2), 775–791.

Hohm, T., Demarsy, E., Quan, C., et al., 2014. Plasma membrane H$^+$-ATPase regulation is required for auxin gradient formation preceding phototropic growth. Mol. Syst. Biol. 10 (9), 751.

Jayakannan, M., Bose, J., Babourina, O., et al., 2013. Salicylic acid improves salinity tolerance in Arabidopsis by restoring membrane potential and preventing salt-induced K + loss via a GORK channel. J. Exp. Bot. 64 (8), 2255–2268.

Kang, N., Cho, C., Kim, N.Y., et al., 2012. Cytokinin receptor-dependent and receptor-independent pathways in the dehydration response of Arabidopsis thaliana. J. Plant Physiol. 169 (14), 1382–1391.

Kang, G., Li, G., Liu, G.Q., et al., 2013. Exogenous salicylic acid enhances wheat drought tolerance by influence on the expression of genes related to ascorbate-glutathione cycle. Biol. Plant. 57, 718–724.

Khan, N., Syeed, S., Masood, A., et al., 2010. Application of salicylic acid increases contents of nutrients and anti-oxidative metabolism in mungbean and alleviates adverse effects of salinity stress. Int. J. Plant. Biol. 1 (1), e1.

Khan, M.I., Asgher, M., Khan, N.A., 2014. Alleviation of salt-induced photosynthesis and growth inhibition by salicylic acid involves glycine betaine and ethylene in mungbean (*Vigna radiata* L.). Plant Physiol. Biochem. 80, 67–74.

Khokon, A.R., Okuma, E., Hossain, M.A., et al., 2011. Involvement of extracellular oxidative burst in salicylic acid-induced stomatal closure in Arabidopsis. Plant Cell Environ. 34 (3), 434–443.

Kilian, J., Whitehead, D., Horak, J., et al., 2007. The AtGenExpress global stress expression data set: protocols, evaluation and model data analysis of UV-B light, drought and cold stress responses. Plant J. 50 (2), 347–363.

Kim, J.I., Baek, D., Park, H.C., et al., 2013. Overexpression of Arabidopsis YUCCA6 in potato results in high-auxin developmental phenotypes and enhanced resistance to water deficit. Mol. Plant. 6 (2), 337–349.

Kodaira, K.S., Qin, F., Tran, L.S., et al., 2011. Arabidopsis Cys2/His2 zinc-finger proteins AZF1 and AZF2 negatively regulate abscisic acid-repressive and auxin-inducible genes under abiotic stress conditions. Plant Physiol. 157 (2), 742–756.

Lambers, H., Scheurwater, I., Mata, C., et al., 1998. Root respiration of fast slow grow plants, as dependent on genotype and N2 supply. A major clue to the function of slow growing plants. In: HLambers, H.Poorter, Van Vuuren, M.M.I. (Eds.), Inherent Variation in Plant Growth. Physiological Mechanism and Ecological Consequences. Leiden in press, Backhuys.

Lee, J., Nam, J., Park, H.C., et al., 2007. Salicylic acid-mediated innate immunity in Arabidopsis is regulated by SIZ1 SUMO E3 ligase. Plant J. 49 (1), 79–90.

Lee, S., Kim, S.G., Park, C.M., 2010. Salicylic acid promotes seed germination under high salinity by modulating antioxidant activity in Arabidopsis. New Phytol. 188 (2), 626–637.

Li, Y.H., Liu, Y.J., Xu, X.L., et al., 2012. Effect of 24-epibrassinolide on drought stress-induced changes in *Chorispora bungeana*. Biol. Plant. 56, 192–196.

Li, J., Besseau, S., Törönen, P., et al., 2013. Defense-related transcription factors WRKY70 and WRKY54 modulate osmotic stress tolerance by regulating stomatal aperture in Arabidopsis. New Phytol. 200 (2), 457–472.

Li, C.H., Wang, G., Zhao, J.L., et al., 2014. The receptor-like kinase SIT1 mediates salt sensitivity by activating MAPK3/6 and regulating ethylene homeostasis in rice. Plant Cell 26 (6), 2538–2553.

Li, W., Yamaguchi, S., Khan, M.A., et al., 2016. Roles of gibberellins and abscisic acid in regulating germination of *Suaeda salsa* dimorphic seeds under salt stress. Front. Plant Sci. 6, 1235.

Ling, F., Su, Q., Jiang, H., et al., 2020. Effects of strigolactone on photosynthetic and physiological characteristics in salt-stressed rice seedlings. Sci. Rep. 10 (1), 6183.

Liu, J., He, H., Vitali, M., et al., 2015. Osmotic stress represses strigolactone biosynthesis in *Lotus japonicus* roots: exploring the interaction between strigolactones and ABA under abiotic stress. Planta 241 (6), 1435–1451.

Lopez-Molina, L., Mongrand, S., Chua, N.H., 2001. A postgermination developmental arrest checkpoint is mediated by abscisic acid and requires the ABI5 transcription factor in Arabidopsis. Proc. Natl. Acad. Sci. U.S.A. 98 (8), 4782–4787.

López-Ráez, J.A., Kohlen, W., Charnikhova, T., et al., 2010. Does abscisic acid affect strigolactone biosynthesis? New Phytol. 187 (2), 343–354.

Lv, S., Zhang, Y., Li, C., et al., 2018. Strigolactone-triggered stomatal closure requires hydrogen peroxide synthesis and nitric oxide production in an abscisic acid-independent manner. New Phytol. 217 (1), 290–304.

Magome, H., Yamaguchi, S., Hanada, A., et al., 2008. The DDF1 transcriptional activator upregulates expression of a gibberellin-deactivating gene, GA2ox7, under high-salinity stress in Arabidopsis. Plant J. 56 (4), 613–626.

Mahesh, K., Balaraju, P., Ramakrishna, B., et al., 2013. Effect of brassinosteroids on germination and seedling growth of radish (*Raphanus sativus* L.) under PEG-6000 induced water stress. Am. J. Plant. Sci. 4, 2305–2313.

Marcińska, I., Czyczyło-Mysza, I., Skrzypek, E., et al., 2013. Alleviation of osmotic stress effects by exogenous application of salicylic or abscisic acid on wheat seedlings. Int. J. Mol. Sci. 14 (7), 13171–13193.

Mason, M., Jha, D., Salt, D., et al., 2010. Type-B response regulators ARR1 and ARR12 regulate expression of AtHKT1;1 and accumulation of sodium in Arabidopsis shoots. Plant J. 64 (5), 753–763.

Mayak, S., Tirosh, T., Glick, B., 2004. Plant growth-promoting bacteria that confer resistance to water stress in tomatoes and peppers. Plant Sci. 166, 525–530.

Merlot, S., Gosti, F., Guerrier, D., et al., 2001. The ABI1 and ABI2 protein phosphatases 2C act in a negative feedback regulatory loop of the abscisic acid signalling pathway. Plant J. 25, 295–303.

Mitchell, J.W., Mandava, N., Worley, J.F., et al., 1970. Brassins-a new family of plant hormones from rape pollen. Nature 225 (5237), 1065–1066.

Miura, K., Okamoto, H., Okuma, E., et al., 2013. SIZ1 deficiency causes reduced stomatal aperture and enhanced drought tolerance via controlling salicylic acid-induced accumulation of reactive oxygen species in Arabidopsis. Plant J. 73 (1), 91–104.

Munemasa, S., Hossain, M.A., Nakamura, Y., et al., 2011. The Arabidopsis calcium-dependent protein kinase, CPK6, functions as a positive regulator of methyl jasmonate signaling in guard cells. Plant Physiol. 155 (1), 553–561.

Munné-Bosch, S., Peñuelas, J., 2003. Photo- and antioxidative protection, and a role for salicylic acid during drought and recovery in field-grown Phillyrea angustifolia plants. Planta 217 (5), 758–766.

Nam, K.H., Li, J., 2002. BRI1/BAK1, a receptor kinase pair mediating brassinosteroid signaling. Cell 110 (2), 203–212.

Nazar, R., Iqbal, N., Syeed, S., et al., 2011. Salicylic acid alleviates decreases in photosynthesis under salt stress by enhancing nitrogen and sulfur assimilation and antioxidant metabolism differentially in two mungbean cultivars. J. Plant Physiol. 168 (8), 807–815.

Nishimura, C., Ohashi, Y., Sato, S., et al., 2004. Histidine kinase homologs that act as cytokinin receptors possess overlapping functions in the regulation of shoot and root growth in Arabidopsis. Plant Cell 16 (6), 1365–1377.

Nishiyama, R., Watanabe, Y., Leyva-González, M.A., et al., 2013. Arabidopsis AHP2, AHP3, and AHP5 histidine phosphotransfer proteins function as redundant negative regulators of drought stress response. Proc. Natl. Acad. Sci. U.S.A. 110, 4840–4845.

Ohta, M., Guo, Y., Halfter, U., et al., 2003. A novel domain in the protein kinase SOS2 mediates interaction with the protein phosphatase 2C ABI2. Proc. Natl. Acad. Sci. U.S.A. 100 (20), 11771–11776.

Park, J.E., Park, J.Y., Kim, Y.S., et al., 2007. GH3-mediated auxin homeostasis links growth regulation with stress adaptation response in Arabidopsis. J. Biol. Chem. 282 (13), 10036–10046.

Peng, J., Li, Z., Wen, X., et al., 2014. Salt-induced stabilization of EIN3/EIL1 confers salinity tolerance by deterring ROS accumulation in Arabidopsis. PLoS Genet. 10 (10), e1004664.

Plaza-Wüthrich, S., Blösch, R., Rindisbacher, A., et al., 2016. Gibberellin deficiency confers both lodging and drought tolerance in small cereals. Front. Plant. Sci. 7, 643.

Qin, X., Zeevaart, J.A.D., 1999. The 9-cis-epoxycarotenoid cleavage reaction is the key regulatory step of abscisic acid biosynthesis in water-stressed bean. Proc. Natl. Acad. Sci. U.S.A. 96 (26), 15354–15361.

Qin, F., Kodaira, K.S., Maruyama, K., et al., 2011. SPINDLY, a negative regulator of gibberellic acid signaling, is involved in the plant abiotic stress response. Plant Physiol. 157, 1900–1913.

Qiu, Q.S., Guo, Y., Dietrich, M.A., et al., 2002. Regulation of SOS1, a plasma membrane Na + /H + exchanger in Arabidopsis thaliana, by SOS2 and SOS3. Proc. Natl. Acad. Sci. U.S.A. 99 (12), 8436–8441.

Ren, H., Gray, W.M., 2015. Saur proteins as effectors of hormonal and environmental signals in plant growth. Mol. Plant 8 (8), 1153–1164.

Sakuraba, Y., Jeong, J., Kang, M.Y., et al., 2014. Phytochrome-interacting transcription factors PIF4 and PIF5 induce leaf senescence in Arabidopsis. Nat. Commun. 5, 4636.

Savchenko, T., Kolla, V.A., Wang, C.Q., et al., 2014. Functional convergence of oxylipin and abscisic acid pathways controls stomatal closure in response to drought. Plant Physiol. 164 (3), 1151–1160.

Sawada, H., Shim, I., Usui, K., 2006. Induction of benzoic acid 2-hydroxylase and salicylic acid biosynthesis-Modulation by salt stress in rice seedlings. Plant Sci. 171, 263–270.

Seo, J.S., Joo, J., Kim, M.J., et al., 2011. OsbHLH148, a basic helix-loop-helix protein, interacts with OsJAZ proteins in a jasmonate signaling pathway leading to drought tolerance in rice. Plant J. 65 (6), 907–921.

Shahzad, A.N., Pitann, B., Ali, H., et al., 2015. Maize genotypes differing in salt resistance vary in jasmonic acid accumulation during the first phase of salt stress. J. Agron. Crop Sci. 201, 443–451.

Shakirova, F., Sakhabutdinova, A., Bezrukova, M., et al., 2003. Changes in the hormonal status of wheat seedlings induced by salicylic acid and salinity. Plant Sci. 164 (3), 317–322.

Shang, Y., Dai, C., Lee, M.M., et al., 2016. BRI1-associated receptor kinase 1 regulates guard cell ABA signaling mediated by Open Stomata 1 in Arabidopsis. Mol. Plant. 9 (3), 447–460.

Sharma, A., Shahzad, B., Kumar, V., et al., 2019. Phytohormones regulate accumulation of osmolytes under abiotic stress. Biomolecules 9 (7), 285.

Shi, C., Qi, C., Ren, H., et al., 2015. Ethylene mediates brassinosteroid-induced stomatal closure via Gα protein-activated hydrogen peroxide and nitric oxide production in Arabidopsis. Plant J. 82 (2), 280–301.

Shi, H., Liu, W., Wei, Y., et al., 2017. Integration of auxin/indole-3-acetic acid 17 and RGA-LIKE3 confers salt stress resistance through stabilization by nitric oxide in Arabidopsis. J. Exp. Bot. 68 (5), 1239–1249.

Shu, K., Liu, X.D., Xie, Q., et al., 2016. Two faces of one seed: hormonal regulation of dormancy and germination. Mol. Plant. 9 (1), 34–45.

Singh, A., Pandey, A., Sribastava, A.K., et al., 2016. Plant protein phosphatase 2C: from genomic diversity to functional multiplicity and importance in stress management. Crit. Rev. Biotechnol. 36 (6), 1023–1035.

Skirycz, A., De Bodt, S., Obata, T., et al., 2010. Developmental stage specificity and the role of mitochondrial metabolism in the response of Arabidopsis leaves to prolonged mild osmotic stress. Plant Physiol. 152 (1), 226–244.

Skirycz, A., Claeys, H., De Bodt, S., et al., 2011. Pause-and-stop: the effects of osmotic stress on cell proliferation during early leaf development in Arabidopsis and a role for ethylene signaling in cell cycle arrest. Plant Cell 23 (5), 1876–1888.

Skoog, F., Miller, C.O., 1957. Chemical regulation of growth and organ formation in plant tissues cultured in vitro. Symp. Soc. Exp. Biol. 11, 118–131.

Spartz, A.K., Ren, H., Park, M.Y., et al., 2014. SAUR inhibition of PP2C-D phosphatases activates plasma membrane H + -ATPases to promote cell expansion in arabidopsis. Plant Cell 26 (5), 2129–2142.

Stintzi, A., Browse, J., 2000. The Arabidopsis male-sterile mutant, opr3, lacks the 12-oxophytodienoic acid reductase required for jasmonate synthesis. Proc. Natl. Acad. Sci. U.S.A. 97 (19), 10625–10630.

Suhita, D., Kolla, V., Vavasseur, A., et al., 2003. Different signaling pathways involved during the suppression of stomatal opening by methyl jasmonate or abscisic acid. Plant Sci. 164, 481–488.

Sun, F., Zhang, W., Hu, H., et al., 2008. Salt modulates gravity signaling pathway to regulate growth direction of primary roots in Arabidopsis. Plant Physiol. 146 (1), 178–188.

Szepesi, A., Csiszár, J., Gémes, K., et al., 2009. Salicylic acid improves acclimation to salt stress by stimulating abscisic aldehyde oxidase activity and abscisic acid accumulation, and increases Na + content in leaves without toxicity symptoms in Solanum lycopersicum L. J. Plant Physiol. 166 (9), 914–925.

Tanaka, Y., Sano, T., Tamaoki, M., et al., 2006. Cytokinin and auxin inhibit abscisic acid-induced stomatal closure by enhancing ethylene production in Arabidopsis. J. Exp. Bot. 57 (10), 2259–2266.

Taylor, I.B., Burbidge, A., Thompson, A.J., 2000. Control of abscisic acid synthesis. J. Exp. Bot. 51 (350), 1563–1574.

Thalmann, M., Pazmino, D., Seung, D., et al., 2016. Regulation of leaf starch degradation by abscisic acid is important for osmotic stress tolerance in plants. Plant Cell 28 (8), 1860–1878.

Thines, B., Katsir, L., Melotto, M., et al., 2007. JAZ repressor proteins are targets of the SCF(COI1) complex during jasmonate signalling. Nature 448 (7154), 661–665.

Tran, L., Urao, T., Qin, F., et al., 2007. Functional analysis of AHK1/ATHK1 and cytokinin receptor histidine kinases in response to abscisic acid, drought, and salt stress in Arabidopsis. Proc. Natl. Acad. Sci. U.S.A. 104, 20623–20628.

Tyler, L., Thomas, S.G., Hu, J., et al., 2004. Della proteins and gibberellin- regulated seed germination and floral development in Arabidopsis. Plant Physiol. 135 (2), 1008–1019.

Urano, K., Maruyama, K., Jikumaru, Y., et al., 2017. Analysis of plant hormone profiles in response to moderate dehydration stress. Plant J. 90 (1), 17–36.

Vardhini, B.V., Anjum, N.A., 2015. Brassinosteroids make plant life easier under abiotic stresses mainly by modulating major components of antioxidant defense system. Front. Environ. Sci. 2, 67.

Vijayan, P., Shockey, J., Lévesque, C.A., et al., 1998. A role for jasmonate in pathogen defense of Arabidopsis. Proc. Natl. Acad. Sci. U.S.A. 95 (12), 7209–7214.

Visentin, I., Vitali, M., Ferrero, M., et al., 2016. Low levels of strigolactones in roots as a component of the systemic signal of drought stress in tomato. New Phytol. 212 (4), 954–963.

Walia, H., Wilson, C., Condamine, P., et al., 2007. Large-scale expression profiling and physiological characterization of jasmonic acid-mediated adaptation of barley to salinity stress. Plant Cell Environ. 30 (4), 410–421.

Wang, C., Li, J., Yuan, M., 2007. Salt tolerance requires cortical microtubule reorganization in Arabidopsis. Plant Cell Physiol. 48 (11), 1534–1547.

Wang, X., Kota, U., He, K., et al., 2008. Sequential transphosphorylation of the BRI1/BAK1 receptor kinase complex impacts early events in brassinosteroid signaling. Dev. Cell 15 (2), 220–235.

Wang, Y., Shen, W., Chan, Z., et al., 2015a. Endogenous cytokinin overproduction modulates ROS homeostasis and decreases salt stress resistance in Arabidopsis thaliana. Front. Plant Sci. 6, 1004.

Wang, Z.Y., Gehring, C., Zhu, J., et al., 2015b. The Arabidopsis vacuolar sorting Receptor1 is required for osmotic stress-induced abscisic acid biosynthesis. Plant Physiol. 167 (1), 137–152.

Wang, J., Song, L., Gong, X., et al., 2020. Functions of jasmonic acid in plant regulation and response to abiotic stress. Int. J. Mol. Sci. 21 (4), 1446.

Watkins, J.M., Chapman, J.M., Muday, G.K., 2017. Abscisic acid-induced reactive oxygen species are modulated by flavonols to control stomata aperture. Plant Physiol. 175 (4), 1807–1825.

Xia, X.J., Gao, C.J., Song, L.X., et al., 2014. Role of H_2O_2 dynamics in brassinosteroid-induced stomatal closure and opening in *Solanum lycopersicum*. Plant Cell Environ. 37 (9), 2036–2050.

Xie, D.X., Feys, B.F., James, S., et al., 1998. COI1: an Arabidopsis gene required for jasmonate-regulated defense and fertility. Science 280 (5366), 1091–1094.

Xiong, L., Zhu, J.K., 2003. Regulation of abscisic acid biosynthesis. Plant Physiol. 133 (1), 29–36.

Xu, Z., Sun, M., Jiang, X., et al., 2018. Glycine betaine biosynthesis in response to osmotic stress depends on jasmonate signaling in watermelon suspension cells. Front. Plant Sci. 9, 1469.

Yang, C., Ma, B., He, S.J., et al., 2015. MAOHUZI6/ETHYLENE INSENSITIVE3-LIKE1 and ETHYLENE INSENSITIVE3-LIKE2 regulate ethylene response of roots and coleoptiles and negatively affect salt tolerance in rice. Plant Physiol. 169 (1), 148–165.

Yoon, J.Y., Hamayun, M., Lee, S.K., et al., 2009. Methyl jasmonate alleviated salinity stress in soybean. J. Crop Sci. Biotechnol. 12, 63–68.

Yoshida, T., Fujita, Y., Sayama, H., et al., 2010. AREB1, AREB2, and ABF3 are master transcription factors that cooperatively regulate ABRE-dependent ABA signaling involved in drought stress tolerance and require ABA for full activation. Plant J. 61, 672–685.

Yu, Y., Wang, J., Shi, H., et al., 2016. Salt stress and ethylene antagonistically regulate nucleocytoplasmic partitioning of COP1 to control seed germination. Plant Physiol. 170 (4), 2340–2350.

Zhang, A., Zhang, J., Zhang, J., et al., 2011a. Nitric oxide mediates brassinosteroid-induced ABA biosynthesis involved in oxidative stress tolerance in maize leaves. Plant Cell Physiol. 52 (1), 181–192.

Zhang, L., Li, Z., Quan, R., et al., 2011b. An AP2 domain-containing gene, ESE1, targeted by the ethylene signaling component EIN3 is important for the salt response in Arabidopsis. Plant Physiol. 157 (2), 854–865.

Zhang, Y., Lv, S., Wang, G., 2018. Strigolactones are common regulators in induction of stomatal closure in planta. Plant Signal. Behav. 13 (3), e1444322.

Zhang, W., Sheng, J., Xu, Y., et al., 2019. Role of brassinosteroids in rice spikelet differentiation and degeneration under soil-drying during panicle development. BMC Plant Biol. 19 (1), 409.

Zhao, Y., 2012. Auxin biosynthesis: a simple two-step pathway converts tryptophan to indole-3-acetic acid in plants. Mol. Plant. 5 (2), 334–338.

Zhao, X.C., Schaller, G.E., 2004. Effect of salt and osmotic stress upon expression of the ethylene receptor ETR1 in *Arabidopsis thaliana*. FEBS Lett. 562 (1–3), 189–192.

Zhao, Y., Christensen, S.K., Fankhauser, C., et al., 2001. A role for flavin monooxygenase-like enzymes in auxin biosynthesis. Science 291 (5502), 306–309.

Zhao, Y., Hull, A.K., Gupta, N.R., et al., 2002. Trp-dependent auxin biosynthesis in Arabidopsis: involvement of cytochrome P450s CYP79B2 and CYP79B3. Genes Dev. 16 (23), 3100–3112.

Zhao, G., Xu, H., Zhang, P., et al., 2017. Effects of 2,4-epibrassinolide on photosynthesis and Rubisco activase gene expression in *Triticum aestivum* L. seedlings under a combination of drought and heat stress. Plant Growth Regul. 81, 377–384.

Role of phytohormones in plant response to drought and salinity stresses

Tanushree Agarwal and Sudipta Ray

Department of Botany, University of Calcutta, Kolkata, West Bengal, India

6.1 Introduction

Plant survival depends on the complex signaling network, comprising genes and regulatory proteins. Phytohormones play a critical role in stress responses. Hormonal crosstalk between different phytohormones can help to ameliorate various stresses, such as drought, salinity, heat, heavy metal stress, or a combination of these. Abiotic stresses pose to be a potential threat to sustainable agricultural productivity. Changing climatic conditions and global warming contribute mainly toward drought and salinity.

Abiotic stress has been reported to drastically affect 70% of crop yield (Mantri et al., 2012). Reports suggest that the salinization of arable land could accelerate a land loss of 30% by the year 2028 and up to 50% by 2050 (Wang et al., 2003). Salinity induces major adverse effects such as ion toxicity, interference in nutrient uptake, impaired stomatal conductance, osmotic stress, changes in biochemical and physiological processes, and an increase in the production of reactive oxygen species (ROS) (Khan et al., 2014). Dehydration stress is caused by limiting water conditions that result due to drought, salinity, or temperature stress. It is by far the single most devastating abiotic stress condition that affects the productivity and yield of crop plants. Drought stress inhibits photosynthesis and stomatal conductance and affects the overall growth of plants (Hasanuzzaman et al., 2013). The physiological and biochemical effect of drought and salinity is related (Llanes et al., 2016; Ma et al., 2020). Both lead to severe osmotic stress, cellular dehydration, production of ROS, damage to the cell membrane, cellular structures, and organelles (Ma et al., 2020).

Phytohormones are small chemical messengers that are instrumental in the regulation of the growth, development, and metabolism of plants. They play a central role in the adaptation of

Plant Hormones in Crop Improvement
DOI: https://doi.org/10.1016/B978-0-323-91886-2.00007-0

109

plants to changing climatic conditions and alleviate stress. The plant growth regulators comprise of abscisic acid (ABA), jasmonic acid (JA), salicylic acid (SA), and ethylene (ET), together with cytokinin (CK), auxin (AUX), gibberellin (GA), brassinosteroids (BRs), and strigolactones (SLs). Hormonal crosstalk involves both synergistic and antagonistic interactions that help to modulate their activities and responses. While the phytohormones endow stress tolerance properties, it sometimes results in reduced growth, so that the plants can redirect their resources toward enduring stress (Skirycz and Inze, 2010).

The synchronization of stress-responsive genes with phytohormone-mediated signal transduction pathways is vital in combating adverse environmental conditions. Transgenic plants overexpressing transcription factors (TFs) or the stress-responsive genes show osmotic stress tolerance. The interactions between the phytohormones, such as ABA, CKs, GA, AUX, SA, and JA SLs, have been considered essential for stress tolerance properties (Lata, 2015; Roy, 2016).

Phytohormones play an important role in regulating plant growth and development. The central role of different phytohormones in drought and salinity has been discussed in the following section.

6.2 Abscisic acid

ABA is a vital phytohormone that regulates plant developmental processes under drought and salinity (Wasilewska et al., 2008). ABA also controls certain developmental and physiological functions in nonstressed conditions, including cell division, enlargement, differentiation, and central metabolism (Brookbank et al., 2021). Besides, ABA plays an important role in seed dormancy, somatic embryogenesis, and fruit ripening (Sano and Marion-Poll, 2021; Salaun et al., 2021). ABA biosynthesis is triggered in response to changing physiological and environmental conditions upon water deficit conditions. ABA metabolism and transport are particularly active upon osmotic stresses, including drought, cold, and high salinity (Fujita et al., 2011). ABA stimulates stomatal closure, changes in gene expression, and results in adaptive physiological responses (Seki et al., 2002; Shinozaki and Yamaguchi-Shinozaki, 2007). Changes in the levels of ROS also enhance ABA production and accumulation (Verslues et al., 2007).

Drought and salinity result in an osmotic imbalance that induces an overlapping network of stress tolerance genes. ABA is present at the starting node of such networks and shares many common elements in the downstream cascade. The transcript accumulation of stress-responsive gene *RD29A* is regulated in both ABA-dependent and ABA-independent manner (Yamaguchi-Shinozaki and Shinozaki, 1993). The salinity stress-induced upregulation of the transcript of pea DNA helicase 45 (*PDH45*) occurs in an ABA-dependent pathway, while calcineurin B-like protein (*CBL*) and CBL-interacting protein kinase (*CIPK*) from pea follow the ABA-independent pathway (Mahajan et al., 2006). The promoters of such stress-responsive genes contain *cis*-elements, such as DRE/CRT (A/GCCGAC), ABRE (PyACGTGGC), MYC recognition sequence (MYCRS; CANNTG), and MYB recognition sequence (MYBRS; C/TAACNA/G), which are regulated by various upstream transcriptional factors (Zhu, 2002). The ABA-dependent stress signaling activates basic leucine zipper TFs called AREB, which binds to the ABRE element to induce the stress-responsive gene (*RD29B*) (Uno et al., 2000). *AREB1/ABF2*, *AREB2/ABF4*, and

ABF3, belonging to the AREB/ABF subfamily of basic leucine zipper (bZIP) TFs in the *Arabidopsis* genome, are induced by both drought stress and ABA (Fujita et al., 2013). TFs like *DREB2A* and *DREB2B* trans-activate the DRE *cis*-element of osmotic stress genes in an ABA-independent manner and help to maintain osmotic stress (Liu et al., 1998). Recent studies have shown that the ABA-dependent proteins *AREB1/ABF2*, *AREB2/ABF4*, and *ABF3* play important roles in the regulation of the drought response by interacting with the ABA-independent proteins *DREB2A*, *DREB1A*, and *DREB2C* (Lee et al., 2010; Kim et al., 2011). This indicates that the two pathways converge at several points during the stress response. ABA leads to the accumulation of numerous proteins, one such major groups are the late embryogenesis abundant (*LEA*) proteins that help in imparting osmotic stress tolerance properties under drought and salinity (Chen et al., 2019). In rice, ABRE motifs were found in all *OsLEA* genes while the *OsDHODH1* gene had both ABRE and DRE/CRT motifs indicating ABA-mediated regulation under osmotic stress (Chen et al., 2016a,b,c). Transgenic *Arabidopsis* expressing wheat *TaRK2.8* showed enhanced tolerance to drought, salt, and cold stress by activating both ABA-dependent genes, such as *ABA1*, *ABA2*, *RD20A*, and *RD29B*, and ABA-independent genes, namely, *CBF1*, *CBF2*, and *CBF3* (Padmalatha et al., 2012). This finding shows that the ABA-dependent pathway and ABA-independent pathway intersect under drought stress conditions.

ABA accumulates in drought-stressed wheat, maize, sorghum, rice, barley, and soybean (Sah et al., 2016). Since ABA is related to water stress conditions, the exposure of plants to salinity also induces an increase in the levels of ABA. The increase in ABA in leaves of salt-stressed *Brassica*, *Phaseolus vulgaris*, and *Zea mays* has been observed (Llanes et al., 2016). ABA can alleviate salt stress in common beans and potatoes (Khadri et al., 2006; Etehadnia et al., 2008). Several genes have been identified in *Arabidopsis* that are induced by ABA upon osmotic stress (Seki et al., 2002). Among the ABA-inducible genes, 54% (133 genes) were induced by high salinity and 63% (155 genes) by drought (Seki et al., 2002). These results indicate the significance of ABA response in abiotic stress signaling pathways induced by drought and high salinity.

The core ABA signaling constituents involve PYRABACTIN RESISTANCE/PYRABACTIN RESISTANCE-LIKE/REGULATORY COMPONENT OF ABA RECEPTOR (PYR/PYL/RCAR) known as ABARs (Ma et al., 2009; Park et al., 2009; Fujii et al., 2009), subclass III SNF1-related protein kinase 2 (SnRK2) (Weiner et al., 2010), and protein phosphatases 2C (PP2Cs). Since then, several reports have provided insight into the dynamics of ABA binding to its receptors. The ABA/RCAR follows a unique gate—latch—lock mechanism. They have an open ligand-binding pocket safeguarded by a gate that closes upon ABA binding. Closure of the gate enables the receptor to interact with a conserved tryptophan residue of PP2C and render it inactive (Singh et al., 2016). PP2Cs also exhibit direct physical interaction with ABA in a complex with PYR/RCAR via its tryptophan residue (Santiago et al., 2012) and can bypass the ABA/PYR regulation. SnRK2s serine—threonine kinases function downstream of PP2Cs in ABA signaling. In the absence of ABA, PP2C blocks the kinase domain of SnRK2 rendering it inactive. Stress results in the production of ABA that bind to PYR/RCAR and ABA/PYR/RCAR interacts with PP2C to remove the inhibition of SnRK2 (Fujii et al., 2009; Umezawa et al., 2009). SnRK2 then phosphorylates downstream components, such as *ABF*, *ABI5*, and *WRKY* in the nucleus, to induce gene expression (Antoni et al., 2011; Miyakawa et al., 2013). SnRK2 also activates slow anion channel-associated 1 (SLAC1) and inhibits potassium channel in *Arabidopsis thaliana* (KAT1), causing stomatal closure (Miyakawa et al., 2013). ABA-induced stomatal closure follows a common signaling pathway that involves receptors, protein kinases, and second messengers, such as calcium and ion flux (Bharath et al., 2021). Among kinases, *OST1* is a primary activating factor of NADPH oxidase and raises the ROS levels of guard cells. An increase in *OST1* kinase is followed by the activation of *RBOH D/F*, which in turn elevates the levels of ROS. Ca^{2+}-dependent CDPKs activate slow anion channel 1 (SLAC1), S-type anion channel 3 (SLAH3), and K^+ out channels to promote ion efflux from guard cells and force stomata to close (Bharath et al., 2021). The SnRK2 activation loop interacts with the PP2C active site, while the Trp residue of PP2C is inserted into the catalytic cleft of SnRK2; thus, both ABAR and SnRK2 utilize a gate−latch−lock mechanism to regulate ABA signaling (Soon et al., 2012). Both PP2C and SnRK2 are critical for regulating the ABA stress signaling cascade. Recent reports suggest that PP2C may confer ABA insensitivity and abiotic stress tolerance in *Arabidopsis* (Singh et al., 2015). Signaling pathways do not exist in a linear or branched form, rather they are a complex network of components of which ABA is a crucial part (Agarwal and Ray, 2020).

6.3 Salicylic acid

SA acts as an endogenous signal molecule that plays a central role in abiotic stress (Devinar et al., 2013). In *A. thaliana*, it enhances seed germination under salinity stress conditions by reducing oxidative damage (Lee et al., 2010). Drought stress results in an increased expression of an SA-responsive gene *PR-1* in chickpea, which shows that pathogenesis-related genes (PR) are induced under both biotic and abiotic stresses (Tiwari et al., 2016). SA-accumulating mutants (*cpr5 and acd6*) in *A. thaliana* exhibited stomatal closure and improved drought tolerance due to the overexpression of PR genes, such as *PR1*, *PR2*, and *PR5* (Liu et al., 2013a,b). SA also regulates the activity of several WRKY TFs, such as *WRKY70* and *WRKY54* (Divi et al., 2010). *WRKY70* and *WRKY54* are implicated in biotic stress response and play a major role as negative regulators of stomatal conductance and osmotic stress tolerance in *Arabidopsis*. Exogenous application of SA in low concentrations improved drought tolerance in tomato and bean plants (Singh and Gautam, 2013). A direct correlation was shown between the water content in the leaves with SA concentration and signaling in plants (Miura and Tada, 2014). Several TFs, such as Apetala 2/Ethylene Responsive Factor (AP2/ERFs), leucine zipper, and Zn fingers, responded to SA under drought and salt stress conditions. These TFs in turn control the expression of stress-responsive genes. The role of SA in salinity stress has been studied in many crops, including *Medicago sativa* (Palma et al., 2013), *Vicia faba* (Azooz, 2009), and *Brassica juncea* (Nazar et al., 2015). SA improved salinity tolerance in *Torreya grandis* by enhancing photosynthesis, chlorophyll content, and antioxidant enzyme activity (Li et al., 2014). SA-deficient *NahG* transgenic of *Arabidopsis* lines showed a reduced activity of antioxidant enzymes (Cao et al., 2009). SA induced significant changes in the expression pattern of glutathione *S*-transferase (GST) gene family members mitigating salinity stress injury in *Solanum lycopersicum* (Csiszar et al., 2014).

Exogenously applied SA (0.5 mM) increased the transcript accumulation of antioxidant genes: *GST1*, *GST2*, *GPX1*, *DHAR*, *GR*, *GPX2*, *MDHAR*, and *GS*. The activity of ascorbate

(AsA)-glutathione (GSH) pathway enzymes was enhanced thereby improving salt tolerance in *Triticum aestivum* (Li et al., 2013). In *A. thaliana*, SA played a major role in improving salinity tolerance by the restoration of membrane potential and prevention of excessive
K^+-loss during salinity stress with the help of the depolarization-activated guard cell
outward-rectifying potassium (GORK) channel (Jayakannan et al., 2013). *Mitragyna speciosa*
showed enhanced expression of stress-responsive genes upon exogenous application of
5.0 µM SA (Jumali et al., 2011). Among the 292 expressed sequence tags (ESTs) analyzed
randomly, genes that encoded antioxidants, chaperone, secondary metabolite biosynthesis,
heat shock proteins (HSPs), and Cytochrome P450 (CYP) responded to SA treatment.

Besides inducing the expression of such stress-responsive genes, SA can also help in the
accumulation of osmolytes like glycine betaine (GB), proline (Pro), soluble sugars, and
amines (Khan et al., 2013). SA can modulate nutrient uptake and improve plant growth
upon abiotic stress (Nazar et al., 2015). SA-mediated changes in the nutrient uptake of
nitrogen (N), phosphorus (P), potassium (K), and calcium (Ca) in *B. juncea* cultivars
resulted in differences in salt tolerance (Syeed et al., 2011). SA also regulates the uptake of
elements, such as copper (C), manganese (Mn), phosphorus (P), calcium (Ca), iron (Fe),
and zinc (Zn), and reduces oxidative stress (Wang et al., 2011). SA modulates the activities
of ROS scavenging enzymes, such as catalase (CAT), peroxidase (POD), superoxide dismutase (SOD), and ascorbate peroxidase (APX), in plants exposed to drought (Saruhan et al.,
2012). SA also helps with the induction of major secondary metabolites, such as terpenes,
phenolics, alkaloids, GSH, phytoalexins, thionins, and defensins (Idrees et al., 2013). SA-
mediated activation of stress-responsive genes requires transcriptional reprograming
where *NPR1* (nonexpressor of pathogenesis-related 1) plays a central role and interacts
with bZIP TFs (Souri et al., 2020). SA enhances the levels of osmolytes, ROS scavenging
enzymes, kinases such as CDPK and MAPK, and TFs thereby regulating the activity of
stress-responsive genes.

6.4 Jasmonic acid

JA is a lipid-based hormone that regulates photosynthesis, growth, and reproductive
development. JA signaling is mainly active in biotic stress and wounding with emerging
evidence of its role in abiotic stress response (Ali and Baek, 2020). JA and its derivatives
[e.g., *cis*-jasmone, jasmonyl isoleucine (JA-Ile), methyl-jasmonate (MeJA), JA-glucosyl ester,
12-hydroxyjasmonic acid sulfate (12-HSO_4-JA), JA-Ile methyl ester, 12-carboxy-JA-IIe,
JA-Ile glucosyl ester, 12-*O*-glucosyl-JA-IIe, and lactones of 12-hydroxy-JA-IIe], are known
as jasmonates (JAs). These are fatty acids derived from cyclopentanones belonging to the
family of oxidized lipids well known as oxylipins (Wasternack and Feussner, 2018). These
are biologically active signaling molecules, ubiquitous in higher groups of plants, abun-
dantly found in flowers and reproductive tissues, but rare in the mature leaves and roots
(Dar et al., 2015; Wasternack and Feussner, 2018). JAs modulate many crucial processes in
plants, such as regulation of cell cycle, vegetative growth, stomatal opening, anthocyanin
biosynthesis, fruit ripening, senescence, rubisco biosynthesis inhibition, glucose transport,
and nitrogen and phosphorus uptake (Ali and Baek, 2020). High levels of JA have been
detected upon drought and salinity, and some plants show a transient increase upon

drought (Llanes et al., 2016). JA increased upon drought conditions in spear tips of *Pinus pinaster* plants, *Asparagus officinalis*, *Carica papaya* seedlings, and *Oryza sativa* leaves and roots (Llanes et al., 2016). Exogenous application of methyl-jasmonate (MeJA) on leaves mitigates the effects of salinity stress in broccoli (*Brassiscaoleracea* L. var. *Italica*). Exposure to sorbitol or mannitol increased the levels of JA precursor 12-oxo-phytodienoic acid (OPDA) and JA metabolite methylJA (MeJA) in tomato, barley, and *A. thaliana* plants (Seltmann et al., 2010). Severe drought conditions in young panicles of the rice plant led to the overexpression of JA carboxyl methyltransferase gene (*AtJMT*) and MeJA accumulation (Kim et al., 2009). MeJA accumulation was also found to cross-link ABA production. Water stress with the application of mannitol in seedlings of two sunflowers (*Helianthus annuus* L.) inbred lines, B59 (drought-sensitive) and B71 (drought-tolerant), resulted in high levels of JA and OPDA (Llanes et al., 2016). In rice, drought induces the interaction of *OsbHLH148* with *OsJAZs*, and its constitutive expression induces *OsDREB1* expression improving drought tolerance (Seo et al., 2011). The overexpression of maize gene *ZmJAZ14* in *Arabidopsis* enhanced tolerance to PEG-induced drought stress (Zhou et al., 2015). In rice leaves and roots, enhanced JA levels under both drought and high salinity led to the activation of stress-related PR and JA-biosynthetic genes (Tani et al., 2008). JA plays an important role in altering the root system architecture and enhancing the acquisition of nutrients from the soil in rice (Deepika and Amarjeet, 2021). These data suggest a significant role of JA in not only plant defense but also upon abiotic stress.

Stress leads to the epimerization of JA to form JA-Ile, and its accumulation in the cytoplasm of the stressed leaves. Under normal conditions, the low levels of JA-Ile prevent the activation of promoters of jasmonate-responsive genes by the different types of TFs. Zhou and Memelink (2016) reported that these TFs are repressed by a series of jasmonate-zinc finger inflorescence meristem (ZIM) domain (JAZ) proteins that act as transcriptional repressors. The JAZ repressors recruit the protein topless (TPL) and the interactor/adapter protein novel interactor of JAZ (NINJA); together, they form an effective transcriptional repression complex JAZ–NINJA–TPL that acts to inhibit the expression of jasmonate-responsive genes. This leads to the changing of the open complex to a closed one through the further recruitment of histone deacetylase 6 (HDA6) and HDA19 (Ali and Baek, 2020).

The formation of JA-Ile in the cytosol and its transportation to the nucleus is upregulated under abiotic stress. JA-Ile facilitates the interaction of JAZ with *COI1* within the SCF complex (Zhai et al., 2015). JAZ proteins are degraded by ubiquitination and subsequent proteasomal degradation with the release of the TFs that trigger the expression of jasmonate-responsive genes. In addition to the TFs, such as ERF, WRKY, and NAC (NAM, ATAF, CUC), JA signaling also activates the calcium channel mitogen-activated protein kinase cascade (Li et al., 2017), and various other processes cross-linking the activity of SA, ABA, and ET to govern plant response to abiotic stresses (Santner and Estelle, 2009).

6.5 Ethylene

ET is responsible for fruit ripening and senescence in plants. ET upon drought and salinity stress conditions may produce either negative or positive effects. ET causes leaf abscission and reduces water loss under drought. Drought enhances the production of

1-aminocyclopropane-1-carboxylic acid (ACC) synthase that regulates a key step in ET biosynthesis (Arraes et al., 2015). ET also regulates different developmental processes of plants during water stress and triggers the activation of TFs and stress-responsive genes such as *ERF1* (Arraes et al., 2015). Exogenous application of ET precursors such as (ACC) or an increase in the production of endogenous ET increases salinity stress tolerance in *Arabidopsis* and maize (Riyazuddin et al., 2020). ET plays an important role in conferring salinity stress tolerance in grapevine, maize, and tomato (Riyazuddin et al., 2020). An increase in *MYB108A*-mediated ET biosynthesis by melatonin also enhanced salinity stress tolerance in grapevines (Riyazuddin et al., 2020). Inhibition of ET biosynthesis and signaling increases the sensitivity of plants to salinity stress. ET confers salinity stress in *Solanum chilense* by maintaining homeostasis, stomatal conductance, and water use efficiency (Riyazuddin et al., 2020). Calcium carbide (CaC_2), a precursor of acetylene that exhibits similar effects to ET, is used to improve seed germination rates and confer tolerance to oxidative stress by increasing solute concentrations and activities of antioxidant enzymes in *Cucumis sativus* (Riyazuddin et al., 2020). Besides the positive regulatory role of ET, it sometimes exhibits a negative effect upon osmotic stress. Transgenic tobacco plants with reduced ET biosynthesis showed elevated salinity tolerance (Riyazuddin et al., 2020). Exogenous application of ET in rice resulted in salinity hypersensitivity. An increase in the production of ET under salinity stress significantly reduced growth, grain filling, and development of spikelets in rice.

The ET signaling and response pathway to the cell nucleus, leading to the expression of stress tolerance genes have been analyzed in the model system *Arabidopsis*. ET is sensed by five receptors localized at the endoplasmic reticulum membrane that are divided into subfamily I (Ethylene Response1 [ETR1] and Ethylene Response Sensor1 [ERS1]) and subfamily II (ETR2, ERS2), and Ethylene Insensitive4 [EIN4] (Lacey and Binder, 2014). The downstream components of ET signaling and response pathway also involve Constitutive Triple Response1 (CTR1), EIN2, EIN3/Ethylene Insensitive-Like Protein1 (EIL1), and ERFs (Stepanova and Alonso, 2009). CTR1 is a negative regulator of ET signaling. The absence of ET triggers the CTR1 kinase activity, which phosphorylates the C-terminal domain of EIN2 and inhibits its nuclear localization. However, in the presence of ET, the receptors and CTR1 are inactive (Muller and Munne-Bosch, 2015). EIN2, which is localized at the ER membrane along with the ET receptors and CTR1, positively regulates ET signaling (Bisson and Groth, 2010). The enhanced transcript accumulation for various ERF genes has also been reported under drought, salt, and cold stresses. Thus, it can be concluded that ET signaling is very important for regulating plant growth and stress responses.

Soybean showed a reduced level of expression of ETR (ET receptors) and CTR genes under drought stress (Eppel and Rachmilevitch, 2016). The *Arabidopsis* ET insensitive mutants, *ein2−5* and *ein3−1*, were found to be more susceptible to drought stress as compared to wild-type Col-0 plants (Cui et al., 2015). In another study, *SlERF5, AtERF5*, and *AtERF6* were established as master regulators after sudden exposure to salt and drought stress (Dubois et al., 2013). Gene expression of these TFs was induced very rapidly in actively growing leaves. *AtERF6* also induces the expression of osmotic stress-responsive genes, such as *STZ, MYB51*, and *WRKY33*.

6.6 Gibberellin

GA is a group of tetracyclic diterpene phytohormones that are crucial for flower and fruit development, germination, leaf expansion, cell elongation, as well as responses to several environmental stresses. GA can influence several photosynthetic enzymes and thereby improve leaf area index, light interception, photosynthetic efficiency, and nutrient use efficiency of plants (Hedden and Thomas, 2016). GA also acts on plant water relations and plays an important role under osmotic stress (Yamaguchi, 2008). A GA-stimulated transcript in *Arabidopsis* encoding *GASA6* has been related to drought and water stress response (Gong et al., 2015).

Gibberellin (GA3) enhances growth, in rice (*O. sativa* L. cv. Nipponbare) plants exposed to salt stress, in a dose-dependent manner. GA3 induces the activity of some salt-induced proteins, such as glutamyl tRNA reductase, enolase, salt stress-induced protein (SALT protein), isoflavone reductase-like protein, and phosphoglucomutase (Wen et al., 2010). Exposure to GA3 also enhances yield in spring wheat (*T. aestivum* L.) by increasing the ion uptake and hormone homeostasis (Iqbal and Ashraf, 2013). GA3 mitigated adverse effects of salt stress in soybean (Hamayun et al., 2010). Plants that showed reduced growth under both drought and salinity may have lower production of GAs or may not respond to the hormone. Modifications of GA levels may affect the source—sink relation by increasing the sucrose levels in source leaves and decreasing photosynthesis to help the plants survive adverse environmental conditions (Iqbal et al., 2011). GA metabolism is directly correlated with the concentration and chemical composition of the salts present in the soil (Llanes et al., 2014).

The core components of the GA signaling pathway have been identified over the past decade. The signaling pathway involves a regulatory loop between GA and DELLA such that DELLA proteins restrain plant growth, whereas the GA promotes growth by overcoming DELLA-mediated growth control (Daviere and Achard, 2016). DELLA proteins play a major role in upregulating or suppressing GA responses under osmotic stress (Plaza-Wuthrich et al., 2016). The GA—GID1—DELLA pathway controls both plant growth and stress responses. In rice and *Arabidopsis*, GA signaling is mediated by binding of GA-to-GA INSENSITIVE DWARF1 (GID1a/b/c) (Yoshida et al., 2018; Jones et al., 2010). The interaction of DELLA proteins with GID1 stimulates the degradation of DELLA (Hedden and Thomas, 2016). *SPINDLY* (*SPY*) gene acts as a negative regulator of GA signaling in plants during abiotic stress probably due to the crosstalk of GA and CKs (Qin et al., 2011). GA biosynthesis and signaling are critical in plant response to abiotic stress.

6.7 Cytokinin

CKs are the phytohormones that have a significant role in cell division, growth, and development in plants. CKs induce stomatal conductance, transpiration, and photosynthesis under water stress. Enhanced syntheses of CKs ameliorate salinity-induced decreases in growth and yield in tomatoes by modifying the shoot hormonal and ionic homeostasis (Tiwari et al., 2017).

Depending on the duration of stress and intensity, CKs show both positive and negative effects on drought tolerance. In transgenic cotton higher levels of endogenous CKs delayed senescence in plants and improved drought tolerance (Tiwari et al., 2017). CKs are thought to be a negative regulator of root growth and branching; degradation of CKs enhances primary root growth and branching upon drought (Tiwari et al., 2017). Drought induces the expression of *AtMYB2* in *Arabidopsis*. *AtMYB2* in turn downregulates isopentenyl transferase (*IPTs*) gene expression, and as a consequence, the endogenous level of CKs decreases. The enhanced expression of *cytokinin oxidase/dehydrogenase1* (*CKX1*) genes in *Arabidopsis* resulted in increased CK degradation promoting both primary root length as well as lateral root formation during drought stress (Tiwari et al., 2017). Further, transgenic barley lines overexpressing CK dehydrogenase displayed partial CK insensitivity and enhanced drought tolerance (Tiwari et al., 2017). Functional analyses of *Arabidopsis* CK receptor mutants revealed that *AHK2*, *AHK3*, and *CRE1/AHK4* function as negative regulators of osmotic stress (Tiwari et al., 2017). Moreover, dehydration, cold, and high salinity activated CKs inducible type A *ARR4* and *ARR5* (Tiwari et al., 2017). Plants react to water-limiting conditions by modulating CK metabolism as in *Arabidopsis*, creeping bent grass, soybean, tobacco, and sunflower or regulating the CK receptors (Pavlu et al., 2018). Ectopic expression of *ipt* accelerates the activity of antioxidant systems during severe drought stress. This prevents excessive ROS accumulation, thereby preserving chloroplast integrity and reducing electrolyte leakage and/or rises in malondialdehyde levels (Pavlu et al., 2018). Ectopic expression of *CKX* in barley activates genes of the flavonoid biosynthesis pathway and flavonoids are effective in drought tolerance (Pavlu et al., 2018). Root-specific overexpression of *CKX* also enhances root growth, nutrient uptake, and drought tolerance (Ramireddy et al., 2018), and helps in the recovery of the plants (Pospisilova et al., 2016).

The two-component CK signaling pathway (TCS) includes membrane-bound (primarily ER) hybrid histidine kinases (HKs) with a CK-binding input domain (CHKs) (Zwack and Rashotte, 2015). The binding of CK triggers the auto-phosphorylation of CHK dimers and activates a phosphorelay cascade involving histidine phosphotransfer proteins (HPts) that carry the signal into the nucleus. The HPTs transfer the phosphoryl group to the receiver domain of two types of response regulator proteins (RRs) (Zwack and Rashotte, 2015). Phosphorylation and activation of B-type RRs (RRBs) containing N-terminal glutamic acid-rich (GARP) domains activates these MYB-like TFs, which regulate the expression of CK-responsive genes. The A-type RRs (RRAs), which are similar to RRBs and only differ due to the absence of GARP domains, are also induced. As such, these proteins act as negative regulators of the CK signal primarily by competing with RRBs for phosphorylation by HPts (Zwack and Rashotte, 2015). Like the RRAs, C-type RRs (RRCs) lack a DNA-binding domain and are also believed to act as negative regulators of CK signaling.

6.8 Auxin

AUX is one of the most important phytohormones that promote the growth of apical meristems and root differentiation. Salinity stress alters the metabolism and distribution of Indole-3-acetic acid (IAA), the precursor of AUX (Schopfer et al., 2002). In *A. thaliana* membrane-bound NAC TF, *NTM2* acts as a molecular link, between the AUX signaling pathway and salt stress (Park et al., 2011).

Drought induces an increase in IAA levels in maize roots and sunflower plants (Llanes et al., 2016). In *Arabidopsis*, drought tolerance was enhanced upon the activation of flavin monooxygenase genes *YUCCA7* and *YUCCA6* (Kim et al., 2013). These genes belong to the tryptophan-dependent AUX biosynthetic pathway, which leads to the rise in endogenous Aux levels. Salinity stress changes the Aux accumulation and distribution pattern in *Arabidopsis* root tip, resulting in the inhibition of primary root growth, reduction in lateral root primordia formation, and a significant increase in the lateral root elongation (Wang et al., 2009). AUX accumulation is also observed in the roots of halophytic *Prosopis strombulifera* under salt treatments (Llanes et al., 2014), which in turn increases the lateral root formation. Exogenous IAA application also induces a similar response. The root and shoot growth of wheat seedlings was also enhanced by IAA application under salt conditions.

The core components of the AUX signaling machinery belong to three protein families: the F-box TRANSPORT INHIBITOR RESPONSE 1/AUXIN SIGNALING F-BOX PROTEIN (*TIR1/AFB*) AUX coreceptors, the Auxin/INDOLE-3-ACETIC ACID (*Aux/IAA*) transcriptional repressors, and the AUXIN RESPONSE FACTOR (*ARF*) TFs. AUX promotes interaction between *TIR1/AFB* and *Aux/IAA* proteins, resulting in a degradation of the *Aux/IAAs* and the release of *ARF* repression (Wang and Estelle, 2014; Salehin et al., 2015). Gene expression upon *ARF* activation enhances tropic responses and the establishment of polarity, as well as embryogenesis and organogenesis in flowering plants, and both gametophyte and sporophyte development in nonflowering plants (Lavy and Estelle, 2016). AUX affects all aspects at the cellular level, such as cell elongation, cell division, and differentiation (Lavy and Estelle, 2016).

6.9 Brassinosteroids

BRs are a group of steroidal phytohormones that are crucial for development as well as in plant responses to abiotic stress. BRs help to induce stress-related gene expression, thereby regulating photosynthesis, antioxidant enzyme activity, accumulation of osmolytes, and induction of other hormonal responses. Treatment with BRs helped improve the biomass reduction in sugar beet plants caused by mild drought. It also increased seedling growth of *Sorghum vulgare* under osmotic stress and improved the drought tolerance in *P. vulgaris* (Llanes et al., 2016). BRs application in rice improved the leaf water economy and CO_2 assimilation that enhanced drought tolerance. Treatment of *A. thaliana* and *Brassica napus* seedlings with BR, 24-epibrassinolide (24-EBR) increased their survival rate upon drought stress (Kagale et al., 2007). Similarly, the exogenous application of BRs can also alleviate the adverse effects of salinity. Foliar spray or root supply of *B. juncea* plants with 28-homobrassinolide (28-HomoBL) after salt treatment of seedlings enhanced the growth and seed yield. However, this compound was found to significantly reduce salt-induced damage by protecting the chloroplast without affecting the cell ultrastructure. It also increased the synthesis of pigment in several species under salinity (Llanes et al., 2016). In wheat, the growth and photosynthetic capacity of salt-tolerant and salt-sensitive are enhanced with the application of 24-EBR (Llanes et al., 2016). Microarray data revealed a strong expression of thioredoxin and monodehydroascorbate reductase and lower expression of *COR78* in BR-deficient *Arabidopsis* plants, while a strong expression of *COR47* was observed in BR-treated plants that indicate that BR regulates cold and drought tolerance (Tiwari et al., 2017). Exogenous BRs have been found very effective in improving the plant resistance to several abiotic stresses, in contrast to endogenous BRs in plant stress responses.

The BRs are initially recognized by brassinosteroid-insensitive 1 (*BRI1*) receptor kinase localized to the plasma membrane as a homodimer (Chakraborty et al., 2015). Exogenously supplemented BRs bind to BRI1 and induce an association with BRI1-associated receptor kinase 1 (*BAK1*) and disassociation of BRI1 kinase inhibitor 1 (*BKI1*). *BRI1* is a multifunction kinase that can phosphorylate Ser, Thr, and Tyr residues, and the phosphorylation of Tyr residue is important for specific aspects of BR signaling in plants (Kour et al., 2021). *BRI1* stimulates a downstream signal network activating a wide array of kinases and phosphatases, which in turn activates a number of TFs that alter the expression of stress-specific genes (Kour et al., 2021).

6.10 Strigolactones

SLs are carotenoid-derived, multifunctional compounds, recognized as a new class of phytohormones that control different processes in plants. They are biosynthesized in the whole plant at low or even undetectable levels (Xie et al., 2003). Different types of SLs have been reported from a wide variety of plant species, including dicots and monocots (Xie et al., 2003). Their mode of action is yet unclear, and so they have been studied mainly through the exogenous application in abiotic stress response. SLs act as a positive regulator under both drought and salt stress (Pandey et al., 2016; Van Ha et al., 2014) and maintain root and shoot architecture. SL biosynthesis is regulated by a set of genes called More axillary growth (*MAX*) genes, namely, *MAX1*, *MAX3*, and *MAX4* (Van Ha et al., 2014). *Arabidopsis* mutants for the max genes were subjected to drought and subsequently to exogenous application of SLs. The drought-sensitive phenotype of the *max3* and *max4* mutants showed improved yield upon exogenous application of SLs. Furthermore, wild-type *Arabidopsis* plants were more tolerant to drought after SL treatment in contrast to the untreated plants (Llanes et al., 2016). Genetic engineering can be used to manipulate the endogenous levels of SLs and improve the drought and salt tolerance of plants.

The signaling pathway of SLs begins with the perception of SLs by α/β-hydrolase that conveys the signal to a leucine-rich-repeat F-box protein (*MAX2* in *Arabidopsis*; *D3* in rice). The pathway has been depicted clearly in a recent report (Faizan et al., 2020) suggesting the involvement of SLs in targeting the expression of stress response genes.

6.11 Crosstalk of phytohormones

During abiotic stress, phytohormone crosstalk involves their synergistic and/or antagonistic action and the coordinated regulation of plant hormone biosynthetic pathways as depicted in Fig. 6.1. Functional genomic approaches in *Arabidopsis* have identified significant cross-linking of phytohormone-mediated signaling under abiotic stresses (Kohli et al., 2013). Drought imposes an increase in endogenous ABA levels in plants. This leads to the activation of ABA biosynthetic genes (Trivedi et al., 2016), accumulation of osmoprotectants, and stomatal closure,

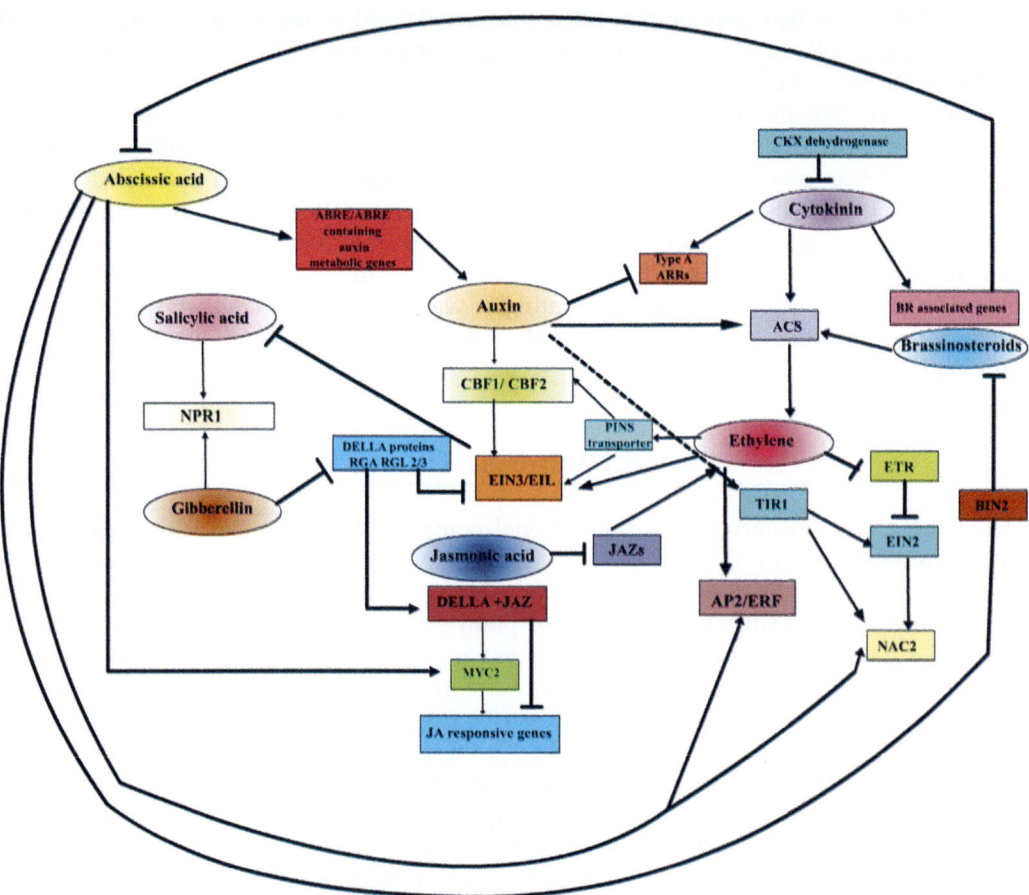

FIGURE 6.1 Crosstalk of phytohormones related to drought and salinity stress in plants. Positive regulation is depicted with arrows and dashed lines and blocked lines depict negative regulation. *Source: Modified from Tiwari, S., Lata, C., Chauhan, P.S., Prasad, V., Prasad, M., 2017. A functional genomic perspective on drought signalling and its crosstalk with phytohormone-mediated signalling pathways in plants. Curr. Genomics 18 (6), 469–482.*

thereby improving drought tolerance (de Ollas and Dodd, 2016). Drought also leads to the accumulation of JA and SA that act to regulate stomatal closure (Miura et al., 2013). JA, GA, and ET interact in an ABA-dependent manner under drought stress. GA also interacts with SA during a stress response. The expression levels of *NPR1* and SA biosynthetic genes are induced upon exogenous application of GA3 (Tiwari et al., 2017). The overexpression of a *Fagus sylvatica* GA-responsive gene *FsGASA4*, a member of the GA3 gene family in transgenic *Arabidopsis*, conferred oxidative stress tolerance linked with enhanced levels of endogenous SA (Tiwari et al., 2017). A comprehensive list of overexpressed genes and their interaction with various phytohormones that leads to stress tolerance has been represented in Table 6.1.

ABA acts through the PP2C family of genes and regulates BR-mediated signaling through *BIN2* or its upstream components (Tiwari et al., 2017). ABA was also found to restrain

TABLE 6.1 A comprehensive list of ectopically expressed genes and their interaction with various phytohormones to induce a stress response in plants.

Sl. no.	Gene	Transgenic	Phytohormone interaction	Stress response	References
1.	CKX1	*Hordeum vulgare*	CK	Drought	Vojta et al. (2016)
2.	MYB12	*Arabidopsis thaliana*	ABA	Drought, salt	Wang et al. (2016)
3.	NAC2	*A. thaliana*	ABA, JA, ET	Drought	Gunapati et al. (2016)
4.	EXPA2	*Nicotiana tabacum*	ABA, JA, SA, GA	Drought	Chen et al. (2016a,b,c)
5.	HDG11	*Triticum aestivum*	ABA	Drought	Li et al. (2016a,b)
6.	HD2D	*A. thaliana*	ABA	Drought, salt, cold	Han et al. (2016)
7.	LEA4−1	*Brassica juncea*	ABA, GA, JA, IAA	Drought, salt	Saha et al. (2016)
8.	DHN	*N. tabacum*	GA, CK, ABA	Drought, salt	Jardak-Jamoussi et al. (2016)
9.	WRKY1	*N. tabacum*	ABA	Drought	Ding et al. (2016)
10.	MYB37	*A. thaliana*	ABA	Drought	Yu et al. (2016)
11.	bZIP23	*Oryza sativa*	ABA	Drought	Dey et al. (2016)
12.	JAZ14	*A. thaliana*	ABA, JA, GA	Drought	Zhou et al. (2015)
13.	CBF3	*A. thaliana*	ABA	Drought, salt	Ma et al. (2015)
14.	DREB2	*N. tabacum*	ABA	Drought	Tan et al. (2015)
15.	ERD4	*A. thaliana*	ABA, SA	Drought	Rai et al. (2015)
16.	YUCCA6	*Poplar*	AUX	Drought	Ke et al. (2015)
17.	EXPB23	*N. tabacum*	ABA, JA, GA, ET, AUX	Drought, salt	Han et al. (2012)
18.	bHLH148	*O. sativa*	JA	Drought	Seo et al. (2011)
19.	FTL1/ DDF1	*A. thaliana*	GA	Drought, cold, heat	Kang et al. (2011)

ABA, Abscisic acid; *AUX*, auxin; *CK*, cytokinin; *ET*, ethylene; *IAA*, indole-3-acetic acid; *JA*, jasmonic acid; *SA*, salicylic acid.

BR-induced stress responses in plants under drought (Divi et al., 2010). Drought stress has generally been found to decrease the production and transport of CKs (Tran et al., 2010). *Arabidopsis* plants when subjected to drought stress showed an increase in the ABA content in roots that correlated with reduced ET production and increased CKs degradation maintaining primary root growth (Tiwari et al., 2017). Recently, CKs have also been shown to be a positive regulator of AUX biosynthesis. Both CKs and IAA signaling are involved in a homeostatic feedback regulatory loop maintaining their proper ratios in developing roots and shoots (Tiwari et al., 2017). The application of AUX regulates the activity of several members of the ACS gene family that encode rate-limiting enzymes in the ET biosynthetic pathway

(Peleg and Blumwald, 2011). BRs and MeJA also activate ACO enzymes, increasing ET production in maize and olive plants (Arraes et al., 2015). In *Arabidopsis*, SLs alter abiotic stress response through both ABA-dependent and ABA-independent signaling pathways (Van Ha et al., 2014). SLs along with ABA play a regulatory role in stomatal closure, senescence of leaves, and stress response, whereas CKs delay the senescence. The crosstalk of phytohormones is vital for the survival of plants under various environmental stresses. However, several components of phytohormone interacting pathways remain to be elucidated.

6.12 Conclusion

From the foregoing discussion, it is evident that plants utilize several phytohormones and elaborate signaling pathways in response to abiotic stress. The signaling interactions among various phytohormones are rather common in controlling the developmental processes in plants. Recent discoveries of the presence of multiple signaling intermediates further help to better understand the intricate web of networks among the phytohormones. Future research involving a complete dissection of the molecular interactions between different phytohormones regulating stress responses will help in the development of stress-resilient crops.

References

Agarwal, T., Ray, S., 2020. Casein kinase2 and its dynamism in abiotic stress management. In: Pandey, G.K. (Ed.), Protein Kinases and Stress Signaling in Plants. https://doi.org/10.1002/9781119541578.ch13.

Ali, M.S., Baek, K.H., 2020. Jasmonic acid signaling pathway in response to abiotic stresses in plants. Int. J. Mol. Sci. 21 (2), 621.

Antoni, R., Gonzalez-Guzman, M., Rodriguez, L., Rodrigues, A., Pizzio, G.A., Rodriguez, P.L., 2011. Selective inhibition of clade A phosphatases type 2C by PYR/PYL/RCAR abscisic acid receptors. Plant. Physiol. 158, 970–980.

Arraes, F.B.M., Beneventi, M.A., de Sa, M.E.L., Paixao, J.F.R., Albuquerque, E.V.S., Marin, S.R.R., et al., 2015. Implications of ethylene biosynthesis and signaling in soybean drought stress tolerance. BMC Plant. Biol. 15, 1.

Azooz, M.M., 2009. Salt stress mitigation by seed priming with salicylic acid in two faba bean genotypes differing in salt tolerance. Intl. J. Agric. Biol. 11, 343–350.

Bharath, P., Gahir, S., Raghavendra, A.S., 2021. Abscisic acid-induced stomatal closure: an important component of plant defense against abiotic and biotic stress. Front. Plant. Sci. 1, 324.

Bisson, M.M., Groth, G., 2010. New insight in ethylene signaling: autokinase activity of ETR1 modulates the interaction of receptors and EIN2. Mol. Plant. 3, 882–889.

Brookbank, B.P., Patel, J., Gazzarrini, S., Nambara, E., 2021. Role of basal ABA in plant growth and development. Genes 12, 1936.

Cao, Y., Zhang, Z.W., Xue, L.W., Du, J.B., Shang, J., Xu, F., et al., 2009. Lack of salicylic acid in *Arabidopsis* protects plants against moderate salt stress. Z. für Naturforschung C. 64 (3–4), 231–238.

Chakraborty, N., Sharma, P., Kanyuka, K., Pathak, R.R., Choudhury, D., Hooley, R., et al., 2015. G-protein α-subunit (GPA1) regulates stress, nitrate and phosphate response, flavonoid biosynthesis, fruit/seed development and substantially shares GCR1 regulation in *A. thaliana*. Plant. Mol. Biol. 89, 559–576.

Chen, H., Liu, L., Wang, L., Wang, S., Cheng, X., 2016a. VrDREB2A, a DREB-binding transcription factor from *Vigna radiata*, increased drought and high-salt tolerance in transgenic *Arabidopsis thaliana*. J. Plant. Res. 129 (2), 263–273.

Chen, S.F., Liang, K., Yin, D.M., Ni, D.A., Zhang, Z.G., Ruan, Y.L., 2016b. Ectopic expression of a tobacco vacuolar invertase inhibitor in guard cells confers drought tolerance in *Arabidopsis*. J. Enzyme Inhib. Med. Chem. 1–5.

Chen, Y., Han, Y., Zhang, M., Zhou, S., Kong, X., Wang, W., 2016c. Overexpression of the wheat expansin gene TaEXPA2 improved seed production and drought tolerance in transgenic tobacco plants. PLoS One 11, e0153494.

Chen, Y., Li, C., Zhang, B., Yi, J., Yang, Y., Kong, C., et al., 2019. The role of the late embryogenesis-abundant (LEA) protein family in development and the abiotic stress response: a comprehensive expression analysis of potato (Solanum tuberosum). Genes 10 (2), 148.

Csiszar, J., Horvath, E., Vary, Z., Galle, A., Bela, K., Brunner, S., et al., 2014. Glutathione transferase supergene family in tomato: salt stress-regulated expression of representative genes from distinct GST classes in plants primed with salicylic acid. Plant. Physiol. Biochem. 78, 15−26.

Cui, M., Lin, Y., Zu, Y., Efferth, T., Li, D., Tang, Z., 2015. Ethylene increases accumulation of compatible solutes and decreases oxidative stress to improve plant tolerance to water stress in Arabidopsis. J. Plant. Biol. 58, 193−201.

Dar, T.A., Uddin, M., Khan, M.M.A., Hakeem, K.R., Jaleel, H., 2015. Jasmonates counter plant stress: a review. Environ. Exp. Bot. 115, 49−57.

Daviere, J.M., Achard, P., 2016. A pivotal role of DELLAs in regulating multiple hormone signals. Mol. Plant. 9, 10−20.

de Ollas, C., Dodd, I.C., 2016. Physiological impacts of ABA-JA interactions under water-limitation. Plant. Mol. Biol. 91 (6), 641−650.

Deepika, D., Amarjeet, S., 2021. Expression dynamics indicate the role of Jasmonic acid biosynthesis pathway in regulating macronutrient (N, P and K^+) deficiency tolerance in rice (Oryza sativa L.). Plant Cell Reports 40, 1495−1521. Available from: https://doi.org/10.1007/s00299-021-02721-5.

Devinar, G., Llanes, A., Masciarelli, O., Luna, V., 2013. Abscisic acid and salicylic acid levels induced by different relative humidity and salinity conditions in the halophyte Prosopis strombulifera. Plant. Growth Regul. 70, 247−256.

Dey, A., Samanta, M.K., Gayen, S., Sen, S.K., Maiti, M.K., 2016. Enhanced gene expression rather than natural polymorphism in coding sequence of the OsbZIP23 determines drought tolerance and yield improvement in rice genotypes. PLoS One 11, e0150763.

Ding, W., Fang, W., Shi, S., Zhao, Y., Li, X., Xiao, K., 2016. Wheat WRKY type transcription factor gene TaWRKY1 is essential in mediating drought tolerance associated with an ABA dependent pathway. Plant. Mol. Biol. Rep. 34, 1111−1126.

Divi, U.K., Rahman, T., Krishna, P., 2010. Brassinosteroid-mediated stress tolerance in Arabidopsis shows interactions with abscisic acid; ethylene and salicylic acid pathways. BMC Plant. Biol. 10, 1.

Dubois, M., Skirycz, A., Claeys, H., Maleux, K., Dhondt, S., De Bodt, S., et al., 2013. Ethylene response factor6 acts as a central regulator of leaf growth under water-limiting conditions in Arabidopsis. Plant. Physiol. 162, 319−332.

Eppel, A., Rachmilevitch, S., 2016. Photosynthesis and photoprotection under drought in the annual desert plant Anastatica hierochuntica. Photosynthetica 54 (1), 143−147.

Etehadnia, M., Waterer, D.R., Tanino, K.K., 2008. The method of ABA application affects salt stress responses in resistant and sensitive potato lines. J. Plant. Growth Regul. 27, 331−341.

Faizan, M., Faraz, A., Sami, F., Siddiqui, H., Yusuf, M., Gruszka, D., et al., 2020. Role of strigolactones: signalling and crosstalk with other phytohormones. Open. Life Sci. 15, 217−228.

Fujii, H., Chinnusamy, V., Rodrigues, A., et al., 2009. In vitro reconstitution of an abscisic acid signalling pathway. Nature 462, 660−664.

Fujita, Y., Fujita, M., Shinozaki, K., Yamaguchi-Shinozaki, K., 2011. ABA mediated transcriptional regulation in response to osmotic stress in plants. J. Plant. Res. 124, 509−525.

Fujita, Y., Yoshida, T., Yamaguchi-Shinozaki, K., 2013. Pivotal role of the AREB/ABF-SnRK2 pathway in ABRE-mediated transcription in response to osmotic stress in plants. Physiol. Plant. 147 (1), 15−27.

Gong, L., Zhang, H., Gan, X., Zhang, L., Chen, Y., Nie, F., et al., 2015. Transcriptome profiling of the potato (Solanum tuberosum L.) plant under drought stress and water stimulus conditions. PLoS One 10, e0128041.

Gunapati, S., Naresh, R., Ranjan, S., Nigam, D., Hans, A., Verma, P.C., et al., 2016. Expression of GhNAC2 from G. herbaceum, improves root growth and imparts tolerance to drought in transgenic cotton and Arabidopsis. Sci. Rep. 6 (1), 1−14.

Hamayun, M., Khan, S.A., Khan, A.L., Shin, J.H., Ahmad, B., Shin, D.H., et al., 2010. Exogenous gibberellic acid reprograms soybean to higher growth and salt stress tolerance. J. Agric. Food Chem. 58 (12), 7226−7232.

Han, Y.Y., Li, A.X., Li, F., Zhao, M.R., Wang, W., 2012. Characterization of a wheat (*Triticum aestivum* L.) expansin gene, TaEXPB23, involved in the abiotic stress response and phytohormone regulation. Plant. Physiol. Biochem. 54, 49-5.

Han, Z., Yu, H., Zhao, Z., Hunter, D., Luo, X., Duan, J., et al., 2016. AtHD2D gene plays a role in plant growth, development, and response to abiotic stresses in *Arabidopsis thaliana*. Front. Plant. Sci. 7, 310.

Hasanuzzaman, M., Nahar, K., Alam, M.M., Roychowdhury, R., Fujita, M., 2013. Physiological, biochemical, and molecular mechanisms of heat stress tolerance in plants. Int. J. Mol. Sci. 14 (5), 9643−9684.

Hedden, P., Thomas, S.G., 2016. Annual plant reviews, The Gibberellins, Vol. 49. John Wiley & Sons.

Idrees, M., Naeem, M., Aftab, T., Khan, M., 2013. Salicylic acid restrains nickel toxicity, improves antioxidant defence system and enhances the production of anticancer alkaloids in *Catharanthus roseus* (L.). J. Hazard. Mater. 252, 367−374.

Iqbal, M., Ashraf, M., 2013. Gibberellic acid mediated induction of salt tolerance in wheat plants: growth, ionic partitioning, photosynthesis, yield and hormonal homeostasis. Environ. Expt. Bot. 86, 76−85.

Iqbal, N., Nazar, R., Khan, M.I.R., Masood, A., Khan, N.A., 2011. Role of gibberellins in regulation of source-sink relations under optimal and limiting environmental conditions. Curr. Sci. 100, 998−1007.

Jardak-Jamoussi, R., Zarrouk, O., Salem, A.B., Zoghlami, N., Mejri, S., Gandoura, S., et al., 2016. Overexpressing *Vitis vinifera* YSK 2 dehydrin in tobacco improves plant performance. Agric. Water Manage. 164, 176−189.

Jayakannan, M., Bose, J., Babourina, O., Rengel, Z., Shabala, S., 2013. Salicylic acid improves salinity tolerance in *Arabidopsis* by restoring membrane potential and preventing salt-induced K$^+$ loss via a GORK channel. J. Exp. Bot. 64, 2255−2268.

Jones, B., Gunneras, S.A., Petersson, S.V., Tarkowski, P., Graham, N., May, S., et al., 2010. Cytokinin regulation of auxin synthesis in *Arabidopsis* involves a homeostatic feedback loop regulated via auxin and cytokinin signal transduction. Plant. Cell 22, 2956−2969.

Jumali, S.S., Said, I.M., Ismail, I., Zainal, Z., 2011. Genes induced by high concentration of salicylic acid in 'Mitragyna speciosa'. Aust. J. Crop. Sci. 5, 296−303.

Kagale, S., Divi, U.K., Krochko, J.E., Keller, W.A., Krishna, P., 2007. Brassinosteroid confers tolerance in *Arabidopsis thaliana* and *Brassica napus* to a range of abiotic stresses. Planta 225, 353−364.

Kang, H.G., Kim, J., Kim, B., Jeong, H., Choi, S.H., Kim, E.K., et al., 2011. Overexpression of FTL1/DDF1, an AP2 transcription factor, enhances tolerance to cold, drought, and heat stresses in *Arabidopsis thaliana*. Plant. Sci. 180, 634−641.

Ke, Q., Wang, Z., Ji, C.Y., Jeong, J.C., Lee, H.S., Li, H., et al., 2015. Transgenic poplar expressing *Arabidopsis* YUCCA6 exhibits auxin-overproduction phenotypes and increased tolerance to abiotic stress. Plant. Physiol. Biochem. 94, 19−27.

Khadri, M., Tejera, N.A., Lluch, C., 2006. Alleviation of salt stress in common bean (*Phaseolus vulgaris*) by exogenous abscisic acid supply. J. Plant. Growth Regul. 25, 110−119.

Khan, M.I.R., Iqbal, N., Masood, A., Per, T.S., Khan, N.A., 2013. Salicylic acid alleviates adverse effects of heat stress on photosynthesis through changes in proline production and ethylene formation. Plant. Signal. Behav. 8 (11), e26374.

Khan, M.I.R., Asgher, M., Khan, N.A., 2014. Alleviation of salt-induced photosynthesis and growth inhibition by salicylic acid involves glycinebetaine and ethylene in mungbean (*Vigna radiata* L.). Plant. Physiol. Biochem. 80, 67−74.

Kim, E.H., Kim, Y.S., Park, S.-H., Koo, Y.J., Choi, Y.D., Chung, Y.-Y., et al., 2009. Methyl jasmonate reduces grain yield by mediating stress signals to alter spikelet development in rice. Plant. Physiol. 149, 1751−1760.

Kim, J.S., Mizoi, J., Yoshida, T., Fujita, Y., Nakajima, J., Ohori, T., et al., 2011. An ABRE promoter sequence is involved in osmotic stress-responsive expression of the DREB2A gene, which encodes a transcription factor regulating drought-inducible genes in *Arabidopsis*. Plant. Cell Physiol. 52, 2136−2146.

Kim, J., Patterson, S.E., Binder, B.M., 2013. Reducing jasmonic acid levels causes Ein2 mutants to become ethylene responsive. FEBS Lett. 587, 226−230.

Kohli, A., Sreenivasulu, N., Lakshmanan, P., Kumar, P.P., 2013. The phytohormone crosstalk paradigm takes center stage in understanding how plants respond to abiotic stress. Plant. Cell Rep. 32, 945−957.

Kour, J., Kohli, S.K., Khanna, K., Bakshi, P., Sharma, P., Singh, A.D., et al., 2021. Brassinosteroid signaling, crosstalk and, physiological functions in plants under heavy metal stress. Front. Plant. Sci. 12.

Lacey, R.F., Binder, B.M., 2014. How plants sense ethylene gas—the ethylene receptors. J. Inorg. Biochem. 133, 58—62.

Lata, C., 2015. Advances in omics for enhancing abiotic stress tolerance in millets. Proc. Indian. Natn. Sci. Acad. 81, 397—417.

Lavy, M., Estelle, M., 2016. Mechanisms of auxin signaling. Development 143 (18), 3226—3229.

Lee, S.J., Kang, J.Y., Park, H.J., Kim, M.D., Bae, M.S., Choi, H.I., et al., 2010. DREB2C interacts with ABF2, a bZIP protein regulating abscisic acid-responsive gene expression, and its overexpression affects abscisic acid sensitivity. Plant. Physiol. 153, 716—727.

Li, G., Peng, X., Wei, L., Kang, G., 2013. Salicylic acid increases the contents of glutathione and ascorbate and temporally regulates the related gene expression in salt-stressed wheat seedlings. Gene 529, 321—325.

Li, T., Hu, Y., Du, X., Tang, H., Shen, C., Wu, J., 2014. Salicylic acid alleviates the adverse effects of salt stress in *Torreya grandis* cv. merrillii seedlings by activating photosynthesis and enhancing antioxidant systems. PLoS One 9 (10), e109492.

Li, L., Zheng, M., Deng, G., Liang, J., Zhang, H., Pan, Z., et al., 2016a. Overexpression of AtHDG11 enhanced drought tolerance in wheat (*Triticum aestivum* L.). Mol. Breed. 36, 1—10.

Li, Z., Zhang, J., Li, J., Li, H., Zhang, G., 2016b. The Functional and regulatory mechanisms of the *Thellungiela salsuginea* ascorbate peroxidase 6 (TsAPX6) in response to salinity and water deficit stresses. PLoS One 11 (4), e0154042.

Li, Y., Qin, L., Zhao, J., Muhammad, T., Cao, H., Li, H., et al., 2017. SlMAPK3 enhances tolerance to tomato yellow leaf curl virus (TYLCV) by regulating salicylic acid and jasmonic acid signaling in tomato (*Solanum lycopersicum*). PLoS One 12 (2), e0172466.

Liu, Q., Kasuga, M., Sakuma, Y., Abe, H., Miura, S., Yamaguchi-Shinozaki, K., et al., 1998. Two transcription factors, DREB1 and DREB2, with an EREBP/AP2 DNA binding domain separate two cellular signal transduction pathways in drought- and low-temperature-responsive gene expression, respectively, in *Arabidopsis*. Plant. Cell 10, 1391—1406.

Liu, P., Xu, Z.S., Pan-Pan, L., Hu, D., Chen, M., Li, L.C., et al., 2013a. A wheat PI4K gene whose product possesses threonine auto-phophorylation activity confers tolerance to drought and salt in *Arabidopsis*. J. Exp. Bot. 64, 2915—2927.

Liu, S., Lv, Z., Liu, Y., Li, L., Zhang, L., 2013b. Network analysis of ABA-dependent and ABA-independent drought responsive genes in *Arabidopsis thaliana*. Genet. Mol. Biol. 41 (3), 624—637.

Llanes, A., Masciarelli, O., Ordoñez, R., Isla, M.I., Luna, V., 2014. Differential growth responses to sodium salts involve different ABA catabolism and transport in the halophyte *Prosopis strombulifera*. Biol. Plant. 58, 80—88.

Llanes, A., Andrade, A., Alemano, S., Luna, V., 2016. Alterations of endogenous hormonal levels in plants under drought and salinity. Am. J. Plant. Sci. 7, 1357—1371.

Ma, Y., Szostkiewicz, I., Korte, A., Moes, D., Yang, Y., Christmann, A., et al., 2009. Regulators of PP2C phosphatase activity function as abscisic acid sensors. Science 324, 1064—1068.

Ma, L.F., Li, Y., Chen, Y., Li, X.B., 2015. Improved drought and salt tolerance of *Arabidopsis thaliana* by ectopic expression of a cotton (*Gossypium hirsutum*) CBF gene. Plant. Cell Tiss. Organ. Cult. 124 (3), 583—598.

Ma, Y., Dias, M.C., Freitas, H., 2020. Drought and salinity stress responses and microbe-induced tolerance in plants. Front. Plant. Sci. 11, 1750.

Mahajan, S., Sopoy, S.K., Tuteja, N., 2006. Cloning and characterization of CBL-CIPK signaling components from a legume (*Pisum sativum*). FEBS J. 273, 907—925.

Mantri, N., Patade, V., Penna, S., Ford, R., Pang, E., 2012. Abiotic stress responses in plants. In: Ahmad, P., Prasad, M.N.V. (Eds.), Abiotic Stress Response in Plants: Metabolism, Productivity and Sustainability. Springer Science, New York, p. 2.

Miura, K., Tada, Y., 2014. Regulation of water, salinity, and cold stress responses by salicylic acid. Front. Plant. Sci. 5, 4.

Miura, K., Okamoto, H., Okuma, E., Shiba, H., Kamada, H., Hasegawa, P.M., et al., 2013. SIZ1 deficiency causes reduced stomatal aperture and enhanced drought tolerance via controlling salicylic acid-induced accumulation of reactive oxygen species in *Arabidopsis*. Plant. J. 73, 91—104.

Miyakawa, T., Fujita, Y., Yamaguchi-Shinozaki, K., Tanokura, M., 2013. Structure and function of abscisic acid receptors. Trends Plant. Sci. 18, 259—266.

Muller, M., Munne-Bosch, S., 2015. Ethylene response factors: a key regulatory hub in hormone and stress signaling. Plant. Physiol. 169 (1), 32–41.

Nazar, R., Umar, S., Khan, N.A., 2015. Exogenous salicylic acid improves photosynthesis and growth through increase in ascorbate-glutathione metabolism and S assimilation in mustard under salt stress. Plant. Signal. Behav. 10 (3), e1003751.

Padmalatha, K.V., Dhandapani, G., Kanakachari, M., Kumar, S., Dass, A., Patil, D.P., et al., 2012. Genome-wide transcriptomic analysis of cotton under drought stress reveal significant down-regulation of genes and pathways involved in fibre elongation and up-regulation of defence responsive genes. Plant. Mol. Biol. 78, 223–246.

Palma, F., Lopez-Gomez, M., Tejera, N.A., Lluch, C., 2013. Salicylic acid improves the salinity tolerance of *Medicago sativa* in symbiosis with *Sinorhizobium meliloti* by preventing nitrogen fixation inhibition. Plant. Sci. 208, 75–82.

Pandey, A., Sharma, M., Pandey, G.K., 2016. Emerging roles of strigolactones in plant responses to stress and development. Front. Plant. Sci. 7, 434.

Park, S.Y., et al., 2009. Abscisic acid inhibits type 2C protein phosphatases via the PYR/PYL family of START proteins. Science 324, 1068–1071.

Park, J., Kim, Y.S., Kim, S.G., Jung, J.H., Woo, J.C., Park, C.M., 2011. Integration of auxin and salt signals by the NAC transcription factor NTM2 during seed germination in *Arabidopsis*. Plant. Physiol. 156 (2), 537–549.

Pavlu, J., Novak, J., Koukalova, V., Luklova, M., Brzobohaty, B., Cerny, M., 2018. Cytokinin at the crossroads of abiotic stress signalling pathways. Int. J. Mol. Sci. 19 (8), 2450.

Peleg, Z., Blumwald, E., 2011. Hormone balance and abiotic stress tolerance in crop plants. Curr. Opin. Plant. Biol. 14, 290–295.

Plaza-Wuthrich, S., Bloesch, R., Rindisbacher, A., Cannarozzi, G., Tadele, Z., 2016. Gibberellin deficiency confers both lodging and drought tolerance in small cereals. Front. Plant. Sci. 7, 643.

Pospisilova, H., Jiskrova, E., Vojta, P., Mrizova, K., Kokas, F., Cudejkova, M.M., et al., 2016. Transgenic barley overexpressing a cytokinin dehydrogenase gene shows greater tolerance to drought stress. N. Biotechnol. 33 (5), 692–705.

Qin, F., Kodaira, K.S., Maruyama, K., Mizoi, J., Tran, L.S., Fujita, Y., et al., 2011. SPINDLY, a negative regulator of gibberellic acid signaling, is involved in the plant abiotic stress response. Plant. Physiol. 157, 1900–1913.

Rai, A.N., Tamirisa, S., Rao, K., Kumar, V., Suprasanna, P., 2015. *Brassica* RNA binding protein ERD4 is involved in conferring salt, drought tolerance and enhancing plant growth in *Arabidopsis*. Plant. Mol. Biol. 90, 375–387.

Ramireddy, E., Hosseini, S.A., Eggert, K., Gillandt, S., Gnad, H., von Wiren, N., et al., 2018. Root engineering in barley: increasing cytokinin degradation produces a larger root system, mineral enrichment in the shoot and improved drought tolerance. Plant. Physiol. 177, 1078–1095.

Riyazuddin, R., Verma, R., Singh, K., Nisha, N., Keisham, M., Bhati, K.K., et al., 2020. Ethylene: a master regulator of salinity stress tolerance in plants. Biomolecules 10 (6), 959.

Roy, S., 2016. Function of MYB domain transcription factors in abiotic stress and epigenetic control of stress response in plant genome. Plant. Signal. Behav. 11 (1), e1117723.

Sah, S.K., Reddy, K.R., Li, J., 2016. Abscisic acid and abiotic stress tolerance in crop plants. Front. Plant. Sci. 7, 571.

Saha, B., Mishra, S., Awasthi, J.P., Sahoo, L., Panda, S.K., 2016. Enhanced drought and salinity tolerance in transgenic mustard [*Brassica juncea* (L.) Czern&Coss.] overexpressing *Arabidopsis* group 4 late embryogenesis abundant gene (AtLEA4-1). Environ. Exp. Bot. 128, 99–111.

Salaun, C., Lepiniec, L., Dubreucq, B., 2021. Genetic and molecular control of somatic embryogenesis. Plants 10, 1467.

Salehin, M., Bagchi, R., Estelle, M., 2015. SCFTIR1/AFB-based auxin perception: mechanism and role in plant growth and development. Plant. Cell 27, 9–19.

Sano, N., Marion-Poll, A., 2021. ABA metabolism and homeostasis in seed dormancy and germination. Int. J. Mol. Sci. 22, 5069.

Santiago, J., Dupeux, F., Betz, K., Antoni, R., Gonzalez-Guzman, M., Rodriguez, L., et al., 2012. Structural insights into PYR/PYL/RCAR ABA receptors and PP2Cs. Plant. Sci. 182, 3–11.

Santner, A., Estelle, M., 2009. Recent advances and emerging trends in plant hormone signalling. Nature 459, 1071–1078.

Saruhan, N., Saglam, A., Kadioglu, A., 2012. Salicylic acid pre-treatment induces drought tolerance and delays leaf rolling by inducing antioxidant systems in maize genotypes. Acta Physiol. Plant. 34, 97–106.

Schopfer, P., Liszkay, A., Bechtold, M., Frahry, G., Wagner, A., 2002. Evidence that hydroxyl radicals mediate auxin-induced extension growth. Planta 214, 821–828.

Seki, M., Ishida, J., Narusaka, M., Fujita, M., Nanjo, T., Umezawa, T., et al., 2002. Monitoring the expression pattern of around 7,000 *Arabidopsis* genes under ABA treatments using a full-length cDNA microarray. Funct. Integr. Genomics 2, 282–291.

Seltmann, M.A., Stingl, N.E., Lautenschlaeger, J.K., Krischke, M., Mueller, M.J., Berger, S., 2010. Differential impact of lipoxygenase 2 and jasmonates on natural and stress-induced senescence in *Arabidopsis*. Plant. Physiol. 152, 1940–1950.

Seo, J.S., Joo, J., Kim, M.J., Kim, Y.K., Nahm, B.H., Song, S., et al., 2011. OsbHLH148, a basic helix-loop-helix protein, interacts with OsJAZ proteins in a jasmonate signaling pathway leading to drought tolerance in rice. Plant. J. 65, 907–921.

Shinozaki, K., Yamaguchi-Shinozaki, K., 2007. Gene networks involved in drought stress response and tolerance. J. Exp. Bot. 58, 221–227.

Singh, P.K., Gautam, S., 2013. Role of salicylic acid on physiological and biochemical mechanism of salinity stress tolerance in plants. Acta Physiol. Plant. 35 (8), 2345–2353.

Singh, A., Jha, S.K., Bagri, J., Pandey, G.K., 2015. ABA inducible rice protein phosphatase 2C confers ABA insensitivity and abiotic stress tolerance in *Arabidopsis*. PLoS One 10 (4), e0125168.

Singh, A., Pandey, A., Srivastava, A.K., Tran, L.S.P., Pandey, G.K., 2016. Plant protein phosphatases 2C: from genomic diversity to functional multiplicity and importance in stress management. Crit. Rev. Biotechnol. 36 (6), 1023–1035.

Skirycz, A., Inze, D., 2010. More from less, plant growth under limited water. Curr. Opin. Biotechnol. 21, 197–203.

Soon, F.F., Ng, L.M., Zhou, X.E., West, G.M., Kovach, A., Tan, M.E., et al., 2012. Molecular mimicry regulates ABA signaling by SnRK2 kinases and PP2C phosphatases. Science 335, 85–88.

Souri, Z., Karimi, N., Farooq, M.A., Akhtar, J., 2020. Phytohormonal signaling under abiotic stress. Plant Life under Changing Environment. Elsevier, Amsterdam, pp. 397–466.

Stepanova, A.N., Alonso, J.M., 2009. Ethylene signaling and response: where different regulatory modules meet. Curr. Opin. Plant. Biol. 12, 548–555.

Syeed, S., Anjum, N.A., Nazar, R., Iqbal, N., Masood, A., Khan, N.A., 2011. Salicylic acid-mediated changes in photosynthesis, nutrients content and antioxidant metabolism in two mustard (*Brassica juncea* L.) cultivars differing in salt tolerance. Acta Physiol. Plant. 33, 877–886.

Tan, D.X., Tuong, H.M., Thuy, V.T.T., Son, L.V., Mau, C.H., 2015. Cloning and overexpression of GmDREB2 gene from a Vietnamese drought-resistant soybean variety. Braz. Arch. Biol. Techn. 58, 651–657.

Tani, T., Sobajima, H., Okada, K., Chujo, T., Arimura, S., Tsutsumi, N., et al., 2008. Identification of the OsOPR7 gene encoding 12-oxophytodienoate reductase involved in the biosynthesis of jasmonic acid in rice. Planta 227, 517–526.

Tiwari, S., Lata, C., Chauhan, P.S., Nautiyal, C.S., 2016. *Pseudomonas putida* attunes morphophysiological, biochemical and molecular responses in *Cicer arietinum* L. during drought stress and recovery. Plant. Physiol. Biochem. 99, 108–117.

Tiwari, S., Lata, C., Chauhan, P.S., Prasad, V., Prasad, M., 2017. A functional genomic perspective on drought signalling and its crosstalk with phytohormone-mediated signalling pathways in plants. Curr. Genomics 18 (6), 469–482.

Tran, L.S.P., Shinozaki, K., Yamaguchi-Shinozaki, K., 2010. Role of cytokinin responsive two component system in ABA and osmotic stress signalings. Plant. Signal. Behav. 5, 148–150.

Trivedi, D.K., Gill, S.S., Tuteja, N., 2016. Abscisic acid (ABA): biosynthesis, regulation, and role in abiotic stress tolerance. In: Tuteja, N.;, Gill, S.S. (Eds.), Abiotic Stress Response in Plants. Wiley-VCH Verlag GmbH & Co. KGaA, Weinheim, Germany, pp. 315–326.

Umezawa, T., Sugiyama, N., Mizoguchi, M., Hayashi, S., Myouga, F., Yamaguchi-Shinozaki, K., et al., 2009. Type 2C protein phosphatases directly regulate abscisic acid-activated protein kinases in *Arabidopsis*. Proc. Natl. Acad. Sci. U.S.A. 106 (41), 17588–17593.

Uno, Y., Furihata, T., Abe, H., Yoshida, R., Shinozaki, K., Yamaguchi-Shinozaki, K., 2000. *Arabidopsis* basic leucine zipper transcription factors involved in an abscisicacid-dependent signal transduction pathway under drought and high-salinity conditions. Proc. Natl. Acad. Sci. U.S.A. 97, 11632–11637.

Van Ha, C., Leyva-Gonzalez, M.A., Osakabe, Y., Tran, U.T., Nishiyama, R., Watanabe, Y., et al., 2014. Positive regulatory role of strigolactone in plant responses to drought and salt stress. Proc. Natl. Acad. Sci. U.S.A. 111 (2), 851–856.

Verslues, P.E., Kim, Y.S., Zhu, J.K., 2007. Altered ABA, proline and hydrogen peroxide in an *Arabidopsis* glutamate: glyoxylate aminotransferase mutant. Plant. Mol. Biol. 64, 205–217.

Vojta, P., Kokas, F., Husickova, A., Gruz, J., Bergougnoux, V., Marchetti, C.F., et al., 2016. Whole transcriptome analysis of transgenic barley with altered cytokinin homeostasis and increased tolerance to drought stress. N. Biotechnol. 33, 676–691.

Wang, R., Estelle, M., 2014. Diversity and specificity: auxin perception and signaling through the TIR1/AFB pathway. Curr. Opin. Plant. Biol. 21, 51–58.

Wang, W., Vinocur, B., Altman, A., 2003. Plant responses to drought, salinity and extreme temperatures: towards genetic engineering for stress tolerance. Planta 218 (1), 1–14.

Wang, H., Liang, X., Wan, Q., Wang, X., Bi, Y., 2009. Ethylene and nitric oxide are involved in maintaining ion homeostasis in *Arabidopsis* callus under salt stress. Planta 230 (2), 293–307.

Wang, C., Zhang, S., Wang, P., Hou, J., Qian, J., Ao, Y., et al., 2011. Salicylic acid involved in the regulation of nutrient elements uptake and oxidative stress in *Vallisneria natans* (Lour.) Hara under Pb stress. Chemosphere 84, 136–142.

Wang, F., Kong, W., Wong, G., Fu, L., Peng, R., Li, Z., et al., 2016. AtMYB12 regulates flavonoids accumulation and abiotic stress tolerance in transgenic *Arabidopsis thaliana*. Mol. Genet. Genomics 294 (4), 1545–1559.

Wasilewska, A., Vlad, F., Sirichandra, C., Redko, Y., Jammes, F., Valon, C., et al., 2008. An update on abscisic acid signaling in plants and more. Mol. Plant. 1, 198–217.

Wasternack, C., Feussner, I., 2018. The oxylipin pathways: biochemistry and function. Annu. Rev. Plant. Biol. 69, 363–386.

Weiner, J.J., Peterson, F.C., Volkman, B.F., Cutler, S.R., 2010. Structural and functional insights into core ABA signaling. Curr. Opin. Plant. Biol. 13, 495–502.

Wen, F.P., Zhang, Z.H., Bai, T., Xu, Q., Pan, Y.H., 2010. Proteomics reveals the effects of gibberellic acid (GA3) on salt stressed rice (*Oryza sativa* L.) shoots. Plant. Sci. 178, 170–175.

Xie, Z., Kasschau, K., Carrington, J., 2003. Negative feedback regulation of Dicer-Like1 in *Arabidopsis* by microRNA-guided mRNA. Curr. Biol. 13 (9), 784–789.

Yamaguchi, S., 2008. Gibberellin metabolism and its regulation. Annu. Rev. Plant. Biol. 59, 225–251.

Yamaguchi-Shinozaki, K., Shinozaki, K., 1993. Arabidopsis DNA encoding two desiccation-responsive rd29 genes. Plant. Physiol. 101 (3), 1119–1120.

Yoshida, H., Tanimoto, E., Hirai, T., Miyanoiri, Y., Mitani, R., Kawamura, M., et al., 2018. Evolution and diversification of the plant gibberellin receptor GID1. Proc. Natl. Acad. Sci. U.S.A. 115 (33), E7844–E7853.

Yu, Y.T., Wu, Z., Lu, K., Bi, C., Liang, S., Wang, X.F., et al., 2016. Overexpression of the MYB37 transcription factor enhances abscisic acid sensitivity, and improves both drought tolerance and seed productivity in *Arabidopsis thaliana*. Plant. Mol. Biol. 90, 267–279.

Zhai, Q., Zhang, X., Wu, F., Feng, H., Deng, L., Xu, L., et al., 2015. Transcriptional mechanism of jasmonate receptor COI1-mediated delay of flowering time in *Arabidopsis*. Plant. Cell 27, 2814–2828.

Zhou, M., Memelink, J., 2016. Jasmonate-responsive transcription factors regulating plant secondary metabolism. Biotechnol. Adv. 34, 441–449.

Zhou, X., Yan, S., Sun, C., Li, S., Li, J., Xu, M., et al., 2015. A maize jasmonate Zim-domain protein, ZmJAZ14, associates with the JA, ABA, and GA signaling pathways in transgenic *Arabidopsis*. PLoS One 10 (3), e0121824.

Zhu, J.K., 2002. Salt and drought stress signal transduction in plants. Annu. Rev. Plant. Biol. 53, 247–273.

Zwack, P.J., Rashotte, A.M., 2015. Interactions between cytokinin signalling and abiotic stress responses. J. Exp. Bot. 66, 4863–4871.

Regulation of plants nutrient deficiency responses by phytohormones

Deepika Deepika, Kamankshi Sonkar and Amarjeet Singh

National Institute of Plant Genome Research, New Delhi, India

7.1 Introduction

Living organisms have the ability to acquire nutrients from their physical environment and subsequently convert them into an energy source for survival and growth. In plants, nutrients are involved in metabolism and physiology either as constituents of metabolites or enzymes for macromolecule biosynthesis. Based on their concentration in plant dry matter, the 14 essential inorganic elements are categorized into macronutrients and micronutrients. Macronutrients include six elements, nitrogen (N), phosphorus (P), potassium (K), sulfur (S), calcium (Ca), and magnesium (Mg), whereas the micronutrients comprised eight elements, chlorine (Cl), iron (Fe), boron (B), manganese (Mn), zinc (Zn), copper (Cu), molybdenum (Mo), and nickel (Ni) (de Bang et al., 2020; Sustr et al., 2019). Nutrients are distributed in a patchy manner in soil due to their variable interactions with spatially and temporally dispersed charged soil particles (Hodge, 2006). This, along with many other factors, leads to either low nutrient concentration in soil or low accessibility for plants. Low availability of a nutrient causes specific deficiency symptoms. But plants usually face multiple nutrient deficiencies simultaneously leading to a complex response and symptoms. Moreover, various biotic and abiotic stress factors, such as pests, pathogens, water deficit, salinity, and light, also interact to cause atypical nutrient deficiency symptoms (Amtmann et al., 2008; Troufflard et al., 2010; Atkinsonand Urwin., 2012).

Globally, nutrient deficiencies are major threats to crop production, causing reduced yields and poor food and feed quality. The world food security challenge is being met by the use of chemical fertilizers (natural and anthropogenic). However, the negative impact of chemical fertilizers on both environment and living beings necessitate improving crop nutrient uptake and use efficiency (Zhang et al., 2015; Gojon, 2017). Nutrient use efficiency (NUE) is defined as the ratio of nutrient elements in the harvested crop output to its input in terms of manure, fertilizer and biological fixation, or any other form. In 2010 NUE for major nutrients like N and P was 0.3 and 0.4, respectively, in India, which should increase to 0.6 and 0.65 by 2050 to feed an additional 273 million (Zhang et al., 2015; Bouwman et al., 2017). The projected increase in NUE could only be achieved with a better understanding of nutrient perception, uptake, transportation, assimilation, and adaptive responses of plant genotypes tolerant to low nutrient supplies.

A huge amount of information about the functional roles of macronutrients in plants is already available. Recent reports highlight the molecular responses of plants to specific nutrients deficiency. Interactions of different nutrient signaling pathways in sensing, uptake, and utilization of multiple nutrients have also been elaborated. An important role of phytohormones has been found in plant responses deficiencies of major macronutrients, including N, P, and K (Schachtman, 2015). Nutrient deficiency signaling and nutrient uptake are regulated by different plant hormones, such as auxin, cytokinin (CK), abscisic acid (ABA), jasmonic acid (JA), gibberellic acid (GA), brassinosteroids (BR), strigolactones (SLs), and ethylene (ET) (Armengaud et al., 2004; Ma et al., 2012; Schachtman, 2015; Vissenberg et al., 2020; Zhang et al., 2017). Additionally, hormones, such as CKs and SLs, are used as relay signals to convey N and P nutrient status across root−shoot−root (Franco-zorrilla et al., 2005; Wang et al., 2006; Umehara et al., 2008). In this chapter, we have described the hormonal regulation of physiological and molecular processes to cope N, P, and K deficiencies in plants.

7.2 Major macronutrients for plants

7.2.1 Nitrogen

N is an essential component of the major macromolecules, such as nucleic acids (DNA, RNA), proteins, and enzymes. One of the most apparent visual symptoms of N deficiency in plants is the yellowing of the whole leaf surface. The deficiency of N affects chlorophyll biosynthesis and causes the breakdown of chlorophyll-binding proteins, resulting in yellow leaves. However, the yellowing of leaves proceeds in an acropetal manner where N-related compounds are broken down in older leaves, and the raw materials, such as amino acids and amines, are transported through phloem to the younger leaves. N levels also directly regulate glutamate, the precursor of chlorophyll biosynthesis. Another prominent symptom of N deficiency is the purple coloration of the shoot, attributed to enhanced anthocyanin production. Anthocyanin biosynthesis is controlled by key TFs, PRODUCTION OF ANTHOCYNANIN PIGMENT1 (PAP1) and (PAP2) (Xu et al., 2015), the expression of which has been shown to be regulated by NO_3^- levels. Elevated miR156 levels during N deficiency enhance anthocyanin production by suppressing the activity of SPL9, a negative regulator of anthocyanin biosynthesis (Cui et al., 2014). However, low NO_3^- levels also reduce anthocyanin production through nitrate signaling involving NLP7 cascade (Rubin et al., 2009) and repressed GA signaling. Majority of N (>98%) exists in environment in its native unreactive (N_2) form that is made usable for plants by the soil microbes through its conversion into different inorganic forms (NH_4^+, NO_2^-, and NO_3^-). Both soil type and plant species determine which N form will be preferably utilized. Cereals grow well in aerobic soils/high pH where NO_3^- is dominantly present, while NH_4^+ is the most preferred form in paddy fields with a reducing soil environment (Gojon, 2017; Zarabi and Jalali, 2012). NO_3^-, being an anion, does not form strong complexes with soil particles and tends to leach out if left unabsorbed in the soil (Jin et al., 2015; Iqbal et al., 2020). Consequently, the concentration of NO_3^- in soil solution varies from 10 to 70 μM (Dechorgnat et al., 2011; Iqbal et al., 2020). Therefore plants are evolved with both high- and low-affinity transport systems (HATS, LATS) for the uptake of NO_3^- from the soil (Wang et al., 2012; Fan et al., 2017).

NO_3^- uptake is an active transport process and occurs against a concentration gradient, as NO_3^- is stored in vacuoles at higher concentration (2–6 mM) compared to soil (0.5–1 mM) (Fan et al., 2017; Iqbal et al., 2020). NO_3^- uptake is initially mediated by the constitutively active low- and high-affinity transporters, cLATS and cHATS, which later increases manifold in a positive feedback loop through an induced expression of high-affinity transporters called iHATS and cHATS. All these transporters have different K_m [cHATS; AtNRT2.6: 2–20 μM; iHATS; AtNRT2.1: 13–79 μM; cLATS; AtNRT1.2/AtNPF4.6: >0.25 mM and iLATS; AtNRT1.1/AtNPF6.3: very high (>0.5 mM)] (Iqbal et al., 2020; Okamoto et al., 2006; Tsay et al., 2007). However, to avoid NO_3^- toxicity, cells also utilize various efflux carriers, such as AtNPF2.7/AtNAXT1 (Segonzac et al., 2007; Kant, 2018). Inside plant cells, NO_3^- is reduced and assimilated either in roots or translocated to shoots via xylem as NO_3^-, NH_4^+, or amino acids for storage. From shoots, it is transported to N sinks, such as developing leaves, fruits, and seeds (Tegeder and Masclaux-Daubresse, 2018).

7.2.2 Phosphorous

Phosphorous is another essential nutrient comprising 0.2% of a plant's dry weight and hence is required in huge amounts. Purple coloration (anthocyanosis) of plant organs in different plants, such as adaxial (maize) or abaxial (tomato) surfaces of leaf, leaf margins, and shoot, is a distinguishing feature of P deficiency (Gould et al., 2018; Hughes and Lev-Yadun, 2015. It is caused by anthocyanin accumulation triggered by an enhanced expression of anthocyanin biosynthetic genes (F'3H and LDOX). Under low Pi conditions, GA levels are reduced, and GA signaling inhibitors DELLAs are activated, which, in turn, activate the expression of anthocyanin biosynthetic genes (Jiang et al., 2007). In soil, P is present in interchangeable states, namely, fixed, active, and solution states. Plants primarily use the inorganic form (Pi, 35%–70%) of soil P. Pi exists as oxyanions, that is, $H_2PO_4^-$, HPO_4^{2-} in soil solution in the concentration range 1–10 μM. Low concentration of Pi is attributed to anions being both highly diffusible and amenable to soil fixation (López-Arredondo et al., 2014; Lambers and Plaxton, 2015; Ham et al., 2018). After its acquisition through roots, Pi transport involves its allocation (root–shoot through xylem), mobilization (from source to sink through phloem, varies in vegetative and reproductive phases), and intracellular transport to various internal organelles, such as mitochondria, plastids, Golgi, and plant vacuoles. Plants acquire, transport, and remobilize soil Pi using both high- and low-affinity transport systems (Raghothama and Karthikeyan, 2005; He et al., 2019). PHT1/PT (phosphate transporter) family proteins comprise both LATS (mM range) and HATS ($1-5\ \mu M$) distinguished by their K_m (Wang et al., 2017). Other Pi transporters in plants belong to SPX-domain containing protein family and PHT2/3/4 families, including intracellular, mitochondrial, plastid, and Golgi-located transporters (Gu et al., 2016; Młodzińska and Zboińska, 2016; Wang et al., 2018). Being a mobile element, P is acquired and transported majorly in the inorganic form throughout plant, except during its remobilization under senescence and P-limiting conditions when its organic sugar phosphates are transported from older to younger leaves. Besides leaves, the reproductive organs, such as flower (both stigma and pollen), embryo, and seeds, act as Pi sink during the reproductive phase, where Pi is further converted into phytate. In seed, P provides fuel for germination and seedling establishment (Gu et al., 2016; Li et al., 2015; Młodzińska and Zboińska, 2016).

Under Low Pi, plants increase the Pi content of the rhizosphere through the secretion of organic acids (OAs) to mobilize the Pi trapped in Al, Fe, and Ca salts. The levels of OAs are increased by induced activity of two key enzymes, Citrate synthase (CS) and PEP carboxylase (PEP-C) of Calvin cycle. OAs have high affinity for metallic ions, such as Al^{3+}, Fe^{3+}, and Ca^{2+}. OAs bind to these ions through ligand exchange, releasing the inorganic Pi from these complex salts (Delhaize et al., 2012).

7.2.3 Potassium

Potassium (K^+) is a monovalent cation, a mobile element of the plant cell and constitutes approximately 10% of the plant's dry weight. Because of its requirement in diverse functions, plant cells maintain a high concentration of K^+ (100 mM). K^+ maintains electroneutrality, osmotic regulation, anion−cation balance, and optimal pH. Additionally, it regulates the stability of >70 enzymes/proteins, such as pyruvate kinase, phosphofructokinase, and starch synthase (Anschütz et al., 2014). In soil, K^+ is present majorly in three forms, dissolved in soil solution (0.1%−0.2%), adsorbed to soil clay particles (1%−2%), and inside crystals of minerals, such as feldspars and mica (96%−99%) (Britzke et al., 2012). Only the soluble form of K^+ is directly available for plants. Hence, despite a significant amount of K^+ present as minerals, its low bioavailability leads to K^+ deficiency; consequently, the application of huge amount of potash fertilizers (K_2O) is required for sustainable agriculture. For example, 72% of the Indian agricultural land requires extensive K^+ fertilizer application that imposes a financial burden on farmers (Yadav and Sidhu, 2016). Moreover, the unused fertilizer leaches down to water bodies and has many serious effects on climate, soil, and human health. The prominent visual symptoms of K^+ deficiency includes chlorosis at the tips of older leaves, which eventually spreads to whole leaf margins, potentially promoting tip senescence (Ueno et al., 2018). Under K^+ deficiency, reduced CO_2 fixation rates result in an excess of electrons, causing enhanced ROS production, which is responsible for leaf tip chlorosis (Cakmak, 2010). However, to overcome K^+ deficiency, plants synthesize polyamines (PAs), such as putrescine, accumulation of which causes additional symptoms like bronzing. PAs function to maintain cellular charge, pH buffering, scavenging of ROS, activation of the antioxidant machinery, alleviation of stress-induced membrane injury, and electrolyte leakage (Pottosin and Shabala, 2016). Another common feature of plants facing K^+ deficiency is slack leaves, primarily caused by reduced turgor pressure in epidermal cells of the stomata. Increased ET content is a major cause for this phenotype, as ET interferes with ABA signaling, which is required to maintain cell turgor. Moreover, under low K^+ conditions, the porosity of xylem walls decreases, resulting in reduced hydraulic conductance and hampered transpiration through leaves (Anschütz et al., 2014).

Several families of membrane protein transporters are involved in K^+ uptake, allocation, and homeostasis. Members of the KT/HAK/KUP protein family transporters, such as HAK1 in rice (Martínez-Cordero et al., 2004), AKT1 and HAK5 in Arabidopsis (Rubin et al., 2009), and KUP1 and KUP2 in barley (Cai et al., 2019), act as HATS and are active at low K^+ concentration (<1 mM). Besides KT/HAK/KUP transporters, Na^+-K^+ transporters, such as HKTs (OsHKT2;1), also help in K^+ uptake (Hamamoto and Uozumi, 2014). However, under normal conditions, more than 50% of the plant K^+ uptake is facilitated through three different types of channels, voltage-gated shaker-type channels (AKT, KAT, SKOR, GORK), voltage-independent two-pore K^+ channels (TPK), and the K^+ inward-rectifier (Kir) channels (Amtmann and Blatt, 2009). Exogenous supply of K^+ mitigates the effects of drought in wheat, rice, and maize (Islam et al., 2015; Islam and Muttaleb, 2016). Moreover, salt stress creates ambient K^+-deficient environment by mimicking Na^+ for uptake (Wang et al., 2013). Hence, plants under K^+ deficiency become more susceptible toward drought and salinity stresses.

7.3 Hormonal regulation of N deficiency responses

7.3.1 Root system architecture

Mild to severe N deficiency has contrasting effects on root growth. Under mild N deficiency, both primary (PR) and lateral roots (LR) elongate to forage N from deeper soil layers. Whereas, severe N deficiency inhibits PR growth (Gruber et al., 2013). NO_3^- uptake, sensing, and signaling are regulated by multiple mechanisms, which mainly include nitrate availability, feedback repression by N status, and hormone signaling. Interestingly, there is a significant correlation between NO_3^- content of the soil and level of auxin in the plant leaves under different N regimes in Arabidopsis (Krouk et al., 2010; Walch-Liu et al., 2006), soybean (Caba et al., 2000) and wheat (Chen et al., 1998). During mild N deficiency, auxin is transported from shoots to roots to assist root foraging while it is removed from the root tips under severe N deficiency. Along with auxin, CLAVATA3/Embryo Surrounding Region-Related (CLE) peptides also inhibit lateral root growth locally during severe N deficiency (Araya et al., 2014) (Fig. 7.1A). There is a two-way connection between N availability and auxin. Presence of nitrate regulates auxin biosynthesis and transport, and accumulated auxin acts as the regulator for both N acquisition and assimilation (Krouk et al., 2011; O'Brien et al., 2016). Mild N deficiency induced LR growth in Arabidopsis which attributed to the function of TAR2 (Tryptophan aminotransferase 2), an auxin biosynthesis gene expressed in vasculature and pericycle of roots (Ma et al., 2014). However, inhibition of LR growth during prolonged N deficiency is controlled by NRT1.1 that facilitates auxin transport in LR epidermal cells and prevents its accumulation in LR tips, inhibiting their growth (Krouk et al., 2010). NRT1.1 is known as a transceptor because it exhibits the dual function of sensing nitrate concentration and transporting auxin based on nitrate availability (Ho et al., 2009). Hence, both the NO_3^-

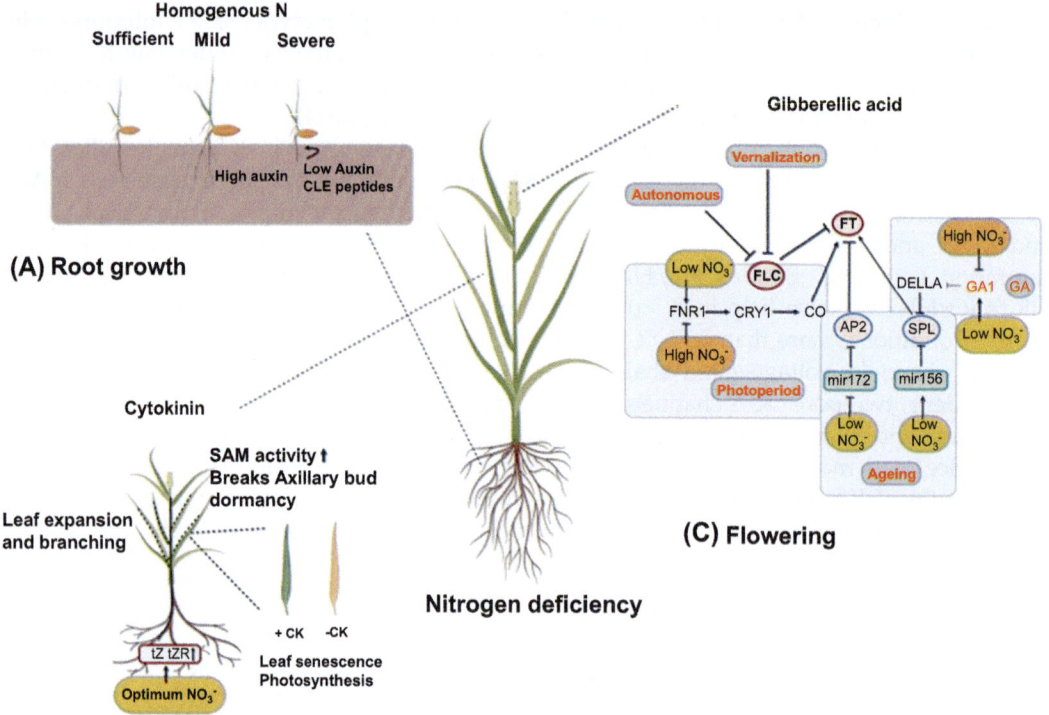

FIGURE 7.1   Regulation of nitrogen deficiency by plant hormones: (A) Mild N deficiency induces a root foraging response through auxin accumulation in root tips, which causes lateral roots to grow faster and elongate to deeper soil layers. Whereas, severe N deficiency quenches auxin from root tips, resulting in shorter roots, which is also an effect of CLAVATA3/Embryo surrounding Region-Related (CLE) peptides; (B) NO$_3^-$ content of soil regulates leaf expansion and shoot branching in plants by modulating CK; (C) NO$_3^-$ levels affect flowering and both are correlated via different pathways: aging, photoperiod, vernalization, autonomous, and GA. Mild N deficiency causes early flowering compared to severe N deficiency and high N levels. All these pathways are modulated by NO$_3^-$ levels and converge at the central protein FT. *CK*, Cytokinin; *GA*, gibberellic acid.

signal and the auxin interact through NRT1.1. However, NRTs are only involved in the local redistribution of auxins. Furthermore, the phosphorylation status of NRT1.1 acts as a regulatory switch between its high- and low-affinity states. In the nonphosphorylated state, NRT1.1 forms homodimer that has low NO$_3^-$ affinity, while NRT1.1^{T101} phosphorylated form is monomeric with high affinity for NO$_3^-$ (Sun et al., 2014). Phosphorylated NRT1.1 induces calcium influx into the cytosol and nucleus, which then phosphorylates NIN-like protein 7 (NLP7) via a calcium kinase cascade. Phosphorylation of NLP7 leads to its translocation into the nucleus where it induces the expression of Arabidopsis Nitrate Response 1 (ANR1) and other nitrate transporters, like NRT2.1 (Castaings et al., 2009; Marchive et al., 2013; Leran et al., 2015; Liu et al., 2017).

PR and LR growths are also regulated by the common effect of NO$_3^-$ availability and auxin signaling via miR393/AFB3 module. Expression of both miR393 and AFB3 (an auxin receptor) are independently induced by nitrate availability, and the miR393 inhibits AFB3 posttranscriptionally (Vidal et al., 2010). Under low NO_3^- conditions, auxin binds to its receptors (members of the SCFTIR1/AFB E3 ubiquitin ligase complex) and promotes recognition and degradation of Aux/IAA repressors via polyubiquitination. This leads to auxin response factors (ARF)-induced auxin-responsive transcription, which, in turn, activates the transcription factor (TF) NAC4, which further induces the expression of, OBP4. The nitrate-AFB3-NAC4-OBP4 signaling is required for the nitrate-dependent LR initiation and emergence (Vidal et al., 2014).

In Arabidopsis, severe N deficiency reduces PR length, and JA is known to be a negative regulator of PR growth in plants suggesting the direct involvement of JA in combating N deficiency (Chen et al., 2011). However, levels of both JA and JA-Ile declined under N deficiency in Arabidopsis (Conesa et al., 2020). Interestingly, many of the JA biosynthesis genes such as *OsLOXs* and *OsAOSs* have been shown to be differentially expressed in rice seedlings at both early and late time points of N deficiency (Deepika and Singh, 2021) hinting toward a different strategy adopted by cereal plants.

7.3.2 Leaf expansion and branching

Nitrogen deficiency leads to stunted growth, reduced branching, and leaf surface area. All these growth defects are a consequence of reduced CK in N-deficient shoots (Rahayu et al., 2005). Under high NO_3^- conditions, CK biosynthesis is promoted in roots. Accumulated CK is then translocated to shoots and regulates shoot growth (Sakakibara et al., 2006) (Fig. 7.1B). NO_3^- augmentation in the N limiting conditions influences CK biosynthesis in Arabidopsis roots, specifically via NO_3^--induced NRT1.1 controlled upregulation of two CK biosynthetic genes, Isopentenyl transferase (IPT3, IPT5) and CYP735A2 (cytochrome P450 monooxygenase) (Woo et al., 2012). The synthesized CK is then translocated from root–shoot as trans-zeatin (tZ) or trans-zeatin-riboside (tZR) and affects shoot apical meristem activity that also affects leaf expansion (Osugi et al., 2017; Landrein et al., 2018). However, branching is promoted by the formation of axillary buds, strongly regulated by NO_3^--induced CK content (Müller et al., 2015). However, the low NO_3^--induced CK accumulation leads to reduced shoot branching through interaction with auxin and SLs (Jong et al., 2014) (Fig. 7.1B).

7.3.3 Early flowering

Nitrogen deficiency promotes transition from vegetative to reproductive stage resulting in earlier flowering compared to N sufficient and prolonged N deficiency conditions (Lin and Tsay, 2017). Many pathways (GA, aging, photoperiod, vernalization, and autonomous) act together to control the N availability and flowering and the level of the central protein FLOWERING LOCUS T(FT). In the GA pathway, low NO_3^- levels induce the expression of the GA biosynthesis gene, GA1, and SUPPRESSOR OF OVEREXPRESSION OF CO 1(SOC1) an integrator of all the endogenous pathways (Liu et al., 2013). Higher GA1 levels repress the activity of DELLA, which in turn modulate the activity of APETALA 2 (AP2) and SQUAMOSA PROMOTER BINDING PROTEIN-LIKE (SPL) TFs of the aging pathway that collectively modulates levels of FT. In the aging pathway, low NO_3^- induces expression of miR156 along with downregulation of miR172. Balance of both these microRNAs inversely affects levels of FT through Tfs AP2 and SPL (Liang et al., 2012; Lin and Tsay, 2017). Low N levels also interact with the photoperiod pathway, where it induces the expression of ferredoxin-$NADP^+$-oxidoreductase (FNR1) that leads to early flowering response through photoreceptor cryptochrome 1 (CRY1) and FT levels (Yuan et al., 2016). However, N status and flowering are also regulated by vernalization and autonomous pathways (Fig. 7.1C).

7.4 Hormonal regulation of P deficiency responses

7.4.1 Root system architecture

Due to limited Pi mobility and its high availability in the topsoil, plant roots experiencing Pi deficiency usually exhibit shorter PRs, longer LRs, and enhanced root hair density (Vance et al., 2003) (Fig. 7.2A). However, PR response to Pi deficiency differs across plant kingdom. In Arabidopsis, primary root length is reduced due to root apical meristem (RAM) exhaustion where root tip cells lose their meristematic activity, and cell expansion in the elongation zone is inhibited (Gutiérrez-Alanís et al., 2018). Interestingly, this root growth defect is caused primarily by the accumulation of Fe^{3+} ions (Ward et al., 2008). Fe^{3+} promotes the exudation of malate through the STOP1-ALMT module in the apoplast of the root tip. The Fe^{3+}-malate complexes causes iron toxicity that leads to ROS production. ROS stimulates callose formation that inhibits root cell elongation; consequently, PR growth is hampered (Balzergue et al., 2017). In cereals such as maize, slight or no change was observed in the PR length, whereas in rice, PR elongated in response to Pi deficiency (Dai et al., 2012; Péret et al., 2014). Transcriptomic studies have shown enhanced expression of PHT1 transporters on root hair tips, suggesting enhanced Pi acquisition through newly developed root–soil interfaces (Ayadi et al., 2015; Remy et al., 2012; Smith et al., 2011).

Interestingly, a vital enzyme of the JA biosynthesis pathway AtOPR3 has been shown to directly inhibit PR growth under Pi deficiency in Arabidopsis without any involvement of JA signaling (Zheng et al., 2016). Moreover, AtOPR3 inhibits PR growth through interaction with GA and ET signaling pathways. Recently, many of the JA biosynthesis genes (*OsLOX6*, *OsAOSs*) were also found to be differentially expressed at both early and late time points of Pi deficiency in rice seedlings suggesting a potential role of JA in mitigating Pi deficiency response. Also, a total of 24 genes from different families of JA biosynthesis genes were strongly upregulated in roots relative to an array of other vegetative and reproductive tissues of rice (Deepika and Singh, 2021). Endogenous auxin is indispensable for RSA modifications under Pi deficiency (Nacry, 2005). Auxin accumulation and its polar transport are essential for LR development. Auxin accumulates in the root tip and inhibits primary root growth in response to low Pi in the soil, irrespective of shoot Pi content. However, auxin accumulation promotes lateral root emergence through auxin/TIR1 signaling, which leads to the activation of an ARF that further promotes lateral root emergence (Bhosale et al., 2018).

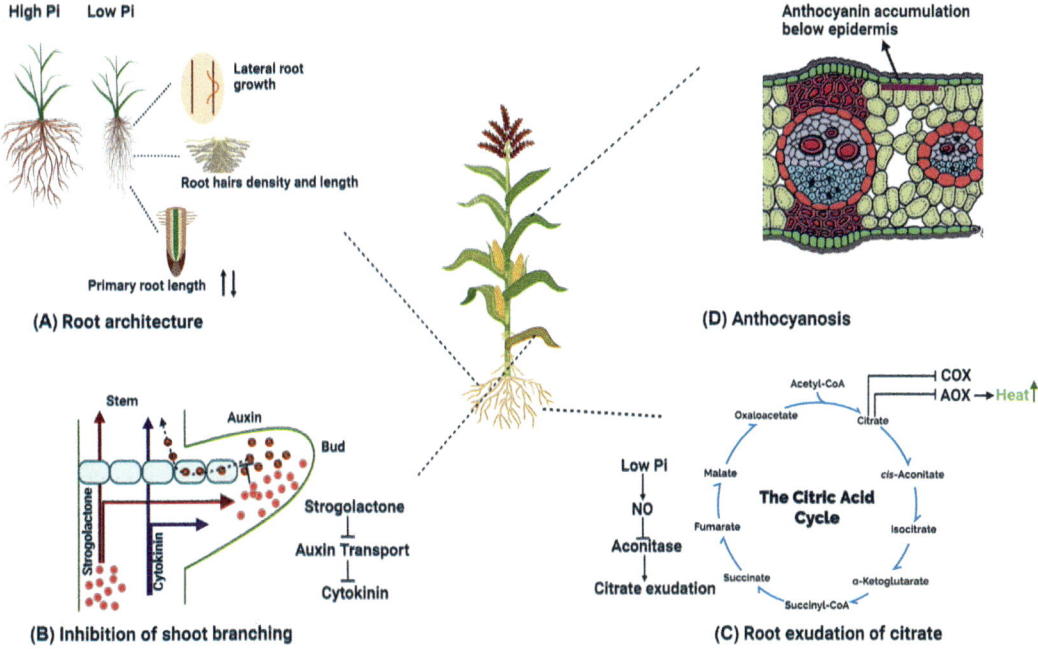

FIGURE 7.2 Regulation of phosphorus deficiency response by plant hormones: (A) RSA is strongly affected by phosphorous (Pi) deficiency. Pi deficiency may enhance or reduce both primary and lateral root growth, whereas it always induces root hair density and length; (B) inhibition of shoot branching; (C) citrate exudation from roots, (D) anthocyanosis. *RSA*, Root system architecture.

Increase in the number of root hairs under Pi deficiency is an integrated outcome of auxin synthesis, transport, and signaling pathways. Root hairs are the epidermal outgrowths, arising from specialized epidermal cells called trichoblasts. Pi deficiency triggers more number of cells to become trichoblasts (Janes et al., 2018) and converts developmentally programed atrichoblasts into potential trichoblasts, which eventually results in more root hairs. Enhanced expression of auxin biosynthesis gene, *TAR1* and *AtAUX1, OsAUX1* leads to enhanced auxin production. Higher auxin level triggers auxin signaling that activates TFs, such as root hair defective (*RHD6*)-like 4(RSL4) that is required for activation of root hair elongation genes (Mangano et al., 2017; Bhosale et al., 2018; Crombez et al., 2019). CK and GA levels decrease along with enhanced ET to modulate RSA under Pi deficiency, but how these hormones interact is still unknown. Plants control the level of CK under low Pi by reducing the expression of the rate-limiting enzyme of CK biosynthesis, IPT3 (Hirose et al., 2008) and CK receptor, AHK4 (Franco-Zorrilla et al., 2002). However, if CKs are applied to plant roots, they inhibit phosphate starvation response (PSR) in both Arabidopsis and Rice (Martín et al., 2000; Shen et al., 2014) and increase intracellular Pi levels. Hence, an exogenous application of CK helps plants survive Pi-deficient conditions by utilizing the internal Pi reserves.

7.4.2 Shoot branching

Shoot branching is attributed to the growth of axillary buds, which is inhibited during Pi deficiency through an integrated hormonal network involving auxin, CKs, and SLs. Pi deficiency induces the biosynthesis of SLs (a peptide hormone) in roots, which is subsequently transported to buds and lead to the degradation of auxin efflux carriers PIN1, thus inhibiting auxin transport from bud to shoots (Kohlen et al., 2011; Kebrom et al., 2013). Consequently, higher auxin levels in the bud decreases the CK content required for bud outgrowth and shoot branching (Fig. 7.2B). In crop plants, Pi deficiency therefore greatly reduces the number of tillers affecting grain yields (Carstensen et al., 2018).

7.4.3 Citrate exudation

Phosphate availability challenges plant survival by affecting two crucial physiological processes, photosynthesis and respiration. Low Pi content severely affects both light and dark reactions of photosynthesis as P is an essential component of ATP and NADPH; Pi deficiency results in reduced ATP levels as Pi along with CO_2 and H_2O is required as the substrate of light reaction. In addition, Pi is essential for the conversion of CO_2 to carbohydrates inside chloroplast stroma (Carstensen et al., 2018). Low Pi also results in a reduced supply of substrates for ATP synthesis in COX pathway, leading to the activation of AOX pathway for optimal electron supply for respiration (Wang et al., 2015)., Furthermore, Pi starvation also results in enhanced nitric oxide (NO) levels in roots, which inhibits aconitase action, leading to citrate accumulation and its exudation to rhizosphere (Wang et al., 2010; Gupta et al., 2013) Enhanced citrate levels further trigger the AOX pathway relative to COX pathway along with citrate exudation to rhizosphere (Fig. 7.2C). Hence, Pi deficiency affects multiple processes from TCA cycle to citrate exudation, and organic acid production.

7.5 Hormonal regulation of K^+ deficiency responses

7.5.1 Root growth

Different plant species and genotypes deploy a combination of diverse mechanisms to cope with K^+ deficiency, such as chemical modifications of the rhizosphere, enhanced K^+ uptake, an active root meristem, and efficient translocation of K^+ between source−sink organs (Chen and Gabelman, 2000, 1995; Sustr et al., 2019). Important crops, such as rice (Jia et al., 2008), maize (Zhao et al., 2016), and *Medicago truncatula*, exhibit proliferation of the roots, whereas in tomato, K^+ uptake from existing root surface is enhanced. RSA alteration mostly involves extensive branching of the root system. In rice, an abundance of fine roots was observed (Klinsawang et al., 2018), while in Arabidopsis and tobacco, the number of first-order lateral roots decreases in K^+ deficiency with an increase in the number of second-order lateral roots (Kellermeier et al., 2014; Song et al., 2018). Moreover, roots undergo significant anatomical alterations under low K^+ such as development of aerenchyma in the cortex for extensive soil search and enhanced endodermal suberization regulated by stress hormone ABA (Barberon et al., 2016; Barberon, 2017). Plants have evolved diverse signaling pathways to respond to low K^+ stress, such as phytohormone signaling, Ca^{2+} signaling, ROS, and miRNAs (Behera et al., 2017). Reduced number of lateral roots under K^+ deficiency is governed by auxin signaling and reduced sucrose transport. Low K^+ content in the soil decreases auxin levels in plants, and degradation of AtPIN1 proteins in the root tip disrupts auxin maxima required for optimal RAM activity. Furthermore, the expression of MYB77, which interacts with ARFs to enhance expression levels of auxin responsive genes for growth of lateral roots, is also reduced (Shin et al., 2007). Under prolonged exposure of K^+ deficiency (6–30 hours), ET promotes the expression of HAK5 (a high affinity K^+ transporter) through ROS−ARF2 cascade. Many studies have shown the induction of JA biosynthesis genes and proteins along with an increased JA content in Arabidopsis, rice, chickpea, and wheat plants under K^+ deficiency (Armengaud et al., 2004; Troufflard et al., 2010; Ma et al., 2012; Li et al., 2017; Deepika and Singh, 2021; Deepika et al., 2022). Also, a significant portion of K^+ deficiency−responsive transcriptome was either replaced/absent in *coi-16* JA receptor mutant that indicates the vital role of JA signaling under K^+ responsein Arabidopsis (Armengaud et al., 2004). This suggests a crucial role of JA in mitigating adverse effects of K^+ deficiency.

7.5.2 Phloem transport

K^+ is required for H^+-coupled sucrose transporters (SUT) driven phloem loading of photoassimilates. The transport of sucrose is accompanied by $H^+ −$ ATPases-mediated influx of H^+ ions from mesophyll apoplast to phloem. K^+ is required to maintain the function of these H^+-ATPases (Wigoda et al., 2014). K^+ deficiency results in the decreased phloem transport of sucrose from source (leaves) to sink (roots). Moreover, the disruption of a metal-binding protein NAKR1 (SODIUM POTASSIUM ROOT DEFECTIVE1) leads to overaccumulated K^+ in the shoots, resulting in impaired phloem loading, a smaller root system, and late flowering. This suggests a close link between K^+ recycling and root growth. Thus K^+-mediated phloem transport ensures proper nutrition supply at the whole plant level.

7.6 Conclusions

Deficiency of each macronutrient has characteristic symptoms that are closely linked to its specific function in the plant. Here, we have described molecular and physiological functions of different macronutrients (N, P, and K^+) along with their specific deficiency symptoms in plants. Different phytohormones act synergistically or antagonistically to regulate the molecular processes underlying these symptoms. Hormones also act as negative regulators of ion uptake in nutrient-abundant conditions. Conversely, nutrients in turn also affect both biosynthesis and signaling of hormones in various conditions. Hence, crosstalk between nutrients and phytohormones is crucial for an efficient maintenance of plant growth and development. However, interactions between nutrient deficiency conditions and several other biotic and abiotic factors are rapidly emerging. The challenge now is to utilize this knowledge to develop sensitive and powerful techniques for early diagnosis and correction of nutrient deficiencies.

Acknowledgment

This work is financially supported by the Science and Engineering Research Board (SERB)—Department of Science and Technology (DST), Govt. of India project (CRG/2021/000694) and core research grant from National Institute of Plant Genome Research (NIPGR) in AS lab. DD and KS are thankful to the council of scientific and industrial research (CSIR), India for research fellowships. Authors are thankful to DBT (Department of Biotechnology)—eLibrary Consortium (DeLCON) for providing access to e-resources.

References

Amtmann, A., Troufflard, S., Armengaud, P., 2008. The effect of potassium nutrition on pest and disease resistance in plants. Physiol. Plant 133, 682–691.

Amtmann, A., Blatt, M.R., 2009. Regulation of macronutrient transport. N. Phytol. 181, 35–52.

Anschütz, U., Becker, D., Shabala, S., 2014. Going beyond nutrition: regulation of potassium homoeostasis as a common denominator of plant adaptive responses to environment. J. Plant. Physiol. 171, 670–687.

Araya, T., Miyamoto, M., Wibowo, J., Suzuki, A., Kojima, S., Tsuchiya, Y.N., et al., 2014. CLE-CLAVATA1 peptide-receptor signaling module regulates the expansion of plant root systems in nitrogen-dependent manner. Proc. Natl. Acad. Sci. U.S.A. 111, 2029–2034.

Armengaud, P., Breitling, R., Amtmann, A., 2004. The potassium-dependent transcriptome of arabidopsis reveals a prominent role of jasmonic acid in nutrient signaling. Plant. Physiol.

Atkinson, N.J., Urwin, P.E., 2012. The interaction of plant biotic and abiotic stresses: from genes to the field. J. Exp. Bot. 63, 3523–3543.

Ayadi, A., David, P., Arrighi, J.-F., Chiarenza, S., Thibaud, M.-C., Nussaume, L., Marin, E., 2015. Reducing the genetic redundancy of Arabidopsis PHOSPHATE TRANSPORTER1 transporters to study phosphate uptake and signaling. Plant Physiol 167, 1511–1526. 25670816.

Balzergue, C., Dartevelle, T., Godon, C., Laugier, E., Meisrimler, C., Teulon, J.M., et al., 2017. Low phosphate activates STOP1-ALMT1 to rapidly inhibit root cell elongation. Nat. Commun. 8.

Barberon, M., 2017. The endodermis as a checkpoint for nutrients. N. Phytol. 213, 1604–1610.

Barberon, M., Vermeer, J.E.M., DeBellis, D., Wang, P., Naseer, S., Andersen, T.G., et al., 2016. Adaptation of root function by nutrient-induced plasticity of endodermal differentiation. Cell 164, 447–459.

Behera, S., Long, Y., Schmitz-Thom, I., Wang, X.P., Zhang, C., Li, H., et al., 2017. Two spatially and temporally distinct Ca2 + signals convey *Arabidopsis thaliana* responses to K + deficiency. N. Phytol. 213, 739–750.

Bhosale, R., Giri, J., Pandey, B.K., Giehl, R.F.H., Hartmann, A., Traini, R., et al., 2018. A mechanistic framework for auxin dependent Arabidopsis root hair elongation to low external phosphate. Nat. Commun. 9, 1–9.

Bouwman, A.F., Beusen, A.H.W., Lassaletta, L., Van Apeldoorn, D.F., Van Grinsven, H.J.M., Zhang, J., et al., 2017. Lessons from temporal and spatial patterns in global use of N and P fertilizer on cropland. Sci. Rep. 7, 1–11.

Britzke, D., da Silva, L.S., Moterle, D.F., dos Santos Rheinheimer, D., Bortoluzzi, E.C., 2012. A study of potassium dynamics and mineralogy in soils from subtropical Brazilian lowlands. J. Soils Sediment. 12, 185–197.

Caba, J.M., Centeno, M.L., Fernández, B., Gresshoff, P.M., Ligero, F., 2000. Inoculation and nitrate alter phytohormone levels in soybean roots: differences between a supernodulating mutant and the wild type. Planta 211, 98–104.

Cai, K., Gao, H., Wu, X., Zhang, S., Han, Z., Chen, X., et al., 2019. The ability to regulate transmembrane potassium transport in root is critical for drought tolerance in barley. Int. J. Mol. Sci. 20.

Cakmak, I., 2010. Plant nutrition research: priorities to meet human needs for food in sustainable ways. Plant Soil 247, 3–24.

Carstensen, A., Herdean, A., Schmidt, S.B., Sharma, A., Spetea, C., Pribil, M., et al., 2018. The impacts of phosphorus deficiency on the photosynthetic electron transport chain1[OPEN]. Plant. Physiol. 177, 271–284.

Carstensen, A., Szameitat, A.E., Frydenvang, J., Husted, S., 2018. Chlorophyll a fluorescence analysis can detect phosphorus deficiency under field conditions and is an effective tool to prevent grain yield reductions in spring barley (*Hordeum vulgare* L.). Plant Soil 434, 79–91.

Castaings, L., Camargo, A., Pocholle, D., Gaudon, V., Texier, Y., Boutet-Mercey, S., et al., 2009. The nodule inception-like protein 7 modulates nitrate sensing and metabolism in Arabidopsis. Plant. J. 57, 426–435.

Chen, J., Gabelman, W.H., 2000. Morphological and physiological characteristics of tomato roots associated with potassium-acquisition efficiency. Sci. Hortic. (Amst.) 83, 213–225.

Chen, J., Gabelman, W.H., 1995. Isolation of tomato strains varying in potassium acquisition using a sand-zeolite culture system. Plant Soil 176, 65−70.

Chen, J.G., Cheng, S.H., Cao, W., Zhou, X., 1998. Involvement of endogenous plant hormones in the effect of mixed nitrogen source on growth and tillering of wheat. J. Plant. Nutr. 21, 87−97.

Chen, Q., Sun, J., Zhai, Q., Zhou, W., Qi, L., Xu, L., et al., 2011. The basic helix-loop-helix transcription factor myc2 directly represses plethora expression during jasmonate-mediated modulation of the root stem cell niche in arabidopsis. Plant Cell 23, 3335−3352.

Conesa, C.M., Saez, A., Navarro-Neila, S., de Lorenzo, L., Hunt, A.G., Sepúlveda, E.B., et al., 2020. Alternative polyadenylation and salicylic acid modulate root responses to low nitrogen availability. Plants 9, 1−16.

Crombez, H., Motte, H., Beeckman, T., 2019. Tackling plant phosphate starvation by the roots. Dev. Cell 48, 599−615.

Cui, L.G., Shan, J.X., Shi, M., Gao, J.P., Lin, H.X., 2014. The miR156-SPL9-DFR pathway coordinates the relationship between development and abiotic stress tolerance in plants. Plant. J. 80, 1108−1117.

Dai, X., Wang, Y., Yang, A., Zhang, W.H., 2012. OsMYB2P-1, an R2R3 MYB transcription factor, is involved in the regulation of phosphate-starvation responses and root architecture in rice. Plant. Physiol. 159, 169−183.

de Bang, T.C., Husted, S., Laursen, K.H., Persson, D.P., Schjoerring, J.K., 2020. The molecular−physiological functions of mineral macronutrients and their consequences for deficiency symptoms in plants. N. Phytol.

Dechorgnat, J., Nguyen, C.T., Armengaud, P., Jossier, M., Diatloff, E., Filleur, S., et al., 2011. From the soil to the seeds: the long journey of nitrate in plants. J. Exp. Bot. 62, 1349−1359.

Deepika, Singh, A., 2021. Expression dynamics indicate the role of Jasmonic acid biosynthesis pathway in regulating macronutrient (N, P and K +) deficiency tolerance in rice (Oryza sativa L.). Plant. Cell Rep. 40 (8), 1495−1512.

Deepika, D., Ankit, Jonwal, S., Mali, K.V., Sinha, A.K., Singh, A., 2022. Molecular analysis indicates the involvement of Jasmonic acid biosynthesis pathway in low-potassium (K +) stress response and development in chickpea (Cicer arietinum). Environ. Exp. Bot. 194, 104753.

Delhaize, E., Ma, J.F., Ryan, P.R., 2012. Transcriptional regulation of aluminium tolerance genes. Trends Plant. Sci. 17, 341−348.

Fan, X., Naz, M., Fan, X., Xuan, W., Miller, A.J., Xu, G., 2017. Plant nitrate transporters: from gene function to application. J. Exp. Bot. 68, 2463−2475.

Franco-Zorrilla, J.M., Martin, A.C., Solano, R., Rubio, V., Leyva, A., Paz-Ares, J., 2002. Mutations at CRE1 impair cytokinin-induced repression of phosphate starvation responses in Arabidopsis. Plant. J. 32, 353−360.

Franco-zorrilla, M., Martı, A.C., Leyva, A., Paz-ares, J., 2005. Interaction between phosphate-starvation, sugar, and cytokinin signaling in arabidopsis and the roles of cytokinin receptors CRE1/AHK4 and AHK3. Plant. Phys. 138, 847−857.

Gojon, A., 2017. Nitrogen nutrition in plants: rapid progress and new challenges. J. Exp. Bot. 68, 2457−2462.

Gould, K.S., Jay-Jllemand, C., Logan, B.A., Baissac, Y., Bidel, L.P.R., 2018. When are foliar anthocyanins useful to plants? Re-evaluation of the photoprotection hypothesis using Arabidopsis thaliana mutants that differ in anthocyanin accumulation. Environ. Exp. Bot. 154, 11−22.

Gruber, B.D., Giehl, R.F.H., Friedel, S., von Wirén, N., 2013. Plasticity of the Arabidopsis root system under nutrient deficiencies. Plant Physiol 163, 161−179. 23852440.

Gu, M., Chen, A., Sun, S., Xu, G., 2016. Complex regulation of plant phosphate transporters and the gap between molecular mechanisms and practical application: what is missing? Mol. Plant. 9, 396−416.

Gupta, K., Dey, A., Gupta, B., 2013. Plant polyamines in abiotic stress responses. Acta Physiol. Plant. 35, 2015−2036.

Gutiérrez-Alanís, D., Ojeda-Rivera, J.O., Yong-Villalobos, L., Cárdenas-Torres, L., Herrera-Estrella, L., 2018. Adaptation to phosphate scarcity: tips from arabidopsis roots. Trends Plant. Sci. 23, 721−730.

Ham, B.K., Chen, J., Yan, Y., Lucas, W.J., 2018. Insights into plant phosphate sensing and signaling. Curr. Opin. Biotechnol. 49, 1−9.

Hamamoto, S., Uozumi, N., 2014. Organelle-localized potassium transport systems in plants. J. Plant. Physiol. 171, 743−747.

He, Q., Wang, F., Wang, Y., Lu, H., Yang, Z., Lv, Q., et al., 2019. Molecular control and genetic improvement of phosphorus use efficiency in rice. Mol. Breed. 39.

Hirose, N., Takei, K., Kuroha, T., Kamada-Nobusada, T., Hayashi, H., Sakakibara, H., 2008. Regulation of cytokinin biosynthesis, compartmentalization and translocation. J. Exp. Bot. 59, 75−83.

Ho, C.H., Lin, S.H., Hu, H.C., Tsay, Y.F., 2009. CHL1 functions as a nitrate sensor in plants. Cell 138, 1184−1194.

Hodge, A., 2006. Plastic plants and patchy soils. J. Exp. Bot. 57, 401−411.

Hughes, N.M., Lev-Yadun, S., 2015. Red/purple leaf margin coloration: potential ecological and physiological functions. Environ. Exp. Bot. 119, 27–39.

Iqbal, A., Qiang, D., Alamzeb, M., Xiangru, W., Huiping, G., Hengheng, Z., et al., 2020. Untangling the molecular mechanisms and functions of nitrate to improve nitrogen use efficiency. J. Sci. Food Agric. 100, 904–914.

Islam, A., Chandrabiswas, J., Sirajul Karim, A.J.M., Salmapervin, M., Saleque, A.M.A., 2015. Effects of Potassium Fertilization on Growth and Yield of Wetland Rice in Grey Terrace Soils of Bangladesh.

Islam, A., Muttaleb, A., 2016. Effect of potassium fertilization on yield and potassium nutrition of Boro rice in a wetland ecosystem of Bangladesh. Arch. Agron. Soil Sci. 62, 1530–1540.

Janes, G., von Wangenheim, D., Cowling, S., Kerr, I., Band, L., French, A.P., et al., 2018. Cellular patterning of arabidopsis roots under low phosphate conditions. Front. Plant. Sci. 9, 1–11.

Jia, Y.b, Yang, X.e, Feng, Y., Jilani, G., 2008. Differential response of root morphology to potassium deficient stress among rice genotypes varying in potassium efficiency. J. Zhejiang Univ. Sci. B 9, 427–434.

Jiang, C., Gao, X., Liao, L., Harberd, N.P., Fu, X., 2007. Phosphate starvation root architecture and anthocyanin accumulation responses are modulated by the gibberellin-DELLA signaling pathway in Arabidopsis. Plant. Physiol. 145, 1460–1470.

Jin, Z., Zhu, Y., Li, X., Dong, Y., An, Z., 2015. Soil N retention and nitrate leaching in three types of dunes in the Mu Us desert of China. Sci. Rep. 5, 14222.

Jong, M.De, Ongaro, V., Ljung, K., 2014. Auxin and strigolactone signaling are required for modulation of arabidopsis shoot branching by nitrogen supply. Plant. Physiol. 166, 384–395.

Kant, S., 2018. Understanding nitrate uptake, signaling and remobilisation for improving plant nitrogen use efficiency. Semin. Cell Dev. Biol. 74, 89–96.

Kebrom, T.H., Spielmeyer, W., Finnegan, E.J., 2013. Grasses provide new insights into regulation of shoot branching. Trends Plant. Sci. 18, 41–48.

Klinsawang, S., Sumranwanich, T., Wannaro, A., Saengwilai, P., 2018. Effects of root hair length on potassium acquisition in rice (Oryza sativa L.). Appl. Ecol. Environ. Res. 16, 1609–1620.

Kellermeier, F., Armengaud, P., Seditas, T.J., Danku, J., Salt, D.E., Amtmann, A., 2014. Analysis of the root system architecture of Arabidopsis provides a quantitative readout of crosstalk between nutritional signals. Plant. Cell 26, 1480–1496.

Kohlen, W., Charnikhova, T., Liu, Q., Bours, R., Domagalska, M.A., Beguerie, S., et al., 2011. Strigolactones are transported through the xylem and play a key role in shoot architectural response to phosphate deficiency in nonarbuscular mycorrhizal host arabidopsis. Plant. Physiol. 155, 974–987.

Krouk, G., Lacombe, B., Bielach, A., Perrine-Walker, F., Malinska, K., Mounier, E., et al., 2010. Nitrate-regulated auxin transport by NRT1.1 defines a mechanism for nutrient sensing in plants. Dev. Cell 18, 927–937.

Krouk, G., Ruffel, S., Gutiérrez, R.A., Gojon, A., Crawford, N.M., Coruzzi, G.M., et al., 2011. A framework integrating plant growth with hormones and nutrients. Trends Plant. Sci. 16, 178–182.

Lambers, H., Plaxton, W.C., 2015. Phosphorus: Back to the Roots. Annu. Plant Rev. Vol. 48, Wiley Online Books.

Landrein, B., Formosa-Jordan, P., Malivert, A., Schuster, C., Melnyk, C.W., Yang, W., et al., 2018. Nitrate modulates stem cell dynamics in Arabidopsis shoot meristems through cytokinins. Proc. Natl. Acad. Sci. U.S.A. 115, 1382–1387.

Leran, S., Garg, B., Boursiac, Y., Corratge-Faillie, C., Brachet, C., Tillard, P., et al., 2015. AtNPF5.5, a nitrate transporter affecting nitrogen accumulation in Arabidopsis embryo. Sci. Rep. 5, 1–7.

Li, G., Wu, Y., Liu, G., Xiao, X., Wang, P., Gao, T., et al., 2017. Large-scale proteomics combined with transgenic experiments demonstrates: an important role of jasmonic acid in potassium deficiency response in wheat and rice. Mol. Cell Proteom. 16, 1889–1905.

Li, Y., Zhang, J., Zhang, X., Fan, H., Gu, M., Qu, H., et al., 2015. Phosphate transporter OsPht1;8 in rice plays an important role in phosphorus redistribution from source to sink organs and allocation between embryo and endosperm of seeds. Plant. Sci. 230, 23–32.

Liang, G., He, H., Yu, D., 2012. Identification of nitrogen starvation-responsive MicroRNAs in Arabidopsis thaliana. PLoS One 7.

Lin, Y.L., Tsay, Y.F., 2017. Influence of differing nitrate and nitrogen availability on flowering control in Arabidopsis. J. Exp. Bot. 68, 2603–2609.

Liu, K.H., Niu, Y., Konishi, M., Wu, Y., Du, H., Sun Chung, H., et al., 2017. Discovery of nitrate-CPK-NLP signalling in central nutrient-growth networks. Nature 545, 311–316.

Liu, T., Li, Y., Ren, J., Qian, Y., Yang, X., Duan, W., et al., 2013. Nitrate or NaCl regulates floral induction in *Arabidopsis thaliana*. Biologia (Bratisl.) 68, 215−222.

López-Arredondo, D.L., Leyva-González, M.A., González-Morales, S.I., López-Bucio, J., Herrera-Estrella, L., 2014. Phosphate nutrition: improving low-phosphate tolerance in crops. Annu. Rev. Plant. Biol. 65, 95−123.

Ma, T., Wu, W., Wang, Y., 2012. Transcriptome analysis of rice root responses to potassium deficiency. BMC Plant. Biol. 12, 1−13.

Ma, W., Li, J., Qu, B., He, X., Zhao, X., Li, B., et al., 2014. Auxin biosynthetic gene TAR2 is involved in low nitrogen-mediated reprogramming of root architecture in Arabidopsis. Plant. J. 78, 70−79.

Mangano, S., Denita-Juarez, S.P., Choi, H.-S., Marzol, E., Hwang, Y., Ranocha, P., et al., 2017. Molecular link between auxin and ROS-mediated polar growth. Proc. Natl. Acad. Sci. U.S.A. 114, 5289−5294.

Marchive, C., Roudier, F., Castaings, L., Bréhaut, V., Blondet, E., Colot, V., et al., 2013. Nuclear retention of the transcription factor NLP7 orchestrates the early response to nitrate in plants. Nat. Commun. 4, 1−9.

Martín, A.C., Pozo, Del, Iglesias, J.C., Rubio, J., Solano, V., De La Peña, R., et al., 2000. Influence of cytokinins on the expression of phosphate starvation responsive genes in Arabidopsis. Plant. J. 24, 559−567.

Martínez-Cordero, M.A., Martínez, V., Rubio, F., 2004. Cloning and functional characterization of the high-affinity K + transporter HAK1 of pepper. Plant. Mol. Biol. 56, 413−421.

Młodzińska, E., Zboińska, M., 2016. Phosphate uptake and allocation − a closer look at *Arabidopsis thaliana* L. and *Oryza sativa* L. Front. Plant. Sci. 7, 1−19.

Müller, D., Waldie, T., Miyawaki, K., Melnyk, J.P.C., Kieber, C.W., Kakimoto, J.J., et al., 2015. Cytokinin is required for escape but not release from auxin mediated apical dominance. Plant. J. 82, 874−886.

Nacry, P., 2005. A role for auxin redistribution in the responses of the root system architecture to phosphate starvation in Arabidopsis. Plant. Physiol. 138, 2061−2074.

O'Brien, J.A.A., Vega, A., Bouguyon, E., Krouk, G., Gojon, A., Coruzzi, G., et al., 2016. Nitrate transport, sensing, and responses in plants. Mol. Plant. 9, 837−856.

Okamoto, M., Kumar, A., Li, W., Wang, Y., Siddiqi, M.Y., Crawford, N.M., et al., 2006. High-affinity nitrate transport in roots of Arabidopsis depends on expression of the NAR2-like gene AtNRT3.1. Plant. Physiol. 140, 1036−1046.

Osugi, A., Kojima, M., Takebayashi, Y., Ueda, N., Kiba, T., Sakakibara, H., 2017. Systemic transport of trans-zeatin and its precursor have differing roles in Arabidopsis shoots. Nat. Plants 3, 1−6.

Péret, B., Desnos, T., Jost, R., Kanno, S., Berkowitz, O., Nussaume, L., 2014. Root architecture responses: in search of phosphate. Plant. Physiol. 166, 1713−1723.

Pottosin, I., Shabala, S., 2016. Transport across chloroplast membranes: optimizing photosynthesis for adverse environmental conditions. Mol. Plant. 9, 356−370.

Raghothama, K.G., Karthikeyan, A.S., 2005. Phosphate acquisition. Plant. Soil 274, 37−49.

Rahayu, Y.S., Walch-Liu, P., Neumann, G., Römheld, V., Von Wirén, N., Bangerth, F., 2005. Root-derived cytokinin s as long-distance signals for NO3−induced stimulation of leaf growth. J. Exp. Bot. 56, 1143−1152.

Remy, E., Cabrito, T.R., Batista, R.A., Teixeira, M.C., Sá-Correia, I., Duque, P., 2012. The Pht1;9 and Pht1;8 transporters mediate inorganic phosphate acquisition by the Arabidopsis thaliana root during phosphorus starvation. New Phytol 195, 356−371. 22578268.

Rubin, G., Tohge, T., Matsuda, F., Saito, K., Scheible, W.R., 2009. Members of the LBD family of transcription factors repress anthocyanin synthesis and affect additional nitrogen responses in arabidopsis. Plant. Cell 21, 3567−3584.

Sakakibara, H., Takei, K., Hirose, N., 2006. Interactions between nitrogen and cytokinin in the regulation of metabolism and development. Trends Plant. Sci. 11, 440−448.

Schachtman, D.P., 2015. The role of ethylene in plant responses to K(+) deficiency. Front Plant Sci 6, 1153. 26734048.

Segonzac, C., Boyer, J.C., Ipotesi, E., Szponarski, W., Tillard, P., Touraine, B., et al., 2007. Nitrate efflux at the root plasma membrane: identification of an Arabidopsis excretion transporter. Plant. Cell 19, 3760−3777.

Shen, C., Yue, R., Yang, Y., Zhang, L., Sun, T., Tie, S., et al., 2014. OsARF16 is involved in cytokinin-mediated inhibition of phosphate transport and phosphate signaling in rice (*Oryza sativa* L). PLoS One 9, 3−12.

Shin, R., Burch, A.Y., Huppert, K.A., Tiwari, S.B., Murphy, A.S., Guilfoyle, T.J., et al., 2007. The Arabidopsis transcription factor MYB77 modulates auxin signal transduction. Plant. Cell Online 19, 2440−2453.

Smith, S.E., Jakobsen, I., Grønlund, M., Smith, F.A., 2011. Roles of arbuscular mycorrhizas in plant phosphorus nutrition: interactions between pathways of phosphorus uptake in arbuscular mycorrhizal roots have important implications for understanding and manipulating plant phosphorus acquisition. Plant Physiol 156, 1050−1057. 21467213.

Song, W., Xue, R., Song, Y., Bi, Y., Liang, Z., Meng, L., et al., 2018. Differential response of first-order lateral root elongation to low potassium involves nitric oxide in two tobacco cultivars. J. Plant. Growth Regul. 37, 114−127.

Sun, J., Bankston, J.R., Payandeh, J., Hinds, T.R., Zagotta, W.N., Zheng, N., 2014. Crystal structure of the plant dual-affinity nitrate transporter NRT1.1. Nature 507, 73−77.

Sustr, M., Soukup, A., Tylova, E., 2019. Potassium in root growth and development. Plants 8.

Tegeder, M., Masclaux-Daubresse, C., 2018. Source and sink mechanisms of nitrogen transport and use. N. Phytol. 217, 35−53.

Troufflard, S., Mullen, W., Larson, T.R., Graham, I.A., Crozier, A., Amtmann, A., et al., 2010. Potassium deficiency induces the biosynthesis of oxylipins and glucosinolates in *Arabidopsis thaliana*. BMC Plant. Biol. 10.

Tsay, Y.F., Chiu, C.C., Tsai, C.B., Ho, C.H., Hsu, P.K., 2007. Nitrate transporters and peptide transporters. FEBS Lett. 581, 2290−2300.

Ueno, H., Maeda, T., Katsuyama, N., Katou, Y., Matsuo, S., Yano, K., et al., 2018. Cation measurements and gene expression analysis suggest tomato leaf marginal necrosis is caused by a jasmonate signal induced by K + starvation in the tip region of leaflets. Hortic. J. 87, 206−213.

Umehara, M., Hanada, A., Yoshida, S., Akiyama, K., Arite, T., Takeda-Kamiya, N., et al., 2008. Inhibition of shoot branching by new terpenoid plant hormones. Nature 455, 195−200.

Vance, C.P., Uhde-Stone, C., Allan, D.L., 2003. Phosphorus acquisition and use: critical adaptations by plants for securing a nonrenewable resource. N. Phytol. 157, 423−447.

Vidal, E.A., Araus, V., Lu, C., Parry, G., Green, P.J., Coruzzi, G.M., et al., 2010. Nitrate-responsive miR393/AFB3 regulatory module controls root system architecture in *Arabidopsis thaliana*. Proc. Natl. Acad. Sci. U.S.A. 107, 4477−4482.

Vidal, E.A., Moyano, T.C., Canales, J., Gutiérrez, R.A., 2014. Nitrogen control of developmental phase transitions in *Arabidopsis thaliana*. J. Exp. Bot. 65, 5611−5618.

Vissenberg, K., Claeijs, N., Balcerowicz, D., Schoenaers, S., 2020. Hormonal regulation of root hair growth and responses to the environment in arabidopsis. J. Exp. Bot. 71 (8), 2412−2427. 31993645.

Walch-Liu, P., Ivanov, I.I., Filleur, S., Gan, Y., Remans, T., Forde, B.G., 2006. Nitrogen regulation of root branching. Ann. Bot. 97, 875−881.

Wang, X., Yi, K., Tao, Y., Wang, F., Wu, Z., Jiang, D., et al., 2006. Cytokinin represses phosphate-starvation response through increasing of intracellular phosphate level. Plant, Cell Environ. 29, 1924−1935.

Wang, B.L., Tang, X.Y., Cheng, L.Y., Zhang, A.Z., Zhang, W.H., Zhang, F.S., et al., 2010. Nitric oxide is involved in phosphorus deficiency-induced cluster-root development and citrate exudation in white lupin. N. Phytol. 187, 1112−1123.

Wang, D., Lv, S., Jiang, P., Li, Y., 2017. Roles, regulation, and agricultural application of plant phosphate transporters. Front. Plant. Sci. 8, 1−14.

Wang, F., Deng, M., Xu, J., Zhu, X., Mao, C., 2018. Molecular mechanisms of phosphate transport and signaling in higher plants. Semin. Cell Dev. Biol. 74, 114−122.

Wang, M., Zheng, Q., Shen, Q., Guo, S., 2013. The critical role of potassium in plant stress response. Int. J. Mol. Sci. 14, 7370−7390.

Wang, Y.Y., Hsu, P.K., Tsay, Y.F., 2012. Uptake, allocation and signaling of nitrate. Trends Plant. Sci. 17, 458−467.

Wang, Z.Q., Huang, H., Deng, J.M., Liu, J.Q., 2015. Scaling the respiratory metabolism to phosphorus relationship in plant seedlings. Sci. Rep. 5, 1−5.

Ward, J.T., Lahner, B., Yakubova, E., Salt, D.E., Raghothama, K.G., 2008. The effect of iron on the primary root elongation of Arabidopsis during phosphate deficiency. Plant. Physiol. 147, 1181−1191.

Wigoda, N., Moshelion, M., Moran, N., 2014. Is the leaf bundle sheath a "smart flux valve" for K + nutrition? J. Plant. Physiol. 171, 715−722.

Woo, J., MacPherson, C.R., Liu, J., Wang, H., Kiba, T., Hannah, M.A., et al., 2012. The response and recovery of the *Arabidopsis thaliana* transcriptome to phosphate starvation. BMC Plant. Biol. 12.

Xu, W., Dubos, C., Lepiniec, L., 2015. Transcriptional control of flavonoid biosynthesis by MYB-bHLH-WDR complexes. Trends Plant. Sci. 20, 176−185.

Yadav, B., Sidhu, A., 2016. Dynamics of Potassium and Their Bioavailability for Plant Nutrition. pp. 187−201.

Yuan, S., Zhang, Z.W., Zheng, C., Zhao, Z.Y., Wang, Y., Feng, L.Y., et al., 2016. Arabidopsis cryptochrome 1 functions in nitrogen regulation of flowering. Proc. Natl. Acad. Sci. U.S.A. 113, 7661−7666.

Zarabi, M., Jalali, M., 2012. Leaching of nitrogen from calcareous soils in western Iran: a soil leaching column study. Environ. Monit. Assess. 184, 7607–7622.

Zhang, X., Davidson, E.A., Mauzerall, D.L., Searchinger, T.D., Dumas, P., Shen, Y., 2015. Managing nitrogen for sustainable development. Nature 528, 51–59.

Zhang, X., Jiang, H., Wang, H., Cui, J., Wang, J., Hu, J., et al., 2017. Transcriptome analysis of rice seedling roots in response to potassium deficiency. Sci. Rep. 7, 5523. 28717149.

Zhao, S., Zhang, M.L., Ma, T.L., Wang, Y., 2016. Phosphorylation of ARF2 relieves its repression of transcription of the K + transporter gene HAK5 in response to low potassium stress. Plant. Cell 28, 3005–3019.

Zheng, H., Pan, X., Deng, Y., Wu, H., Liu, P., Li, X., 2016. AtOPR3 specifically inhibits primary root growth in Arabidopsis under phosphate deficiency. Sci. Rep. 6, 24778. 27101793.

Extended role of auxin: reconciliation of growth and defense responses under biotic stress

Gyöngyi Major and Gábor Jakab

Department of Plant Biology, Faculty of Sciences, University of Pécs, Pécs, Hungary

The regulatory role of auxin in plant growth and development has been acknowledged and studied for a long time. Recent studies, however, provide new insights into the role of auxin in plant defense responses indicating a trade-off between defense and growth. In this review, we will discuss the diverse ways of auxin counteracting with plant defense regulation through interaction with other hormones, like salicylic acid (SA) and jasmonic acid (JA) with well-characterized functions in plant pathogenesis.

8.1 Diversity of auxin biosynthesis

In plants, the major natural form of auxin is indole-3-acetic acid (IAA). In addition, several endogenous auxinic compounds, such as 4-chloroindole-3-acetic acid (4-Cl-IAA), indole-3-butyric acid (IBA), and phenylacetic acid (PAA) with auxin-like activity, have been identified (Korasick et al.,2013; Paque and Weijers, 2016). Auxinic compounds exist in both free and conjugated forms in plants. Free IAA is the active form of auxin, and the conjugated auxins are considered storage forms or intermediates of the degradation pathways (Korasick et al., 2013; Woodward and Bartel, 2005). Free IAA can be released from IAA conjugates, such as IAA esters, IAA-sugars, and IAA-amino acid conjugates, by hydrolysis (Hangarter and Good, 1981; Korasick et al., 2013; Ludwig-Müller, 2011; Yu et al., 2014). Free IAA can also be produced from IBA by a process similar to fatty acid β-oxidation in the peroxisomes (Frick and Strader, 2018; Simon and Petrášek, 2011). Pathways for IAA biosynthesis are traditionally defined by their distinctive central intermediates and subdivided into two categories: the Trp-independent and Trp-dependent

147

pathways. Auxin biosynthesis occurs mainly in the aerial parts of the plant, especially in young developing leaves and meristems, from where it is transported to the rest of the plant (Zhao, 2010, 2014). However, local auxin biosynthesis can also occur in other tissues, such as the meristematic region of the primary root or the tips of emerged lateral roots (Brumos et al., 2018; Zhao, 2018).

The Trp-independent pathway was first identified in orange pericarp. The biosynthetic pathway was further characterized in a Trp-auxotroph *Zea mays* mutant defective in Trp synthase b activity. This mutant has low levels of Trp but accumulates 50-fold higher amounts of IAA if compared to the wild-type (Wright et al., 1991). Indole, the last intermediate of the Trp biosynthetic pathway, has been suggested as the precursor for this pathway (Wang et al., 2015). A cytosol-localized indole synthase (INS) is the key enzyme in Trp-independent IAA biosynthesis pathways, and auxin generated this way has an important role in apical—basal pattern formation during early embryogenesis in *Arabidopsis* (Wang et al., 2015).

The tryptophan-dependent pathways were named after their major intermediates. They can be divided into the indole-3-acetaldoxime (IAOx), indole-3-acetamide (IAM), tryptamine (TAM), and indole-3-pyruvic acid (IPA) pathways. The IAOx pathway operates in few plant species that have CYP79B family members to convert Trp to IAOx. IAOx was identified in *Arabidopsis* (Bak et al., 2001). CYP79B2 and CYP79B3 are cytochrome P450 monooxygenases that catalyze the conversion of Trp into IAOx (Hull et al., 2000; Zhao et al., 2002). IAOx is then converted to IAA, camalexin (CL), and indole glucosinolates (IGs). Both CL and IG have antimicrobial activities. CL is an indolic phytoalexin that is thought to damage fungal cell membranes (Sanchez-Vallet et al., 2010) and is mostly effective against necrotrophic fungi such as *Alternaria brassicicola* (Schlaeppi et al., 2010). In contrast, IGs are toxic to a wide range of bacteria (Kim et al., 2015).

The IAM pathway has been well characterized in bacteria (Manulis et al., 1994; Li et al., 2018; Zhang et al., 2019) and seems to be also conserved in plants (Lehmann et al., 2010). This pathway exists widely in plants, but it is still unclear exactly how IAM is produced. The conversion of IAM to IAA by *Arabidopsis* AMIDASE 1 (AMI1) has been demonstrated (Pollmann et al., 2003). The physiological significance of the IAM pathway in plants is under investigation. The research group of Gao suggests that IAMH1 and IAMH2 (two homologous IAM HYDROLASE genes) are the main enzymes responsible for converting IAM into IAA in *Arabidopsis* (Gao et al., 2020). Genetic and biochemical studies have finally revealed the importance of the WEAK ETHYLENE INSENSITIVE 8 (WEI8)/ TRYPTOPHAN AMINOTRANSFERASE OF ARABIDOPSIS1 (TAA1) gene in the IPA pathway, encoding for the Trp-aminotransferase that converts Trp to IPA (Stepanova et al., 2008; Mano and Nemoto, 2012). The YUCCA (YUC) gene family encodes flavin monooxygenase—like proteins and was originally associated with the TAM pathway (Zhao et al., 2002; Tivendale et al., 2014). Recent experimental evidence suggests that the YUC proteins function downstream of the TAA1 in the IPA pathway (Kriechbaumer et al., 2017; Blakeslee et al., 2019). The TAA/YUC pathway converts Trp to IAA in two consecutive steps. First, Trp is metabolized into IPA by the TAA family of aminotransferases. Then, IPA undergoes oxidative decarboxylation catalyzed by the YUC family to produce IAA. YUC proteins catalyze a rate-limiting step of the IPA pathway, which is the main IAA biosynthesis pathway in *Arabidopsis* and maize (Mashiguchi et al., 2011).

8.2 Disease resistance or growth

In a changing environment, the best defensive strategy is not necessarily the construction of strong constitutive defense systems. The assignment of metabolites and proteins to resistance may limit other plant physiological processes. The effective harmonization provided by the hormone network allows plants to respond quickly to environmental changes, thereby using limited nutrient resources in a cost-effective manner. The role of many phytohormones in the regulation of resistance reactions of plants to biotic and abiotic stress indicates that there is a close relationship between two physiological processes: development and adaptation to environmental influences (Walters and Heil, 2007; Kempel et al., 2011).

In their natural environments, plants are under a constant pressure of different biotic stressors (e.g., bacteria, fungi, viruses, *Oomycetes*, and insects) that endangers plant survival and its reproduction (Fig. 8.1). Plants have a multilayered immune system that is fully capable of perceiving and stopping the invading pathogens (Wang et al., 2019). Despite the host-mediated immune barriers, virulent pathogens still can manage to cause infection in susceptible host plants. The plant's innate immune system relies on the specific detection of the microbe/pathogen-associated molecular patterns (M/PAMPs) by pattern recognition receptors (PRRs), which are relatively conserved molecules (Jones and Dangl, 2006). This interaction between the PRRs and the PAMPs activates a type of basal resistance called PAMP-triggered immunity (PTI), which prevents the growth of many microbes and nonhost pathogens (Jones and Dangl, 2006; Couto and Zipfel, 2016). Successful pathogens secrete effector proteins that deactivate PTI. To counteract this, plant resistance (R) proteins recognize effectors or their activity and turn on effector-triggered immunity (ETI) (Yuan et al., 2021).

The essential roles of SA and ET/JA-mediated signaling pathways in resistance to pathogens are well characterized (Denancé et al., 2013; Čarná et al., 2014; Yang et al., 2015; Yuan et al., 2021). SA signaling positively regulates plant defense against biotrophic pathogens, which are only capable of completing their life cycle in living tissues. For resistance to herbivorous pests and necrotrophic pathogens, that degrade plant tissue during infection ET/JA pathways are usually required (Thomma et al., 1998; Checker et al., 2018; Bürger and Chory, 2019). It is usually considered that inducible defense was developed to save energy in enemy-free environment (Van Loon et al., 2006). Therefore, a cost−benefit balance of defense and growth may improve plant fitness. So these phytohormones act in concert to regulate many developmental processes and are also capable to modulate plant−pathogen interactions (Robert-Seilaniantz et al., 2011a; Huang et al., 2020).

8.3 Counteracting effects of salicylic acid and auxin

SA is a phenolic hormone and affects plant defense against biotrophic and hemibiotrophic pathogens, and the establishment of systemic acquired resistance (SAR) (Durrant and Dong, 2004). It can control growth, development, senescence, and stress responses in plants (Raskin, 1992; Rivas-San Vicente and Plasencia, 2011; van Butselaar and Van den

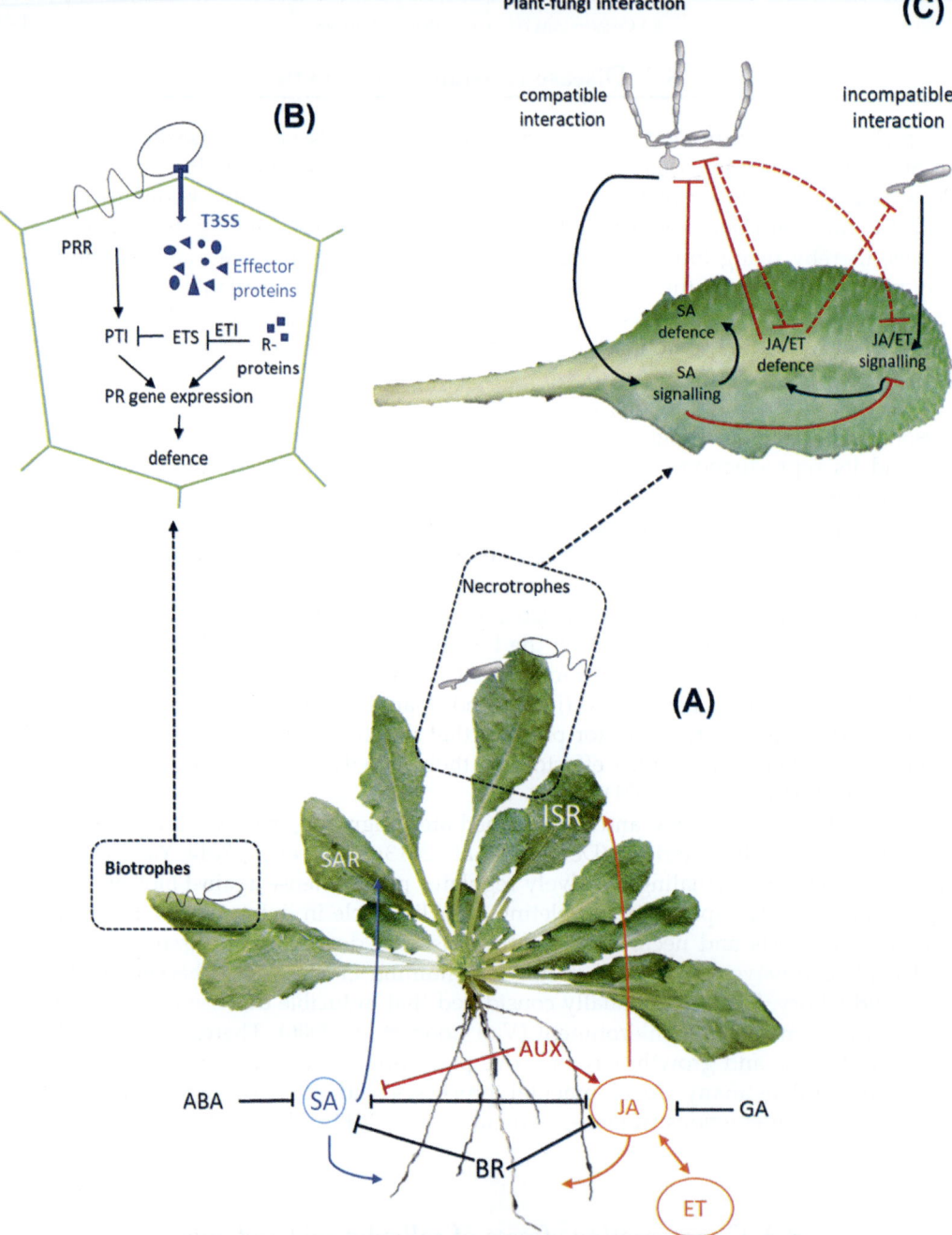

FIGURE 8.1 (A) SAR is established against biotrophic pathogens and is controlled by SA. ISR requires JA and ET signaling and is found as a response triggered by necrotrophic pathogens or mutualistic organisms. (B) Typical components and steps preceding the establishment of SAR. (C) Although only incompatible PM—host interactions elicit JA/ET-mediated defense, JA/ET-induced defense responses are effective against virulent PM fungi if stimulated constitutively, artificially (e.g., JA treatment) or systemically (Piriformospora indica root colonization). These findings suggest that virulent fungi suppress JA/ET signaling during compatible interactions. This suppression might involve the antagonistic action of SA signaling (Kuhn et al., 2016). Solid lines indicate experimentally supported impacts, while dashed lines indicate speculative connections. *ET*, ethylene; *ISR*, Induced systemic resistance; *JA*, jasmonic acid; *PM*, powdery mildew; *SA*, salicylic acid; *SAR*, systemic acquired resistance. Source: *(A) and (B) are modified from Bürger and Chory (2019), (C) is based on Kuhn et al. (2016).*

Ackerveken, 2020). In the following, we present how mutual interactions between auxin and SA affect both growth and defense in plants. Many plant pathogens can either produce auxin themselves or manipulate host auxin biosynthesis to interfere with the host's normal developmental processes (Naseem et al., 2015; Kunkel and Johnson, 2021). In response, plants repress auxin signaling during infection as a defense strategy. Auxin induces gene expression through direct physical interaction with the TIR1-like F box proteins, which in turn remove the Aux/IAA family of transcriptional repressors (Checker et al., 2018; Bürger and Chory, 2019). In *Arabidopsis*, SA treatment stabilizes the Aux/IAA by the repression of TIR1 receptor gene, leading to the downregulation of the expression of auxin-related genes (Wang et al., 2007). Wang et al. found that coinoculation of *Pseudomonas syringae* pv. *maculicola* ES4326 (*Psm*4326) with the synthetic auxin NAA (1-naphthaleneacetic acid) not only enhanced disease symptoms but also promoted pathogen growth. They observed that auxin application could inhibit the full induction of SA-mediated *PR-1* expression. These results support the hypothesis that auxin might downregulate host defense responses (Yamada, 1993; Naseem and Dandekar, 2012; Djami-Tchatchou et al., 2020). Moreover, the enhanced susceptibility to *Psm*4326 of plants expressing the *NahG* gene (encoding a bacterial salicylate hydroxylase that degrades SA) is partially reverted by the *axr2−1* mutation, which disrupts auxin signaling, further indicating that auxin signaling is part of the SA-induced resistance signaling pathway (Wang et al., 2007).

Interaction between SA and auxins was further clarified by the characterization of the regulatory pattern of the GH3.5 gene, which is involved in auxin homeostasis by the negative feedback regulation of endogenous auxin levels in *Arabidopsis* plants (Park et al., 2007). GH3 enzymes have been proposed to regulate endogenous auxin content by conjugating amino acids to IAA (Westfall et al., 2010). In *Arabidopsis*, the major part of the endogenous IAA pool is present either in amide linkages with amino acids and peptides or in ester linkages with sugars (Hangarter and Good, 1981; Ludwig-Müller, 2011; González-Lamothe et al., 2012). Only a small fraction is present as free IAA, indicating that conjugated forms play an important role in auxin homeostasis (Bajguz and Piotrowska, 2009). Lines overexpressing GH3.5 have lower levels of Aux/IAA proteins, overexpression of SA signaling pathway, and enhanced resistance to *P. syringae*. In *Arabidopsis*, six GH3 proteins, AtGH3.1, AtGH3.2/YDK1, AtGH3.5/WES1, AtGH3.6/DFL1, AtGH3.9, and AtGH3.17/VAS2, are linked to IAA activity in plant growth and development (Khan and Stone, 2007; Park et al., 2007; Westfall et al., 2010). Of these proteins, AtGH3.5 is also implicated in SA-linked pathogen responses. Studies with the loss-of-function mutant *wes1* and two gain-of-function mutants *wes1-D* and *gh3.5−1D* indicate that AtGH3.5 contributes to both IAA and SA responses (Park et al., 2007). The gain-of-function lines displayed both low auxin phenotypes along with increased disease resistance. The *Arabidopsis* mutant, *wes1-D* (in which a GH3 gene WES1 is activated by nearby insertion of the 35S enhancer), exhibited auxin-deficient traits, such as reduced growth and altered leaf shape. WES1 is also induced by SA (biotic stress) and abscisic acid (abiotic stress). Thus *wes1-D* was resistant to both biotic and abiotic stresses, and stress-responsive genes, such as PR and CBF genes, were upregulated in this mutant. In contrast, a T-DNA insertional mutant showed reduced stress resistance. They proposed that GH3-mediated growth suppression directs the reallocation of metabolic resources to resistance establishment and indicates the fitness costs of induced resistance (Park et al., 2007). Moreover, these transgenic lines also displayed enhanced resistance to abiotic stress and induction of the ABA regulatory pathway (Park et al., 2007). The inhibitory effect of SA on auxin homeostasis has also been investigated in detail (Wang et al., 2007). Auxin is one among the core signals for plant cellular developmental programs, and disruption of auxin signaling has negative implications for normal plant development (Stone et al., 2008; Spiess et al., 2014; Schaller et al., 2015). The SA-regulated auxin homeostasis has interesting implications for the trade-off between growth and defense (van Butselaar and Van den Ackerveken, 2020; Wang et al., 2007).

During the analysis of an activation-tagged mutant gh3.5—1D, Zhang et al. found that the overexpression of the GH3.5 gene resulted in an elevated accumulation of SA and an increased expression of the marker gene PR-1 in local and systemic tissues in response to avirulent pathogens (Zhang et al., 2007). In contrast, the knockdown of GH3.5 by T-DNA insertional mutation partially compromised SAR and was associated with diminished PR-1 expression in systemic tissues. After pathogen infection, the gh3.5—1D mutant also accumulated high levels of free IAA and was impaired in different resistance-gene-mediated responses. Similar responses were also observed in the mutant dfl1-D, activation-tagged in the GH3.6 gene impacted the auxin pathway, indicating an important role of GH3.5/GH3.6 in disease susceptibility (Zhang et al., 2007). Their microarray analysis confirmed that the SA and auxin pathways were simultaneously augmented in gh3.5—1D after infection with the avirulent pathogen P. syringae pv. tomato DC3000. The SA pathway was amplified by GH3.5 through inducing SA-responsive genes and basal defense components, whereas the auxin pathway was derepressed through upregulating IAA biosynthesis and downregulating auxin repressor genes (Zhang et al., 2007). Zhang et al. proposed a model describing the role of GH3.5 in the Arabidopsis—P. syringae interaction. In the compatible interaction, GH3.5 is activated to modulate the auxin pathway resulting in enhanced disease susceptibility through increasing IAA biosynthesis and derepressing auxin signaling. IAA further induces GH3.5 to augment those processes. In addition, GH3.5 might also function as an IAA-amidosynthetase to regulate IAA homeostasis. In the compatible/incompatible interactions, GH3.5 positively modulates the SA pathway to enhance plant defense response through elevating SA biosynthesis, activating SA-induced genes, WRKYs, and basal defense—related genes. Feedback regulation of GH3.5 by SA amplifies those effects. In this case, GH3.5 might also synthesize SA-Asp with unknown functions during the interactions (Zhang et al., 2007).

Reciprocally, SA-mediated defenses are attenuated by auxin. The increased auxin signaling by the overexpression of the auxin-receptor gene AFB1 leads to a significant reduction in SA accumulation in Arabidopsis after pathogen infection (Robert-Seilaniantz et al., 2011b). PAMPs such as bacterial flagellin-peptide flg22 induce an Arabidopsis microRNA (miR393) that leads to the mRNAs of receptor genes (TIR1, AFB2, and AFB3) being targeted for cleavage, resulting in the suppression of auxin signaling and increased resistance to the bacterium P. syringae (Navarro et al., 2006). In contrast, AFB1 is partially resistant to miR393-mediated degradation. Therefore, miR393 is not able to suppress auxin signaling (Robert-Seilaniantz et al., 2011b). Robert-Seilaniantz et al. found that miR393 redirects secondary metabolite biosynthesis away from camalexin (which is more effective against necrotrophic fungi) and toward glucosinolates (which are implicated in biotrophic resistance). In their study, Robert-Seilaniantz et al. overexpressed miR393, and they proved that plants are more resistant to biotroph pathogens as did Navarro et al. before (Navarro et al., 2006), and they found that plants are more susceptible to necrotrophic pathogens. The action of miR393 reduces the plant sensitivity to auxin and prevents the activation of ARF1 and ARF9. Inhibition of auxin signaling by miR393 redirects the metabolic flow of the tryptophan metabolic pathway; thus, the plant does not produce auxins and antimicrobial IGs and camalexin. The *Arabidopsis thaliana* microRNA miR167 controls the patterns of expression of its target genes AUXIN RESPONSE FACTOR6 (ARF6), ARF8, and IAA-Ala RESISTANT3 to regulate diverse processes, including flower development, root development, and response to osmotic stress (Wu et al., 2006; Su et al., 2016). Caruana et al. (2020) found that miR167 also modulates defense against pathogens through ARF6 and ARF8. miR167 is differentially expressed in response to *P. syringae* infections, and the overexpression of miR167 confers very high levels of resistance (Li., et al., 2011; Zhang et al., 2011). This resistance appears to be due to suppression of auxin responses and is partially dependent upon SA signaling and also depends upon altered stomatal behavior in these plants. Plants overexpressing miR167 constitutively maintain small stomatal apertures, resulting in very high resistance when the pathogen is inoculated onto the leaf surface (Caruana et al., 2020). Since the closure of stomata upon the detection of *P. syringae* is an important aspect of the basal defense response preventing bacterial entering into the leaves (Zeng et al., 2011). Additionally, the SAR response is severely compromised in plants overexpressing miR167, in agreement with previous work (Wang et al., 2007) indicating that the activation of SAR requires intact auxin signaling responses.

The auxin signaling pathway is able to suppress the SA pathway. However, YUC1-overexpressing plants (YUCCA 1 auxin biosynthesis enzyme) with higher endogenous auxin levels showed increased susceptibility without any effect on the SA response (Mutka et al., 2013). Mutka et al. observed that plants overexpressing YUC1 gene leading to elevated IAA levels are more susceptible to the phytopathogen *P. syringae* DC3000. This study suggests that the increased susceptibility is not primarily due to suppression of SA-mediated defenses. Instead, high IAA levels promote pathogenesis via a different mechanism. To investigate plant defense responses as a result of elevated IAA levels, wild-type and IAA-overproducing plants were inoculated with *P. syringae* DC3000 expressing the type-III effector protein AvrRpm1. Bacterial growth was higher in IAA-overproducing plants treated with the bacteria compared with wild-type plants. This implies that elevated IAA levels promote pathogen growth (Mutka et al., 2013).

Taken together, these results signify the importance of the GH3.5 enzyme in the SA- and auxin-mediated growth−defense trade-off, in which higher SA levels reduce the pool of active IAA, and thus, defense is prioritized overgrowth. However, in the absence of pathogens, auxin-mediated suppression of SA responses can be redirected to facilitate growth programs. GH3.5 might also synthesize SA-Asp and IAA-Asp, allowing it to control the growth−defense trade-off (Fig. 8.2). Localized high levels of auxin are associated with meristematic cells, organ initiation, and polar growth, and GH3.5 is specifically expressed in these cells. For leaves, younger developing leaves exhibit moderate levels of IAA and high Asp. As leaves mature and senesce and/or are infected by a pathogen, the concentrations of IAA and Asp decrease. Senescence and particularly (hemi)-biotrophic pathogens, in contrast, increase SA accumulation in plants. GH3.5 is also induced by these

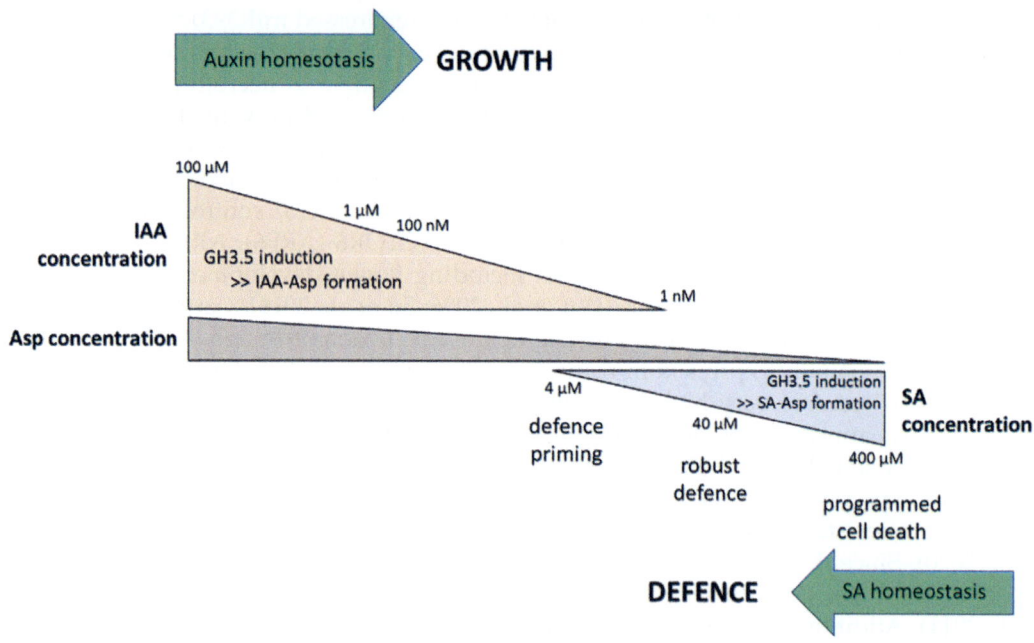

FIGURE 8.2   Model for GH3.5 function as a mediator of growth versus defense. *Source: Data from Mackelprang, R., Okrent, R.A., Wildermuth, M.C., 2017. Preference of Arabidopsis thaliana GH3.5 acyl amido synthetase for growth versus defense hormone acyl substrates is dictated by concentration of amino acid substrate aspartate. Phytochemistry 143, 19−28. https://doi.org/10.1016/J.PHYTOCHEM.2017.07.001.*

pathogens concordant with SA accumulation. The kinetics of GH3.5 dependence on Asp show a dramatic preference for IAA when Asp is high, consistent with GH3.5 function to regulate IAA homeostasis via conversion of IAA to inactive IAA-Asp during growth/development when both IAA and Asp are high. Decreasing Asp due to aging and/or stress creates more favorable conditions for GH3.5 conjugation of SA to SA-Asp, thereby controlling SA homeostasis and defense. Thus GH3.5 fine-tunes local growth versus defense responses and hormone homeostasis (Mackelprang et al., 2017). Pasternak et al. investigated the role of SA on plant growth and development. They found that low-concentration SA plays an important role in shaping root meristem structure and root system architecture, whereas high concentration (greater than 50 μM) of SA inhibited all growth processes in the root. These effects are due to changes in auxin synthesis and transport (Pasternak et al., 2019). Thus the crosstalk between auxin and SA has interesting biotechnological implications for higher yield and better protection from pests.

8.4 Interactions in jasmonic acid- and auxin-regulated plant responses

Jasmonates are a class of lipid-derived defense signaling hormone that regulates plant immune responses against necrotrophic pathogens and insect herbivores (Pieterse et al., 2012).

The perception of both auxin and JA involves the formation of coreceptor complexes in which hormone-specific E3-ubiquitin ligases of the SKP1-Cullin-F-box protein (SCF) type interact with specific repressor proteins (Pérez and Goossens, 2013). Across the plant kingdom, the Aux/IAA and the JASMONATE-ZIM DOMAIN (JAZ) proteins correspond to the auxin- and JA-specific repressors, respectively. In the absence of the hormones, these repressors form a complex with transcription factors (TFs) specific for to both pathways. They also recruit several proteins, among which the general corepressor TOPLESS (TPL), and thereby prevent the TFs from activating gene expression (Pérez and Goossens, 2013). NINJA recruits to the corepressor TPL and TPL-related proteins (TPRs) to the JAZ/MYC/ ± COI complex (Pauwels et al., 2010). TPL also represses the expression of auxin-responsive genes (Szemenyei et al., 2008). Thus TPL and TPRs appear to be at the apex of hormonal crosstalk, linking JA signaling to auxin signaling and possibly other hormone pathways (Pérez and Goossens, 2013). The hormone-mediated interaction between the SCF and the repressors targets the latter for 26S proteasome-mediated degradation, which in turn releases the TFs to allow modulating hormone-dependent gene expression. One point of crosstalk between the two hormones occurs at the level of hormone perception because both IAA and JA-Ile are perceived by SCF E3-ligases (Williams et al., 2019). Auxin-JA crosstalk has also been reported to occur through the interaction of auxin- and JA-related regulators of gene expression. Nagpal et al. showed that the expression of JAZ1/TIFY10A protein is inducible by JA and auxin as well. Moreover, this expression is independent of the JA signaling pathway and is controlled by Aux/IAA ARF (Grunewald et al., 2009). In their study, they used auxin receptor loss-of-function mutants, and they observed a decrease in JA concentration. They demonstrated that JAZ1 is induced by auxin treatment; however, the stability of JAZ proteins is not affected by auxin treatment (Grunewald et al., 2009). Qi et al. investigated the effects of JA on the auxin pathway during the necrotrophic pathogen *A. brassicicola* infection (Qi et al., 2012). They used different *A. thaliana* mutants defective in distinct aspects of the auxin pathway because these plants are more susceptible to the necrotrophic pathogens. They found that during infection, free IAA level was elevated due to the overexpression of the auxin biosynthetic genes, whereas the auxin transport genes were downregulated. These effects of the infection led to enhanced auxin response in host plants. When they used exogenous IAA and MeJA together, the expressions of PDF1.2 (PLANT DEFENSIN 1.2) and HEL (HEVEIN-LIKE) defense marker genes were induced. Taken together these results, JA and auxin interact positively in plant defense during infection of *A. brassicicola*, and enhancement of JA-dependent defense response may be part of the auxin-mediated defense mechanism (Qi et al., 2012). Llorente et al. (2008) investigated the role of auxin during the infection of the necrotrophic fungi *Plectosphaerella cucumerina*. They have worked with *Arabidopsis* auxin signaling mutants *axr1*, *axr2*, and *axr6* that have defects in the auxin-stimulated SCF (Skp1−Cullin−F-box) ubiquitination pathway, which exhibit increased susceptibility to the necrotrophic pathogens. Ubiquitin-mediated proteolysis by the proteasome contributes to the restriction of the fungal diseases. Stabilization of the auxin transcriptional repressor AXR3 that is normally targeted for removal by the SCF-ubiquitin/proteasome machinery occurs upon *P. cucumerina* infection. SGT1b (one of two *Arabidopsis* SGT1 genes encoding HSP90/HSC70 co-chaperones) promotes the functions of SCF E3-ubiquitin ligase complexes in auxin and JA responses and resistance conditioned by certain resistance genes to biotrophic pathogens. They found that *sgt1b* mutants are as resistant to *P. cucumerina* as wild-type plants. Both the synthetic inhibitor of auxin transport TIBA

(Geldner et al., 2001) and the proteasome activity inhibitor MG132 (Ramos et al., 2001) pheno-copy the signaling defects of *axr1−12* and *axr2−1* mutants, leading to an enhanced suscepti-bility to necrotrophic fungi (Llorente et al., 2008). Thus they proved that the repression of auxin signaling, either through mutations in the auxin pathway or by pharmacological inter-ference with the auxin response, increases *Arabidopsis* susceptibility to necrotrophic fungi.

8.5 Pathogens produce and degrade auxins

Many pathogenic microbes and plant growth promoting *Rhizobacteria* have evolved complete pathways for auxin biosynthesis, with tryptophan as the main precursor (Spaepen et al., 2007). The impact of exogenous auxin produced by the pathogen on plant development ranges from positive to negative effects. The consequence for the plant is usually a function of (1) the amount of IAA produced that is available to the plant and (2) the sensitivity of the plant tissue to changes in IAA concentration.

8.6 Stimulation of plant cell growth

Based on the role of auxins in promoting plant cell division and expansion, apparently IAA plays an important role in diseases caused by tumorigenic plant pathogens, such as *Agrobacterium tumefaciens, Pseudomonas savastanoi* (formerly *P. syringae* pv. *savastanoi*), and *Pantoea agglomerans* (Dodueva et al., 2007; Lee et al., 2009). In the case of *A. tumefaciens*, the main source of the IAA involved in disease development is not synthesized directly by the pathogen but rather is produced by plant cells that have been genetically transformed by the *A. tumefaciens* T-DNA element (Binns, 1988). During infection, the T-DNA is delivered into the host cell nucleus via a complex process involving a large number of virulence genes that are regulated by a highly evolved signaling process (Narasimhulu et al., 1996; Veena et al., 2003).

8.7 Auxin as signaling molecule

Auxin can promote the virulence of PtoDC3000 through two different mechanisms: (1) activating host auxin signaling to suppress SA-mediated plant defenses and (2) directly impacting the pathogen by modulating virulence gene expression (Djami-Tchatchou et al., 2020) (Fig. 8.3). IAA acts as a signaling molecule, coordinating the expression of virulence genes required during different phases of pathogenesis. Upon PtoDC3000 infection, the detection of MAMPs induces the expression of basal host defense responses mediated by SA. Early during pathogenesis, bacteria colonizing the apoplast assemble the T3SS and secrete type III effector (T3E) proteins into host cells to suppress MAMP-induced defenses (Wei et al., 2018; Xin et al., 2018). The T3SS is employed by most Gram-negative plant-pathogenic bacteria as an essential virulence factor due to a critical role of T3Es in patho-genesis through the promotion of bacterial growth and virulence in host plants mainly by interfering with PTI signaling networks (Büttner, 2016). Once that is accomplished, the

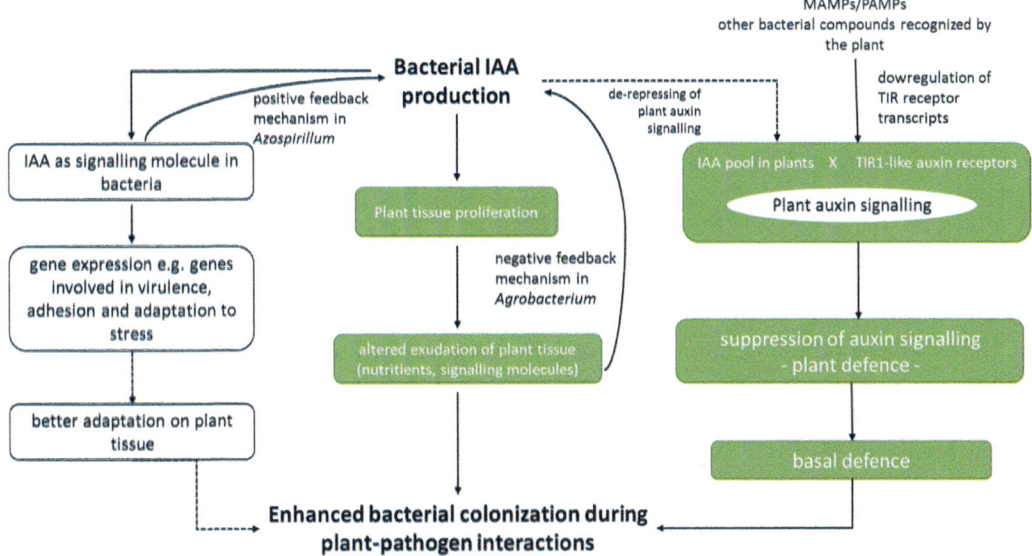

FIGURE 8.3 IAA in pathogenic and beneficial microorganism—plant interactions. Signaling taking place in the plant is indicated in green boxes; signaling taking place in bacterial cells is indicated in white boxes. Full lines indicate demonstrated links; dashed lines indicate hypothesized links. *IAA*, Indole-3-acetic acid. *Source: Data from Spaepen, S., Vanderleyden, J., Remans, R., 2007. Indole-3-acetic acid in microbial and microorganism-plant signaling. FEMS Microbiol. Rev. 31 (4), 425—448. https://doi.org/10.1111/J.1574-6976.2007.00072.X.*

bacteria can obtain water and nutrients and multiply to high levels in the apoplast. PtoDC3000 infection results in elevated IAA levels in infected tissue, possibly due to auxin synthesis by both the host and PtoDC3000 (via activity of AldA) (McClerklin et al., 2018). IAA also promotes the growth of PtoDC3000 independently of the suppression of SA-mediated defenses, by regulating the expression of pathogen virulence genes. They found that IAA downregulates T3SS genes after they are no longer needed (e.g., 24 hours post infection) and activates virulence genes, such as tvrR (encodes a TF to be required for PtoDC3000 virulence), that are required during the intermediate or late stages of infection (Djami-Tchatchou et al., 2020).

8.8 Modulation of defense responses

Elevated levels of auxin promote disease susceptibility in several pathogenic interactions (Kidd et al., 2011; Robert-Seilaniantz et al., 2011a). McClerklin et al. (2018) recently demonstrated that the PtoDC3000 *aldA* mutant exhibits reduced virulence on *A. thaliana* plants, suggesting that auxin synthesized by the pathogen is a virulence factor. They also observed that SA-mediated defenses were elevated in *A. thaliana* plants infected with the *aldA* mutant, and that growth of the *aldA* was restored to normal levels in *sid2 A. thaliana* mutant (which is impaired in SA synthesis). These results suggest that pathogen-derived auxin promotes virulence by suppressing SA-mediated defenses (McClerklin et al., 2018).

There is discrepancy between the studies about the role of auxin during pathogenesis. This could be because (1) auxin promotes DC3000 virulence via multiple different mechanisms, and (2) pathogen-derived and plant-derived auxins play different roles during pathogenesis. The data of McClerklin et al. suggests that the stimulatory effect of AldA-dependent DC3000-synthesized IAA on virulence acts via suppressing SA-mediated defense signaling, while auxin produced by the plant (e.g., YUC1-dependent) promotes pathogen growth via a mechanism that acts independently or downstream of SA-mediated defenses. Another possible role for IAA during pathogenesis is through a direct effect on the pathogen, for example, by regulating virulence gene expression (McClerklin et al., 2018). Future studies examining the impact of the source, the targets, and possibly also the form of auxin during pathogenesis will provide important insight into the roles of auxin in promoting disease development by DC3000.

These findings are consistent with several earlier studies indicating that auxin suppresses SA-mediated defense responses (Kazan and Manners, 2009; Park et al., 2007; Robert-Seilaniantz et al., 2011a). In another study, Chen et al. examined the phenotypes of uninfected transgenic *A. thaliana* plants expressing the *P. syringae* type III effector protein, AvrRpt2, to gain insight into the effect of this virulence factor on host physiology (Chen et al., 2007). They found that the presence of AvrRpt2 during PstDC3000 infection is correlated with an increased level of free IAA in infected plants, and that application of auxin during infection promoted disease development in infected plants. As a result, AvrRpt2 may promote pathogen virulence by altering host auxin physiology. A local increase in free IAA levels at the site of *P. syringae* infection could facilitate pathogen entry and bacterial growth, for example, loosening the cell wall, stomatal opening. Auxin also upregulates the production of ethylene (Yi et al., 1999), which has been proposed to mediate pathogen-induced tissue damage and thus promote disease symptom production (Yamada, 1993). Cui et al. also examined the effects of AvrRpt2 (Cui et al., 2013). They reported that AvrRpt2 promotes auxin response by stimulating the turnover of Aux/IAA proteins. They found that AvrRpt2 acts additively with auxin to stimulate Aux/IAA turnover, suggesting distinct, yet proteasome-dependent, mechanisms operated by AvrRpt2 and auxin to control Aux/IAA stability. AvrRpt2 is a Cys protease that is active in plant cells (Whalen et al., 1991). Cui et al. (2013) tested AvrRpt2 mutants for the ability to promote auxin response and AXR2 protein stability by mutating the conserved residues comprising the catalytic triad of the Cys protease (Cys-122, His-208, and Asp-226) have been mutated, and they found that this activity is required for AvrRpt2 caused susceptibility. Importantly, transgenic plants expressing the dominant *axr2−1* mutation recalcitrant to AvrRpt2-mediated degradation ameliorated the virulence functions of AvrRpt2 but did not alter the avirulent function mediated by the corresponding RPS2 resistance protein. Thus promoting auxin response via modulating the stability of the key transcription repressors Aux/IAA is a mechanism used by the bacterial AvrRpt2 to promote pathogenicity (Cui et al., 2013).

An example of an SA-independent defense mechanism that appears to be modulated by auxin signaling is production of IGs that have antimicrobial activities. In *Arabidopsis*, an expression of the basal defense−elicited miRNA miR393 downregulates auxin signaling (Navarro et al., 2006), resulting in an increased accumulation of several IGs (Robert-Seilaniantz et al., 2011b).

In plants, IAA is typically conjugated to amino acids such as IAA-Asp. This conjugate has been reported to promote *Botrytis cinerea* and *Pst*DC3000 pathogenesis during plant infection (González-Lamothe et al., 2012). In *Arabidopsis*, the GH3.2 gene, which plays a role in synthesizing IAA-Asp, was induced during pathogen infection (Jahn et al., 2013). González-Lamothe et al. (2012) observed that synthetic auxin was not able to promote bacterial growth. This could be due to the substrate specificity of GH3 proteins (Westfall et al., 2012). When they used the *gh3.2* mutant, the amount of IAA-Asp decreased, making the plants less susceptible to infection. While no effect could be observed on pathogens grown in culture media, IAA-Asp clearly contributed to pathogen growth and virulence gene activation when the measure was made on inoculated plants (González-Lamothe et al., 2012). The BcatrB gene of *B. cinerea* encodes an ABC transporter that promotes virulence by increasing tolerance to the phytoalexin camalexin. This gene is also known to be induced by camalexin (Stefanato et al., 2009). Interestingly, their results show that BcatrB expression is more highly induced in Col-0 than in *gh3.2* plants 48 hpi, but camalexin contents remain unchanged under the same conditions. These observations suggest that the reduced susceptibility to *B. cinerea* observed in *gh3.2* plants and the induction of BcatrB mediated by GH3.2 and IAA-Asp are not related to a modification of the camalexin biosynthesis pathway (González-Lamothe et al., 2012). A similar observation was made in Pst DC3000 for the virulence genes AvrPto and HopAO1. These genes encode effectors secreted through the T3SS and contribute to pathogen virulence by suppressing host basal defense responses (Lee et al., 2019). The virulence induced by IAA-Asp occurs during plant infection. This induction might require a direct contact between the pathogen and the host or a component of the host. It is still to be investigated.

8.9 Conclusion

Plants are under continuous biotic stress caused by different attackers (e.g., bacteria, fungi, viruses, *Oomycetes*, and insects) that endanger plant survival and offspring. Crosstalk between defense signaling hormones such as SA and JAs and growth regulator auxin play a significant role in mediating the trade-off between growth and defense in plants. Here we summarized recent studies on how auxin counteracts with plant defense processes regulation through interaction with JA and SA signaling during pathogenesis. IAA can be secreted by the pathogen, and/or it can be produced by the manipulated host. The source and the concentration as well as the plant tissue can alter the effects of IAA. The pathogen can use it to promote susceptibility and to suppress host defense by inhibiting the SA pathway that is effective in protecting plants against biotrophs. By contrast, an increase in auxin content can lead to augmented JA levels, thereby helping defense against necrotrophs. Furthermore, IAA serves as signaling molecule for the pathogen to regulate virulence genes during infection.

References

Bajguz, A., Piotrowska, A., 2009. Conjugates of auxin and cytokinin. Phytochemistry 70 (8), 957–969. Available from: https://doi.org/10.1016/J.PHYTOCHEM.2009.05.006.

Bak, S., Tax, F.E., Feldmann, K.A., Galbraith, D.W., Feyereisen, R., 2001. CYP83B1, a cytochrome P450 at the metabolic branch point in auxin and indole glucosinolate biosynthesis in *Arabidopsis*. Plant. Cell 13 (1), 101–111. Available from: https://doi.org/10.1105/TPC.13.1.101.

Binns, A., 1988. Cell biology of agrobacterium infection and transformation of plants. Annu. Rev. Microbiol. 42 (1), 575–606. Available from: https://doi.org/10.1146/annurev.micro.42.1.575.

Blakeslee, J.J., Spatola Rossi, T., Kriechbaumer, V., 2019. Auxin biosynthesis: spatial regulation and adaptation to stress. J. Exp. Bot. 70 (19), 5041–5049. Available from: https://doi.org/10.1093/JXB/ERZ283.

Brumos, J., Robles, L.M., Yun, J., Vu, T.C., Jackson, S., Alonso, J.M., et al., 2018. Local auxin biosynthesis is a key regulator of plant development. Dev. Cell 47 (3), 306–318. Available from: https://doi.org/10.1016/J.DEVCEL.2018.09.022. e5.

Bürger, M., Chory, J., 2019. Stressed out about hormones: how plants orchestrate immunity. Cell Host Microbe 26 (2), 163–172. Available from: https://doi.org/10.1016/J.CHOM.2019.07.006.

Büttner, D., 2016. Behind the lines—actions of bacterial type III effector proteins in plant cells. FEMS Microbiol. Rev. 40 (6), 894. Available from: https://doi.org/10.1093/FEMSRE/FUW026.

Čarná, M., Repka, V., Skůpa, P., Šturdík, E., 2014. Auxins in defense strategies. Biologia 69 (10), 1255–1263. Available from: https://doi.org/10.2478/S11756-014-0431-3.

Caruana, J.C., Dhar, N., Raina, R., 2020. Overexpression of *Arabidopsis* microRNA167 induces salicylic acid-dependent defense against *Pseudomonas syringae* through the regulation of its targets ARF6 and ARF8. Plant. Direct 4 (9), 1–16. Available from: https://doi.org/10.1002/pld3.270.

Checker, V.G., Kushwaha, H.R., Kumari, P., Yadav, S., 2018. Role of phytohormones in plant defense: signaling and cross talk BT – molecular aspects of plant-pathogen interaction. Mol. Asp. Plant-Pathogen Interact. 159–184. Available from: https://doi.org/10.1007/978-981-10-7371-7_7.

Chen, Z., Agnew, J.L., Cohen, J.D., He, P., Shan, L., Sheen, J., et al., 2007. *Pseudomonas syringae* type III effector AvrRpt2 alters *Arabidopsis thaliana* auxin physiology. Proc. Natl. Acad. Sci. U.S.A. 104 (50), 20131–20136. Available from: https://doi.org/10.1073/pnas.0704901104.

Couto, D., Zipfel, C., 2016. Regulation of pattern recognition receptor signalling in plants. Nat. Rev. Immunol. 16 (9), 537–552. Available from: https://doi.org/10.1038/nri.2016.77.

Cui, F., Wu, S., Sun, W., Coaker, G., Kunkel, B., He, P., et al., 2013. The *Pseudomonas syringae* type III effector AvrRpt2 promotes pathogen virulence via stimulating *Arabidopsis* auxin/indole acetic acid protein turnover. Plant. Physiol. 162 (2), 1018–1029. Available from: https://doi.org/10.1104/PP.113.219659.

Denancé, N., Sánchez-Vallet, A., Goffner, D., Molina, A., 2013. Disease resistance or growth: the role of plant hormones in balancing immune responses and fitness costs. Front. Plant. Sci. 0 (MAY), 155. Available from: https://doi.org/10.3389/FPLS.2013.00155.

Djami-Tchatchou, A.T., Harrison, G.A., Harper, C.P., Wang, R., Prigge, M.J., Estelle, M., et al., 2020. Dual role of auxin in regulating plant defense and bacterial virulence gene expression during *Pseudomonas syringae* PtoDC3000 pathogenesis. Mol. Plant-Microbe Interact. 33 (8), 1059–1071. Available from: https://doi.org/10.1094/MPMI-02-20-0047-R.

Dodueva, I., Frolova, N., Lutova, L., 2007. Plant tumorigenesis: different ways for shifting systemic control of plant cell division and differentiation. Transgenic Plant. J. 1 (1), 17–38.

Durrant, W.E., Dong, X., 2004. Systemic acquired resistance. Annu. Rev. Phytopathol. 42, 185–209. Available from: https://doi.org/10.1146/annurev.phyto.42.040803.140421.

Frick, E.M., Strader, L.C., 2018. Roles for IBA-derived auxin in plant development. J. Exp. Bot. 69 (2), 169–177. Available from: https://doi.org/10.1093/JXB/ERX298.

Gao, Y., Dai, X., Aoi, Y., Takebayashi, Y., Yang, L., Guo, X., et al., 2020. Two homologous INDOLE-3-ACETAMIDE (IAM) HYDROLASE genes are required for the auxin effects of IAM in *Arabidopsis*. J. Genet. Genomics 47 (3), 157–165. Available from: https://doi.org/10.1016/j.jgg.2020.02.009.

Geldner, N., Friml, J., Stierhof, Y.D., Jürgens, G., Palme, K., 2001. Auxin transport inhibitors block PIN1 cycling and vesicle trafficking. Nature 413 (6854), 425–428. Available from: https://doi.org/10.1038/35096571. *2001 413:6854*.

González-Lamothe, R., El Oirdi, M., Brisson, N., Bouarab, K., 2012. The conjugated auxin indole-3-acetic acid-aspartic acid promotes plant disease development. Plant. Cell 24 (2), 672–777. Available from: https://doi.org/10.1105/tpc.111.095190.

Grunewald, W., Vanholme, B., Pauwels, L., Plovie, E., Inzé, D., Gheysen, G., et al., 2009. Expression of the *Arabidopsis* jasmonate signalling repressor JAZ1/TIFY10A is stimulated by auxin. EMBO Rep. 10 (8), 923–928. Available from: https://doi.org/10.1038/embor.2009.103.

Hangarter, R.P., Good, N.E., 1981. Evidence that IAA conjugates are slow-release sources of free IAA in plant tissues. Plant. Physiol. 68 (6), 1424–1427. Available from: https://doi.org/10.1104/PP.68.6.1424.

Huang, S., Zhang, X., Fernando, W.G.D., 2020. Directing trophic divergence in plant-pathogen interactions: antagonistic phytohormones with NO doubt? Front. Plant. Sci. 11, 1850. Available from: https://doi.org/10.3389/FPLS.2020.600063/BIBTEX.

Hull, A.K., Vij, R., Celenza, J.L., 2000. *Arabidopsis* cytochrome P450s that catalyze the first step of tryptophan-dependent indole-3-acetic acid biosynthesis. Proc. Natl. Acad. Sci. U.S.A. 97 (5), 2379. Available from: https://doi.org/10.1073/PNAS.040569997.

Jahn, L., Mucha, S., Bergmann, S., Horn, C., Staswick, P., Steffens, B., et al., 2013. The clubroot pathogen (*Plasmodiophora brassicae*) influences auxin signaling to regulate auxin homeostasis in *Arabidopsis*. Plants (Basel, Switz.) 2 (4), 726–749. Available from: https://doi.org/10.3390/PLANTS2040726.

Jones, J.D. G., & Dangl, J.L. 2006. The plant immune system. In Nature (444, 7117, pp. 323–329). Nature Publishing Group. Available from: https://doi.org/10.1038/nature05286.

Kazan, K., Manners, J.M., 2009. Linking development to defense: auxin in plant-pathogen interactions. Trends Plant. Sci. 14 (7), 373–382. Available from: https://doi.org/10.1016/j.tplants.2009.04.005.

Kempel, A., Schädler, M., Chrobock, T., Fischer, M., Van Kleunen, M., 2011. Tradeoffs associated with constitutive and induced plant resistance against herbivory. Proc. Natl. Acad. Sci. U.S.A. 108 (14), 5685–5689. Available from: https://doi.org/10.1073/PNAS.1016508108/-/DCSUPPLEMENTAL.

Khan, S., Stone, J., 2007. *Arabidopsis thaliana* GH3.9 in auxin and jasmonate cross talk. Plant. Signal. Behav. 2 (6), 483–485. Available from: https://doi.org/10.4161/psb.2.6.4498.

Kidd, B.N., Kadoo, N.Y., Dombrecht, B., Tekeoğlu, M., Gardiner, D.M., Thatcher, L.F., et al., 2011. Auxin Signaling and Transport Promote Susceptibility to the Root-Infecting Fungal Pathogen *Fusarium oxysporum* in *Arabidopsis*. 24(6), 733–748. https://doi.org/10.1094/MPMI-08-10-0194.

Kim, J.I., Dolan, W.L., Anderson, N.A., Chapple, C., 2015. Indole glucosinolate biosynthesis limits phenylpropanoid accumulation in *Arabidopsis thaliana*. Plant. Cell 27 (5), 1529–1546. Available from: https://doi.org/10.1105/TPC.15.00127.

Korasick, D.A., Enders, T.A., Strader, L.C., 2013. Auxin biosynthesis and storage forms. J. Exp. Bot. 64 (9), 2541–2555. Available from: https://doi.org/10.1093/JXB/ERT080.

Kriechbaumer, V., Botchway, S.W., Hawes, C., 2017. Localization and interactions between *Arabidopsis* auxin biosynthetic enzymes in the TAA/YUC-dependent pathway. J. Exp. Bot. 68 (12), 4195–4207. Available from: https://doi.org/10.1093/JXB/ERW195.

Kuhn, H., Kwaaitaal, M., Kusch, S., Acevedo-Garcia, J., Wu, H., Panstruga, R., 2016. Biotrophy at its best: novel findings and unsolved mysteries of the *Arabidopsis*-powdery mildew pathosystem. Arabidopsis Book 2016 (14), e0184. Available from: https://doi.org/10.1199/TAB.0184.

Kunkel, B.N., Johnson, J.M.B., 2021. Auxin plays multiple roles during plant-pathogen interactions. Cold Spring Harb. Perspect. Biol. 13 (9). Available from: https://doi.org/10.1101/CSHPERSPECT.A040022.

Lee, C.W., Efetova, M., Engelmann, J.C., Kramell, R., Wasternack, C., Ludwig-Müller, J., et al., 2009. *Agrobacterium tumefaciens* promotes tumor induction by modulating pathogen defense in *Arabidopsis thaliana*. Plant. Cell 21 (9), 2948–2962. Available from: https://doi.org/10.1105/tpc.108.064576.

Lee, J.H., Kim, H., Chae, W.B., Oh, M.H., 2019. Pattern recognition receptors and their interactions with bacterial type III effectors in plants. Genes. Genomics 41 (5), 499–506. Available from: https://doi.org/10.1007/s13258-019-00801-1.

Lehmann, T., Hoffmann, M., Hentrich, M., Pollmann, S., 2010. Indole-3-acetamide-dependent auxin biosynthesis: a widely distributed way of indole-3-acetic acid production. Eur. J. Cell Biol. 89 (12), 895–905. Available from: https://doi.org/10.1016/j.ejcb.2010.06.021.

Li, M., Guo, R., Yu, F., Chen, X., Zhao, H., Li, H., et al., 2018. Indole-3-acetic acid biosynthesis pathways in the plant-beneficial bacterium *Arthrobacter pascens* ZZ21. Int. J. Mol. Sci. 19 (2), 443. Available from: https://doi.org/10.3390/IJMS19020443. 2018, Vol. 19, Page 443.

Li, Y., Wang, W., Zhou, J.M., 2011. Role of small RNAs in the interaction between *Arabidopsis* and *Pseudomonas syringae*. Front. Biol. 6 (6), 462–467. Available from: https://doi.org/10.1007/S11515-011-1169-8.

Llorente, F., Muskett, P., Sánchez-Vallet, A., López, G., Ramos, B., Sánchez-Rodríguez, C., et al., 2008. Repression of the auxin response pathway increases *Arabidopsis* susceptibility to necrotrophic fungi. Mol. Plant. 1 (3), 496–509. Available from: https://doi.org/10.1093/MP/SSN025.

Ludwig-Müller, J., 2011. Auxin conjugates: their role for plant development and in the evolution of land plants. J. Exp. Bot. 62 (6), 1757–1773. Available from: https://doi.org/10.1093/JXB/ERQ412.

Mackelprang, R., Okrent, R.A., Wildermuth, M.C., 2017. Preference of *Arabidopsis thaliana* GH3.5 acyl amido synthetase for growth versus defense hormone acyl substrates is dictated by concentration of amino acid substrate aspartate. Phytochemistry 143, 19–28. Available from: https://doi.org/10.1016/J.PHYTOCHEM.2017.07.001.

Mano, Y., Nemoto, K., 2012. The pathway of auxin biosynthesis in plants. J. Exp. Bot. 63 (8), 2853–2872. Available from: https://doi.org/10.1093/JXB/ERS091.

Manulis, S., Shafrir, H., Epstein, E., Lichter, A., Barash, I., 1994. Biosynthesis of indole-3-acetic acid via the indole-3-acetamide pathway in *Streptomyces* spp. Microbiology 140 (5), 1045–1050. Available from: https://doi.org/10.1099/13500872-140-5-1045.

Mashiguchi, K., Tanaka, K., Sakai, T., Sugawara, S., Kawaide, H., Natsume, M., et al., 2011. The main auxin biosynthesis pathway in *Arabidopsis*. Proc. Natl. Acad. Sci. U.S.A. 108 (45), 18512–18517. Available from: https://doi.org/10.1073/PNAS.1108434108.

McClerklin, S.A., Lee, S.G., Harper, C.P., Nwumeh, R., Jez, J.M., Kunkel, B.N., 2018. Indole-3-acetaldehyde dehydrogenase-dependent auxin synthesis contributes to virulence of *Pseudomonas syringae* strain DC3000. PLoS Pathog. 14 (1), e1006811. Available from: https://doi.org/10.1371/JOURNAL.PPAT.1006811.

Mutka, A.M., Fawley, S., Tsao, T., Kunkel, B.N., 2013. Auxin promotes susceptibility to *Pseudomonas syringae* via a mechanism independent of suppression of salicylic acid-mediated defenses. Plant. J. 74 (5), 746–754. Available from: https://doi.org/10.1111/TPJ.12157.

Narasimhulu, S.B., Deng, X.B., Sarria, R., Gelvin, S.B., 1996. Early transcription of Agrobacterium T-DNA genes in tobacco and maize. Plant. Cell 8 (5), 873–886. Available from: https://doi.org/10.1105/TPC.8.5.873.

Naseem, M., Dandekar, T., 2012. The role of auxin-cytokinin antagonism in plant-pathogen interactions. PLoS Pathog. 8 (11), e1003026. Available from: https://doi.org/10.1371/journal.ppat.1003026.

Naseem, M., Srivastava, M., Tehseen, M., Ahmed, N., 2015. Auxin crosstalk to plant immune networks: a plant-pathogen interaction perspective. Curr. Protein Peptide Sci. 16 (5), 389–394. Available from: https://doi.org/10.2174/1389203716666150330124911.

Navarro, L., Dunoyer, P., Jay, F., Arnold, B., Dharmasiri, N., Estelle, M., et al., 2006. A plant miRNA contributes to antibacterial resistance by repressing auxin signaling. Science 312 (5772), 436–439. Available from: https://doi.org/10.1126/SCIENCE.1126088.

Paque, S., Weijers, D., 2016. Q&A: auxin: the plant molecule that influences almost anything. BMC Biol. 14 (1), 1–5. Available from: https://doi.org/10.1186/S12915-016-0291-0. *2016 14:1*.

Park, J.E., Park, J.Y., Kim, Y.S., Staswick, P.E., Jeon, J., Yun, J., et al., 2007. GH3-mediated auxin homeostasis links growth regulation with stress adaptation response in *Arabidopsis*. J. Biol. Chem. 282 (13), 10036–10046. Available from: https://doi.org/10.1074/JBC.M610524200/ATTACHMENT/9BBD7493-C4D2-4CE9-AB10-F6D47BF3C20E/MMC1.PDF.

Pasternak, T., Groot, E.P., Kazantsev, F.V., Teale, W., Omelyanchuk, N., Kovrizhnykh, V., et al., 2019. Salicylic acid affects root meristem patterning via auxin distribution in a concentration-dependent manner. Plant. Physiol. 180 (3), 1725–1739. Available from: https://doi.org/10.1104/PP.19.00130.

Pauwels, L., Barbero, G.F., Geerinck, J., Tilleman, S., Grunewald, W., Pérez, A.C., et al., 2010. NINJA connects the co-repressor TOPLESS to jasmonate signalling. Nature 464 (7289), 788. Available from: https://doi.org/10.1038/NATURE08854.

Pérez, A.C., Goossens, A., 2013. Jasmonate signalling: a copycat of auxin signalling? Plant, Cell & Environ. 36 (12), 2071–2084. Available from: https://doi.org/10.1111/PCE.12121.

Pieterse, C.M.J., Van Der Does, D., Zamioudis, C., Leon-Reyes, A., Van Wees, S.C.M., 2012. Hormonal modulation of plant immunity. Annu. Rev. Cell Dev. Biol. 28, 489–521. Available from: https://doi.org/10.1146/ANNUREV-CELLBIO-092910-154055.

Pollmann, S., Neu, D., Weiler, E.W., 2003. Molecular cloning and characterization of an amidase from *Arabidopsis thaliana* capable of converting indole-3-acetamide into the plant growth hormone, indole-3-acetic acid. Phytochemistry 62 (3), 293–300. Available from: https://doi.org/10.1016/S0031-9422(02)00563-0.

Qi, L., Yan, J., Li, Y., Jiang, H., Sun, J., Chen, Q., et al., 2012. *Arabidopsis thaliana* plants differentially modulate auxin biosynthesis and transport during defense responses to the necrotrophic pathogen *Alternaria brassicicola*. N. Phytol. 195 (4), 872–882. Available from: https://doi.org/10.1111/j.1469-8137.2012.04208.x.

Ramos, J.A., Zenser, N., Leyser, O., Callis, J., 2001. Rapid degradation of auxin/indoleacetic acid proteins requires conserved amino acids of domain II and is proteasome dependent. Plant. Cell 13 (10), 2349–2360. Available from: https://doi.org/10.1105/TPC.010244.

Raskin, I., 1992. Role of salicylic acid in plants. Annu. Rev. Plant. Physiol. Plant Mol. Biol. 43 (1), 439–463. Available from: https://doi.org/10.1146/annurev.pp.43.060192.002255.

Rivas-San Vicente, M., Plasencia, J., 2011. Salicylic acid beyond defence: its role in plant growth and development. J. Exp. Bot. 62 (10), 3321–3338. Available from: https://doi.org/10.1093/JXB/ERR031.

Robert-Seilaniantz, A., Grant, M., Jones, J.D.G., 2011a. Hormone crosstalk in plant disease and defense: more than just JASMONATE-SALICYLATE antagonism. Annu. Rev. Phytopathol. 49, 317–343. Available from: https://doi.org/10.1146/ANNUREV-PHYTO-073009-114447.

Robert-Seilaniantz, A., MacLean, D., Jikumaru, Y., Hill, L., Yamaguchi, S., Kamiya, Y., et al., 2011b. The microRNA miR393 re-directs secondary metabolite biosynthesis away from camalexin and towards glucosinolates. Plant. J. 67 (2), 218–231. Available from: https://doi.org/10.1111/j.1365-313X.2011.04591.x.

Sanchez-Vallet, A., Ramos, B., Bednarek, P., López, G., Piślewska-Bednarek, M., Schulze-Lefert, P., et al., 2010. Tryptophan-derived secondary metabolites in *Arabidopsis thaliana* confer non-host resistance to necrotrophic *Plectosphaerella cucumerina* fungi. Plant. J. 63 (1), 115–127. Available from: https://doi.org/10.1111/J.1365-313X.2010.04224.X.

Schaller, G.E., Bishopp, A., Kieber, J.J., 2015. The Yin-Yang of hormones: cytokinin and auxin interactions in plant development. Plant. Cell 27 (1), 44–63. Available from: https://doi.org/10.1105/TPC.114.133595.

Schlaeppi, K., Abou-Mansour, E., Buchala, A., Mauch, F., 2010. Disease resistance of *Arabidopsis* to phytophthora brassicae is established by the sequential action of indole glucosinolates and camalexin. Plant. J. 62 (5), 840–851. Available from: https://doi.org/10.1111/J.1365-313X.2010.04197.X.

Simon, S., Petrášek, J., 2011. Why plants need more than one type of auxin. Plant. Sci. 180, 454–460. Available from: https://doi.org/10.1016/j.plantsci.2010.12.007.

Spaepen, S., Vanderleyden, J., Remans, R., 2007. Indole-3-acetic acid in microbial and microorganism-plant signaling. FEMS Microbiol. Rev. 31 (4), 425–448. Available from: https://doi.org/10.1111/J.1574-6976.2007.00072.X.

Spiess, G.M., Hausman, A., Yu, P., Cohen, J.D., Rampey, R.A., Zolman, B.K., 2014. Auxin input pathway disruptions are mitigated by changes in auxin biosynthetic gene expression in *Arabidopsis*. Plant. Physiol. 165 (3), 1092–1104. Available from: https://doi.org/10.1104/PP.114.236026.

Stefanato, F.L., Abou-Mansour, E., Buchala, A., Kretschmer, M., Mosbach, A., Hahn, M., et al., 2009. The ABC transporter BcatrB from *Botrytis cinerea* exports camalexin and is a virulence factor on *Arabidopsis thaliana*. Plant. J. 58 (3), 499–510. Available from: https://doi.org/10.1111/J.1365-313X.2009.03794.X.

Stepanova, A.N., Robertson-Hoyt, J., Yun, J., Benavente, L.M., Xie, D.Y., Doležal, K., et al., 2008. TAA1-mediated auxin biosynthesis is essential for hormone crosstalk and plant development. Cell 133 (1), 177–191. Available from: https://doi.org/10.1016/J.CELL.2008.01.047.

Stone, B.B., Stowe-Evans, E.L., Harper, R.M., Brandon Celaya, R., Ljung, K., Sandberg, G., et al., 2008. Disruptions in AUX1-dependent auxin influx alter hypocotyl phototropism in *Arabidopsis*. Mol. Plant. 1 (1), 129–144. Available from: https://doi.org/10.1093/MP/SSM013.

Su, Y.H., Liu, Y.B., Zhou, C., Li, X.M., Zhang, X.S., 2016. The microRNA167 controls somatic embryogenesis in *Arabidopsis* through regulating its target genes ARF6 and ARF8. Plant. Cell, Tissue Organ. Cult. 124 (2), 405–417. Available from: https://doi.org/10.1007/S11240-015-0903-3/TABLES/3.

Szemenyei, H., Hannon, M., Long, J.A., 2008. TOPLESS mediates auxin-dependent transcriptional repression during *Arabidopsis* embryogenesis. Science 319 (5868), 1384–1386. Available from: https://doi.org/10.1126/SCIENCE.1151461/SUPPL_FILE/SZEMENYEI.SOM.PDF.

Thomma, B.P.H.J., Eggermont, K., Penninckx, I.A.M.A., Mauch-Mani, B., Vogelsang, R., Cammue, B.P.A., et al., 1998. Separate jasmonate-dependent and salicylate-dependent defense-response pathways in *Arabidopsis* are essential for resistance to distinct microbial pathogens. Proc. Natl. Acad. Sci. U.S.A. 95, 25.

Tivendale, N.D., Ross, J.J., Cohen, J.D., 2014. The shifting paradigms of auxin biosynthesis. Trends Plant. Sci. 19 (1), 44–51. Available from: https://doi.org/10.1016/J.TPLANTS.2013.09.012.

van Butselaar, T., Van den Ackerveken, G., 2020. Salicylic acid steers the growth–immunity tradeoff. Trends Plant. Sci. 25 (6), 566–576. Available from: https://doi.org/10.1016/J.TPLANTS.2020.02.002.

Van Loon, L.C., Rep, M., Pieterse, C.M.J., 2006. Significance of inducible defense-related proteins in infected plants. Annu. Rev. Phytopathol. 44, 135–162. Available from: https://doi.org/10.1146/annurev.phyto.44.070505.143425.

Veena, Jiang, H., Doerge, R.W., Gelvin, S.B., 2003. Transfer of T-DNA and Vir proteins to plant cells by *Agrobacterium tumefaciens* induces expression of host genes involved in mediating transformation and suppresses host defense gene expression. Plant. J. 35 (2), 219–236. Available from: https://doi.org/10.1046/J.1365-313X.2003.01796.X.

Walters, D., Heil, M., 2007. Costs and trade-offs associated with induced resistance. Physiol. Mol. Plant. Pathol. 71 (1–3), 3–17. Available from: https://doi.org/10.1016/J.PMPP.2007.09.008.

Wang, B., Chu, J., Yu, T., Xu, Q., Sun, X., Yuan, J., et al., 2015. Tryptophan-independent auxin biosynthesis contributes to early embryogenesis in *Arabidopsis*. Proc. Natl. Acad. Sci. U.S.A. 112 (15), 4821–4826. Available from: https://doi.org/10.1073/PNAS.1503998112.

Wang, D., Pajerowska-Mukhtar, K., Culler, A.H., Dong, X., 2007. Salicylic acid inhibits pathogen growth in plants through repression of the auxin signaling pathway. Curr. Biol. 17 (20), 1784–1790. Available from: https://doi.org/10.1016/J.CUB.2007.09.025.

Wang, Y., Tyler, B.M., Wang, Y., 2019. Defense and counterdefense during plant-pathogenic oomycete infection. Annu. Rev. Microbiol. 73, 667–696. Available from: https://doi.org/10.1146/annurev-micro-020518-120022.

Wei, H.L., Zhang, W., Collmer, A., 2018. Modular study of the type III effector repertoire in *Pseudomonas syringae* pv. tomato DC3000 reveals a matrix of effector interplay in pathogenesis. Cell Rep. 23 (6), 1630–1638. Available from: https://doi.org/10.1016/J.CELREP.2018.04.037.

Westfall, C.S., Herrmann, J., Chen, Q., Wang, S., Jez, J.M., 2010. Modulating plant hormones by enzyme action the GH3 family of acyl acid amido synthetases. Plant. Signal. Behav. 5 (12), 1607–1612. Available from: https://doi.org/10.4161/psb.5.12.13941.

Westfall, C.S., Zubieta, C., Herrmann, J., Kapp, U., Nanao, M.H., Jez, J.M., 2012. Structural basis for prereceptor modulation of plant hormones by gh3 proteins. Science 336 (6089), 1708–1711. Available from: https://doi.org/10.1126/SCIENCE.1221863/SUPPL_FILE/WESTFALL.SM.PDF.

Whalen, M.C., Innes, R.W., Bent, A.F., Staskawicz, B.J., 1991. Identification of *Pseudomonas syringae* pathogens of *Arabidopsis* and a bacterial locus determining avirulence on both *Arabidopsis* and soybean. Plant. Cell 3 (1), 49–59. Available from: https://doi.org/10.1105/TPC.3.1.49.

Williams, C., Fernández-Calvo, P., Colinas, M., Pauwels, L., Goossens, A., 2019. Jasmonate and auxin perception: how plants keep F-boxes in check. J. Exp. Bot. 70 (13), 3401–3414. Available from: https://doi.org/10.1093/JXB/ERZ272.

Woodward, A.W., Bartel, B., 2005. Auxin: regulation, action, and interaction. Ann. Bot. 95 (5), 707–735. Available from: https://doi.org/10.1093/AOB/MCI083.

Wright, A.D., Sampson, M.B., Neuffer, M.G., Michalczuk, L., Slovin, J.P., Cohen, J.D., 1991. Indole-3-acetic acid biosynthesis in the mutant maize orange pericarp, a tryptophan auxotroph. Science 254 (5034), 998–1000. Available from: https://doi.org/10.1126/SCIENCE.254.5034.998.

Wu, M.F., Tian, Q., Reed, J.W., et al., 2006. Arabidopsis microRNA167 controls patterns of ARF6 and ARF8 expression, and regulates both female and male reproduction. Development 133 (21), 4211–4218. Available from: https://doi.org/10.1242/dev.02602.

Xin, X.F., Kvitko, B., He, S.Y., 2018. *Pseudomonas syringae*: what it takes to be a pathogen. Nat. Rev. Microbiol. 16 (5), 316–328. Available from: https://doi.org/10.1038/nrmicro.2018.17.

Yamada, T., 1993. The role of auxin in plant-disease development, Annual Review of Phytopathology, Vol. 31. Annual Reviews Inc, pp. 253–273. Available from: https://doi.org/10.1146/annurev.py.31.090193.001345.

Yang, Y.X., Ahammed, G.J., Wu, C., Fan, S.Y., Zhou, Y.H., et al., 2015. Crosstalk among Jasmonate, Salicylate and Ethylene signaling pathways in plant disease and immune responses. Current Protein & Peptide Science 16 (5), 450–461. Available from: https://doi.org/10.2174/1389203716666150330141638.

Yi, H.C., Joo, S., Nam, K.H., Lee, J.S., Kang, B.G., Kim, W.T., 1999.). Auxin and brassinosteroid differentially regulate the expression of three members of the 1-aminocyclopropane-1-carboxylate synthase gene family in mung bean (*Vigna radiata* L.). Plant. Mol. Biol. 41 (4), 443–454. Available from: https://doi.org/10.1023/A:1006372612574.

Yu, P., Hegeman, A.D., Cohen, J.D., 2014. A facile means for the identification of indolic compounds from plant tissues. Plant. J. 79 (6), 1065–1075. Available from: https://doi.org/10.1111/TPJ.12607.

Yuan, M., Ngou, B.P.M., Ding, P., Xin, X.F., 2021. PTI-ETI crosstalk: an integrative view of plant immunity. Curr. Opin. Plant. Biol. 62, 102030. Available from: https://doi.org/10.1016/j.pbi.2021.102030.

Yuan, M., Pok, B., Ngou, M., Ding, P., Xin, X.-F., 2021. PTI-ETI crosstalk: an integrative view of plant immunity this review comes from a themed issue on biotic interactions. Curr. Opin. Plant. Biol. 2021, 102030. Available from: https://doi.org/10.1016/j.pbi.2021.102030.

Zeng, W., Brutus, A., Kremer, J.M., Withers, J.C., Gao, X., Da Jones, A.D., et al., 2011. A genetic screen reveals *Arabidopsis* stomatal and/or apoplastic defenses against *Pseudomonas syringae* pv. tomato DC3000. PLoS Pathog. 7 (10), e1002291. Available from: https://doi.org/10.1371/JOURNAL.PPAT.1002291.

Zhang, P., Jin, T., Sahu, S.K., Xu, J., Shi, Q., Liu, H., et al., 2019. The distribution of tryptophan-dependent indole-3-acetic acid synthesis pathways in bacteria unraveled by large-scale genomic analysis. Molecules 24 (7), 1411. Available from: https://doi.org/10.3390/MOLECULES24071411. *2019, Vol. 24, Page 1411.*

Zhang, W., Gao, S., Zhou, X., Chellappan, P., Chen, Z., Zhou, X., et al., 2011. Bacteria-responsive microRNAs regulate plant innate immunity by modulating plant hormone networks. Plant. Mol. Biol. 75 (1), 93–105. Available from: https://doi.org/10.1007/S11103-010-9710-8/FIGURES/6.

Zhang, Z., Li, Q., Li, Z., Staswick, P.E., Wang, M., Zhu, Y., et al., 2007. Dual regulation role of GH3.5 in salicylic acid and auxin signaling during *Arabidopsis-Pseudomonas syringae* interaction. Plant. Physiol. 145 (2), 450–464. Available from: https://doi.org/10.1104/PP.107.106021.

Zhao, Y., 2010. Auxin biosynthesis and its role in plant development. Annu. Rev. Plant. Biol. 61, 49–64. Available from: https://doi.org/10.1146/ANNUREV-ARPLANT-042809-112308.

Zhao, Y., 2014. Auxin biosynthesis. Arabidopsis Book./Am. Soc. Plant. Biologists 12, e0173. Available from: https://doi.org/10.1199/TAB.0173.

Zhao, Y., 2018. Essential roles of local auxin biosynthesis in plant development and in adaptation to environmental changes. Annu. Rev. Plant. Biol. Available from: https://doi.org/10.1146/annurev-arplant-042817.

Zhao, Y., Hull, A.K., Gupta, N.R., Goss, K.A., Alonso, J., Ecker, J.R., et al., 2002. Trp-dependent auxin biosynthesis in *Arabidopsis*: involvement of cytochrome P450s CYP79B2 and CYP79B3. Genes. Dev. 16 (23). Available from: https://doi.org/10.1101/GAD.1035402.

Further reading

Bouzroud, S., Gouiaa, S., Hu, N., Bernadac, A., Mila, I., Bendaou, N., Zouine, M., et al., 2018. Auxin response factors (ARFs) are potential mediators of auxin action in tomato response to biotic and abiotic stress (Solanum lycopersicum). PloS one 13 (2). Available from: https://doi.org/10.1371/journal.pone.0193517.

Fan, S., Chang, Y., Liu, G., Shang, S., Tian, L., Shi, H., et al., 2020. Molecular functional analysis of auxin/indole-3-acetic acid proteins (Aux/IAAs) in plant disease resistance in cassava. Physiol Plantarum 168, 88–97. Available from: https://doi.org/10.1111/ppl.12970.

Kacprzyk, J., Burke, R., Schwarze, J., McCabe, P.F., 2022. Plant programmed cell death meets auxin signalling. FEBS J 289, 1731–1745. Available from: https://doi.org/10.1111/febs.16210.

Korver, R.A., Koevoets, I.T., Testerink, C., 2018. Out of shape during stress: a key role for auxin. Trends in plant science 23 (9), 783–793. Available from: https://doi.org/10.1016/j.tplants.2018.05.011.

Singh, D., Dhiman, V.K., Pandey, H., Dhiman, V.K., Pandey, D., 2022. Crosstalk Between Salicylic Acid and Auxins, Cytokinins and Gibberellins Under Biotic Stress. In: Aftab, T. (Ed.), Auxins, Cytokinins and Gibberellins Signaling in Plants. Signaling and Communication in Plants. Springer, Cham. Available from: https://doi.org/10.1007/978-3-031-05427-3_11.

Zhang, Q., Gong, M., Xu, X., Li, H., Deng, W., 2022.). Roles of Auxin in the Growth, Development, and Stress Tolerance of Horticultural Plants. Cells 11 (17). Available from: https://doi.org/10.3390/cells11172761.

9

Jasmonic acid biosynthesis pathway and its functional role in plants

Ankit Ankit, Saravanappriyan Kamali and Amarjeet Singh

National Institute of Plant Genome Research, New Delhi, India

9.1 Introduction

Plants encounter various abiotic and biotic stresses including drought, heat, cold, salinity, osmotic stress, fungal infection, herbivore attacks in their natural habitat. In these unavoidable circumstances various phytohormones play crucial roles in regulating plant growth and development (Khan et al., 2019, 2020a, 2020b; Nazir et al., 2019, 2021, 2022; Poór et al., 2021). Jasmonic acid (JA) is a lipid-derived phytohormone which acts as a signal as well as regulator in various physiological processes and stress conditions. Methyl ester of JA (MeJA) is the first active jasmonate which was detected and isolated as an odorant from *Jasminium grandiflorum* flowers (Demole et al., 1962). Among the conjugates of JA, JA-Ile is the most biologically active form (Fonseca et al., 2009). Recently, *cis*-(+)-12-oxophytodienoic acid (OPDA) an intermediate in the lipoxygenases (LOX) pathway for JA biosynthesis has been shown to be functional signaling molecule instead of JA in lower plants, such as *Marchantia polymorpha* (liverworts), *Physcomitrella patens* (moss) and *Selaginella martensii* (spikemoss) (Ogorodnikova et al., 2015; Stumpe et al., 2010; Yamamoto et al., 2015). Apart from bryophytes, fungus species such as *Fusarium oxysporum* have JA and/or JA-Ile conjugate (Miersch et al., 1999). Although, JA and its derivatives are distributed among bryophytes and fungi, most of the homologs of JA biosynthesis enzymes are present in major lineages of land plants (Han, 2017). In last decade, studies have been performed in both monocotyledons as well as dicotyledons plants to better understand the JA biosynthesis mechanism. In *Arabidopsis*, JA biosynthesis mainly occurs in chloroplast, peroxisome and cytoplasm (Ruan et al., 2019). In chloroplast, OPDA is synthesized from unsaturated fatty acid α-linolenic acid (α-LeA) derived from the chloroplast membrane, followed by its conversion into JA in peroxisome. The conversion of JA into different functional and structural metabolites takes place in the cytoplasm. JA and its other derivatives like MeJA and JA-Ile are collectively known as jasmonates. Major enzymes which are

Plant Hormones in Crop Improvement
DOI: https://doi.org/10.1016/B978-0-323-91886-2.00008-2

involved at different steps of JA biosynthesis pathway belong to different families like galactolipases [phospholipase A1 (PLA1)], LOXs, allene oxide synthase (AOS), allene oxide cyclase (AOC) and OPDA reductases (OPRs) (Deepika et al. 2022). Apart from these, enzymes like acyl-CoA oxidase, L-3-ketoacyl CoA thiolase and 4-coumarateCoA ligase-like enzymes catalyze the β-oxidation of pentenyl side chain of JA intermediates (Castillo et al., 2004; Koo et al., 2006; Schilmiller et al., 2007). This chapter illustrates the participation of these enzymes in the JA biosynthesis pathway. In addition, their subcellular localization, structure, expression analysis and recent functional aspects in response to stress as well as their regulation in plants was also discussed.

9.2 Release of linolenic acid for jasmonic acid biosynthesis

The fatty acid, α-LeA (18:3) is the primary precursor of JA biosynthesis which is hydrolyzed from chloroplast membrane by the activity of a galactolipase enzyme. PLA1 hydrolyzes the galactolipids with *sn-1* specificity. Earliest evidence of JA biosynthesis initiation by α-LeA as precursor were provided by a study on DEFECTIVE IN ANTHER DEHISCENCE1 (DAD1) in *Arabidopsis* flower stamen filaments (Ishiguro et al., 2001). *DAD1* gene encodes phospholipase A1 which catalyze the initial step of JA biosynthesis, i.e., hydrolysis of α-LeA. In *Arabidopsis*, *dad1* mutant display defects in anther dehiscence, pollen maturation, and flower opening which could be reversed by exogenous JA or α-LeA. DAD1 protein was found to be chloroplast localized and it contains conserved lipase active site. It's enzymatic PLA1 activity in *Escherichia coli* further confirmed its role in initiation of JA biosynthesis in chloroplast (Ishiguro et al., 2001).

DONGLE (DGL), a DAD1 homolog is chloroplast localized and is responsible for maintaining basal JA level under conditions in leaves and regulates vegetative growth (Hyun et al., 2008). AtPLA1 is an acylhydrolase which is involved in basal JA production and provide resistance to neurotrophic fungus *Botrytis cinerea* (Yang et al., 2007). Glycerolipase GLA1 is required to supply trienoic fatty acid for JA biosynthesis in leaves and roots after wounding and stimulated herbivory but not in response to infection with *Phytophothora parasitica* in *Nicotiana attenuate* (Bonaventure et al., 2011). *Arabidopsis* DAF [DEFECTIVE IN ANTHER DEHISCENCE1 (DAD1)-Activating Factor] functions as a RING-finger E3 ubiquitin ligase protein that control anther dehiscence by positively regulating the expression of DAD1 in the JA biosynthesis pathway (Peng et al., 2013). 35S:DAF RNAi/antisense plants shows significant reduction in the expression of DAD1 whereas, 35S:DAF failed to rescue the *dad1* mutant. This depicts the involvement of DAF upstream of DAD1 in the JA biosynthesis pathway (Peng et al., 2013). The *Arabidopsis* floral homeotic gene AGAMOUS (AG) directly binds to the 5′ coding region of DAD1 and induces its transcriptional expression in the late stamen development stage (Ito et al., 2007). A splice variant (ARF8.2) of the auxin response factor gene ARF8 is responsible for increased expression of the DAD1 gene, which controls stomium opening, the last step in anther dehiscence (Ghelli et al., 2018). Thus, DAF, AG and ARF8 may be independently involved in the JA biosynthesis pathway through regulation of DAD1. Plastid lipase *PLIP2* and *PLIP3* are the two ABA and abiotic stress responsive genes. Plants overexpressing PLIP2 and PLIP3 exhibit overall reduced growth which is accompanied by an accumulation of oxylipins, including JA, its derivatives. Both of these chloroplastic enzymes contribute in the release of fatty acids from chloroplast membrane lipids for JA biosynthesis (Wang et al., 2018). Induction of *PLIP1* and *PLIP2* in response to ABA and abiotic stresses, as well as their involvement in JA biosynthesis indicate that they may integrate plant abiotic and biotic stress responses (Wang et al., 2018).

9.3 Lipoxygenases

LOXs (EC 1.13.11.12) are important class of enzymes of oxylipin and JA biosynthesis pathway. LOXs have been found to be involved in various physiological processes under abiotic and biotic stress conditions in plants. LOXs belong to multigene family which is present both, in plants and animals. They are non-heme, non-sulfur, iron containing dioxygenases which catalyze the oxidation of fatty acids like linoleic acid, α-linolenic acid, arachidonic acid etc. at C9 or C13 positions stereo-specifically (Wasternack and Feussner, 2018). Structurally, LOX proteins consist of two conserved PLAT (polycystin-1, lipoxygenase, and α-toxin) domain at N-terminus and a lipoxygenase domain at C-terminus. PLAT domain has eight-stranded antiparallel β-barrel whereas, lipoxygenase domain has a much larger α-helical subunit with five conserved histidine residues which are involved in iron binding (Newcomer and Brash, 2015). Based on the positional specific of oxygenation at carbon backbone (either C-9/C-13) of α-LeA, they are categorized into 9-LOXs and 13-LOXs (Feussner and Wasternack, 2002). Only 13-LOXs are involved in the Vick and Zimmermann pathway of JA biosynthetic pathway in plants. An overview of JA biosynthesis pathway is presented in Fig. 9.1.

Furthermore, 13-LOX are classified as type I and type II subgroups. Members of type I subgroup exhibit high sequence similarity (>75%) among themselves and are localized at cytoplasm whereas, member of type II subgroup have moderate sequence similarity (> 35%), they contain N-terminal transit peptide and localized at chloroplast (Feussner and Wasternack, 2002). LOX family has been identified in different plant species. In *Arabidopsis*, there are six LOX family members, out of which four are 13-LOXs (AtLOX2, AtLOX3, AtLOX4, AtLOX6) and two are 9-LOXs (AtLOX1, AtLOX5) (Bannenberg et al., 2009). Tomato genome encodes nine 9-LOX members and five 13-LOX members (Upadhyay et al., 2019). In cucumber there are 23 predicted *LOX* genes in its genome (Yang et al., 2012). Similarly, 16 LOXs are identified in rice (Deepika and Singh, 2021), 8 LOXs in pepper (Lim et al., 2015), 11 LOXs in *Raphanus sativus* (Wang et al., 2019) and 18 in chickpea *(Cicer arietinum)* (Deepika et al., 2022). Cotton has a total 64 putative *LOX* genes that are identified from 4 cotton species (*Gossypium hirsutum*, *Gossypium barbadense*, *Gossypium arboreum*, and *Gossypium raimondii*), and in the phylogenetic tree they are clustered in three categories (9-LOX, 13-LOX type I, and 13-LOX type II) (Shaban et al., 2018). After the release of α-LeA (18:3) acid from chloroplast membrane, oxygenation at its C-13 position is catalyzed by 13-LOX. This results into the production of 13-HPOT (13S-hydroperoxyoctadecatrienoic acid) which is subsequently converted into JA and its derivatives. In *Arabidopsis*, all four 13-LOXs (AtLOX2, AtLOX3, AtLOX4, and AtLOX6) are known to be involved in JA biosynthesis. LOX2 is the main enzyme of JA biosynthesis in leaf and xylem, especially in response to wounding, osmotic stress and during senescence

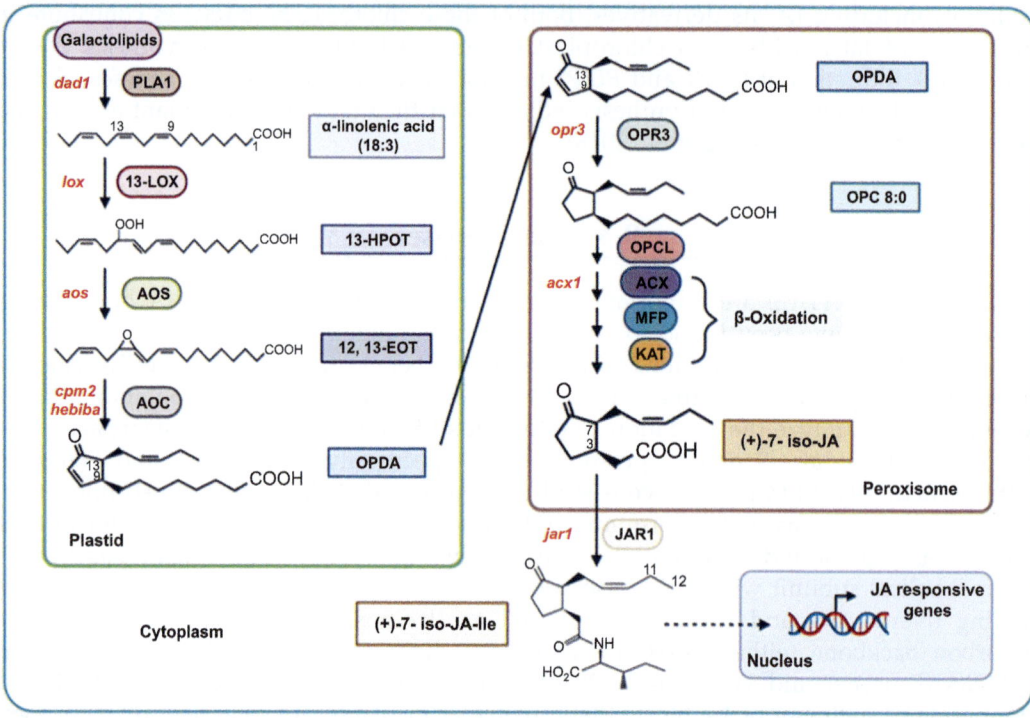

FIGURE 9.1 **Overview of jasmonic acid (JA) biosynthesis pathway in plants.** JA/JA-Ile biosynthesis initiation occurs in plastid where α-LeA (α-linolenic acid) is produced by the action of PLA1on galactolipids. 13-LOX utilizes α-LeA to generate 13-HPOT (13S)-hydroperoxyoctadecatrienoic acid). AOS and AOC enzymatic action leads the formation of OPDA (cis-(+)-12-oxophytodienoic acid). OPDA is then converted into OPC-8 (3-oxo-2-(2-pentenyl)-cyclopentane-1-octanoic acid) by OPR3 in peroxisomes. OPCL (OPC-8:0:CoA ligase) action on OPC-8 followed by β-oxidation of pentenyl side chain of JA intermediates by acx1, MFP and KATgenerates (+)-7-iso-JA in peroxisomes. In cytosol, (+)-7-iso-JA is conjugated with amino acid by JAR1 to produce (+)-7-iso-JA-Ile. JA and its conjugates are associated and regulate the expression of various JA responsive genes in the nucleus. All the mutant enzymes in *Arabidopsis* are shown in red.

(Bell et al., 1995; Glauser et al., 2009; Seltmann et al., 2010). LOX3 and LOX4 have been found to regulate male fertility as *lox3 lox4* double mutant are male sterile (Caldelari et al., 2011). The male sterility of *lox3 lox4* double mutant could be restored upon complementing with LOX3/LOX4 cDNA or exogenous treatment with JA (Caldelari et al., 2011). These findings confirm the involvement of LOX3 and LOX 4 in JA biosynthesis in *Arabidopsis*. Moreover, LOX3 and LOX4 exhibit differential activity in response to root-knot nematode (*Meloidogyne javanica*) and cyst nematode (*Heterodera schachtii*), where LOX4 plays major role in controlling plant defense against nematode infection (Ozalvo et al., 2014). In roots, AtLOX6, is involved in the oxylipin formation in response to osmotic and drought stress, as well as it is also essential for JA production in response to wounding (Grebner et al., 2013). The *lox6* knockout mutants show high sensitivity to drought, and attracts detritivores crustacean. This suggests that LOX6 mediate JA production in response to abiotic and biotic factors (Grebner et al., 2013). LOX3 has been found to induce in response to salinity stress. The *lox3* mutant exhibits hypersensitivity to salt treatment and MeJA treatment improves the salinity induced phenotype of the *lox3* mutant (Ding et al., 2016), indicating that LOX3 is responsible for salinity induced JA production. Although all four 13-LOXs are involved in JA biosynthesis upon wounding, but LOX6 contributes to the rapid accumulation of JA and JA-Ile in distal leaves upon wounding (Chauvin et al., 2013). In rice, infection by fungus *Magnaporthe oryzae* leads to the induction of miR319b. Interestingly, rice OsLOX2 and OsLOX5 were specifically suppressed by miR319b overexpression or *M. oryzae* infection. Therefore, *M. oryzae* defeats host immunity and induce pathogenicity by suppressing JA biosynthesis pathway genes through miR319b. OsLOX2/ 5 induction has also been proposed to provide resistance against rice blast (Zhang et al., 2018). The genome of melon encodes five 13-LOXs (CmLOX08, CmLOX10, CmLOX12, CmLOX13 and CmLOX18) with chloroplast transit peptide, and they all are readily induced by wounding and H_2O_2 treatment. In addition, these 13-LOXs respond differentially to phytohormones such as, ABA, SA, JA and abiotic stresses such as cold and heat (Liu et al., 2016). These findings suggests the diverse functional role of LOXs in abiotic and biotic stresses and plant defense.

9.4 Allene oxide synthase

After LOXs, the AOS is the first enzyme involved in JA biosynthesis in chloroplast. The, cloning, purification and characterization of AOS was done in early 90's from flax seeds for the first time (Song et al., 1993, 1991). AOS belongs to the CYP74 family of cytochrome P450 enzymes in plants (Hughes et al., 2009). AOS enzymes catalyze the production of intermediate components like EOT [(13S)-12,13-epoxy-octadecanoic acid] and EHT [(11S)-10,11-epoxy-hexadecatrienoic acid] from peroxides synthesized by lipoxygenases LOXs in JA biosynthetic pathway. These intermediates are highly unstable with a half-life of few seconds in aqueous solution (Wasternack and Feussner, 2018). *Arabidopsis* has a single copy gene that encodes AOS, protein which is chloroplast localized. AtAOS contains membrane binding domain and heme-binding domain. A knockout mutation in *AOS* results in male sterility and defect in wound signal transduction due to impaired JA biosynthesis (Park et al., 2002). In rice, AOS interacts with 9- and 13-hydroxyperoxides where, Phe347 is crucial for AOS functionality and Asn278 residue in catalytic site is important for maintaining stability of complexes (Tyagi et al., 2016). Analysis of AtAOS crystal shows the involvement of an unusual active site in regulating the reaction of epoxyallylic radical and its cation by means of interactions with an aromatic π-system. AOS activity was reduced when the amino acid from this active site are replaced with non-polar residues (Lee et al., 2008). Another crystal structure of guayule (*Parthenium argentatum*) AOS and its complex analog 13(*S*)-hydroxyoctadeca-9Z,11-*E*-dienoic acid have been determined. This structure has a classic P450 fold with an unusually long heme binding loop and a unique I-helix (Li et al., 2008). Asn-276 residue in the substrate binding site may interact with hydroperoxy group of its substrate whereas, Lys-282 at the entrance may control substrate access and binding. Thus, Asn-276 and Lys-282 residue play an important role in catalysis mechanism of AOS (Li et al., 2008). In tea (*Camellia sinensis*) flower, only AOS2 gene was found to be upregulated upon exposure to insect (*Thrips hawaiiensis*) attacks (Peng et al., 2018).

Functional characterization of grapevine (*Vitis vinifera* L. *Sauvignon blanc*) VvAOS suggests its role in plant growth and development. VvAOS could complement male sterile phenotype in *Arabidopsis* JA biosynthesis mutants, which confirmed its role in JA biosynthesis (Dumin et al., 2018). Expression of *Castanea crenata* CcAOS in *Arabidopsis* result into increase in JA level, delay in pathogen (*Phytophthora cinnamomi*) progression, and transgenic plants exhibit vigorous growth even in presence of pathogen (Serrazina et al., 2021). In rice, both OsAOS1 and OsAOS2 are involved in herbivore-induced JA biosynthesis. The transcript levels of both genes are enhanced by mechanical wounding, herbivore attack (*Chilo suppressalis, Niaparvata lugens*) and in response to JA treatment (Zeng et al., 2021).

9.5 Allene oxide cyclase

AOC participates in JA biosynthesis pathway by catalyzing the stereospecific conversion of 12,13(*S*)-epoxy-9(Z),11,15(Z)-octadecatrienoic acid (12,13-EOT) to OPDA (Yoeun et al., 2018). *Arabidopsis* AOC2 contains eight-stranded antiparallel β-barrel with a C-terminal partial helical extension, along with a hydrophobic binding cavity with two distinct polar patches (Hofmann et al., 2006). Crystallography and structure analysis revealed that in crystalline state AOC exists as a trimer (Hofmann et al., 2006; Neumann et al., 2012; Otto et al., 2016). *Arabidopsis* contains four functional members in AOC family (AOC1−4) express in tissue and organ specific manner (Stenzel et al., 2012). GUS bases promoter analysis showed that in fully developed leaves, AOC1, AOC2 and AOC3 are active in entire leaf tissue whereas, AOC4 shows vascular bundle specific activity. In addition, AOC3 and AOC4 promoters show root specific activity whereas, AOC1 and AOC4 have partial activity during flower development (Stenzel et al., 2012). Protein-protein interaction analysis also revealed *in vivo* heterodimerization among AOCs (Otto et al., 2016; Stenzel et al., 2012). These findings indicate JA biosynthesis could be regulated by differential expression and heteromerization among AOCs in *Arabidopsis*. In rice, only single copy gene encodes for chloroplast localized AOC, which is predicted to have a core β-barrel. Structural analysis demonstrates that OsAOC exists as a mixture of multimers in the solution, out of which homodimer appears to be its preferred confirmation (Yoeun et al., 2018). In rice, two photomorphogenic mutants, *cpm2* and *hebiba* were found to be defective in the *OsAOC* gene (Riemann et al., 2013). Both mutants had significantly lower jasmonates levels and were more susceptible to blast fungus *M. oryzae* than WT plants (Riemann et al., 2013). Recent advances in technology made it possible to generate JA deficient mutants of rice by targeting single copy AOC gene using CRISPR/Cas9 system (Nguyen et al., 2020). In sugarcane, ScAOC1 exhibits differential expression in response to various abiotic stresses like cold, salt, PEG etc. and phytohormones treatment of MeJA, ABA and SA. Transient overexpression of *ScAOC1* in *Nicotiana benthamiana* leaves confers resistance against *Ralstonia solanacearum* and *Fusarium solani* var. *coeruleum* pathogens. Therefore, ScAOC1 may have an important role in response to various biotic and abiotic stresses in sugarcane (Sun et al., 2020). Herbivory-induced stunted growth of *Nicotiana attenuate* caused by GA deficiency shows regulatory roles of JA-Ile and AOC on GA metabolism. JA-Ile positively regulates the GA catabolic genes whereas, expression of GA biosynthetic genes is controlled by AOC derivatives and not by JA-Ile (Yang et al., 2020).

9.6 OPDA reductase

OPDA produced in chloroplast is transported to peroxisome where, rest of the reaction of JA biosynthesis occur. However, the transport mechanism of OPDA across two membranes has not been identified yet. A putative transport mechanism involving intrinsic acyl Co-A thioesterase activity of a fatty acid-transporting peroxisomal ABC transporter is proposed. In peroxisome, conversion of OPDA to JA involves the reduction of 10,11 double bond of OPDA to yield 3-oxo-2-(2′-pentenyl)-cyclopentane-1-octanoic acid (OPC-8:0). This reaction is catalyzed by OPR (Schaller et al., 2000). Biesgen and Weiler initially reported that *OPR2* gene is involved in JA biosynthesis during pollen development (Biesgen and Weiler, 1999).

Later, characterization of the *opr3*mutantswhich show male sterile phenotype revealed that the *OPR3* gene product is the only isozyme that is involved in JA production in anther tissue, which is essential for pollen development and its release (Stintzi, 2000). Only OPR3 from OPR family could selectively reduce (9*S*,13*S*)-OPDA (a natural precursor of JA), rather than (9*R*,13*R*)-OPDA. Whereas both OPR1 and OPR2 function in removal of (9*R*,13*R*)-OPDA due to spontaneous cyclization of 12,13-epoxytrienoic acid (Laudert and Weiler, 1998). Therefore, the OPR family in plants is categorized into two subgroups: subgroup I which includes OPR1 and OPR2, subgroup II including OPR3 (Stintzi, 2000). OsOPR7, AtOPR3 and SlOPR3 proteins fall in same phylogenetic clade and belong to subgroup II which hints towards their conserved biological role in JA biosynthesis.

Structural studies of OPR show that OPR belongs to Old Yellow Enzyme family protein. OPR3 contains the usual observed $(\beta/\alpha)8$-barrel fold, where cylindrical β sheet which is made up of eight β strands is surrounded by eight α-helices. The αB helix make a common phosphate-binding motif which anchors the phosphate group of the FMN (Wilmanns et al., 1991). The substrate is reduced by hydride transfer from $FMNH^-$ N to C^β and protonated at C^α by Y-190. Y-246 and Y-78 of OPR1 above the FMN form a narrow entrance, resulting in a high degree of substrate stereo selectivity. The corresponding amino acids of OPR3 are H-244 and F-74, which turns away from the active site, forming wide cavity. These differences in geometry accounts for "relaxed" specificity of OPR3 (Breithaupt et al., 2006).

In plants, OPR3 is preferentially expressed in phloem and is involved in various physiological processes like lateral root development, seed germination inhibition, flower development, biotic and abiotic stress tolerance. The functional disruption of OPR3 leads to reduction in overall performance of the plants. In *Arabidopsis*, *opr3* mutants were found to have more lateral roots associated with reduced expression of an auxin biosynthesis gene *ASA1*. This suggests that JA together with auxin is involved in root architecture modeling (Li et al., 2013). Transgenic wheat plants with high *AtOPR3* expression levels showed delayed germination, slower growth, late flowering and senescence, and improved tolerance to short-term freezing (Pigolev et al., 2018). In tomato, silencing *SiOPR3* resulted in lower JA and OPDA levels and increased susceptibility to pathogens. Moreover, pathogen induced callose deposition was reduced in *SiOPR3* mutants (Scalschi et al., 2015). In wheat, RNAi lines of *TaOPR3* were susceptible to heat stress, while its over-expression lines were tolerant to heat stress (Pigolev et al., 2018). Application of MeJA to *TaOPR3* RNAi lines reduced the leakage of electrolytes due to heat stress. This indicates the crucial role of TaOPR3 mediated JA accumulation in heat stress tolerance.

9.7 Activation of OPC8:0 before peroxisomal β-oxidation

Plant genomes are comprised of large family of acyl-activating enzymes (AAE). The *AAE* superfamily in *Arabidopsis thaliana*, for example, is comprised of 63 members which are involved in the activation of various carboxylic acid substrates (Shockey et al., 2003). These include 4-CoA Ligases (4-CL) and adenylating enzymes that metabolize phytohormones. The large number of AAEs in higher plants presumably reflects the diversity of carboxylic acid-containing compounds involved in primary and secondary metabolism. These AAEs are also involved in conversion of OPC-8:0 into OPC-8:0-CoA esters before entering the peroxisomal β-oxidation cycle. An acyl-CoA oxidase (ACX) that metabolizes OPC-8:0-CoA and OPDA-CoA but not the respective un-esterified compounds was found to be involved in JA production in tomato leaves (Li et al., 2005). Also, two 4CL-like AAEs from *Arabidopsis* were shown to have the capacity to activate C18 and C16 precursors of JA in vitro (Koo et al., 2006). Reduced levels of JA and increased levels of OPC-8:0 in *opcl1−1* and *opcl1−2* knock-out mutants indicated the involvement of OPCL1 in JA biosynthesis. However, the *opcl1−1* and *opcl1−2* mutants had accumulated nearly 50% of JA, which suggest that other members of the peroxisome-targeted group of 4CL-like proteins contributes to the activation of JA precursors (Schilmiller et al., 2007).

9.8 β-Oxidation

Vick and Zimmermann investigated the metabolism of ^{18}O-labeled OPDA, OPC8:0 and 15,16-dihydro-OPC8:0 and observed the formation of JA and dh-JA, with two intermediates having acyl side chain shortened by two and four carbons. These intermediates are indicatives of three rounds of β-oxidation during formation of JA from OPC8:0 (Vick and Zimmerman, 1987). In plants, fatty acid β-oxidation generally occurs in peroxisomes. But the location of β-oxidation involved in JA biosynthesis was not assured, and it was assumed to be compartmentalized in peroxisome due to the presence of OPR3 in peroxisomes. This was further concluded with the discovery of three *Arabidopsis* 4-coumarate: CoA ligase-(4CL)-like proteins which activate the OPC-8:0 for further oxidation, and these were localized in peroxisomes (Schilmiller et al., 2007). β-oxidation process takes place via three enzymes (1) acyl-CoA oxidase, (2) multifunctional protein (MFP) and (3) L-3-ketoacyl-CoA thiolase (KAT). Together these proteins catalyze the cleavage of acetate units from acyl-CoAs. Each of these three enzymes are encoded by a small gene family in *Arabidopsis* (Li et al., 2005).

9.9 Acyl-CoA oxidase

ACX (EC 1.3.3.6) is the first enzyme involved in fatty acid β-oxidation. It catalyzes the oxidation of fatty acyl-CoA into 2-trans-enoyl-CoA. During this reaction, the flavine adenine dinucleotide containing ACX donates electron to oxygen in substrate, forming hydrogen peroxide (Schilmiller et al., 2007). In plants, ACX forms a gene family of isoenzymes. In *Arabidopsis*, ACXs family is comprised of six enzymes with overlapping specificity for acyl-CoA substrates of various chain lengths. These exhibits activity with medium—to long—chain fatty acy-CoAs, which is (C_{12} to C_{16})—and having maximum activity for C_{16} acyl-CoAs. ACXs contains flavin mononucleotide binding motif (CGGHGY) and are conserved across the plant species. AtACX1 contain a tripeptide sequence ([SAC]-[KHR]-[LM]) at its C-terminal end, corresponding to the peroxisomal targeting signal (PTS1), which is usually found in plant peroxisomal enzymes such as isocitrate lyase and malate synthase. While, AtACX2, AtACX3 and AtACX4 contain a pre-sequence at their N terminal, which corresponds to PTS2 found in malate dehydrogenase and citrate synthase (Xin et al., 2019). Therefore, plant ACXs have either PTS1 or PTS2 for import into peroxisomes. The reaction of ACX is catalyzed in a two-step process. First, the double bond formation between C_α–C_β of the substrate by oxidation. Second, the formation of hydrogen peroxide by introduction of hydride ion from FAD to the molecular oxygen. Among five expressed *ACX* gene family members (*ACX1*–5) in *Arabidopsis*, only *ACX1* is known to be involved in JA production. However, recently, double *acx1/5* mutant were found to have reduced JA formation in *A. thaliana* which suggests possible role of ACX5 also in JA biosynthesis (Schilmiller et al., 2007) The expression level of *ACX1* was high during flower opening and decreased during petal senescence in *Camelia sinensis*, this correlates with the pattern to that of JA and JA-Ile accumulation (Xin et al., 2019). In *Brassica campestris*, mutation in *ACX1* leads to petal degeneration (Peng et al., 2019). Similar to AtACX1, OsACX1 and SlACX1 are the Acyl-CoA oxidases involved in fatty acid β-oxidation in rice and tomato respectively (Wasternack, 2007).

9.10 Peroxisomal multifunction protein

MFP is another enzyme involved in β-oxidation of OPC8:0 into JA. MFP adds water to 2-*trans*-enoyl-CoA double bond and oxidizes the resultant L-3-hydroxyacyl-CoA into 3-keto-acyl-CoA, using NAD^+. MFP contains 2-*trans*-enoyl-CoA hydratase and 3-hydroxyacyl-CoA dehydrogenase activity, Δ^3,Δ^2-Enoyl-CoA isomerase and D-3-hydroxyacyl-CoA epimerase activities (Ohya et al., 2008). Structural studies showed that MFP is 725-aa long protein organized into two structural entities separated by an α-linker. The two structural entities are the ECH domain and HACD domain. AtMFP-ECH catalyzes the *syn* addition of water across the double bond of an α, β-unsaturated thioester converting 2-*trans*-enoyl-CoA to L-3-hydroxyacyl-CoA. Whether the addition of water proceeds via a concerted reaction or a stepwise mechanism is not clear. MFP-ECH can also catalyze the isomerization of β,γ-unsaturated CoA thioesters to form the α,β-unsaturated thioesters for further β-oxidation (Willadsen and Eggerer, 1975). MFP2-HACD catalyzes the oxidation of the L-3-hydroxyacyl-CoA hydroxyl group to a keto group while reducing NAD^+ to NADH (Noyes et al., 1974). The HACD entity can be further divided into two domains $HACD_N$ and $HACD_C$ with interdomain interaction, dominated by nonpolar residues (Arent et al., 2010). *Arabidopsis* has two peroxisomal MFPs, AIM1 and MFP2. A complete block in β-oxidation in *ataim1/atmfp2*double mutant results in impaired embryo development (Rylott et al., 2006). *AtMFP2* is induced during germination and expressed primarily during post-germinative growth, whereas *AtAIM1* is expressed predominantly in siliques, flowers, and seedlings. Mutations in the *AtMFP2*leads to a sucrose-dependent seedling phenotype

(Richmond and Bleecker, 1999), whereas *AtAIM1* disruption causes abnormal inflorescence development resulting in low fertility. AtAIM1 seems to have higher affinity for short chain acyl-CoA, although *aim1* plants have elevated levels of C18:1 and C18:2 unsaturated fatty acids. The different phenotypes and substrate specificities of two MFPs reveal their diverse physiological roles.

9.11 Keto-acyl thiolase

The last reaction in of the β-oxidation pathway is catalyzed by KAT which leads to the formation of (+)-7-iso-JA in peroxisomes. A thiolase cleaves acetyl-CoA from the fatty acyl chain, resulting in the formation of fatty acyl-CoA, which is 2-C shorter than the original substrate (Wasternack, 2007). KAT exists in dimeric form, each subunit consists of three domains; the N-domain, C-domain, and Loop domain. *Arabidopsis* genome has three *KAT* genes that encodes three peroxisome KAT isoforms: *AtKAT1*, *AtKAT2*, and *AtKAT5* and one cytosolic form. The three peroxisomal KAT have been reported to be involved in JA biosynthesis. AtKAT2 belongs to type I 3-ketoacyl-CoA thiolase (Footitt et al., 2007). The *atkat2* mutant was found to accumulate reduced level of JA in wounded cotyledons and leaves, while only the cotyledons accumulate 3-oxo-2-(pent-2′-enyl)-cyclopentane-1-octa-noic acid (OPC-8:0). This indicates that a defect in one of the thiolase isoenzymes impairs β-oxidation of OPC-8:0 to JA (Afitlhile et al., 2005). AtKAT2 is a very sensitive redox switch capable of regulating β-oxidation even at minor perturbations of the peroxisomal redox environment.

9.12 Jasmonate resistance 1 (JAR1)

JA released from peroxisome into cytoplasm and gets conjugated with various amino acids. JA conjugate with isoleucine forms bioactive JA-Ile which is extensively studied as functional signaling component. This conjugation reaction is catalyzed by an ATP-dependent JA-amido synthetase encoded by the gene *JAR1* (Meesters et al., 2014). Structurally, JAR1 are members of firefly luciferase family proteins. Among different oxylipins, JAR1 has highest substrate specificity for JA. The *Arabidopsis jar1−1* mutants had very low amount of JA-Ile and plants had reduced root growth and were sensitive to ozone and pathogens (Staswick and Tiryaki, 2004). In tobacco, leaf wounding markedly increased the amount of JA−Ile and transient silencing of two tobacco orthologues of *Arabidopsis* JAR1; *NaJAR4* and *NaJAR6*, suppressed the wounding induce JA−Ile accumulation. In addition, silenced plants were susceptible to *Manduca sexta* larvae and induction of a TPI was hampered. While, treatment of silenced plants with JA−Ile reversed the defects in plant defense against insect (Suza and Staswick, 2008). Similarly, rice *OsJAR1* mutants were susceptible to herbivores (Fukumoto et al., 2013). Another major role of JAR1 was found to be in integrating the far-red light and JA signaling and makes the plants shade tolerant (Wen et al., 2017). The summary of various enzymes and their role are given in Table 9.1.

TABLE 9.1   Summary of various JA biosynthetic enzymes and their roles in plants.

Enzyme	Plant	Localization	Role	References
DAD1	Arabidopsis thaliana	Chloroplast	Pollen maturation and flower opening	Ishiguro et al. (2001)
GLA1	Nicotiana attenuate	Chloroplast	Wounding	Bonaventure et al. (2011)
PLIP2 and PLIP3	A. thaliana	Chloroplast	ABA, cold, drought	Wang et al. (2018)
LOX2	A. thaliana	Chloroplast	Wounding, osmotic stress and senescence in leaves	Bell et al. (1995), Glauser et al. (2009), Seltmann et al. (2010)
LOX3	A. thaliana	Chloroplast	Male sterility, nematode infection, salinity	Caldelari et al. (2011), Ozalvo et al. (2014), and Ding et al. (2016)
LOX4	A. thaliana	Chloroplast	Male sterility, nematode infection	Caldelari et al. (2011) and Ozalvo et al. (2014)
LOX6	A. thaliana	Chloroplast	Wounding, smotic and drought stress	Grebner et al. (2013) and Chauvin et al. (2013)
TomLOXD	Solanum lycopersicum	Chloroplast	Wounding, neurotropic pathogen infections	
OsLOX2 and OsLOX5	Oryza sativa	Chloroplast	Fungal infection	Zhang et al. (2018)
PgLOX4 and PgLOX5	Panax ginseng	Chloroplast	Wounding	Bae et al. (2016)
CmLOX(08, 10,12, 13, 18)	Cucumis melo	Chloroplast	Wounding	Liu et al. (2016)
AOS	A. thaliana	Chloroplast	Wounding, male sterility	Park et al. (2002)
CsAOS2	Camellia sinensis	Chloroplast	Insect attack	Peng et al. (2018)
VvAOS	Vitis vinifera	Chloroplast	Male sterility, wounding, fruit ripening	Dumin et al. (2018)
CcAOS	Castanea crenata	Chloroplast	Pathogen infection	Serrazina et al. (2021)
OsAOS1 and OsAOS2	O. sativa	Chloroplast	Wounding, herbivore attack	Zeng et al. (2021)
OsAOC	O. sativa	Chloroplast	Salinity, fungal infection	Riemann et al. (2013) and Hazman et al. (2015)
ScAOC1	Saccharum officinarum	Chloroplast	Cold, salinity, pathogen infection	Sun et al. (2020)
OPR3	A. thaliana	Peroxisome	Flower development and lateral root formation	Li et al. (2013)

(Continued)

TABLE 9.1   (Continued)

Enzyme	Plant	Localization	Role	References
Sl*OPR3*	*Solanum lycopersicum*	Peroxisome	Pollen development and pathogen resistance	
Ta*OPR3*	*Triticum aestivum*	Peroxisome	Heat and pathogen resistance	Pigolev et al. (2018)
ACX1	*A. thaliana*	Peroxisome	Wound response	Castillo et al. (2004)
Cs*ACX1*	*Camellia sinensis*	Peroxisome	Flower opening	Xin et al. (2019)
Os*ACX1*	*O. sativa*	Peroxisome	Wound response	Kim et al. (2007)
Os*ACX2* Os*ACX3*	*O. sativa*	Peroxisome	Seed germination	Kim et al. (2007)
MFP2	*A. thaliana*	Peroxisome	Flower and seedling development	Rylott et al. (2006)
AIM1	*A. thaliana*	Peroxisome	Flower development	Richmond and Bleecker (1999)
KAT1, KAT2, KAT5	*A. thaliana*	Peroxisome	Flower development, pathogen resistance and root development	Afitlhile et al. (2005)
Bn*KAT2*	*Brassica napus*	Peroxisome	Seed germination	Wiszniewski et al. (2012)
Bn*KAT5*	*B. napus*	Peroxisome	Flower development	Wiszniewski et al. (2012)
JAR1	*A. thaliana*	Cytosol	Wound response and pathogen resistance	Suza and Staswick (2008)
Na*JAR4* Na*JAR5*	*Nicotiana tobaccum*	Cytosol	Wound response	Suza and Staswick (2008)
Os*JAR1*	*O. sativa*	Cytosol	Wound response	Wen et al. (2017)

9.13 Conclusions

JA is one of the major phytohormones already known to be involved in various physiological processes along with its response against a variety of stresses. JA biosynthesis pathway consists of various enzymes encoded by different gene family members that in a cascade convert galactolipids into JA and its amino acid conjugates. Mutants and overexpression analysis of various key enzymes depicts their importance and contribution in the JA biosynthesis pathway. Expression analysis suggests the role of JA and its biosynthesis enzymes in response to different stresses. Studies which are discussed above help to rectify the significant role of each enzyme in the JA biosynthesis pathway and improve understanding of intracellular JA response. Although, recent studies provide useful insights on JA biosynthesis, but still there are various gaps including the regulatory mechanism of the JA biosynthesis pathway in response to changing environmental conditions that require further investigation.

Acknowledgment

This work is financially supported by the Science and Engineering Research Board (SERB)—Department of Science and Technology (DST), Govt. of India project (CRG/2021/000694) and core research grant from National Institute of Plant Genome Research (NIPGR) in AS lab. SK is thankful to council of scientific and industrial research (CSIR), India for research fellowships. Authors are thankful to DBT (Department of Biotechnology)-eLibrary Consortium (DeLCON) for providing access to e-resources.

References

Afitlhile, M.M., Fukushige, H., Nishimura, M., Hildebrand, D.F., 2005. A defect in glyoxysomal fatty acid β-oxidation reduces jasmonic acid accumulation in *Arabidopsis*. Plant. Physiol. Biochem. 43, 603–609.

Arent, S., Christensen, C.E., Pye, V.E., Nørgaard, A., Henriksen, A., 2010. The multifunctional protein in peroxisomal β-oxidation: structure and substrate specificity of the *Arabidopsis thaliana* protein MFP2. J. Biol. Chem. 285, 24066–24077.

Bae, K.S., Rahimi, S., Kim, Y.J., Devi, B.S.R., Khorolragchaa, A., Sukweenadhi, J., et al., 2016. Molecular characterization of lipoxygenase genes and their expression analysis against biotic and abiotic stresses in *Panax ginseng*. Eur. J. Plant. Pathol. 145, 331–343.

Bannenberg, G., Martínez, M., Hamberg, M., Castresana, C., 2009. Diversity of the enzymatic activity in the lipoxygenase gene family of *Arabidopsis thaliana*. Lipids 44, 85–95.

Bell, E., Creelman, R.A., Mullet, J.E., 1995. A chloroplast lipoxygenase is required for wound-induced jasmonic acid accumulation in *Arabidopsis*. Proc. Natl. Acad. Sci. U. S. A. 92, 8675–8679.

Biesgen, C., Weiler, E.W., 1999. Structure and regulation of OPR1 and OPR2, two closely related genes encoding 12-oxophytodienoic acid-10,11-reductases from *Arabidopsis thaliana*. Planta 208, 155–165.

Bonaventure, G., Schuck, S., Baldwin, I.T., 2011. Revealing complexity and specificity in the activation of lipase-mediated oxylipin biosynthesis: a specific role of the *Nicotiana attenuata* GLA1 lipase in the activation of jasmonic acid biosynthesis in leaves and roots. Plant, Cell Env. 34, 1507–1520.

Breithaupt, C., Kurzbauer, R., Lilie, H., Schaller, A., Strassner, J., Huber, R., et al., 2006. Crystal structure of 12-oxophytodienoate reductase 3 from tomato: self-inhibition by dimerization. Proc. Natl. Acad. Sci. U. S. A. 103 (39), 14337–14342.

Caldelari, D., Wang, G., Farmer, E.E., Dong, X., 2011. *Arabidopsis* lox3 lox4 double mutants are male sterile and defective in global proliferative arrest. Plant. Mol. Biol. 75, 25–33.

Castillo, M.C., Martínez, C., Buchala, A., Métraux, J.P., Léon, J., 2004. Gene-specific involvement of β-oxidation in wound-activated responses in *Arabidopsis*. Plant. Physiol. 135, 85–94.

Chauvin, A., Caldelari, D., Wolfender, J.L., Farmer, E.E., 2013. Four 13-lipoxygenases contribute to rapid jasmonate synthesis in wounded *Arabidopsis thaliana* leaves: a role for lipoxygenase 6 in responses to long-distance wound signals. N. Phytol. 197, 566–575.

Deepika, Singh, A., 2021. Expression dynamics indicate the role of Jasmonic acid biosynthesis pathway in regulating macronutrient (N, P and K$^+$) deficiency tolerance in rice (*Oryza sativa* L.). Plant. Cell Rep. 40 (8), 1495–1512.

Deepika, D., Ankit, Jonwal, S., Mali, K.V., Sinha, A.K., Singh, A., 2022. Molecular analysis indicates the involvement of Jasmonic acid biosynthesis pathway in low-potassium (K$^+$) stress response and development in chickpea (*Cicer arietinum*). Environ. Exp. Bot. 194 (June 2021), 104753.

Demole, E., Lederer., E., Mercier, D., 1962. Isolation and determination of the structure of methyl jasmonate, an odorous constituent characteristic of jasmine essence. Helv. Chim. Acta 45 (2), 675–685.

Ding, H., Lai, J., Wu, Q., Zhang, S., Chen, L., Dai, Y.S., et al., 2016. Jasmonate complements the function of *Arabidopsis* lipoxygenase3 in salinity stress response. Plant. Sci. 244, 1–7.

Dumin, W., Rostas, M., Winefield, C., 2018. Identification and functional characterisation of an allene oxide synthase from grapevine (*Vitis vinifera* L. Sauvignon blanc). Mol. Biol. Rep. 45, 263–277.

Feussner, I., Wasternack, C., 2002. The lipoxygenase pathway. Annu. Rev. Plant. Biol. 53, 275–297.

Fonseca, S., Chini, A., Hamberg, M., Adie, B., Porzel, A., Kramell, R., et al., 2009. (+)-7-iso-Jasmonoyl-L-isoleucine is the endogenous bioactive jasmonate. Nat. Chem. Biol. 5, 344–350.

Footitt, S., Cornah, J.E., Pracharoenwattana, I., Bryce, J.H., Smith, S.M., 2007. The *Arabidopsis* 3-ketoacyl-CoA thiolase-2 (kat2-1) mutant exhibits increased flowering but reduced reproductive success. J. Exp. Bot. 58, 2959–2968.

Fukumoto, K., Alamgir, K.M., Yamashita, Y., Mori, I.C., Matsuura, H., Galis, I., 2013. Response of rice to insect elicitors and the role of OsJAR1 in wound and herbivory-induced JA-Ile accumulation. J. Integr. Plant. Biol. 55 (8), 775–784.

Ghelli, R., Brunetti, P., Napoli, N., De Paolis, A., Cecchetti, V., Tsuge, T., et al., 2018. A newly identified flower-specific splice variant of AUXIN RESPONSE FACTOR8 regulates stamen elongation and endothecium lignification in *Arabidopsis*. Plant Cell 30, 620–637.

Glauser, G., Dubugnon, L., Mousavi, S.A.R., Rudaz, S., Wolfender, J.L., Farmer, E.E., 2009. Velocity estimates for signal propagation leading to systemic jasmonic acid accumulation in wounded *Arabidopsis*. J. Biol. Chem. 284, 34506–34513.

Grebner, W., Stingl, N.E., Oenel, A., Mueller, M.J., Berger, S., 2013. Lipoxygenase6-dependent oxylipin synthesis in roots is required for abiotic and biotic stress resistance of *Arabidopsis*. Plant. Physiol. 161, 2159–2170.

Han, G.Z., 2017. Evolution of jasmonate biosynthesis and signalling mechanisms. J. Exp. Bot. 68, 1323–1331.

Hazman, M., Hause, B., Eiche, E., Nick, P., Riemann, M., 2015. Increased tolerance to salt stress in OPDA-deficient rice ALLENE OXIDE CYCLASE mutants is linked to an increased ROS-scavenging activity. J. Exp. Bot. 66, 3339–3352.

Hofmann, E., Zerbe, P., Schaller, F., 2006. The crystal structure of *Arabidopsis thaliana* allene oxide cyclase: insights into the oxylipin cyclization reaction. Plant Cell 18, 3201–3217.

Hughes, R.K., Domenico, De, Santino, A., S., 2009. Plant cytochrome CYP74 family: biochemical features, endocellular localisation, activation mechanism in plant defence and improvements for industrial applications. ChemBioChem 10, 1122–1133.

Hyun, Y., Choi, S., Hwang, H.J., Yu, J., Nam, S.J., Ko, J., et al., 2008. Cooperation and functional diversification of two closely related galactolipase genes for jasmonate biosynthesis. Dev. Cell 14, 183–192.

Ishiguro, S., Kawai-Oda, A., Ueda, J., Nishida, I., Okada, K., 2001. The defective in anther dehiscence1 gene encodes a novel phospholipase A1 catalyzing the initial step of jasmonic acid biosynthesis, which synchronizes pollen maturation, anther dehiscence, and flower opening in *Arabidopsis*. Plant Cell 13, 2191–2209.

Ito, T., Ng, K.H., Lim, T.S., Yu, H., Meyerowitz, E.M., 2007. The homeotic protein AGAMOUS controls late stamen development by regulating a jasmonate biosynthetic gene in *Arabidopsis*. Plant Cell 19, 3516–3529.

Khan, M.I.R., Jahan, B., Alajmi, M.F., Rehman, M.T., Khan, N.A., 2019. Exogenously-sourced ethylene modulates defense mechanisms and promotes tolerance to zinc stress in mustard (*Brassica juncea* L.). Plants 8 (12), 540.

Khan, M.I.R., Jahan, B., AlAjmi, M.F., Rehman, M.T., Khan, N.A., 2020a. Ethephon mitigates nickel stress by modulating antioxidant system, glyoxalase system and proline metabolism in Indian mustard. Physiol. Mol. Biol. Plants 26 (6), 1201–1213.

Khan, M.I.R., Trivellini, A., Chhillar, H., Chopra, P., Ferrante, A., Khan, N.A., et al., 2020b. The significance and functions of ethylene in flooding stress tolerance in plants. Environ. Exp. Bot. 179, 104188.

Kim, M.C., Kim, T.H., Park, J.H., Moon, B.Y., Lee, C.H., Cho, S.H., 2007. Expression of rice acyl-CoA oxidase isoenzymes in response to wounding. J. Plant Physiol. 164 (5), 665–668.

Koo, A.J.K., Hoo, S.C., Kobayashi, Y., Howe, G.A., 2006. Identification of a peroxisomal acyl-activating enzyme involved in the biosynthesis of jasmonic acid in *Arabidopsis*. J. Biol. Chem. 281, 33511–33520.

Laudert, D., Weiler, E.W., 1998. Allene oxide synthase: a major control point in *Arabidopsis thaliana* octadecanoid signalling. Plant J. 15 (5), 675–684.

Lee, D.S., Nioche, P., Hamberg, M., Raman, C.S., 2008. Structural insights into the evolutionary paths of oxylipin biosynthetic enzymes. Nature 455, 363–368.

Li, C., Schilmiller, A.L., Liu, G., Lee, G.I., Jayanty, S., Sageman, C., et al., 2005. Role of β-oxidation in jasmonate biosynthesis and systemic wound signaling in tomato. Plant Cell 17 (3), 971–986.

Li, S., Ma, J., Liu, P., 2013. OPR3 is expressed in phloem cells and is vital for lateral root development in *Arabidopsis*. Can. J. Plant Sci. 93, 165–170.

Li, L., Chang, Z., Pan, Z., Fu, Z.Q., Wang, X., 2008. Modes of heme binding and substrate access for cytochrome P450 CYP74A revealed by crystal structures of allene oxide synthase. Proc. Natl. Acad. Sci. U. S. A. 105, 13883–13888.

Lim, C.W., Han, S.W., Hwang, I.S., Kim, D.S., Hwang, B.K., Lee, S.C., 2015. The pepper lipoxygenase CaLOX1 plays a role in osmotic, drought and high salinity stress response. Plant Cell Physiol. 56, 930–942.

Liu, J.Y., Zhang, C., Shao, Q., Tang, Y.F., Cao, S.X., Guo, X.O., et al., 2016. Effects of abiotic stress and hormones on the expressions of five 13-CmLOXs and enzyme activity in oriental melon (*Cucumis melo var. makuwa Makino.* J. Integr. Agric. 15, 326–338.

Meesters, C., Mönig, T., Oeljeklaus, J., Krahn, D., Westfall, C.S., Hause, B., et al., 2014. A chemical inhibitor of jasmonate signaling targets JAR1 in *Arabidopsis thaliana.* Nat. Chem. Biol. 10, 830–836.

Miersch, O., Bohlmann, H., Wasternack, C., 1999. Jasmonates and related compounds from *Fusarium oxysporum.* Phytochemistry 50, 517–523.

Nazir, F., Hussain, A., Fariduddin, Q., 2019. Interactive role of epibrassinolide and hydrogen peroxide in regulating stomatal physiology, root morphology, photosynthetic and growth traits in *Solanum lycopersicum* L. under nickel stress. Environ. Exp. Bot. 162, 479–495.

Nazir, F., Hussain, A., Fariduddin, Q., Tanveer, A.K., 2021. Brassinosteroid and hydrogen peroxide improve photosynthetic efficiency and maintain chloroplast ultrastructure, stomatal movement, root morphology, cell viability and reduce Cu-triggered oxidative burst in tomato. Ecotoxicol. Environ. Saf. 207, 111081.

Nazir, F., Fariduddin, Q., Tanveer, A.K., 2022. Interaction between brassinosteroids and hydrogen peroxide networking signal molecules in plants. Brassinosteroids Signalling: Intervention with Phytohormones and Their Relationship in Plant Adaptation to Abiotic Stresses. pp. 59–79.

Neumann, P., Brodhun, F., Sauer, K., Herrfurth, C., Hamberg, M., Brinkmann, J., et al., 2012. Crystal Structures of *physcomitrella patens* AOC1 and AOC2: insights into the enzyme mechanism and differences in substrate specificity. Plant Physiol. 160, 1251–1266.

Newcomer, M.E., Brash, A.R., 2015. The structural basis for specificity in lipoxygenase catalysis. Protein Sci. 24, 298–309.

Nguyen, T.H., Mai, H.T.T., Moukouanga, D., Lebrun, M., Bellafiore, S., Champion, A., 2020. CRISPR/Cas9-mediated gene editing of the jasmonate biosynthesis OsAOC gene in rice. In: Champion, A., Laplaze, L. (Eds.), Jasmonate in Plant Biology. Methods in Molecular Biology, 2085. Humana, New York, NY.

Noyes, B.E., Glatthaar, B.E., Garavelli, J.S., Bradshaw, R.A., 1974. Structural and functional similarities between mitochondrial malate dehydrogenase and L3 hydroxyacyl CoA dehydrogenase. Proc. Natl. Acad. Sci. U. S. A. 71, 1334–1338.

Ogorodnikova, A.V., Mukhitova, F.K., Grechkin, A.N., 2015. Oxylipins in the spikemoss *Selaginella martensii*: detection of divinyl ethers, 12-oxophytodienoic acid and related cyclopentenones. Phytochemistry 118, 42–50.

Ohya, H., Ogata, A., Nakamura, K., Chung, K.M., Sano, H., 2008. A stress-responsive multifunctional protein involved in β-oxidation in tobacco plants. Plant Biotechnol. 25, 503–508.

Otto, M., Naumann, C., Brandt, W., Wasternack, C., Hause, B., 2016. Activity regulation by heteromerization of *Arabidopsis* allene oxide cyclase family members. Plants 5, 1021–1058.

Ozalvo, R., Cabrera, J., Escobar, C., Christensen, S.A., Borrego, E.J., Kolomiets, M.V., et al., 2014. Two closely related members of *Arabidopsis* 13-lipoxygenases (13-LOXs), LOX3 and LOX4, reveal distinct functions in response to plant-parasitic nematode infection. Mol. Plant Pathol. 15, 319–332.

Park, J.H., Halitschke, R., Kim, H.B., Baldwin, I.T., Feldmann, K.A., Feyereisen, R., 2002. A knock-out mutation in allene oxide synthase results in male sterility and defective wound signal transduction in *Arabidopsis* due to a block in jasmonic acid biosynthesis. Plant J. 31, 1–12.

Peng, S., Huang, S., Liu, Z., Feng, H., 2019. Mutation of ACX1, a jasmonic acid biosynthetic enzyme, leads to petal degeneration in chinese cabbage (*Brassica campestris ssp. pekinensis*). Int. J. Mol. Sci. Artic. 20 (9), 2310.

Peng, Q., Zhou, Y., Liao, Y., Zeng, L., Xu, X., Jia, Y., et al., 2018. Functional characterization of an allene oxide synthase involved in biosynthesis of Jasmonic acid and its influence on metabolite profiles and ethylene formation in tea (*Camellia sinensis*) flowers. Int. J. Mol. Sci. 19 (8).

Peng, Y.J., Shih, C.F., Yang, J.Y., Tan, C.M., Hsu, W.H., Huang, Y.P., et al., 2013. A RING-type E3 ligase controls anther dehiscence by activating the jasmonate biosynthetic pathway gene DEFECTIVE in ANTHER DEHISCENCE1 in *Arabidopsis*. Plant J. 74, 310–327.

Pigolev, A., V, Miroshnichenko, D.N., Pushin, A.S., Terentyev, V., V, Boutanayev, A.M., et al., 2018. Overexpression of *Arabidopsis* OPR3 in hexaploid wheat (*Triticum aestivum* L.) alters plant development and freezing tolerance. Int. J. Mol. Sci. Artic. 19 (2), 3989.

Poór, P., Nawaz, K., Gupta, R., Ashfaque, F., Khan, M.I.R., 2021. Ethylene involvement in the regulation of heat stress tolerance in plants. Plant Cell Rep. 1–24.

Richmond, T.A., Bleecker, A.B., 1999. A defect in β-oxidation causes abnormal inflorescence development in *Arabidopsis*. Plant Cell. 11 (10), 1911–1923.

Riemann, M., Haga, K., Shimizu, T., Okada, K., Ando, S., Mochizuki, S., et al., 2013. Identification of rice Allene Oxide Cyclase mutants and the function of jasmonate for defence against *Magnaporthe oryzae*. Plant J. 74, 226–238.

Ruan, J., Zhou, Y., Zhou, M., Yan, J., Khurshid, M., Weng, W., et al., 2019. Jasmonic acid signaling pathway in plants. Int. J. Mol. Sci. 20.

Rylott, E.L., Eastmond, P.J., Gilday, A.D., Slocombe, S.P., Larson, T.R., Baker, A., et al., 2006. The *Arabidopsis thaliana* multifunctional protein gene (MFP2) of peroxisomal β-oxidation is essential for seedling establishment. Plant J. 45 (6), 930–941.

Scalschi, L., Sanmartín, M., Camañes, G., Troncho, P., Sánchez-Serrano, J.J., García-Agustín, P., et al., 2015. Silencing of OPR3 in tomato reveals the role of OPDA in callose deposition during the activation of defense responses against *Botrytis cinerea*. Plant J. 81 (2), 304–315.

Schaller, F., Biesgen, C., Müssig, C., Altmann, T., Weiler, E.W., 2000. 12-Oxophytodienoate reductase 3 (OPR3) is the isoenzyme involved in jasmonate biosynthesis. Planta 210 (6), 979–984.

Schilmiller, A.L., Koo, A.J.K., Howe, G.A., 2007. Functional diversification of acyl-coenzyme A oxidases in jasmonic acid biosynthesis and action. Plant Physiol. 143, 812–824.

Seltmann, M.A., Hussels, W., Berger, S., 2010. Jasmonates during senescence signals or products of metabolism. Plant Signal. Behav. 5, 1–5.

Serrazina, S., Machado, H., Costa, R.L., Duque, P., Malhó, R., 2021. Expression of Castanea crenata allene oxide synthase in *Arabidopsis* improves the defense to *Phytophthora cinnamomi*. Front. Plant Sci. 12, 1–15.

Shaban, M., Ahmed, M.M., Sun, H., Ullah, A., Zhu, L., 2018. Genome-wide identification of lipoxygenase gene family in cotton and functional characterization in response to abiotic stresses. BMC Genomics 19, 1–13.

Shockey, J.M., Fulda, M.S., Browse, J., 2003. *Arabidopsis* contains a large superfamily of acyl-activating enzymes. Phylogenetic and biochemical analysis reveals a new class of acyl-coenzyme A synthetases. Plant Physiol. 132 (2), 1065–1076.

Song, W.C., Brash, A.R., 1991. Purification of an allene oxide synthase and identification of the enzyme as a cytochrome P-450. Science 253 (5021), 781–784.

Song, W.-C., Funk, C.D., Brash, A.R., 1993. Molecular cloning of an allene oxide synthase: a cytochrome P450 specialized for the metabolism of fatty acid hydroperoxides (jasmonic acid/lipoxygenase/chloroplast tnsdt peptide/flaxseed). Proc. Natl. Acad. Sci. U. S. A. 90, 8519–8523.

Staswick, P.E., Tiryaki, I., 2004. The oxylipin signal jasmonic acid is activated by an enzyme that conjugates it to isoleucine in *Arabidopsis*. Plant Cell 16 (8), 2117–2127.

Stenzel, I., Otto, M., Delker, C., Kirmse, N., Schmidt, D., Miersch, O., et al., 2012. ALLENE OXIDE CYCLASE (AOC) gene family members of *Arabidopsis thaliana*: tissue- and organ-specific promoter activities and in vivo heteromerization. J. Exp. Bot. 63 (17), 6125–6138.

Stintzi, A., 2000. The *Arabidopsis* male-sterile mutant, opr3, lacks the 12-oxophytodienoic acid reductase required for jasmonate synthesis. Proc. Natl Acad. Sci. U. S. A. 97 (19), 10625–10630.

Stumpe, M., Göbel, C., Faltin, B., Beike, A.K., Hause, B., Himmelsbach, K., et al., 2010. The moss *Physcomitrella patens* contains cyclopentenones but no jasmonates: mutations in allene oxide cyclase lead to reduced fertility and altered sporophyte morphology. N. Phytol. 188, 740–749.

Sun, T., Cen, G., You, C., Lou, W., Wang, Z., Su, W., et al., 2020. ScAOC1, an allene oxide cyclase gene, confers defense response to biotic and abiotic stresses in sugarcane. Plant Cell Rep. 39, 1785–1801.

Suza, W.P., Staswick, P.E., 2008. The role of JAR1 in jasmonoyl-L-isoleucine production during *Arabidopsis* wound response. Planta 227, 1221–1232.

Tyagi, C., Singh, A., Singh, I.K., 2016. Mechanistic insights into mode of action of rice allene oxide synthase on hydroxyperoxides: an intermediate step in herbivory-induced jasmonate pathway. Comput. Biol. Chem. 64, 227–236.

Upadhyay, R.K., Handa, A.K., Mattoo, A.K., 2019. Transcript abundance patterns of 9- and 13-lipoxygenase subfamily gene members in response to abiotic stresses (heat, cold, drought or salt) in tomato (*Solanum lycopersicum* L.) highlights member-specific dynamics relevant to each stress. Genes 10 (9), 683.

Vick, B.A., Zimmerman, D.C., 1987. Oxidative systems for modification of fatty acids: the lipoxygenase pathway. Lipids: Structure and Function. Elsevier, pp. 53–90.

Wang, J., Hu, T., Wang, W., Hu, H., Wei, Q., Wei, X., et al., 2019. Bioinformatics analysis of the lipoxygenase gene family in radish (*Raphanus sativus*) and functional characterization in response to abiotic and biotic stresses. Int. J. Mol. Sci. 20.

Wang, K., Guo, Q., Froehlich, J.E., Hersh, H.L., Zienkiewicz, A., Howe, G.A., et al., 2018. Two abscisic acid-responsive plastid lipase genes involved in jasmonic acid biosynthesis in *Arabidopsis thaliana*. Plant. Cell 30, 1006–1022.

Wasternack, C., 2007. Jasmonates: an update on biosynthesis, signal transduction and action in plant stress response, growth and development. Ann. Bot. 100 (4), 681–697.

Wasternack, C., Feussner, I., 2018. The oxylipin pathways: biochemistry and function. Annu. Rev. Plant. Biol. 69, 363–386.

Wen, C.K., Swain, S., Jiang, H.W., Hsieh, H.L., 2017. FAR-RED INSENSITIVE 219/JAR1 contributes to shade avoidance responses of *Arabidopsis* seedlings by modulating key shade signaling components. Front. Plant Sci. 8, 1901.

Willadsen, P., Eggerer, H., 1975. Substrate stereochemistry of the enoyl-CoA hydratase reaction. Eur. J. Biochem. 54 (1), 247–252.

Wilmanns, M., Hyde, C.C., Davies, D.R., Kirschner, K., Jansonius, J.N., 1991. Structural conservation in parallel beta/alpha-barrel enzymes that catalyze three sequential reactions in the pathway of tryptophan biosynthesis. Biochemistry 30 (38), 9161–9169.

Wiszniewski, A.A., Smith, S.M., Bussell, J.D., 2012. Conservation of two lineages of peroxisomal (type I) 3-ketoacyl-CoA thiolases in land plants, specialization of the genes in Brassicaceae, and characterization of their expression in *Arabidopsis thaliana*. J. Exp. Bot. 63 (17), 6093–6103.

Xin, Z., Chen, S., Ge, L., Li, X., Sun, X., 2019. The involvement of a herbivore-induced acyl-CoA oxidase gene, CsACX1, in the synthesis of jasmonic acid and its expression in flower opening in tea plant (*Camellia sinensis*). Plant. Physiol. Biochem. 135, 132–140.

Yamamoto, Y., Ohshika, J., Takahashi, T., Ishizaki, K., Kohchi, T., Matusuura, H., et al., 2015. Functional analysis of allene oxide cyclase, MpAOC, in the liverwort *Marchantia polymorpha*. Phytochemistry 116, 48–56.

Yang, F., Tang, J., Yang, D., Yang, T., Liu, H., Luo, W., et al., 2020. Jasmonoyl-L-isoleucine and allene oxide cyclase-derived jasmonates differently regulate gibberellin metabolism in herbivory-induced inhibition of plant growth. Plant Sci. 300.

Yang, W., Devaiah, S.P., Pan, X., Isaac, G., Welti, R., Wang, X., 2007. AtPLAI is an acyl hydrolase involved in basal jasmonic acid production and *Arabidopsis* resistance to Botrytis cinerea. J. Biol. Chem. 282, 18116–18128.

Yang, X.Y., Jiang, W.J., Yu, H.J., 2012. The expression profiling of the lipoxygenase (LOX) family genes during fruit development, abiotic stress and hormonal treatments in cucumber (*cucumis sativus L.*). Int. J. Mol. Sci. 13, 2481–2500.

Yoeun, S., Cho, K., Han, O., 2018. Structural evidence for the substrate channeling of rice allene oxide cyclase in biologically analogous nazarov reaction. Front. Chem. 6, 1–11.

Zeng, J., Zhang, T., Huangfu, J., Li, R., Lou, Y., 2021. Both allene oxide synthases genes are involved in the biosynthesis of herbivore-induced jasmonic acid and herbivore resistance in rice. Plants 10, 1–14.

Zhang, X., Bao, Y., Shan, D., Wang, Zhihui, Song, X., Wang, Zhaoyun, et al., 2018. *Magnaporthe oryzae* induces the expression of a microrna to suppress the immune response in rice. Plant Physiol. 177, 352–368.

10

Brassinosteroids in plant growth and development

Ritesh Kumar Yadav*, Loitongbam Lorinda Devi* and Amar Pal Singh

National Institute of Plant Genome Research, New Delhi, India

10.1 Introduction

Brassinosteroids (BRs) are involved in modulating diverse plant processes, including photomorphogenesis, reproduction, cell division and elongation, xylem differentiation, and stress responses. Initially, BRs were isolated from *Brassica napus* pollen (Grove et al., 1979). Later, by the genetic screening approach in *Arabidopsis*, several BRs biosynthesis and signaling components were identified with a diverse role in plant growth—related processes (Clouse et al., 1996; Li and Chory, 1999; Sreeramulu et al., 2013; Szekeres et al., 1996). BR-deficient mutants showed several pleiotropic effects, such as dwarfism, delayed flowering, altered leaf morphology, and compromised cell elongation and division processes in both above- and underground plant tissues (Clouse, 2011). In contrast, exogenous high BRs signaling activity represses BRs biosynthesis through feedback mechanism and alters the essential genes involved in defining cellular processes. (Chaiwanon and Wang, 2015; Nolan et al., 2017; Singh and Savaldi-Goldstein, 2015; Vragović et al., 2015; Zhu et al., 2013; Ackerman-Lavert et al., 2020).

Unlike animal steroid receptors, which are restricted to the nucleus and function as a transcription factor, BRs bind with the plasma membrane (PM)—confined receptor, BR insensitive1 (BRI1), and its homologs, BRI1-Like1 and 3 (BRL1 and BRL3) in complex with the BRI1-associated receptor kinase 1/Somatic embryogenesis receptor kinase 3 (BAK1/ SERK3) (Belkhadir and Chory, 2006). In the absence of BRs, the BRI1 kinase inhibitor 1 (BKI1) binds with the carboxyl terminus of the BRI1 and represses BRI1 receptor signaling activity (Wang and Chory, 2006; Jaillais et al., 2011; Jiang et al., 2015).

* Equal contribution.

Plant Hormones in Crop Improvement
DOI: https://doi.org/10.1016/B978-0-323-91886-2.00004-5

In contrast, BRs availability leads to BKI1 phosphorylation by BRI1 thus causing a dissociation of BKI1 from the PM (Wang et al., 2008). After a series of transphosphorylation events, BRI1 (receptor) and BRI1-associated kinase1 (BAK1; coreceptor) together acts as a functional receptor complex that phosphorylates the BR signaling kinases (BSKs) and Constitutive differential growth1 (CDG1) to activate phosphatase BRI1-suppressor1 (BSU1) (Wang and Chory, 2006; Tang et al., 2008; Wang et al., 2008; Kim et al., 2011). Activated BSU1 inhibits the negative regulator of BR signaling kinase, BR insensitive2 (BIN2) by dephosphorylating it (Kim and Wang, 2010). Although BIN2 has been well explored as a negative regulator of the BR signaling pathway, it also phosphorylates BSK3 to enhance BR signaling, thus making a feedback loop (Ren et al., 2019). Dephosphorylation of BIN2 leads to its proteasome-dependent degradation, thus releasing its inhibitory effect on BRs signaling transcriptional effectors, BRI1-EMS suppressor 1/Brassinazole-resistant1 (BES1/ BZR1) (Wang et al., 2001; Nam and Li, 2002; Yin et al., 2002; He et al., 2005; Yin et al., 2005; Vert and Chory, 2006; Hao et al., 2010). Further, these BRs signaling transcription factors regulate the plethora of genes that are accountable for various plant developmental and stress-related processes.

In this chapter, we intend to highlight the recent findings in the BRs field and discuss the latest studies related to BRs action in regulating the cellular processes. We highlight the roles of BR during hypocotyl elongation, leaf and stomata development, and root development.

10.2 Brassinosteroids and development of above-ground plant organs

BR signaling plays an important role in cell proliferation in the shoot apical meristems (SAMs). BRs regulate the controlled growth of organ boundaries involved in the formation of new organs from SAM (Reddy et al., 2004; Barton, 2010). BRs induce the differentiation of vascular tissues in the stem. The BR signaling mutants (*bri1−116*, *bri1−301*, and *bri1brl1brl3*) and BRs biosynthesis mutant such as *cpd* showed fewer vascular bundles. However, enhanced BR signaling/response lines (BRI1-GFP overexpression, *bes1-D*, *bzr1-D*, or *DWF4*-OX) developed an increased number of vascular bundles compared to wild-type plants (Ibañes et al., 2009). Different aspects of BRs-mediated aerial part development in plants, such as hypocotyl elongation, leaf development, regulation of flower development, and senescence, are elaborated under the following sub-headings.

10.2.1 BRs in cell size determination during hypocotyl elongation

Hormonal and environmental signaling cues modulate postembryonic growth that involves cell division, proliferation, and elongation followed by differentiation (Zhang et al., 2009; Heyman et al., 2013; Sun et al., 2015; Kang et al., 2017). Interconnected hormonal and environmental signaling networks, such as auxin, gibberellins (GAs), light, and temperature, play a crucial role during hypocotyl elongation in *Arabidopsis* (Bai et al., 2012). Tremendous progress has been made in understanding the genetic network, and it was demonstrated that several BRs-regulated genes were coregulated by auxin and GA to govern the cell elongation using *Arabidopsis* as a model system (Tanaka et al., 2003). In-line with GA and BRs interaction, the hypocotyl elongation mediated by GAs was suppressed in BR-deficient mutants, whereas BR response was largely normal in GA-deficient mutants. The mechanistic framework connecting BR−GA and light *via* Phytochrome-interacting factor4 (PIF4) has been shown to modulate cell elongation during hypocotyl development (Gallego-Bartolomé et al., 2012; Oh et al., 2014).

In this loop, transcription factor BZR1 forms a heterodimer with the PIF4 transcription factor to regulate central genes related to the cell elongation process. BZR1−PIF4 module activates the Paclobutrazol resistance (PRE) family transcription factors. PACLOBUTRAZOL RESISTANT, ILI1 BINDING bHLH PROTEIN1, and HOMOLOG OF BEE2 INTERACTING WITH IBH1 are downstream targets for BZR1−PIF4−ARF6 complex to regulate cell elongation process (Oh et al., 2012). In-line with these findings, DELLA proteins interact with BZR1 and PIF4 to repress their activity, thereby repressing the downstream genes and ultimately inhibiting the cell elongation. Thus hormonal and environmentally controlled transcriptional networks (BZR1−PIF4−DELLA) determine the cell elongation during hypocotyl growth (Li et al., 2012) (Fig. 10.1).

A genome-wide study using hypocotyl growth as an experimental system reveals that Auxin response factor6 (ARF6), BZR1, and PIF4 have shared common target genes that are involved in hypocotyl growth. Further, DELLA proteins interact with the ARF6 and abolishe its DNA binding activity, thus hampering the hypocotyl growth (Oh et al., 2014). Taken together, these studies elucidated a molecular circuit among the growth hormones auxin−BR−GA and the regulator of environmental factors (light and temperature) PIF4, that together constitute a loop for defining the hypocotyl growth.

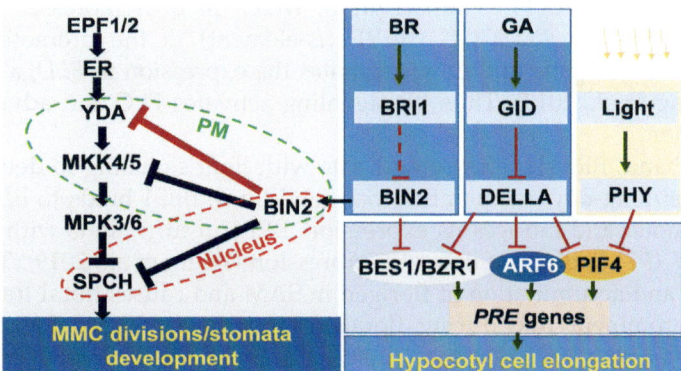

FIGURE 10.1 **BR-regulated stomata and hypocotyl development/elongation in *Arabidopsis thaliana*.** BIN2 plays a dual role in stomata development based on its subcellular localization and in response to environmental and developmental cues. BIN2 in the nucleus suppresses the SPCH and decreases stomata number. BIN2 forms a complex with BASL and Polar localization during asymmetric division and redistribution (POLAR) protein and moves toward the plasma membrane and inhibits YDA and MKK4/5. SPCH gets activated and promotes MMC divisions leading to an increase in stomata number. BR−GA−auxin and light-modulated signaling cascades integrate to modulate the hypocotyl elongation/growth. BASL, Breaking of asymmetry in the stomatal lineage; BRs, brassinosteroids; GA, gibberellin; MMC, meristemoid mother cell.

10.2.2 Role of brassinosteroids in regulating organ boundary and floral organ development

BRs regulate plant reproductive organ development by governing the different aspects of flower development in plants including *Arabidopsis* and *Prunus avium*. Exogenous BR treatment promotes pollen tube growth *in vitro* in *Prunus avium* suggesting essential role of BRs in reproductive organ development (Hewitta et al., 1985). The reduced fertility in BR-deficient mutant, *cpd* (*Constitutive photomorphogenic dwarf*), might be associated with the reduced growth rate of the pollen tube (Hewitta et al., 1985; Szekeres et al., 1996). Later, it was found that the male sterility in BR-deficient and insensitive mutants was due to the abnormal development of tapetum and microspores (Ye et al., 2010). In addition, BES1 binds with the promoters and modulates the expression of key genes, such as *Tapetum determinant1* (*TDF1*) and *Male sterility1&2* (*MS1/MS2*), which are necessary for anther and pollen development (Ye et al., 2010).

Flowering time is a necessary developmental switch that controls the life cycle of a plant. Mutants either defective in BRs signaling (*bri1*) or biosynthesis, *deetiolated2* (*det2*), *dwarf4* (*dwf4*), and *cpd* show late-flowering phenotype (Li et al., 2010). The *bri1* shows an increased level of floral repressor gene, *Flowering Locus C* (*FLC*), which may delay the flowering time in this mutant (Domagalska et al., 2007). On contrary, by considering the developmental criteria of the floral transition from the first leaf instead of growth rate or delayed flowering, the *bri1* and BR biosynthesis mutants show early flowering phenotype (Li et al., 2018).

BRs facilitate the BZR1 and BES1-interacting myc-likes (BIMs) mediated expression of FLC and represses floral transition (Li et al., 2018; Li and He, 2020). Constitutively active BR signaling mutant, *bzr1−1D*, shows delayed flowering in *Arabidopsis*. Both BZR1 and BIM1 bind with CGTGTG motif (a *cis*-regulatory BRRE element) located at first intron in *FLC* gene and upregulated FLC transcription, which in turn represses floral transition. BZR1 also interacts with CGTGGG (BRRE *cis*-element) in the promoter region of the *Flowering Locus D* (*FLD*) gene and downregulates the expression of *FLD*, a negative regulator of *FLC* (Zhang et al., 2013). Thus BR signaling activates FLC and adversely influences the floral transition.

On the other hand, BRs signaling coordinate with light signaling to determine the flowering time. Under long-day/temperature rise conditions, BES1 binds to *BEE1* (*BR enhanced expression1*) promoter and induces its expression. BEE1 in turn binds with the promoter of *Flowering Locus T* (*FT*) and upregulates *FT* expression (Wang et al., 2019). The FT promotes the biosynthesis and accumulation of florigen in SAM and causes floral transition.

BRs play a prominent role during floral organogenesis. The deformity in the plant organ boundary (such as inflorescence, stem, and flower) has been observed in constitutively active BRs signaling mutants, *bzr1-1D* and *bes1-D*, and the compromised (reduced) BR mutants, *bri1-5*, *bin2-1*, and *det2-1* (Espinosa-Ruiz et al., 2017). The transcription factor, LOB (Lateral organ boundaries), binds to the promoter (GCGGCG *cis*-element: LBD motif) region of *BAS1* and increases the biosynthesis of C-26 hydroxylated BR derivatives (inactive BR) in organ boundary and reduces the level of active BRs (Bell et al., 2012; Li and He, 2020). Additionally, BZR1 binds to the promoters of organ boundary identity genes, such as *Cup-shaped cotyledons* (*CUC1, CUC2, CUC3*) and *Lateral organ fusion1* (*LOF1*) and represses them (Hibara et al., 2006; Lee et al., 2009). The *CUC* genes are involved in the regulation of floral organogenesis and the development of lateral organs (Vroemen et al., 2003). BES1 binds with the general repressor TOPLESS (TPL) and regulates *CUCs* expression (Espinosa-Ruiz et al., 2017). Hence, BZR1 and BES1 have the potential to interact with a wide range of organ boundary identity genes to regulate floral organogenesis (Sun et al., 2010; Oh et al., 2012).

10.2.3 Brassinosteroids in leaf development

Reduced BR response and biosynthesis mutants, such as *bri1*, *det2*, and *cpd*, show cabbage-shaped dark green leaves with shorter petiole length (Wang et al., 2001; Li et al., 2002; Yin et al., 2002). In contrast, elevated BR response, signaling, and biosynthesis (*BRI1−GFP* overexpression lines and dominant BR signaling gene, *bes1-D*) show elongated leaves with longer petioles (Oh et al., 2011). The smaller leaf size of the BR-deficient mutants (*det2* and *dwarf1*) was shown to be associated with the reduced cell numbers in the leaf blade that suggested the plausible effect of BR in modulating the cell division to regulate the leaf morphology (Nakaya et al., 2002). Zhiponova et al. (2013) showed the role of BRs in leaf epidermal as well as mesophyll cell proliferation and expansion. The BR biosynthesis and signaling was elucidated to differentially control the cell division and differentiation processes in the leaf (Zhiponova et al., 2013). The BR biosynthesis mutant, *cpd*, treated with BR regained its size due to enhanced cell expansion compared with the division rate. The ratio of BRI1 receptor and BR ligand concentrations stimulated the BR responses and play a key role in determining the cell fate. Additionally, *det2* mutant in *Arabidopsis* develop small and round leaves, while the exogenous application of BR recovered the leaf size/shape by promoting cellular features like cell division and elongation (Nakaya et al., 2002).

BRs play an essential role in stomata development (Lau and Bergmann, 2012). In *Arabidopsis* leaves, stomata were separated with at least one pavement cell, while BR-deficient mutants show stomata clusters. Additionally, BR treatment represses stomata density, thus demonstrating the role of BR in stomata development. The BR-insensitive mutant, *bri1−116*, and plants overexpressing the negative regulator of BR signaling kinase, BIN2, show an increased stomatal density along with the increased meristemoids and guard mother cells (Kim et al., 2012). BIN2 has been shown to differentially affect stomata development depending on the environmental conditions and plant developmental stage. It has been shown that stomata development is genetically controlled by MAPK kinase kinase (MAPKKK), YODA (YDA), and ERECTA (ER) family of receptors (ER, ERL1, and ERL2). The Erecta family receptors activate YDA leading to the phosphorylation of MAPKKs (MKK4/MKK5). The MKK4/MKK5 activates MAPKs (MPK3/MPK6) that ultimately inhibits the key transcription factor SPEECHLESS (SPCH) that regulates the activity of stomata precursor cells. The BIN2 kinase has been shown to alter and phosphorylate YDA/MKK4 as well as the SPCH based on subcellular localization to regulate stomatal development and clustering. In the nucleus, BIN2 suppresses the SPCH and decreases stomata number (Gudesblat et al., 2012). BIN2 forms a complex with Breaking of asymmetry in the stomatal lineage (BASL) and Polar localization during asymmetric division and redistribution (POLAR) protein and moves to the PM. At the PM, BIN2 phosphorylates YDA and MKK4/5 and inhibits their activity (Khan et al., 2013; Luschnig and Vert, 2014; Zhang et al., 2015; Houbaert et al., 2018). In this case, SPCH gets activated and promotes meristemoid mother cell (MMC) division leading to increased stomata number (Fig. 10.1).

10.3 Brassinosteroids in root development

Studies using BR biosynthesis and signaling mutants revealed the importance of BR in root development. Roots are the hidden organ of the plant, which are responsible for the acquisition of nutrients and water as well as providing anchorage to the plant. Recently, many studies have shown the role of BRs in regulating root system architecture (RSA) and its spatiotemporal activity in mediating root development and nutrient foraging responses (Singh et al., 2014; Wei and Li, 2016; Singh et al., 2018; Devi et al., 2020; Pandey et al., 2020; Devi et al., 2022; Gupta et al., 2022).

Root growth patterning is determined by the spatial balance of cell division, elongation, and differentiation (Chaiwanon and Wang, 2015; Singh and Savaldi-Goldstein, 2015). The root is divided into different developmental zones. These zones include (1) the meristematic zone where the cells undergo active cellular divisions, (2) the elongation zone, in which the cells cease dividing and start increasing their length and volume, and (3) the differentiation zone, where the cells start differentiating into root hairs and lateral roots (LRs) (Dolan et al., 1993; Beemster and Baskin, 1998; Ubeda-Tomás et al., 2012). The meristematic zone is further divided into the apical and basal meristem. In the apical meristem, the cell exhibits a high rate of proliferation, while in the basal meristem, the division slows down and cells start increasing in size.

At the base of the root apical meristem (RAM), there is a small group of stem cells, which comprises the root stem cell niche. The stem cell niche functions as the precursors of all specialized cells. These cells are pivotal for sustaining root growth as they replace damaged tissues (Sabatini et al., 2003; Heyman et al., 2014). The quiescent center (QC) constitutes the core of the stem cell niche. The QC cells are characterized by very low mitotic activity and maintain the undifferentiated state of the surrounding stem cells (Sarkar et al., 2007). These cells also function as a reservoir of cells by providing new cells to damaged areas of plant roots (Heyman et al., 2014; Vilarrasa-Blasi et al., 2014). BRs are shown to regulate cell elongation, RAM maintenance, vascular development, LR, and root hair formation. In the following sections, we will be discussing the regulatory roles of BR in governing these processes.

10.3.1 Brassinosteroid in root meristem maintenance

The aggregate of cellular division and cellular expansion rate determine the total root length of a plant. BRs regulate a plethora of genes involved in cellular divisions and elongation in the meristematic zone. The BR signaling transcription factors, BZR1 and BES1, either represses or induces their target genes by direct binding to the E-BOX and BRRE motifs present in their promoters (He et al., 2005). BRs orchestrate root growth in concentration-dependent manner. A high level of active form of BR, brassinolide (BL) (above 0.04 nM BL), restricts primary root growth due to abnormal cell cycle, while a lower BL dose (≤ 0.04 nM) promotes primary root growth (González-García et al., 2011). The root growth of BR biosynthesis mutant, *dwf4*, can be recovered by an exogenous application of low concentration of BL. Thus an optimal level of BR is critical for promoting root growth. At the root tip, the QC cells, which function as a reservoir for new cells, constitute the stem cell niche, and BRs have been shown to regulate the QC cells' activity by modulating the *Brassinosteroids at vascular and organizing center* (*BRAVO*) (Sarkar et al., 2007; Heyman et al., 2014; Vilarrasa-Blasi et al., 2014).

BRAVO, expressed in QC and vascular initials, is a transcription factor belonging to the R2R3-MYB family (Vilarrasa-Blasi et al., 2014). Activation of BR signaling causes the repression of *BRAVO* transcripts by BES1. BES1 also heterodimerizes with *BRAVO*, inhibiting it at the protein level. The transcription factors, BES1 and BZR1, contain a conserved ERF-associated amphiphilic repression (EAR) motif in their C-terminal. TOPLESS (TPL) and TOPLESS-related (TPR) bind to the EAR motifs of various transcription factors and regulate their downstream targets. It has been shown that BES1 represses *BRAVO* expression by interacting with TPL at the EAR domain. Thus BES1−TPL interaction is necessary for promoting QC cell division (Planas-Riverola et al., 2019). Hence, QC cell identity is maintained by *BRAVO* through the repression of BR-mediated QC divisions. In addition, *Ethylene response factor115* (*ERF115*) functions as a rate-limiting factor for QC cell division. The ERF115 promotes QC cell division by activating phytosulfokine5 (PSK5) peptide hormone. The expression of *ERF115* in the QC is enhanced by the exogenous application of BL (Heyman et al., 2013). Both *bes1-D* and *bzr1-D* induce the expression of *ERF115* while inhibiting the *BRAVO* function. Thus BR inhibits *BRAVO* and promotes *ERF115* in the QC to balance the QC division rates.

Wuschel-related homeobox5 (*WOX5*) is an important transcription factor regulating stem cell fate. *WOX5* expression is restricted in the QC and functions in maintaining stem cell identity (Sarkar et al., 2007). The auxin response factors, ARF10 and ARF16, repress *WOX5* expression. BZR1-mediated QC cell division reactivation involves the expression of *WOX5*. The *WOX5* expression is significantly increased by exogenous BR and in the constitutively active BR signaling mutant, *bes1-D* (González-García et al., 2011). Hence, BRs promote *WOX5* transcription, which in turn promotes QC cell divisions (Fig. 10.2). BRs have been reported to regulate the columella stem cells (CSCs) differentiation in a dose-dependent manner. Low concentrations of BRs restrict stem cell differentiation, while higher levels of them promote stem cell differentiation (Lee et al., 2015; Planas-Riverola et al., 2019). *bzr1-D* and *bes1-D* have opposing effects in CSC differentiation (Lee et al., 2015). BES1 promotes CSC differentiation into starch-filled columella cells (CC) while BZR1 inhibits this differentiation processes.

BRs transcriptionally regulate the expression of many genes involved in the root developmental process in a BZR1-dependent manner (Chaiwanon and Wang, 2015). The root cap is necessary for directional root growth in response to gravity and surrounding environmental cues (Blancaflor et al., 1998). The root cap development is under the control of various transcription factors. The stem cell activity of the root cap is regulated by the transcription factors FEZ and SMB (Sombrero) (Willemsen et al., 2008). The FEZ and SMB antagonistically regulate each other. FEZ promotes periclinal root cap forming cell

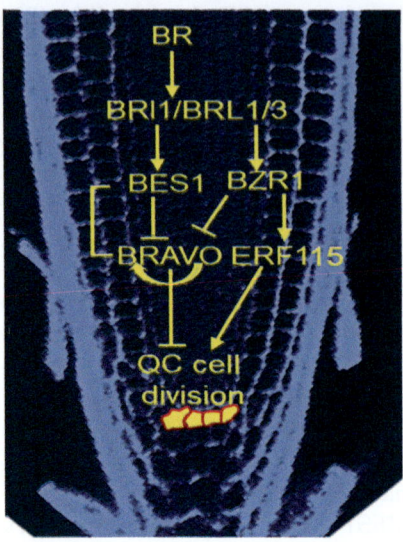

FIGURE 10.2 **Brassinosteroid regulated QC cell division.** Major components regulated by BR signaling during QC cell divisions in *Arabidopsis* root. BR perception from the membrane-bound receptors BRI1/BRL1/BRL3 activates the BR signaling transcription factors, BES1 and BZR1. BES1/BZR1 represses *BRAVO* that inhibits QC cell division (Vilarrasa-Blasi et al., 2014). However, BZR1 activates *ERF115* to promote QC cell divisions (Heyman et al., 2013). QC, quiescent center.

divisions, while SMB inhibits FEZ activity (Willemsen et al., 2008; Bennett et al., 2010). It was found that BL treatment enhanced the expression of *SMB*, while it repressed the *FEZ* expression. Additionally, BR receptor mutant, *bri1−116* shows reduced *SMB* expression. This shows that CSC differentiation is repressed in a BZR1-dependent manner. Bearskin1 (BRN1) and BRN2 along with SMB functions in regulating LR cap (LRC) development (Bennett et al., 2010; Lee et al., 2015). The *BRN1* and 2 levels are upregulated by BL treatment, and *BRN1* expression is enhanced in *bes1-D*. Further, molecular and genetic interaction studies are needed to illustrate how does BR regulate these genes in controlling root cap development (Lee et al., 2015).

BRs interact with other phytohormones in regulating RAM maintenance. One such major phytohormone is auxin. The RAM development and maintenance involved the synchronous activity of auxin. BRs have been shown to regulate auxin distribution at the root tip by regulating auxin transporters, PINs (*PIN3*, *PIN4*, and *PIN7*) expression, and their polar localization (Lee et al., 2015). A low concentration of BL treatment reduced the expression of *PIN3*, *PIN4*, and *PIN7* as in *bzr1-D*. However, under high BL treatment, *PIN7* expression is enhanced as in *bes1-D*. The PIN4 protein expression is enhanced in *bri1−116* plants as compared to control plants, while in *bzr1-D*, its expression is significantly reduced (Lee et al., 2015). Thus BR regulates the complex interplay of auxin distribution by PINs, which in turn mediate the cellular meristematic activity and cell elongation.

BR-controlled root growth relies on a fine balance of its level and signaling in a tissue-specific manner (Chaiwanon and Wang, 2015; Vragović et al., 2015; Ackerman-Lavert et al., 2020). For example, BRs responsive genes are induced by BRs mediated signals originating from the epidermal cell layer, while stelar BRs signaling repressed the expression of BR responsive genes. Translatome mapping of *Arabidopsis* roots revealed that targeted BR signaling from the epidermal cell layer promotes the induction of genes involved in auxin biosynthesis and transport, such as *Tryptophan aminotransferase of arabidopsis1* (*TAA1*) and *Pinformed2* (*PIN2*) (Hacham et al., 2011; Singh and Savaldi-Goldstein, 2015; Vragović et al., 2015). Thus, BRs from the epidermis induce auxin biosynthesis and translocation in the stem cells leading to cell divisions, thereby delaying the onset of cell differentiation. On the other hand, BRs from inner cell layers promote early differentiation. BR perception by BRI1 and its homologous receptors, BRL1 and BRL3, in the stele tissues buffered the epidermal effect on meristem size (Vragović et al., 2015). The stem cell niche is characterized by low BR activity due to a reduced level of nuclear-localized BZR1 (Chaiwanon and Wang, 2015). BR reduced the expression of auxin-induced genes, such as *Plethora 1,2,4/ Baby boom* (*PLT1, PLT2, PLT4/BBM*), at the root tip, which are necessary for stem cell identity (Aida et al., 2004; Chaiwanon and Wang, 2015). Nonetheless, a higher level of nuclear-localized BZR1 in the transition−elongation zone enhances BR-induced genes expression. This shows that BR repressed the expression of genes at the root meristem, while the transition−elongation zone is defined by BR-induced genes expression. Auxin−BR functions antagonistically in root cell elongation. High levels of auxin inhibit cell elongation while promoting lateral cell expansion. In contrast, high BRs promote cell elongation by modulating the genes involved in cell wall remodeling (Chaiwanon and Wang, 2015). BRs also function in a cell-specific manner. BRI1 expression in the epidermal hair cells promotes cell elongation, while its expression in nonhair cells inhibits it (Fridman et al., 2014). The phenotypic defects of *bri1brl1brl3* triple mutants are recovered by BR signaling from the developing protophloem. Hence, BR signaling and response from the inner cell to the outer layers drive root growth in a noncell-autonomous manner (Kang et al., 2017).

BRs have limited mobility, therefore, distribution of enzymes involved in their biosynthesis is are critical for optimal BR activity. Distribution of nuclear BES1 and BZR1 along the longitudinal root axis directs the gradient of BR response with higher and reduced response in the transition−elongation zone and the meristem zone, respectively (Chaiwanon and Wang, 2015). Recently, it is shown that BR biosynthesis largely occurs in the elongation zone and lesser in the meristematic zone of the root. This indicated that local BR biosynthesis maintains BR gradients along the longitudinal root axis ensuring optimal cellular elongation and cellular division (Vukašinović et al., 2021). The genes involved in BR biosynthesis are localized in different cell layers. *CPD* is expressed in procambial and central CC, *Brassinosteroid-6-oxidase1* and *2* (*BR6OX1* and *BR6OX2*) in the endodermis and pericycle, *DWF4*, and *DET2* in the apical meristem with *DWF4* having maximal expression in the epidermis. These spatiotemporal and tissue-specific expressions of BR biosynthesis genes suggested that the intermediates of the BR biosynthesis pathways have to be exchanged between cells for synchronized growth (Vukašinović et al., 2021).

The UPBEAT1 (UPB1) belongs to basic helix−loop−helix (bHLH) transcription factor. This bHLH transcription factor regulates the balance of cell division and differentiation and maintains root meristem zone activity. UPB1 interacts with PRE2/3 and inhibits their transcription (Li et al., 2020). Reduction of *UPB1* expression promotes root meristem development and acts downstream of BIN2. BIN2 phosphorylates UPB1 and enhanced its protein stability and transcriptional activity, while BES1 represses the *UPB1* transcription level (Li et al., 2020). In addition, Rapid alkalinization factor (RALF) is a peptide hormone, which represses cell elongation by activating a cell surface receptor FER (Feronia). This peptide hormone and BRs act oppositely. It was found that *AtRALF1* overexpression causes reduced cell size, whereas silencing *AtRALF1* enhances cell elongation. Also, *AtRALF1* overexpressed lines are less sensitive to BL. RALF1 further induces BR downregulated genes, such as *CPD* and *DWF4* (Bergonci et al., 2014; Wei and Li, 2016). In brief, BR and RALF regulate the cell elongation processes in an opposing manner where BRs promote cell elongation, while RALF inhibits the elongation process. The regulatory mechanism between the two hormones needs to be elucidated and is open for more investigation (Bergonci et al., 2014). Major genes involved in root growth are summarized in Table 10.1.

TABLE 10.1 Brassinosteroid regulated genes involved in root apical meristem development.

Gene	Functions	Effect of brassinosteroid (BRs)	Expression site/tissue	References
BRAVO	*BRAVO* represses QC divisions by opposing BR-dependent cell divisions in QC cells. BES1 represses the expression of *BRAVO*	Down	QC and vascular initials	Vilarrasa-Blasi et al. (2014)
PLT1	Belongs to Apetella2 (AP2) class transcription factor. Functions in maintaining stem cell activity and QC identity. Transcribed in response to auxin response factors	Down	QC	Chaiwanon and Wang (2015)
PLT2	Belongs to Apetella2 (AP2) class transcription factor. Functions in maintaining stem cell activity and QC identity. Transcribed in response to auxin response factors	Down	QC	Chaiwanon and Wang (2015)
UPB1	Modulates balance between root cell proliferation and differentiation	Down	Elongation zone	Li et al. (2020)
FEZ	FEZ promotes periclinal root cap forming cell divisions	Down	Root cap stem cells	Willemsen et al. (2008); Bennett et al. (2010); Lee et al. (2015)
WOX5	Belongs to the WUSCHEL (WUS) homeodomain transcription factor. Functions in maintaining QC identity and repressed the CSCs differentiation	Down	QC	Sarkar et al. (2007); González-García et al. (2011); Chaiwanon and Wang (2015)
ERF115	AP2 transcription factor. This ERF115 promotes QC cell division by activating Phytosulfokine PSK5 peptide hormone	Up	QC	Heyman et al. (2013)
PIN2	Auxin efflux transporter and maintain auxin maxima	Up	Localized in the epidermis and lateral root cap	Hacham et al. (2011); Vragović et al. (2015)

(Continued)

TABLE 10.1   (Continued)

Gene	Functions	Effect of brassinosteroid (BRs)	Expression site/tissue	References
PIN3	Auxin efflux carrier, root auxin maxima maintenance, mediate root gravitropism	Down	Columella cells, basal side of vascular cells	Ganguly et al. (2012); Lee et al. (2015)
PIN4	Auxin efflux carrier, maintenance of auxin gradients	Down	QC and provascular cells	Lee et al. (2015)
SMB	Involved in lateral root cap development. Acts in a feedback loop mechanism with *FEZ*. *FEZ* activates *SMB*, while *SMB* represses *FEZ* expression	Up		Bennett et al. (2010)
BRN1	Bearskin1 (*BRN1*) and *BRN2* together with *SMB* functions in regulating LRC development	Up	Root cap	Bennett et al. (2010)
BRN2	Bearskin1 (*BRN1*) and *BRN2* together with *SMB* functions in regulating LRC development	Up	Root cap	Bennett et al. (2010)
TAA1	Auxin biosynthesis. *TAA1* functions in maintaining root meristem size and growth of root hair	Up	QC	Stepanova et al. (2008); Vragović et al. (2015)

CSC, Columella stem cell; *LRC*, lateral root cap; *QC*, quiescent center.

10.3.2 Brassinosteroids in root vasculature development

The hormonal and environmental signals together regulate the organization of the vascular system. BRs have been reported to play roles in regulating vascular cells development within the root. The receptor-like protein 44 (RLP44) forms a complex with BRI1 and BAK1 and functions in regulating cell wall modification (Wolf et al., 2014; Holzwart et al., 2018). RLP44 is expressed in the root vasculature tissue. The *bri1brl1brl3* triple mutants have an increased number of differentiated xylem cells. Control of xylem cell numbers requires both BR and RLP44 activity. It has been found that activating BRI1 downstream signaling molecules does not affect metaxylem numbers. This suggests that BRI1 is critical for xylem cell fate determination through RLP44. The RLP44 functions as a link between BR and the phytosulfokine (PSK) peptide signaling pathways. The PSK is a small secreted peptide hormone that is perceived by Phytosulfokine receptor-1 and receptor-2 (Sauter, 2015). PSK functions in maintaining the procambial identity. RLP44 stabilizes the complex formation of PSKR1−BAK1 and BRI1−BAK1. Thus BRI1 negatively affects xylem cell fate and maintains procambial activity through its interaction with RLP44 and PSKR1 (Holzwart et al., 2018).

Brevis radix (BRX) and Octopus (OPS) are positive regulator of protophloem differentiation. Loss of function mutants, *ops* and *brx*, has impaired protophloem differentiation. BIN2 has been reported to play a role in protophloem differentiation by interacting with OPS. OPS at the PM recruits BIN2 and inhibits its binding to BZR1 and BES1. Thus, OPS enhances the BRs signaling output by antagonizing BIN2 (Anne et al., 2015). Exogenous BR application partially recovered the differentiation of the protophloem sieve elements in the *ops* mutant. In addition, *brxbri1* double mutant displayed almost no differentiating phloem sieve elements (Kang et al., 2017). Moreover, *brxbri1* double mutant has only a single protophloem strand. The targeted BRI1 signaling using *Cotyledon vascular pattern2* (*CVP2*) and *Membrane-associated kinase regulator5* (*MAKR5*) promoters in the developing protophloem recovered all phenotypic defects of *bri1brl1brl3* triple mutants highlighting the roles of BRs in protophloem differentiation (Kang et al., 2017).

10.3.3 Brassinosteroids in lateral root development

LRs constitute an important component of RSA. LRs originate from the pericycle founder cell, which is located opposite the xylem poles. These pericycle founder cells give rise to LR primordium. The LR primordium emerges out from the parental root by undergoing cell expansion (Dolan et al., 1993; Péret et al., 2009). Several hormones regulate LR initiation and growth. BRs are essential for regulating LR development along with the auxin. The *bri1* mutant has a reduced number of LRs. Exogenous application of BL rescued this defect (Bao et al., 2004). In addition, exogenous BL application (1–100 nM) enhances LR formation in wild-type roots. It is also observed that the auxin transport inhibitor naphthylphthalamic acid (NPA) inhibits LR emergence. However, BL treatment recovered NPA inhibited LR primordia (LRP) development. This illustrated that BR promotes LR development by enhancing acropetal transport of auxin in the root. Additionally, treatment with both BL and IAA enhances LR initiation indicating that these hormones work synergistically in LR development (Bao et al., 2004). Moreover, histochemical GUS analysis of *DR5-GUS* (auxin reporter) revealed that the expression of auxin-inducible genes are enhanced by BR (Bao et al., 2004). Several transcription factors control LR development (Perotti et al., 2019). *A. thaliana—activating factor2* (*ATAF2*) belongs to the NAC family transcription factor. This transcription factor contains NAC domain at the N terminal. The NAC domain is responsible for binding to DNA and protein, while the C terminal regulates the expression of downstream target genes. The LR development is reported to be regulated by *ATAF2* in a BR-dependent manner. Also, the loss of function mutant *ataf2* has more LRs than the wild type (Peng et al., 2015). Nevertheless, LR growth in *ataf2* mutant is reduced by inhibiting BR biosynthesis using brassinazole (BRZ). Thus, *ATAF2* acts as a repressor of LR development. Further research is needed to determine the molecular mechanism behind this observation (Peng et al., 2015). Recently, it has been reported that LR organogenesis is regulated by BIN2 *via* ARF7 and ARF19 phosphorylation (Cho et al., 2013). A higher concentration of BR (1 μM BL) inhibits LR development. This inhibition is due to the induction of *Auxin/indole-3 acetic acid* (*AUX/IAA*) genes expressions, such as *Auxin resistant2/ Indole-3-acetic acid7* (*AXR2/ IAA7*), *Solitary root/Indole-3-acetic acid14* (*SLR/IAA14*), *IAA28*, and *AXR3/IAA17*. The high AUX/IAA levels then inhibit auxin signaling resulting in reduced LR development and growth (Kim et al., 2006).

10.3.4 Brassinosteroids in root hairs development

Root hairs are tubular extensions that arise from the epidermal cell layer in the differentiation zone. They play a crucial role in increasing the surface area of the root, that enhances water and nutrient absorption from the soil (Wei and Li, 2016). In plants, development of root hairs has been divided into three types. In type I, root hair arises randomly from any epidermal cell, such as in rice. In the case of type II, as in *Brachypodium*, the epidermal cell undergoes asymmetrical division, wherein the smaller cell gives rise to root hairs. In *Arabidopsis*, the type III system of root hair development takes place where root hairs develop from the cell adjacent to the two cortical cells (H cells), while the cell attached to the single cortical cell developed into nonhair cells (N cells) (Dolan, 1996).

The *Glabra3* (*GL3*) and *Enhancer of GL3* (*EGL3*), members of bHLH transcription factors family, *Werewolf* (*WER*), a MYB transcription factor family, and *Transparent testa* (*TTG*), a *WD-40* repeat transcription factor family are central in root hair development and forms a complex. This complex activates the expression of *Glabra2* (*GL2*) and *Caprice* (*CPC*) (Lee and Schiefelbein, 1999; Bernhardt et al., 2003). *Root hair defective6* (*RHD6*), a bHLH transcription factor, positively regulates root hair development (Lin et al., 2015; Masucci and Schiefelbein, 1994). *RHD6* expression is inhibited by GL2, thus repressing root hair development. CPC, on the other hand, moves to the hair cells, competes, displaces WER from the complex, and binds to GL3/EGL3-TTG complex. This resulted in a reduction of *GL2* expression. In the absence of GL2, *RHD6* expression is promoted and induces root hair development. Inhibition of *RHD6* expression by GL2 in the epidermal cell leads to nonhair cell fate. However, the WER-GL3/EGL3-TTG complex enhances CPC. This CPC forms a different complex with GL3/EGL3-TTG by displacing WER, thus promoting hair−cell fate (Lee and Schiefelbein, 2002; Ryu et al., 2005; Wei and Li, 2018). BRs have been reported to be essential for determining the fate of root epidermal cells. *WER* and *GL2* expression are enhanced by BRs, and *bri1* mutant shows reduced expression (Kuppusamy et al., 2009; Cheng et al., 2014).

Recently, it has been reported that BRs have a role in regulating hair (H) and nonhair (N) cell fate. BRs repress H-cell fate and promote N-cell fate. Mutants with BR deficiency have more hair numbers than wild-type. In contrast, mutant having enhanced BR-signaling shows reduced hair numbers compared with wild type. Expression analysis of *pGL2-GUS* in the background of BR-related mutants revealed that BR biosynthesis mutant (*det2−1*) and signaling mutant (*bri1−116*) lacked *GL2* expression in their N cells of root epidermis. However, enhanced BR signaling in *BRI1-OX* and *bin2−3bil1bil2*, ectopically expressed *GL2* in H cells (Cheng et al., 2014). Additionally, biochemical studies revealed that BIN2 phosphorylates EGL3 and TTG. Thus phosphorylation by BIN2 inactivates WER-GL3/EGL3-TTG complex thereby reducing GL2 expression in nonhair cells. The phosphorylation of EGL3 by BIN2 at T399 and T209/T213 regulates the movement of EGL3 from H cells to N cells (Cheng et al., 2014; Wei and Li, 2018). All these studies together revealed that BRs regulate cell fate determination by regulating the key transcriptional regulators involved in root hair patterning and development.

10.4 Conclusions and future perspective

BRs play a crucial role in plant developmental processes in both above- and below-ground organs. In the above-ground organs, BRs regulate the development of leaf, stomata, hypocotyl, and flower from SAM. These processes are mediated through the BR signaling. BIN2 is directly involved in the stomata development processes, while hypocotyl elongation and flower development are monitored through BES1/BZR1. BR signaling components also interact with other hormones and external stimuli to monitor the developmental processes. The BR—GA signaling together promotes hypocotyl elongation. BR signaling and light signaling together regulate floral transition. For instance, BR signaling inhibits flowering, but in association with light signaling, it causes floral transition from SAM. In addition, plant root growth is also regulated by BRs in a dose-dependent manner. High BR concentrations inhibit root growth, while lower concentrations promote it. Although much knowledge of the BR signaling pathway has been illustrated, however, the transport system of BR is still a mystery and open for future investigation. Further, the tissue-specific regulation of BR with auxin along the root developmental zones is critical for stem cell activity and cellular division, which in turn mediate root growth (Chaiwanon and Wang, 2015; Vragović et al., 2015). However, deeper knowledge and direct interaction studies are required for the BRs-auxin module in RSA modification under normal as well as various environmental cues.

The discovery of spatiotemporal activity of BR in regulating plant growth is recently emerging (Hacham et al., 2011; Kang et al., 2017). Detailed studies with a focus on identifying regulatory genes could be a future target for developing plants with robust adaptations to changing environments. It is also known that BR from epidermis promotes auxin biosynthesis gene, *TAA1* expression and *PIN2* accumulation (Hacham et al., 2012; Singh and Savaldi-Goldstein, 2015; Vragović et al., 2015). This could be an important initiation point for a better understanding between BR-auxin modules during normal as well as stress condition. Here, we summarized the molecular mechanism associated with BR signaling during different plant organ developmental processes in a comprehensive manner with future possibilities.

Acknowledgments

The authors acknowledge the members of APS lab. APS acknowledges DST-SERB and the Department of Biotechnology (DBT), India for the grant (ECR/2018/000526) and Har Govind Khorana—Innovative Young Biotechnologist Award Grant (BT/11/IYBA/2018/02), respectively. RKY and LLD acknowledge CSIR for RA and SRF fellowship, respectively. The authors are thankful to the DBT e-library Consortium (DeLCON) for providing access to e-resources.

References

Ackerman-Lavert, M., Fridman, Y., Matosevich, R., Khandal, H., Friedlander, L., Vragović, K., et al., 2020. Auxin requirements for a meristematic state in roots depend on a dual brassinosteroid function. Curr. Biol. 31, 4462—4472.

Aida, M., Beis, D., Heidstra, R., Willemsen, V., Blilou, I., Galinha, C., et al., 2004. The PLETHORA genes mediate patterning of the Arabidopsis root stem cell niche. Cell 119, 109—120.

Anne, P., Azzopardi, M., Gissot, L., Beaubiat, S., Hématy, K., Palauqui, J.C., 2015. OCTOPUS negatively regulates BIN2 to control phloem differentiation in *Arabidopsis thaliana*. Curr. Biol. 25, 2584—2590.

Bai, M.-Y., Shang, J.-X., Oh, E., Fan, M., Bai, Y., Zentella, R., et al., 2012. Brassinosteroid, gibberellin and phytochrome impinge on a common transcription module in Arabidopsis. Nat. Cell Biol. 2012, 810—817. 148 14.

Bao, F., Shen, J., Brady, S.R., Muday, G.K., Asami, T., Yang, Z., 2004. Brassinosteroids interact with auxin to promote lateral root development in arabidopsis. Plant. Physiol. 134, 1624−1631.

Barton, M.K., 2010. Twenty years on: the inner workings of the shoot apical meristem, a developmental dynamo. Dev. Biol. 341, 95−113.

Beemster, G.T.S., Baskin, T.I., 1998. Analysis of cell division and elongation underlying the developmental acceleration of root growth in *Arabidopsis thaliana*. Plant. Physiol. 116, 1515−1526.

Belkhadir, Y., Chory, J., 2006. Brassinosteroid signaling: a paradigm for steroid hormone signaling from the cell surface. Science (80-) 314, 1410−1411.

Bell, E.M., Lin, W., Husbands, A.Y., Yu, L., Jaganatha, V., Jablonska, B., et al., 2012. Arabidopsis LATERAL ORGAN BOUNDARIES negatively regulates brassinosteroid accumulation to limit growth in organ boundaries. Proc. Natl. Acad. Sci. U.S.A. 109, 21146−21151.

Bennett, T., van den Toorn, A., Sanchez-Perez, G.F., Campilho, A., Willemsen, V., Snel, B., et al., 2010. SOMBRERO, BEARSKIN1, and BEARSKIN2 regulate root cap maturation in Arabidopsis. Plant. Cell 22, 640−654.

Bergonci, T., Ribeiro, B., Ceciliato, P.H.O., Guerrero-Abad, J.C., Silva-Filho, M.C., Moura, D.S., 2014. *Arabidopsis thaliana* RALF1 opposes brassinosteroid effects on root cell elongation and lateral root formation. J. Exp. Bot. 65, 2219−2230.

Bernhardt, C., Lee, M.M., Gonzalez, A., Zhang, F., Lloyd, A., Schiefelbein, J., 2003. The bHLH genes GLABRA3 (GL3) andENHANCER OF GLABRA3(EGL3) specify epidermal cell fate in the Arabidopsis root. Development 130, 6431−6439.

Blancaflor, E.B., Fasano, J.M., Gilroy, S., 1998. Mapping the functional roles of cap cells in the response of Arabidopsis primary roots to gravity. Plant. Physiol. 116, 213−222.

Chaiwanon, J., Wang, Z.Y., 2015. Spatiotemporal brassinosteroid signaling and antagonism with auxin pattern stem cell dynamics in Arabidopsis roots. Curr. Biol. 25, 1031−1042.

Cheng, Y., Zhu, W., Chen, Y., Ito, S., Asami, T., Wang, X., 2014. Brassinosteroids control root epidermal cell fate via direct regulation of a MYB-bHLH-WD40 complex by GSK3-like kinases. eLife 3, 2525.

Cho, H., Ryu, H., Rho, S., Hill, K., Smith, S., Audenaert, D., et al., 2013. A secreted peptide acts on BIN2-mediated phosphorylation of ARFs to potentiate auxin response during lateral root development. Nat. Cell Biol. 2013, 66−76. 161 16.

Clouse, S.D., 2011. Brassinosteroid signal transduction: from receptor kinase activation to transcriptional networks regulating plant development. Plant. Cell 23, 1219−1230.

Clouse, S.D., Langford, M., Mcmorris, T.C., 1996. A Brassinosteroid-lnsensitive Mutant in *Arabidopsis thaliana* Exhibits Multiple Defects in Growth and Development. Plant. Physiol. 111 (3), 671−678.

Devi, L.L., Pandey, A., Gupta, S., Singh, A.P., 2022. The interplay of auxin and brassinosteroid signaling tunes root growth under low and different nitrogen forms. Plant. Physiol. 189 (3), 1757−1773. Available from: https://doi.org/10.1093/plphys/kiac157.

Devi, L.L., Pandey, A., Singh, A.P., 2020. Root plasticity under low phosphate availability: A physiological and molecular approach to plant adaptation under limited phosphate availability. Improv. Abiotic Stress. Toler. Plants 13−30.

Dolan, L., 1996. Pattern in the root epidermis: an interplay of diffusible signals and cellular geometry. Ann. Bot. 77, 547−553.

Dolan, L., Janmaat, K., Willemsen, V., Linstead, P., Poethig, S., Roberts, K., et al., 1993. Cellular organisation of the *Arabidopsis thaliana* root. Development 119, 71−84.

Domagalska, M.A., Schomburg, F.M., Amasino, R.M., Vierstra, R.D., Nagy, F., Davis, S.J., 2007. Attenuation of brassinosteroid signaling enhances FLC expression and delays flowering. Development 134, 2841−2850.

Espinosa-Ruiz, A., Martínez, C., de Lucas, M., Fàbregas, N., Bosch, N., Caño-Delgado, A.I., et al., 2017. TOPLESS mediates brassinosteroid control of shoot boundaries and root meristem development in *Arabidopsis thaliana*. Development 144, 1619−1628.

Fridman, Y., Elkouby, L., Holland, N., Vragović, K., Elbaum, R., Savaldi-Goldstein, S., 2014. Root growth is modulated by differential hormonal sensitivity in neighboring cells. Genes. Dev. 28, 912−920.

Gallego-Bartolomé, J., Minguet, E.G., Grau-Enguix, F., Abbas, M., Locascio, A., Thomas, S.G., et al., 2012. Molecular mechanism for the interaction between gibberellin and brassinosteroid signaling pathways in Arabidopsis. Proc. Natl. Acad. Sci. U.S.A. 109, 13446−13451.

Ganguly, A., Lee, S.H., Cho, H.T., 2012. Functional identification of the phosphorylation sites of Arabidopsis PIN-FORMED3 for its subcellular localization and biological role. Plant. J. 71, 810–823.

González-García, M.P., Vilarrasa-Blasi, J., Zhiponova, M., Divol, F., Mora-García, S., Russinova, E., et al., 2011. Brassinosteroids control meristem size by promoting cell cycle progression in Arabidopsis roots. Development 138, 849–859.

Grove, M.D., Spencer, G.F., Rohwedder, W.K., Mandava, N., Worley, J.F., Warthen, J.D., et al., 1979. Brassinolide, a plant growth-promoting steroid isolated from Brassica napus pollen. Nature 1979, 216–217. 2815728 281.

Gudesblat, G.E., Schneider-Pizoń, J., Betti, C., Mayerhofer, J., Vanhoutte, I., van Dongen, W., et al., 2012. SPEECHLESS integrates brassinosteroid and stomata signalling pathways. Nat. Cell Biol. 2012, 548–554. 145 14.

Gupta, S., Devi, L.L., Singh, A.P., 2022. Nitric oxide. Nitric Oxide Plants 230–247.

Hacham, Y., Holland, N., Butterfield, C., Ubeda-Tomas, S., Bennett, M.J., Chory, J., et al., 2011. Brassinosteroid perception in the epidermis controls root meristem size. Development 138, 839–848.

Hacham, Y., Sela, A., Friedlander, L., Savaldi-Goldstein, S., 2012. BRI1 activity in the root meristem involves post-transcriptional regulation of PIN auxin efflux carriers. Plant Signaling & Behavior. 7, 68–70.

Hao, F., Zhao, S., Dong, H., Zhang, H., Sun, L., Miao, C., 2010. Nia1 and Nia2 are involved in exogenous salicylic acid-induced nitric oxide generation and stomatal closure in Arabidopsis. J. Integr. Plant. Biol. 52, 298–307.

He, J.X., Gendron, J.M., Sun, Y., Gampala, S.S.L., Gendron, N., Sun, C.Q., et al., 2005. BZR1 is a transcriptional repressor with dual roles in brassinosteroid homeostasis and growth responses. Science (80-) 307, 1634–1638.

Hewitta, F.R., Hough, B.T., O'neillb, P., Sassea, J.M., Williamsb, E.G., Rowana, K.S., 1985. Effect of brassinolide and other growth regulators on the germination and growth of pollen tubes of Prunus avium using a multiple hanging-drop assay. Aust. J. Plant. Physiol. 12, 201–202.

Heyman, J., Cools, T., Vandenbussche, F., Heyndrickx, K.S., Leene, J.V., Vercauteren, I., et al., 2013. ERF115 controls root quiescent center cell division and stem cell replenishment. Science (80-) 342, 860–863.

Heyman, J., Kumpf, R.P., De Veylder, L., 2014. A quiescent path to plant longevity. Trends Cell Biol. 24, 443–448.

Hibara, K., Karim, M.R., Takada, S., Taoka, K., Furutani, M., Aida, M., et al., 2006. Arabidopsis CUP-SHAPED COTYLEDON3 regulates postembryonic shoot meristem and organ boundary formation. Plant. Cell 18, 2946–2957.

Holzwart, E., Huerta, A.I., Glöckner, N., Gómez, B.G., Wanke, F., Augustin, S., et al., 2018. BRI1 controls vascular cell fate in the Arabidopsis root through RLP44 and phytosulfokine signaling. Proc. Natl. Acad. Sci. U.S.A. 115, 11838–11843.

Houbaert, A., Zhang, C., Tiwari, M., Wang, K., de Marcos Serrano, A., Savatin, D.V., et al., 2018. POLAR-guided signalling complex assembly and localization drive asymmetric cell division. Nature 563, 574–578.

Ibañes, M., Fàbregas, N., Chory, J., Caño-Delgado, A.I., 2009. Brassinosteroid signaling and auxin transport are required to establish the periodic pattern of Arabidopsis shoot vascular bundles. Proc. Natl. Acad. Sci. U.S.A. 106, 13630–13635.

Jaillais, Y., Hothorn, M., Belkhadir, Y., Dabi, T., Nimchuk, Z.L., Meyerowitz, E.M., et al., 2011. Tyrosine phosphorylation controls brassinosteroid receptor activation by triggering membrane release of its kinase inhibitor. Genes. Dev. 25, 232–237.

Jiang, J., Wang, T., Wu, Z., Wang, J., Zhang, C., Wang, H., et al., 2015. The intrinsically disordered protein BKI1 is essential for inhibiting BRI1 signaling in plants. Mol. Plant. 8, 1675–1678.

Kang, Y.H., Breda, A., Hardtke, C.S., 2017. Brassinosteroid signaling directs formative cell divisions and protophloem differentiation in Arabidopsis root meristems. Development 144, 272–280.

Khan, M., Rozhon, W., Bigeard, J., Pflieger, D., Husar, S., Pitzschke, A., et al., 2013. Brassinosteroid-regulated GSK3/shaggy-like kinases phosphorylate mitogen-activated protein (MAP) kinase kinases, which control stomata development in Arabidopsis thaliana. J. Biol. Chem. 288, 7519–7527.

Kim, H., Park, P.-J., Hwang, H.-J., Lee, S.-Y., Oh, M.-H., Kim, S.-G., 2006. Brassinosteroid signals control expression of the AXR3/IAA17 gene in the cross-talk point with auxin in root development. Biosci. Biotechnol. Biochem. 70, 768–773.

Kim, T.-W., Michniewicz, M., Bergmann, D.C., Wang, Z.-Y., 2012. Brassinosteroid regulates stomatal development by GSK-mediated inhibition of a MAPK pathway. Nature 2012, 419–422. 4827385 482.

Kim, T.W., Guan, S., Burlingame, A.L., Wang, Z.Y., 2011. The CDG1 kinase mediates brassinosteroid signal transduction from BRI1 receptor kinase to BSU1 phosphatase and GSK3-like kinase BIN2. Mol. Cell 43, 561–571.

Kim, T.-W., Wang, Z.-Y., 2010. Brassinosteroid Signal Transduction from Receptor Kinases to Transcription Factors. Annu. Rev. Plant. Biol. 61, 681–704.

Kuppusamy, K.T., Chen, A.Y., Nemhauser, J.L., 2009. Steroids are required for epidermal cell fate establishment in Arabidopsis roots. Proc. Natl. Acad. Sci. U.S.A. 106, 8073–8076.

Lau, O.S., Bergmann, D.C., 2012. Stomatal development: a plant's perspective on cell polarity, cell fate transitions and intercellular communication. Development 139, 3683–3692.

Lee, D.-K., Geisler, M., Springer, P.S., 2009. LATERAL ORGAN FUSION1 and LATERAL ORGAN FUSION2function in lateral organ separation and axillary meristem formation in Arabidopsis. Development 136, 2423–2432.

Lee, H.-S., Kim, Y., Pham, G., Kim, J.W., Song, J.-H., Lee, Y., et al., 2015. Brassinazole resistant 1 (BZR1)-dependent brassinosteroid signalling pathway leads to ectopic activation of quiescent cell division and suppresses columella stem cell differentiation. J. Exp. Bot. 66, 4835–4849.

Lee, M.M., Schiefelbein, J., 1999. WEREWOLF, a MYB-related protein in Arabidopsis, is a position-dependent regulator of epidermal cell patterning. Cell 99, 473–483.

Lee, M.M., Schiefelbein, J., 2002. Cell pattern in the Arabidopsis root epidermis determined by lateral inhibition with feedback. Plant. Cell 14, 611–618.

Li, J., Chory, J., 1999. Brassinosteroid actions in plants. J. Exp. Bot. 50, 275–282.

Li, J., Wen, J., Lease, K.A., Doke, J.T., Tax, F.E., Walker, J.C., 2002. BAK1, an Arabidopsis LRR receptor-like protein kinase, interacts with BRI1 and modulates brassinosteroid signaling. Cell 110, 213–222.

Li, L., Ye, H., Guo, H., Yin, Y., 2010. Arabidopsis IWS1 interacts with transcription factor BES1 and is involved in plant steroid hormone brassinosteroid regulated gene expression. Proc. Natl. Acad. Sci. U.S.A. 107, 3918–3923.

Li, Q.-F., Wang, C., Jiang, L., Li, S., Sun, S.S.M., He, J.-X., 2012. An interaction between BZR1 and DELLAs mediates direct signaling crosstalk between brassinosteroids and gibberellins in Arabidopsis. Sci. Signal. 5, ra72.

Li, T., Lei, W., He, R., Tang, X., Han, J., Zou, L., et al., 2020. Brassinosteroids regulate root meristem development by mediating BIN2-UPB1 module in Arabidopsis. PLoS Genet. 16, e1008883.

Li, Z., He, Y., 2020. Roles of brassinosteroids in plant reproduction. Int. J. Mol. Sci. 21, Page 872 21: 872.

Li, Z., Ou, Y., Zhang, Z., Li, J., He, Y., 2018. Brassinosteroid signaling recruits histone 3 lysine-27 demethylation activity to FLOWERING LOCUS C chromatin to inhibit the floral transition in Arabidopsis. Mol. Plant. 11, 1135–1146.

Lin, Q., Ohashi, Y., Kato, M., Tsuge, T., Gu, H., Qu, L.-J., et al., 2015. GLABRA2 Directly Suppresses Basic Helix-Loop-Helix Transcription Factor Genes with Diverse Functions in Root Hair Development. Plant Cell 27, 2894–2906.

Luschnig, C., Vert, G., 2014. The dynamics of plant plasma membrane proteins: PINs and beyond. Development 141, 2924–2938.

Masucci, J.D., Schiefelbein, J.W., 1994. The rhd6 mutation of *Arabidopsis thaliana* alters root-hair initiation through an auxin- and ethylene-associated process. Plant. Physiol. 106, 1335–1346.

Nakaya, M., Tsukaya, H., Murakami, N., Kato, M., 2002. Brassinosteroids control the proliferation of leaf cells of *Arabidopsis thaliana*. Plant. Cell Physiol. 43, 239–244.

Nam, K.H., Li, J., 2002. BRI1/BAK1, a receptor kinase pair mediating brassinosteroid signaling. Cell 110, 203–212.

Nolan, T.M., Brennan, B., Yang, M., Chen, J., Zhang, M., Li, Z., et al., 2017. Selective autophagy of BES1 mediated by DSK2 balances plant growth and survival. Dev. Cell 41 (33–46), e7.

Oh, E., Zhu, J.-Y., Wang, Z.-Y., 2012. Interaction between BZR1 and PIF4 integrates brassinosteroid and environmental responses. Nat. Cell Biol. 2012, 802–809. 148 14.

Oh, E., Zhu, J.Y., Bai, M.Y., Arenhart, R.A., Sun, Y., Wang, Z.Y., 2014. Cell elongation is regulated through a central circuit of interacting transcription factors in the Arabidopsis hypocotyl. eLife. Available from: https://doi.org/10.7554/ELIFE.03031.

Oh, M.-H., Sun, J., Oh, D.H., Zielinski, R.E., Clouse, S.D., Huber, S.C., 2011. Enhancing Arabidopsis leaf growth by engineering the BRASSINOSTEROID INSENSITIVE1 receptor kinase. Plant. Physiol. 157, 120–131.

Pandey, A., Devi, L.L., Singh, A.P., 2020. Review: emerging roles of brassinosteroid in nutrient foraging. Plant. Sci. 110474.

Peng, H., Zhao, J., Neff, M.M., 2015. ATAF2 integrates Arabidopsis brassinosteroid inactivation and seedling photomorphogenesis. Development 142, 4129–4138.

Péret, B., De Rybel, B., Casimiro, I., Benková, E., Swarup, R., Laplaze, L., et al., 2009. Arabidopsis lateral root development: an emerging story. Trends Plant. Sci. 14, 399–408.

Perotti, M.F., Ribone, P.A., Cabello, J.V., Ariel, F.D., Chan, R.L., 2019. AtHB23 participates in the gene regulatory network controlling root branching, and reveals differences between secondary and tertiary roots. Plant. J. 100, 1224–1236.

Planas-Riverola, A., Gupta, A., Betegoń-Putze, I., Bosch, N., Ibanęs, M., Cano-Delgado, A.I., 2019. Brassinosteroid signaling in plant development and adaptation to stress. Development 146, 151894. Available from: https://doi.org/10.1242/dev.151894.

Reddy, G.V., Heisler, M.G., Ehrhardt, D.W., Meyerowitz, E.M., 2004. Real-time lineage analysis reveals oriented cell divisions associated with morphogenesis at the shoot apex of *Arabidopsis thaliana*. Development 131, 4225−4237.

Ren, H., Willige, B.C., Jaillais, Y., Geng, S., Park, M.Y., Gray, W.M., et al., 2019. BRASSINOSTEROID-SIGNALING KINASE 3, a plasma membrane-associated scaffold protein involved in early brassinosteroid signaling. PLoS Genet. 15, e1007904.

Ryu, K.H., Kang, Y.H., Park, Y., Hwang, I., Schiefelbein, J., Lee, M.M., 2005. The WEREWOLF MYB protein directly regulates CAPRICEtranscription during cell fate specification in the Arabidopsis root epidermis. Development 132, 4765−4775.

Sabatini, S., Heidstra, R., Wildwater, M., Scheres, B., 2003. SCARECROW is involved in positioning the stem cell niche in the Arabidopsis root meristem. Genes. Dev. 17, 354−358.

Sarkar, A.K., Luijten, M., Miyashima, S., Lenhard, M., Hashimoto, T., Nakajima, K., et al., 2007. Conserved factors regulate signalling in *Arabidopsis thaliana* shoot and root stem cell organizers. Nature 2007, 811−814. 4467137 446.

Sauter, M., 2015. Phytosulfokine peptide signalling. J. Exp. Bot. 66, 5161−5169.

Singh, A.P., Fridman, Y., Friedlander-Shani, L., Tarkowska, D., Strnad, M., Savaldi-Goldstein, S., 2014. Activity of the brassinosteroid transcription factors BRASSINAZOLE RESISTANT1 and BRASSINOSTEROID INSENSITIVE1-ETHYL METHANESULFONATE-SUPPRESSOR1/ BRASSINAZOLE RESISTANT2 blocks developmental reprogramming in response to low phosphate availability. Plant. Physiol. 166, 578−588.

Singh, A.P., Fridman, Y., Holland, N., Ackerman-Lavert, M., Zananiri, R., Jaillais, Y., et al., 2018. Interdependent nutrient availability and steroid hormone signals facilitate root growth plasticity. Dev. Cell 46 (59−72), e4.

Singh, A.P., Savaldi-Goldstein, S., 2015. Growth control: brassinosteroid activity gets context. J. Exp. Bot. 66, 1123−1132.

Sreeramulu, S., Mostizky, Y., Sunitha, S., Shani, E., Nahum, H., Salomon, D., et al., 2013. BSKs are partially redundant positive regulators of brassinosteroid signaling in Arabidopsis. Plant. J. 74, 905−919.

Stepanova, A.N., Robertson-Hoyt, J., Yun, J., Benavente, L.M., Xie, D.Y., Doležal, K., et al., 2008. TAA1-mediated auxin biosynthesis is essential for hormone crosstalk and plant development. Cell 133, 177−191.

Sun, S., Chen, D., Li, X., Qiao, S., Shi, C., Li, C., et al., 2015. Brassinosteroid signaling regulates leaf erectness in *Oryza sativa* via the control of a specific U-Type cyclin and cell proliferation. Dev. Cell 34, 220−228.

Sun, Y., Fan, X.Y., Cao, D.M., Tang, W., He, K., Zhu, J.Y., et al., 2010. Integration of brassinosteroid signal transduction with the transcription network for plant growth regulation in Arabidopsis. Dev. Cell 19, 765−777.

Szekeres, M., Németh, K., Koncz-Kálmán, Z., Mathur, J., Kauschmann, A., Altmann, T., et al., 1996. Brassinosteroids rescue the deficiency of CYP90, a cytochrome P450, controlling cell elongation and de-etiolation in Arabidopsis. Cell 85, 171−182.

Tanaka, K., Nakamura, Y., Asami, T., Yoshida, S., Matsuo, T., Okamoto, S., 2003. Physiological roles of brassinosteroids in early growth of arabidopsis: brassinosteroids have a synergistic relationship with gibberellin as well as auxin in light-grown hypocotyl elongation. J. Plant. Growth Regul. 2003, 259−271. 223 22.

Tang, W., Kim, T.-W., Oses-Prieto, J.A., Sun, Y., Deng, Z., Zhu, S., et al., 2008. BSKs mediate signal transduction from the receptor kinase BRI1 in Arabidopsis. Science (80-) 321, 557−560.

Ubeda-Tomás, S., Beemster, G.T.S., Bennett, M.J., 2012. Hormonal regulation of root growth: integrating local activities into global behaviour. Trends Plant. Sci. 17, 326−331.

Vert, G., Chory, J., 2006. Downstream nuclear events in brassinosteroid signalling. Nature 2006, 96−100. 4417089 441.

Vilarrasa-Blasi, J., González-García, M.P., Frigola, D., Fàbregas, N., Alexiou, K.G., López-Bigas, N., et al., 2014. Regulation of plant stem cell quiescence by a brassinosteroid signaling module. Dev. Cell 30, 36−47.

Vragović, K., Sela, A., Friedlander-Shani, L., Fridman, Y., Hacham, Y., Holland, N., et al., 2015. Translatome analyses capture of opposing tissue-specific brassinosteroid signals orchestrating root meristem differentiation. Proc. Natl. Acad. Sci. U.S.A. 112, 923−928.

Vroemen, C.W., Mordhorst, A.P., Albrecht, C., Kwaaitaal, M.A.C.J., de Vries, S.C., 2003. The CUP-SHAPED COTYLEDON3 gene is required for boundary and shoot meristem formation in Arabidopsis. Plant. Cell 15, 1563−1577.

Vukašinović, N., Wang, Y., Vanhoutte, I., Fendrych, M., Guo, B., Kvasnica, M., et al., 2021. Local brassinosteroid biosynthesis enables optimal root growth. Nat. Plants 2021, 619−632. 75 7.

Wang, S.-L., Viswanath, K.K., Tong, C.-G., An, H.R., Jang, S., Chen, F.-C., 2019. Floral induction and flower development of orchids. Front. Plant. Sci. 0, 1258.

Wang, X., Chory, J., 2006. Brassinoteroids regulate dissociation of BKI1, a negative regulator of BRI1 signaling, from the plasma membrane. Science (80-) 313, 1118−1122.

Wang, X., Kota, U., He, K., Blackburn, K., Li, J., Goshe, M.B., et al., 2008. Sequential transphosphorylation of the BRI1/BAK1 receptor kinase complex impacts early events in brassinosteroid signaling. Dev. Cell 15, 220−235.

Wang, Z.Y., Seto, H., Fujioka, S., Yoshida, S., Chory, J., 2001. BRI1 is a critical component of a plasma-membrane receptor for plant steroids. Nature 410, 380−383.

Wei, Z., Li, J., 2016. Brassinosteroids regulate root growth, development, and symbiosis. Mol. Plant. 9, 86−100.

Wei, Z., Li, J., 2018. Receptor-like protein kinases: key regulators controlling root hair development in *Arabidopsis thaliana*. J. Integr. Plant. Biol. 60, 841−850.

Willemsen, V., Bauch, M., Bennett, T., Campilho, A., Wolkenfelt, H., Xu, J., et al., 2008. The NAC domain transcription factors FEZ and SOMBRERO control the orientation of cell division plane in Arabidopsis root stem cells. Dev. Cell 15, 913−922.

Wolf, S., Does, D.van der, Ladwig, F., Sticht, C., Kolbeck, A., Schürholz, A.-K., et al., 2014. A receptor-like protein mediates the response to pectin modification by activating brassinosteroid signaling. Proc. Natl. Acad. Sci. U.S.A. 111, 15261−15266.

Ye, Q., Zhu, W., Li, L., Zhang, S., Yin, Y., Ma, H., et al., 2010. Brassinosteroids control male fertility by regulating the expression of key genes involved in Arabidopsis anther and pollen development. Proc. Natl. Acad. Sci. U.S.A. 107, 6100−6105.

Yin, Y., Vafeados, D., Tao, Y., Yoshida, S., Asami, T., Chory, J., 2005. A new class of transcription factors mediates brassinosteroid-regulated gene expression in Arabidopsis. Cell 120, 249−259.

Yin, Y., Wang, Z.Y., Mora-Garcia, S., Li, J., Yoshida, S., Asami, T., et al., 2002. BES1 accumulates in the nucleus in response to brassinosteroids to regulate gene expression and promote stem elongation. Cell 109, 181−191.

Zhang, L.-Y., Bai, M.-Y., Wu, J., Zhu, J.-Y., Wang, H., Zhang, Z., et al., 2009. Antagonistic HLH/bHLH transcription factors mediate brassinosteroid regulation of cell elongation and plant development in rice and Arabidopsis. Plant. Cell 21, 3767.

Zhang, Y., Li, B., Xu, Y., Li, H., Li, S., Zhang, D., et al., 2013. The cyclophilin CYP20-2 modulates the conformation of BRASSINAZOLE-RESISTANT1, which binds the promoter of FLOWERING LOCUS D to regulate flowering in Arabidopsis. Plant. Cell 25, 2504−2521.

Zhang, Y., Wang, P., Shao, W., Zhu, J.K., Dong, J., 2015. The BASL polarity protein controls a MAPK signaling feedback loop in asymmetric cell division. Dev. Cell 33, 136−149.

Zhiponova, M.K., Vanhoutte, I., Boudolf, V., Betti, C., Dhondt, S., Coppens, F., et al., 2013. Brassinosteroid production and signaling differentially control cell division and expansion in the leaf. N. Phytol. 197, 490−502.

Zhu, J.-Y., Sae-Seaw, J., Wang, Z.-Y., 2013. Brassinosteroid signalling. Development 140, 1615−1620.

11

Understanding the role of phytohormones in governing heat, cold, and freezing stress response

Mohan Sharma[1,2],*, Harshita B. Saksena[1],*,
Halidev Krishna Botta[1],* and Ashverya Laxmi[1]

[1]Lab 203, National Institute of Plant Genome Research, New Delhi, Delhi, India [2]Institute of Biology III, Albert-Ludwigs-University Freiburg, Schänzlestraße, Freiburg im Breisgau, Germany

11.1 Introduction

Plants are surrounded by several enemies, such as abiotic and biotic stresses, in their environment and due to their sedentary nature, they cannot escape to the more favorable locations. Instead, they adopt several adaptive strategies that help them cope with stress in a better way. One such program is large-scale transcriptome reprogramming of stress-related genes. Over the past two decades, significant advances have been made that suggest that phytohormones are important players in helping plants adapt to stressful conditions. Traditionally, auxins, cytokinin (CK), brassinosteroids (BRs), gibberellic acid (GA), and strigolactones (SLs) are known for their essential roles in plant growth and development (Mishra et al., 2009; Petrášek and Friml, 2009; Zhao, 2010; Smith, 2014; Belkhadir and Jaillais, 2015; Kieber and Schaller, 2018). On the other hand, jasmonic acid (JA), salicylic acid (SA), and ethylene are mainly involved in regulating the defense of plants to biotic stress (Yang et al., 2015; Verma et al., 2016). Furthermore, abscisic acid (ABA) is known for its profound role in plant responses to abiotic stresses (Vishwakarma et al., 2017). However, emerging evidence has shown that phytohormones that are primary regulators of defense responses, such as JA, SA, and ethylene, are involved in plant growth and plasticity as well as abiotic stress responses

* Authors Mohan Sharma, Harshita B. Saksena, and Halidev Krishna Botta contributed equally to this chapter.

(Kazan, 2015; Dubois et al., 2018), while growth-related hormones, such as auxin, CK, BR, and GA, are seen for their role in plant responses to many abiotic stresses, including heat, cold, and cold stress (Kazan, 2013; Colebrook et al., 2014; Planas-Riverola et al., 2019; Prerostova et al., 2020a; Kothari and Lachowiec, 2021). In this chapter, we have discussed mechanistic insights into the function of phytohormones in counteracting plant responses to heat, cold, and freezing stress.

11.2 Regulation of high and low temperature responses by growth promoting hormones

11.2.1 Role of auxin in regulating ambient temperature and heat stress response

Temperature is one of the major environmental factors that affect crop productivity. Every 1°C changes in environmental temperature affects crop yield by 10% (Zhao et al., 2017). A large number of studies assessed the basal as well as acquired thermotolerance, when plants encountered detrimental lethal temperature (ranges between 35°C and 45°C), resulting into plant death (Lämke et al., 2016; Ohama et al., 2017; Sharma et al., 2019). However, a mild increase in atmospheric temperature in the range of 27°C–32°C causes growth and developmental changes in plants including *Arabidopsis* without resulting into plant death. The growth and developmental changes in plants due to high ambient temperature are termed as thermomorphogenesis (Quint et al., 2016; Casal and Balasubramanian, 2019). Auxin is a central regulatory player in the plant responses to ambient temperature. Inside plant tissues, auxin distribution is regulated by its biosynthesis, conjugation and transport (Zazímalová et al., 2010; Zhao, 2010). The core components of the auxin signaling pathway include the TRANSPORT INHIBITOR RESPONSE 1/AUXIN SIGNALING F-BOX PROTEIN (TIR1/AFB) auxin coreceptors, the Auxin/INDOLE-3-ACETIC ACID (Aux/IAA) transcriptional repressors, and the AUXIN RESPONSE FACTOR (ARF) transcription factors. Presence of auxin promotes the interaction between TIR/AFB and Aux/IAA protein, resulting in the proteasomal degradation of Aux/IAA, relieving the Aux/IAA mediated ARF repression (Salehin et al., 2015). Auxin is produced in the leaves and cotyledons upon exposure to high ambient temperature that acts as a mobile signal in controlling hypocotyl elongation and leaf thermonastic movements (Bellstaedt et al., 2019). The elongation of hypocotyls and upward movement of the leaves provide plants an adaptive feature that maximize the cooling capacity of the plants by moving away from the high soil surface temperature and to get better access for the cool air (Quint et al., 2016; Casal and Balasubramanian, 2019). In response to high ambient temperatures, PHYTOCHROME INTERACTING FACTOR4 (PIF4) promotes auxin biosynthesis in the cotyledon and leaves. PIF4 regulates the transcription of genes involved in auxin biosynthesis, such as *TRYPTOPHAN AMINOTRANSFERASE OF ARABIDOPSIS 1 (TAA1), CYTOCHROME P450, FAMILY 79, SUBFAMILY B, POLYPEPTIDE 2 (CYP79B2), YUCCA8,* and *YUCCA9,* by binding directly to their promoters (Franklin et al., 2011; Sun et al., 2012). Auxin then travels from the cotyledon to hypocotyl where it activates the growth promoting hormone BR biosynthesis and signaling (Ibañez et al., 2018; Bellstaedt et al., 2019). Auxin response factors are also induced in response to warm temperature that activates the genes involved in cell expansion in the hypocotyl epidermis (Reed et al., 2018). Auxin signaling is also enhanced by the stabilization of auxin receptor TIR1 at warm temperature through an HEAT SHOCK PROTEIN 90 (HSP90)-dependent mechanism (Wang et al., 2016b). Besides, auxin biosynthesis and signaling, PIN-FORMED (PIN)-mediated auxin redistribution also plays significant role in leaf thermonastic movements. PIF4 governs the expression of *PINOID* (*PID*), a protein kinase which regulates the phosphorylation-dependent PIN polarization to regulate polar auxin transport (Park et al., 2019). Moreover, auxin interacts with regulators of diverse endogenous (such as phytohormone) and exogenous signals (such as light and temperature) to control temperature mediated responses.

In recent years, significant progress has been made which suggests the crucial involvement of auxin in stress responses. Among other stresses, high temperature above the optimum range affects plant physiology and crop productivity negatively. Recent reports suggest that auxin plays a very significant role in controlling plant responses to heat stress. In rice, a sharp decline in spikelet fertility was found under high temperature stress, which was reversed when plants were subjected to exogenous 1-naphthaleneacetic acid treatment. During high temperature, a sharp increase in auxin levels was also observed in a rice variety along with the increased expression of auxin biosynthesis *YUCCA* genes (Zhang et al., 2018). In *Nicotiana tabacum*, overexpression of a phosphate transporter *OsPT8* enhanced tobacco plant tolerance to heat stress through the modulation of root architecture and increased antioxidant ability. Transcript profiling suggests that overexpressing *OsPT8* in tobacco significantly increased the auxin biosynthesis (*NtYUCCA 6, 8*), transport (*NtPIN 1,2*), as well as signaling (*NtARF1 and NtARF2*) genes under high temperature stress (Song et al., 2019). A proteome and transcriptome analysis of *Lentinula edodes* under heat stress showed upregulation of several HSP proteins as well as key enzymes involved in tryptophan and IAA metabolism. Also, exogenous application of IAA protected plants from heat stress (Wang et al., 2018). NEEDLE1, a mitochondria localized ATP-dependent metalloprotease, regulates heat stress response through the alterations of auxin signaling. *needle1* was found to be a temperature sensitive mutant, which showed strong genetic interaction with the components of the auxin signaling (Liu et al., 2019). A mutation in eukaryotic translation initiation factor showed thermosensitive and pleiotropic growth defects due to the perturbation of translational efficiency of auxin-regulated, ribosome-related, and electron transport genes (Zhang et al., 2017). However, a large number of studies suggest high-throughput data in supporting the role of auxin in plant protection to lethal environmental temperatures, a detailed candidate based functional characterization of genes will be helpful in elucidating the molecular mechanism of auxin-mediated thermotolerance response. All these studies indicate a favorable role of auxin in thermal protection (Fig. 11.1A). In contrast, a few studies suggest that auxin concentration is reduced in response to temperature stress (Table 11.1). As auxin is a major regulator of stamen and pollen development, loss of function of auxin biosynthesis genes showed severe defects in flower development and had either reduced or no pollen in flowers (Cheng et al., 2006). In *Arabidopsis* and Barley, heat stress reduced the endogenous auxin levels in the developing anthers. This reduced level of auxin can be correlated with the reduced expression of *YUCCA2* and *YUCCA6* in the stamens of *Arabidopsis* and Barley (Sakata et al., 2010; Higashitani, 2013; Ozga et al., 2017). In another study, heat stress caused accumulation of reactive oxygen species (ROS) and hydrogen peroxide, which suppressed auxin-

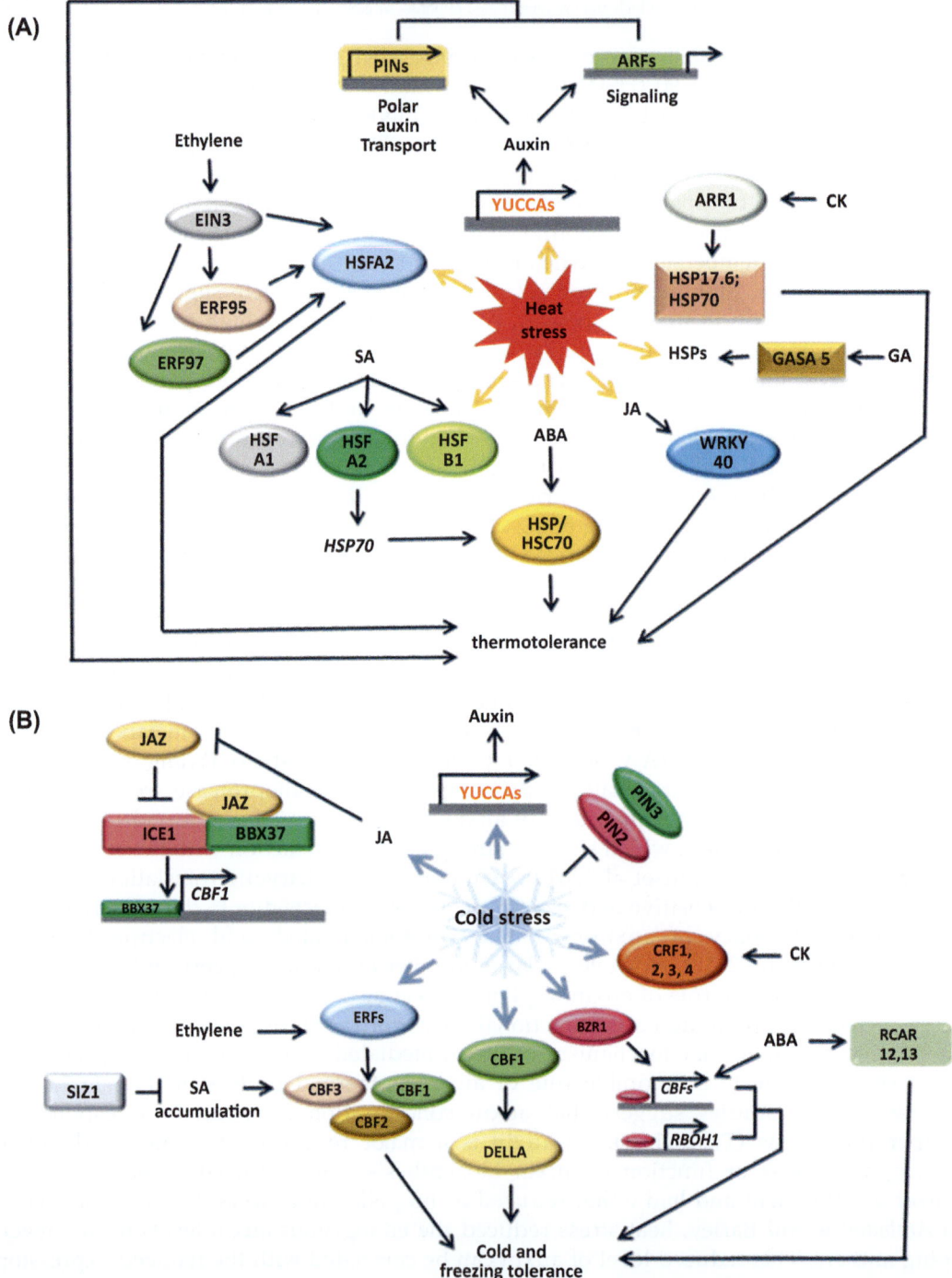

FIGURE 11.1 Role of various phytohormones in regulating (A) thermotolerance and (B) cold and freezing stress response in plants.

TABLE 11.1 Summarising regulation of the different phytohormone levels by heat, cold and freezing stresses in various plant species.

Hormone	Organism	References
Heat stress response		
Auxin		
1. Auxin accumulation.	All tissues of *Arabidopsis*.	Prerostova et al. (2020c)
2. Increase of auxin biosynthesis	Seed and pericarp of *Pisum sativum L*.	Kaur et al. (2021)
3. Increase of IAA levels.	Heat tolerant plant of *L. edodes*.	Wang et al. (2018)
4. Increase of auxin biosynthesis genes	In heat tolerance rice.	Sharma et al. (2021)
5. Change in auxin distribution	*B. napus* during embryogenesis.	Dubas et al. (2014)
6. Affects auxin signaling	Potato.	Trapero-Mozos et al. (2018)
Cytokinin		
1. Increase in CK levels.	*Arabidopsis*.	Dobrá et al. (2015) and Skalák et al. (2016)
	Heat tolerant ryegrass.	Li et al. (2020b)
2. Increase in *cis-zeatin*	HS recovered *Arabidopsis*.	Prerostova et al. (2020b)
3. Maintained CK levels	Heat tolerant high yielding rice.	Wu et al. (2016)
4. Increased CK transportation to aerial parts	Heat tolerance rice.	Wu et al. (2017)
5. Decreased *trans-zeatin* levels	HS recovered *Arabidopsis*.	Prerostova et al. (2020c)
Abscisic acid		
1. Increased ABA levels	In heat treated as well as heat treated primed *Arabidopsis*.	Prerostova et al. (2020c)
	Heliotropium thermophilum.	Sezgin Muslu and Kadıoğlu (2021)
Jasmonic acid		
1. Increased levels of JA	In maize ears of heat tolerant variety.	Wang et al. (2020a)
2. Increased levels of JA, 12-oxophytodienoic acid (OPDA) and a JA-isoleucine (JA-Ile) conjugate.	*Arabidopsis*.	Clarke et al. (2009)
Gibberillic acid		
1. Sustained GA levels	Heat tolerant ryegrass.	Li et al. (2020b)
2. Sustained GA1 levels	High heat tolerant rice.	Wu et al. (2016)
Cold and freezing stress response		
Auxin		
1. Increased IAA levels	Crown tissues of Wheat.	Kosová et al. (2012)
	Cold tolerance winter wheat.	Majláth et al. (2012)
2. Affect auxin-regulated genes.	*Arabidopsis*.	Hannah et al. (2005)
	Rice.	Jain and Khurana (2009)
Cytokinin		
1. Increased levels of *cis-zeatin*	*Lolium perenne L*.	Prerostova et al. (2021b)
2. Increase of *trans-zeatin* levels in primed plants	*Lolium perenne L*.	Prerostova et al. (2021b)
3. Decreased CK levels	*Arabidopsis*.	Prerostova et al. (2021a)
4. Early peak in *cis-zeatin* followed by increase of active *trans-zeatin*	Winter wheat.	Kosová et al. (2012)
Abscisic acid		
1. Increased ABA levels	*Arabidopsis*.	Prerostova et al. (2021a)
	Lolium perenne L.	Prerostova et al. (2021b)
Jasmonic acid		
1. Increased JA levels	Rice.	Du et al. (2013)
Salicylic acid		
1. Increased SA biosynthesis and SA levels	In cold primed winter wheat	Wang et al. (2021)

responsive promoter activation in the *Arabidopsis* mesophyll cells (Dong et al., 2007). These results also suggest that inhibition of cell expansion and cell cycle at high temperature might be due to inhibition of auxin signaling (Beard et al., 2012) (Fig. 11.1A).

11.3 Role of auxin in regulating cold and freezing stress response

A first potential link between the auxin and cold stress came into picture by the study of plant gravistimulation under cold response. Inflorescence gravitropism, which is majorly regulated by auxin, is actually inhibited by cold (Fukaki et al., 1996; Wyatt et al., 2002). Using a combination of elegant experiments, Shibasaki et al. (2009) showed that cold stress primarily affects intracellular auxin transport. Cold stress inhibits the asymmetric redistribution and cycling of PIN2 (Shibasaki et al., 2009). Also, the lateral distribution of PIN3, which is an early marker of root gravity response, was also inhibited by cold stress (Shibasaki et al., 2009). Another report suggests that endosomal trafficking of auxin efflux carriers was blocked by cold stress and this response was mediated by GNOM ARF-GEF (Ashraf and Rahman, 2019). Cold stress was found to be involved in the regulation of biologically active auxin levels (IAA), but this effect was dependent on the plant species and developmental context. A 24-hour exposure of cold temperature induced IAA levels in rice, which was also corroborated with the increased expression of *YUCCA* genes and decreased expression of *GH3*, which are involved in the conjugation, and therefore inactivation of the biologically active auxins (Du et al., 2013). Moreover, in cold-tolerant rice varieties, cold treatment induced the expression of auxin pathway regulated genes than their cold sensitive counterparts (Zhao et al., 2015; Guan et al., 2019). Global transcriptome profiling of *Arabidopsis* in response to cold showed upregulation of auxin pathway genes in an INDUCER OF CBF EXPRESSION 1 (ICE1)-dependent manner (Lee et al., 2005). Lately, it has been suggested that auxin signaling mediates crosstalk with the miRNA machinery in regulating cold stress response in the *Arabidopsis* roots. A genome wide transcriptome profiling of wild type and *solitary root* (*slr1*) mutants showed abundance of 180 known and 71 novel cold-responsive microRNAs. Furthermore, target gene expression shows the plausible involvement of miR169/NF-YA module in regulating Aux/IAA14-mediated cold stress response (Aslam et al., 2020). Furthermore, it has been observed that hydrogen sulfide levels are increased in response to cold, which regulates downstream auxin signaling, leading to chilling tolerance in cucumber (Zhang et al., 2020a). Apart from cold stress tolerance, auxin has been found to be involved in the postharvest and fruit ripening of avocados (Vincent et al., 2020) (Fig. 11.1B).

11.3.1 Role of cytokinin in regulating heat stress response

CKs are one of the well-known phytohormones whose signaling is vital in numerous aspects throughout plant growth and development. CKs are known to act by promoting cell division and showing antisenescent properties. In addition, its signaling is required to regulate various abiotic stress responses (Liu et al., 2020). Along with their role in regulating abiotic stress tolerance, CKs also regulate various biotic stresses (Ma, 2008; Jameson and Song, 2016; Cortleven et al., 2019).

CK plays an important role in conferring heat stress tolerance to plants. Various studies confirm that external application of CK or upregulation of endogenous CK levels reduces crop losses due to heat stress. External application of CK helps in reducing the side effects of high heat (Wu et al., 2016). Furthermore, SY63, a rice variety with a higher transport rate of root-derived CK in aerial parts during heat stress, is associated with increased heat tolerance (Wu et al., 2017). Similarly, exogenous application of CK also resulted in better growth and yield under heat stress (Yang et al., 2016; Liu and Huang, 2018). Use of INCYDE-F further enhances thermotolerance in *Arabidopsis* by inhibiting CK catabolism during heat stress (Prerostova et al., 2020b). In *Arabidopsis*, it was found that more than 70% of heat stress induced HSPs were similarly induced by CK. Furthermore, the increase in CK signaling through *ARR1* overexpression increased plant's thermotolerance via regulating Hsp17.6 and Hsp70 during heat stress (Karunadasa et al., 2022) (Fig. 11.1A).

Increased heat tolerance is also observed in plants in which endogenous CK levels are sustained or increased during heat stress. For instance, overexpression of adenine *ISOPENTENYL TRANSFERASE* (*IPT*) with the *SENESCENCE-ACTIVATED12* promoter-*SAG12:IPT* and *HSP18* promoter-*HSP18:IPT* in *Agrostis stolonifera* showed increased thermotolerance (Xu et al., 2010). Furthermore, the CYTOKININ OXIDASE/ DEHYDROGENASE (CKX) activity was lower in the heat-tolerant rice variety than the heat susceptible one (Wu et al., 2017). The lower level of CKX activity maintains the active CK levels. The role of downstream CK transcription factors, such as TYPE-B RESPONSE REGULATORS OF ARABIDOPSIS (B-type ARR) and CYTOKININ RESPONSE FACTOR (CRF), toward heat tolerance responses remains to be known. In conclusion, all these works suggest a connection between CKs and heat stress response (Fig. 11.1A).

11.3.2 Role of cytokinin in regulating cold and freezing stress response

CK also plays an important role in regulating cold stress tolerance. External supplementation of CK to the media attenuated cold stress-associated damage to *Arabidopsis* (Xia et al., 2009). Furthermore, cold stress tolerance was compromised in *Arabidopsis* plants with DEX::CKX, which overexpress *HvCKX2*, a CK catabolic gene. (Prerostova et al., 2021a). In addition, cold tolerance is observed in sugarcane when *IPT* is overexpressed during cold stress through the cold-inducible promoter of the *Arabidopsis thaliana COLD-REGULATED 15A* (*AtCOR15A*) gene (Belintani et al., 2012). *Arabidopsis* plants overexpressing *B-type ARR1* showed better survival during cold stress. Furthermore, the *B-type arr1* mutant has compromised cold tolerance compared to its wild type (Jeon and Kim, 2013). In addition, CK-regulated *SlCRF1* of *Solanum lycopersicum* L. was strongly upregulated by cold stress (Shi et al., 2014). In addition, CK modulates CRF2 and CRF3 in *Arabidopsis* to confer better survival potential under cold stress, regulating lateral root development (Jeon et al., 2016). Moreover, CRF4 is induced by cold and has been found to be a positive regulator for cold tolerance (Zwack et al., 2016). So, it is clear that CRFs controlled by CK signaling have a role in regulating abiotic stress tolerance (Fig. 11.1B).

11.3.3 Role of brassinosteroids in regulating high ambient temperature and heat stress response

BRs are steroid phytohormones, which were first discovered in the pollens of *Brassica napus* (Mitchell et al., 1970; Planas-Riverola et al., 2019). Apart from the role of BR in plant growth and development, several reports also suggest its contribution in adaptation to environmental stresses. The loss of function *bri1-emssuppressor 1* (*bes1*) mutant exhibited high heat sensitivity with reduced PSII efficiency and increased photoinhibition and photooxidative stress. However, the heat sensitivity phenotype of *bes1* was alleviated in the presence of 24-epibrassinolide (24-EBR) (Setsungnern et al., 2020). Similarly, exogenous application of EBR led to the alleviation of heat stress induced reduction in photosynthesis and enhanced the activity of antioxidant enzymes in tomato (Ogweno et al., 2008; Ahammed et al., 2020). Exogenous foliar application of EBR enhanced the antioxidant and osmoprotectant system in wheat thus resulting in improved growth and photosynthetic efficacy under heat stress (Hussain et al., 2019; Ahammed et al., 2020). BR could ameliorate the adverse effects of high temperature stress in rice, which resulted in enhanced photosynthesis and reduced lipid peroxidation and proline content during heat stress (Quintero-Calderón et al., 2021).

External application of BR or BR biosynthesis inhibitors proved the role of BR in mediating the effect of heat stress on the pistil activity during anthesis in PTSGMS rice lines (Chen et al., 2021). At a molecular level, EBR treatment resulted in enhanced accumulation of heat shock proteins (HSPs) by restricting the loss of components of translational machinery during heat stress in *B. napus* (Dhaubhadel et al., 2002). In contrast, BR-deficient mutants, *de-etiolated 2* (*det2-1*) and *dwarf4* (*dwf4*), were not impaired in heat induced HSP synthesis (Kagale et al., 2007), thus suggesting that HSP synthesis is not indispensable for BR-induced heat stress tolerance (Ahammed et al., 2020) (Fig. 11.1A).

BR has also been shown to have a role in controlling thermomorphogenesis. Under high ambient temperature, BRASSINAZOLE-RESISTANT 1 (BZR1) regulates thermomorphogenesis via directly functioning downstream to PIF4 and auxin to induce elongation or by binding to the *PIF4* promoter to induce its transcription (Ibañez et al., 2018). Elevated temperatures resulted in PIF4 accumulation and caused PIF4-BES1 dimerization to be dominant over BES1 dimerization. Limited availability of BES1 homodimers leads to derepression of transcription of BR biosynthesis genes and feedback inhibition of BR signaling. Meanwhile, abundance of PIF4-BES1 complex activates auxin-related, cell wall modifying and BR biosynthesis genes required for thermomorphogenesis (Martínez et al., 2018; Planas-Riverola et al., 2019).

11.3.4 Role of brassinosteroids in regulating cold and freezing stress response

Several studies have also mentioned the role of BR in providing protection against low temperature stress including chilling and freezing stress. In tomato, BR deficiency caused sensitivity to chilling stress while overexpression of BR biosynthetic *DWARF* (*DWF*) gene or application of EBR diminished the oxidative damages caused by chilling stress. In addition, BR enhanced the expression of *RESPIRATORY BURST OXIDASE HOMOLOG1* (*RBOH1*) and *GLUTAREDOXIN* (*GRX*) genes and increased the ratio of reduced/oxidized

2-cysteine peroxiredoxin (2-Cys Prx) to mediate chilling tolerance in tomato (Xia et al., 2018). Heidari et al. (2021) documented an increase in the antioxidant enzymes and reduction in oxidative damage in cold-tolerant and cold-sensitive varieties of tomato on EBR treatment (Heidari et al., 2021). Plants employ the phenomenon of photoprotection to prevent photoinhibition under cold stress. Under chilling stress, BRs facilitate photoprotection via activating BZR1, which regulates the transcription of *RBOH1* and apoplastic production of H_2O_2, thus contributing to chilling tolerance in tomato (Fang et al., 2019). BR also has a role in modulating the cell wall enzymes known as pectin methylesterases during chilling stress (Qu et al., 2011). Overexpression of BR biosynthesis gene, *DWF4* in *Arabidopsis* led to high expression of cold inducible gene, *COR15A*, and enhanced cold tolerance (Divi and Krishna, 2010). On the contrary, Kim et al. (2010) showed that *bri1* exhibited increased tolerance to cold and *BRI1*-overexpressing *Arabidopsis* plants were sensitive to cold stress. These opposite phenotypes were a result of higher basal expression of stress inducible genes in *bri1−9* mutant (Kim et al., 2010). A recent study denotes that cold stress- and BR-induced BZR1 was bound to the promoters of *C-REPEAT BINDING FACTOR 1* (*CBF1*), *C-REPEAT BINDING FACTOR 3* (*CBF3*), and *RBOH1* and promoted their transcription. *RBOH1*-dependent H_2O_2 synthesis was essential for the accumulation of BZR1 and the expression of *CBFs* via regulating glutathione homeostasis. Thus interplay of BR and redox signaling was responsible for regulating *CBF* transcription during cold stress in tomato (Fang et al., 2021) (Fig. 11.1B).

BR also participates in enhancing freezing tolerance. BR induces freezing tolerance via posttranslational modification of basic helix loop helix transcription factor CESTA, which controls *COR* gene expression in a CBF-dependent and -independent manner (Eremina et al., 2016). Cold-induced accumulation of unphosphorylated BZR1 activated the expression of *CBF1* and *CBF2* along with other stress inducible genes of CBF independent pathway to confer freezing tolerance in *Arabidopsis*. Moreover, mutant of *GLYCOGEN SYNTHASE KINASE 3* (*GSK3*) like kinases, *bin2−3*, *bil1*, and *bil2*, exhibited enhanced freezing tolerance while overexpression of *BRASSINOSTEROID INSENSITIVE2* (*BIN2*) showed hypersensitivity to freezing tolerance (Li et al., 2017). Under prolonged cold stress, BIN2 negatively regulates INDUCER OF CBF EXPRESSION1 (ICE1) through phosphorylation, thereby leading to its degradation via HIGH EXPRESSION OF OSMOTICALLY RESPONSIVE GENE1 (HOS1), thus causing attenuation of *CBF* expression to balance growth and stress responses in *Arabidopsis* (Ye et al., 2019).

11.3.5 Role of gibberellic acid in regulating heat stress response

GAs are natural diterpenoids which are one of the essential plant hormones. Heat stress results in inhibition of seed germination and seedling growth, but upon exogenous application of GA, there is an increase in seed germination and growth in Arabidopsis (Alonso-Ramirez et al., 2009). Also, during heat stress, the external application of GA showed positive effects by increasing the sugar and protein content of wheat (Asthir and Bhatia, 2014). GA_3 along with silicon reduced heat-induced oxidative stress in Date Palm (*Phoenix dactylifera* L.). They together promoted growth and development by reducing the ABA and increasing SA accumulation (Khan et al., 2020). GA regulates *GIBBERELLIC*

ACID-STIMULATED IN ARABIDOPSIS5 (*GASA5*), which then modulates heat response through regulating HSPs (Zhang and Wang, 2011). It was found that *GASA4* of maize is upregulated by heat stress. Furthermore, overexpression of maize *GASA4* in *Arabidopsis* leads to the overall increase in thermotolerance as well as increased germination of heat-stressed seeds (Ko et al., 2007). GA contributes positively in modulating plant growth during heat stress early via SA regulation (Fig. 11.1A).

The role of GA is shown toward the regulation of anticipated phenotypic changes during thermomorphogenesis. GA-mediated DELLAs may negatively regulate thermomorphogenesis by binding with *PIF4*, *ARF6* and *BZR1* (Koini et al., 2009; Choi and Oh, 2016; Li et al., 2018). It was shown that in *ga1−3* mutant, a strong suppression of heat induced flowering gives an idea about the role of GA in early flowering response (Balasubramanian et al., 2006). Furthermore, the SHORT VEGETATIVE PHASE (SVP) represses GA biosynthesis. But upon flowering inductive photoperiods, *SVP* expression reduces, which results in acceleration of GA biosynthesis at shoot apex (Andrés et al., 2014). GA enhances the PIF4 activity under heat stress posttranslationally (Stavang et al., 2009). PIF4 activation by GA, which in turn activates the FLOWERING LOCUS T (FT) and shows a positive role of GA in early flowering (Kumar et al., 2012). ACTIN-RELATED PROTEIN 6 (ARP6) is a chromatin remodeling factor, which regulates the accumulation of histone variant H2A.Z and is thought to regulate the temperature depended expression of *FT* by PIF4 (Deal et al., 2007; SV and PA, 2010). It is found that early flowering phenotype of *arp6* can only be complemented with GA inhibitors to inhibit GA biosynthesis. This resulted in severe attenuation of early flowering in *arp6*. This conveys that GA signaling plays an upstream regulatory role for flowering. Furthermore, exogenous GA application was able to induce flowering even under noninductive short-day photoperiod and low temperatures (15°C) independently of *FT*, *TSF*, and *PIF* genes (Galvão et al., 2015).

11.3.6 Role of gibberellic acid in regulating cold stress response

The GAs are found to be more abundant in cold-treated seeds, which help in germination. Both cold and GA promote the germination of seeds (Fu et al., 2014; Tuttle et al., 2015). A recent study showed that GA_3 maintained pollen viability under cold stress, whereas lower GA levels due to the application of paclobutrazol, an inhibitor of GA biosynthesis, reduced the pollen number as well as their viability in almond (*Amygdalus communis* L.) (Li et al., 2021).

Cold stress resulting in the increased expression of *CBF1/DREB1B* in plants, leading to DELLA accumulation (Achard et al., 2008a). Furthermore, DELLA accumulation resulted in growth retardation but promoted survival ability during cold stress (Achard et al., 2008b). (Fig. 11.1B). Application of GA degraded the DELLA accumulation, thereby promoting plant growth during cold stress (Achard et al., 2008a). A genome-wide study on characterization of GIBBERELLIC ACID INSENSITIVE, REPRESSOR OF GAI, AND SCARECROW (GRAS) family of *Cucumis sativus* L. found that most of the *CsaGRAS* genes were regulated by cold (Lu et al., 2020). This study can open new perspectives of understanding GA-mediated cold tolerance responses by characterizing candidate genes to provide mechanistic insights of cold tolerance.

11.4 Regulation of high and low temperature responses by defense hormones

11.4.1 Role of abscisic acid in regulating heat stress response

ABA role is well established in various aspects of growth and development as well as in numerous stress responses (Nakashima et al., 2014). Reduced levels of ABA signaling in ABA-deficient *aba1* mutant compromises the heat tolerance (Suzuki et al., 2016). ABA-dependent heat stress response was also shown in walnut (*Juglans regia*), through *JrWRKY6* and *JrWRKY53* (Yang et al., 2017). In another report, ABA regulates the transcription of *HSP70* in H_2O_2-regulated manner (Li et al., 2014) (Fig. 11.1A). ABA reduces the desiccation by controlling the stomatal conductance during various stresses. Rice mutants having semirolled leaves with lowered stomatal conductance, lower transpiration, higher H_2O_2, and higher respiration rate, an increase in ABA levels results in hypersensitivity toward the heat stress (Li et al., 2020a). Similarly, the negative role of ABA toward heat stress was studied in citrus plants Cleopatra mandarin and Carrizo citrange, where low levels of ABA facilitated increased transpiration in Carrizo citrange, thus reducing the heat stress effects (Zandalinas et al., 2016). ABA is thought to regulate energy homeostasis during the stress response. Recurrent heat stress increases the plant's thermomemory and also has increased levels of ABA in *Arabidopsis* (Prerostova et al., 2020b). Furthermore, ABA-mediated SNF1-RELATED PROTEIN KINASE 2 (SnRK2) and TARGET OF RAPAMYCIN (TOR) negatively regulate each other in a yin yang fashion. Contradicting the fact that TOR expression is lowered during stress conditions, the latest works of (Sharma et al., 2020) showed that short-term thermomemory is regulated through TOR mediated pathway during heat stress recovery having high energy status.

11.4.2 Role of abscisic acid in regulating cold stress response

A transcriptome analysis showed that out of 245 ABA-induced genes, 25 genes were also inducible by cold proving that ABA can directly regulate cold response (Seki et al., 2002). Together with low-temperature induction, ABA enhanced the expression of *CBF/DREB1* transcription factors providing better cold stress tolerance in grapevine buds (Rubio et al., 2019). Coexpression of stylo *9-CIS-EPOXYCAROTENOID DIOXYGENASE* (*SgNCED1*) and yeast *D-ARABINONO-1,4-LACTONE OXIDASE* (*ALO*) genes in *N. tabacum* L. and *Stylosanthes guianensis*, a forage legume, resulted in improved performance to both drought and cold stress by upregulating ABA and ascorbate levels (Bao et al., 2016). Enhanced ABA perception through overexpression of *PYR1-LIKE10* (*PYL10*) in rice plants showed an increase in cold tolerance of Indica rice variety (Verma et al., 2019). In addition, Zhang et al. (2019) showed that overexpression of ABA receptors, *REGULATORY COMPONENTS OF ABA RECEPTOR 12* (*RCAR12*) or *RCAR13*, in *Arabidopsis* showed positive roles in regulating both heat and cold stress tolerances (Zhang et al., 2019) (Fig. 11.1B). Likewise, overexpression of *OsPYL3* and *OsPYL10* showed enhanced tolerance to both cold and drought tolerance of rice (Tian et al., 2015; Lenka et al., 2018; Verma et al., 2019). This can show light toward vegetation and improving food security in cold deserts. Along with other stresses like high salt and drought stress, ABA-dependent bZIP transcription factor *TabZIP60* of wheat was strongly induced by cold and positively regulates multiple abiotic stress (Zhang et al., 2015).

11.4.3 Role of ethylene in regulating heat stress response

Ethylene is a gaseous hormone involved in many stresses as well as developmental responses. Pretreatment of *Arabidopsis* seedlings with ethylene precursor, aminocyclopropane-1-carboxylic acid (ACC) provided protection against heat stress and reduced the oxidative damage of the membranes. Genetic evidence showed that ethylene insensitive mutant, *ethylene response 1 (etr1)* was sensitive to heat stress and exhibited high oxidative damage (Larkindale and Knight, 2002). Similarly, exogenous application of ACC prior to heat stress in creeping bentgrass (*Agrostis stolonifera* var. palustris) improved heat stress tolerance. Also, pretreatment with ACC led to increased activity of antioxidant enzymes under heat stress (Larkindale and Huang, 2004). In tomato, a semidominant mutation in the ethylene response sensor-like ethylene receptor [*Never ripe (Nr)* mutant] exhibited lower number of viable pollen grains in comparison to wild type when exposed to heat stress. Also, prior treatment with an ethylene releaser known as ethephon during pollen maturation stage led to a significant number of viable pollen grains upon exposure to heat stress while treatment with ethylene biosynthesis inhibitor, aminoethoxyvinylglycine (AVG) led to decrease in the germinating pollen grains under heat stress (Firon et al., 2012; Poór et al., 2021). Treatment with ethephon altered the heat stressed pollen proteome of tomato and led to increase in proteins involved in translation, degradation, tricarboxylic acid cycle, and RNA regulation and antioxidant enzymes (Jegadeesan et al., 2018). In RNA-seq analysis of leaves of reproductive stage soybean grown under heat, 12 *ERFs* were found to be highly upregulated thus suggesting an important role of ethylene signaling in heat stress (Xu et al., 2019). Overexpression of *ETHYLENE RESPONSE FACTOR 1* (*ERF1*) led to increased thermotolerance and upregulation of *AtHSFA3* and *HSPs* in *Arabidopsis* (Cheng et al., 2013). Application of 10 μM ACC in rice seedlings protected them from membrane damage and improved the fresh weight as well as chlorophyll *a/b* ratio under heat stress. Moreover, ACC treatment in the presence of heat enhanced the expression of heat shock factors, such as *HSFA1a* and *HSFA2a, c, d, e,* and *f,* in comparison to rice seedlings treated with heat without ACC (Wu and Yang, 2019). Recently, Huang et al. (2021) have documented the role of ERF95 and ERF97 in positively regulating heat stress tolerance (Fig. 11.1A). Overexpression of *ERF95* resulted in better heat stress tolerance along with reduced membrane damage whereas quadruple mutant (*erfq*), lacking *ERF95, ERF96, ERF97,* and *ERF98* exhibited lesser survival and higher membrane damage (Huang et al., 2021). A root endophyte *Enterobacter* sp. SA187 confers thermotolerance by *HSFA2*-dependent H3K4 trimethylation of chromatin at heat stress memory gene loci, wherein *HSFA2* was regulated by ethylene signaling via ETHYLENE-INSENSITIVE3 (EIN3) (Shekhawat et al., 2021).

11.4.4 Role of ethylene in cold and freezing stress response

Application of continuous ethylene in peach fruit could protect it from chilling injury and membrane damage by increasing phenolic content and lipid metabolism, including elevation of phospholipid and sphingolipid content to confer membrane stability (Chen et al., 2021). On the contrary, Garcia et al. (2020) observed that ethylene-insensitive *etr2b* mutant in zucchini fruit showed reduced accumulation of H_2O_2 and better tolerance to chilling injury (García et al., 2020). Ethylene could also induce the expression of *CBF1* in tomato, which was evident in the correlated peak expression of

Lycopersicon esculentum CBF1 and endogenous ethylene level at 16-hour cold treatment suggesting that ethylene is important for protection against chilling injury in fruits (Zhao et al., 2009).

Overexpression of *CdERF1* from *Cynodon dactylon* (L.) Pers in *Arabidopsis* improved the cold tolerance via activation of antioxidant enzymes and stress-related genes, such as *CBF2, EARLY ARABIDOPSIS ALUMINUM INDUCED 1 (EARLI1), PEROXIDASE 71 (PER71),* and *LIPID TRANSFER PROTEIN (LTP)* (Hu et al., 2020). *ERF057* was highly induced by cold in grapevine and overexpression of *ERF057* enhanced the cold tolerance via increasing the activity of antioxidant enzymes and upregulation of several stress-induced genes including *CBF1, CBF2,* and *CBF3* through binding to the GCC box and DRE motifs (Sun et al., 2016).

Illgen et al. (2020) characterized the four closely related members of the ERF family namely *ERF102, ERF103, ERF104,* and *ERF105* in cold stress (Illgen et al., 2020). *erf105* mutant showed reduced freezing tolerance before and after cold acclimation. The authors also found that high order mutants of *ERF102* and *ERF103* were freezing sensitive and displayed reduced expression of cold induced *CBFs* and *COR* genes (Illgen et al., 2020). Overexpression of tomato *ERF* gene, *TERF2/LeERF2* in tobacco protected them from membrane damage and increased the expression of cold induced genes thereby enhancing the freezing tolerance. Blocking of ethylene biosynthesis or ethylene signaling pathway using inhibitors led to reduction in the freezing tolerance in *TERF2/LeERF2* overexpressing tobacco (Zhang and Huang, 2010) (Fig. 11.1B).

Apart from several reports suggesting positive role of ethylene in mediating cold and freezing tolerance, some studies also direct its role as a negative regulator in cold and freezing tolerance. Through chemical and genetic evidence, Shi et al. (2012) reported that application of ACC as well as *ethylene overproducer1 (eto1)* decreased the freezing tolerance in Arabidopsis while treatment with ethylene biosynthesis or perception inhibitors increased freezing tolerance (Shi et al., 2012). Inhibition of the ethylene signaling pathway in soybean resulted in a significant increase in *GmDREB1A;1* and *GmDREB1A;2* transcripts, while promotion caused reduction in these transcripts. Application of silver nitrate, an ethylene signaling inhibitor led to increased freezing tolerance in acclimated and nonacclimated conditions. Taken together, ethylene signaling via EIN3 negatively regulates CBF/DREB1 mediated freezing tolerance in soybean (Robison et al., 2019). Likewise, another study in legume, *Medicago truncatula* revealed that exogenous application of ethylene biosynthesis inhibitors enhanced freezing tolerance while ethylene treatment alleviated freezing tolerance (Zhao et al., 2014).

11.4.5 Role of jasmonic acid in regulating ambient and heat stress response

Jasmonates are induced when plants are attacked by some herbivorous and pathogens or in response to wounding. Apart from this, jasmonates are also involved in a variety of physiological processes that include root growth inhibition, trichrome development, anthocyanin accumulation, male sterility, senescence, and response to various biotic and abiotic stresses (Wasternack et al., 2013; Goossens et al., 2016; Sharma and Laxmi, 2016; Yang et al., 2019; Jang et al., 2020; Wang et al., 2020b).

JA positively regulates heat stress tolerance. Under natural environment, high temperature is accompanied with high light intensities. A recent report showed that JA is indispensable for a combination of high temperature and high light. Combination of HL and HS caused increased accumulation of JA and JA-Ile, together with the increased expression of over 2000 stress-associated genes many of them belong to JA signaling. Also, a mutant compromised in JA biosynthesis (allene oxide synthase) showed higher sensitivity to combined HL + HS, suggesting crucial role of JA signaling in regulating HL + HS response (Balfagón et al., 2019). In tomato, priming with two plant-derived materials named as eugenol and anise oil showed enhanced thermotolerance which was found to be due to initiation of JA and SA mediated defense response (Tsai et al., 2019).

Heat stress triggered the expression of JA signaling genes, which induced the accumulation of endogenous JA, and the accumulation of agarwood sesquiterpene in *Aquilaria sinensis* while block of JA signaling using nordihydroguaiaretic acid could inhibited the JA pathway and therefore JA accumulation, leading to decreased sesquiterpene compounds (Xu et al., 2016). A mutant of *constitutive expresser of PR1 proteins* (*cpr5−1*), which exhibits constitutive activation of SA, JA, and ethylene signaling, showed increased tolerance to heat stress. When a mutant of JA signaling, *jasmonate resistant 1−1* (*jar1−1*) was crossed with *cpr5−1*, the double mutants showed reduced basal thermotolerance similar to *opda reductase3* (*opr3*) and *coronatine insensitive 1−1* (*coi1−1*) (Clarke et al., 2009). Furthermore, exogenous application of JA regulates the transcription of WRKY transcription factor genes, such as *Capsicum annuum CaWRKY40*, which in turn regulates the plant responses to high temperature. Tobacco plants overexpressing *CaWRKY40* showed derepression of JA biosynthetic gene *N. tabacum LIPOXYGENASE 1* (*NtLOX1*) by heat stress. *CaWRKY40* overexpressing plants showed increased tolerance whereas RNAi plants showed susceptibility to heat stress and therefore WRKY40 facilitates the crosstalk between JA and heat stress signaling (Dang et al., 2013) (Fig. 11.1A).

11.4.6 Role of jasmonic acid in regulating cold and freezing tolerance

In the past decade, results shown by Hu et al. (2013) suggest that JA also governs tolerance to cold stress in *A. thaliana*. JA regulates cold stress response by directly targeting the ICE-CBF pathway. At the molecular level, several JAZ repressors physically interacted with the ICE transcription factors and therefore inhibited its transcriptional activity. Also, *JAZ1/JAZ4* overexpression plants through increased repression of *ICE1* showed reduced tolerance to freezing stress. Altogether, these results indicate that JA signaling works upstream to ICE1-CBF pathway in regulating freezing stress tolerance (Hu et al., 2013).

Besides *Arabidopsis*, studies have shown the regulatory role of JA in cold and freezing stress tolerance in other plant species, such as rice and tomato. Several JA biosynthetic genes, such as *DEFECTIVE IN ANTHER DEHISCENCE1* (*OsDAD1*), *OsLOX2*, *ALLENE OXIDE CYCLASE* (*OsAOC*), *ALLENE OXIDE SYNTHASE* (*OsAOS1*), *OsAOS2*, *12-OXOPHYTODIENOATE REDUCTASE 1* (*OsOPR1*), and *OsOPR7* were found to be upregulated in response to cold stress (Du et al., 2013). In tomato (*S. lycopersicum*), light signals modulate the levels of endogenous phytohormones, such as JA and ABA, in regulating cold stress response. Moreover, this report suggested that the *suppressor of prosystemin-mediated responses 2* (*spr2*) mutant, which shows partial deficiency of JA displayed increased cold sensitivity (Wang et al., 2016a). In another report, MeJA application induced the expression of *SIMYC2*, which confers chilling tolerance possibly through the increase in proline and lycopene content and improving the antioxidant enzymes activities (Min et al., 2018a). Very recently, it has also been observed that MeJA reduced chilling injury in tomato by promoting polyamine biosynthesis. MeJA signaling transcription factor SIMYC2 could bind directly on the promoters of genes involved in polyamines biosynthesis and activate their transcription (Min et al., 2021). In *Poncirus trifoliate*, PtrMYC2 facilitates the merging of JA signal to modulate cold tolerance. PtrMYC2 regulates cold induced glycine betaine accumulation through direct regulation of *PtrBADH-l*, a betaine aldehyde dehydrogenase (BADH)-like gene. BADH interacts with PtrMYC2 and PtrMYC2 also occupied the *cis*-regulatory G-box elements in the promoter region of BADH and activated its transcription, ultimately leading to cold tolerance (Ming et al., 2021).

Besides, cold and freezing tolerance, JA signaling components have been shown to play roles during the cold storage of several tropical or subtropical fruits, such as banana, mango, papaya, and avocado (Sayyari et al., 2011; Zhao et al., 2013; Ba et al., 2016). In case of Banana, the expression of two MYC2 homologs, *MaMYC2a* and *MaMYC2b* were induced rapidly after MeJA treatment during cold storage. Through protein—protein interactions, authors showed that MaMYC2s interacted with cold responsive transcription factor protein MaICE1. Also, the expression of several cold-responsive genes was induced by MeJA application (Zhao et al., 2013). A *LATERAL ORGAN BOUNDARY 5* (*LBD5*) gene also confers chilling tolerance in Banana fruit. The expression of *MaLBD5* was induced in response to cold and MeJA. MaLBD5 interacted with MaJAZ1 repressor, thereby inhibiting the MaLBD5-mediated transactivation of a jasmonate biosynthesis gene, termed *MaAOC2* (Ba et al., 2016). In peach fruits, application of MeJA promotes the adjustments of fatty acids in phospholipids, leading to MeJA-induced alleviation of chilling injury via induction of JA-mediated CBF pathway (Chen et al., 2019a). An Apple B-box protein BBX37 links JA-mediated cold stress tolerance via JAZ-BBX37-ICE1-CBF pathway. BBX37 interacts with MdICE1 to enhance its transcriptional activity on MdCBF1, thus promoting cold tolerance. Two JAZ proteins MdJAZ1 and MdJAZ2 interacted with MdBBX37 and this interaction not only repressed the transcriptional activity on MdCBF1 and MdCBF4, it also interfered with the interaction between MdBBX37 and MdICE1, thus negatively regulates JA-mediated cold tolerance response (An et al., 2021). (Fig. 11.1B). In the rubber tree (*Hevea brasiliensis*), an ICE-like transcription factor HbICE2 is involved in the cold tolerance response through JA signaling. HbJAZ1 and HbJAZ12 physically interact with the HbICE2 through protein—protein interaction and repressed the transcriptional activity of HbICE2 (Chen et al., 2019b).

11.4.7 Role of salicylic acid in heat stress

Several studies have shown the importance of SA in biotic stress; however, many reports also provide evidence for its role in abiotic stresses like high and low temperature stress. Foliar application of 1 mM SA reduced membrane damage as well as H_2O_2 accumulation and increased antioxidant enzymes in cucumber seedlings under heat stress and recovery (Shi et al., 2006). Supplementation of 10 µmol/L SA in the growth medium led to enhanced heat stress tolerance in tobacco along with an increase in the antioxidant enzyme activity (Dat et al., 2000). In tomato, application of 1 mM SA leads to an increase in quantum yield of photosystem II, increased antioxidant enzyme activity, reduced membrane damage, lower lipid peroxidation and H_2O_2 contents, and an increase in proline content under heat stress, thus protecting the plants from the harmful effects of high temperature stress (Shah Jahan et al., 2019). Another study in Tri-genomic *B. napus* L. revealed that prior treatment with SA improved leaf and root biomass and exhibited increase in antioxidant enzyme activity under heat stress (Ghani et al., 2021). SA treatment in wheat resulted in an increase in proline synthesis under heat stress. Also, SA treatment suppressed ethylene biosynthesis under heat stress leading to better proline metabolism, N assimilation, and photosynthesis (Khan et al., 2013). In rice, exogenous treatment of SA led to ROS accumulation in anthers and prohibited the programmed cell death and degradation of tapetum under heat stress. SA treatment also led to significant increase in the pollen viability under heat stress in comparison to non-SA treated or paclobutrazol (SA inhibitor) treated rice plants (Feng et al., 2018). Snyman and Cronjé (2008) deciphered the effects of SA on HSPs and HSFs in tomato. Treatment with SA resulted in activation of DNA binding of HSFs. However, transcription of *HSP70* mRNA was not induced. Combined treatment of SA and heat shock exhibited increased HSF–DNA binding, enhanced transcription of HSP70 mRNA and gene expression of *HSFA1*, *HSFA2*, and *HSFB1*, leading to heightening of Hsp/Hsc70 levels. Thus SA-mediated accumulation of HSP70 is due to regulation of HSFs by SA (Snyman and Cronjé, 2008) (Fig. 11.1A).

11.4.8 Role of salicylic acid in regulating cold and freezing stress response

Exogenous foliar application of acetylsalicylic acid (ASA) in *Phaseolus vulgaris* improved the growth and photosynthetic parameters, increased total soluble sugars and proline accumulation under chilling stress. ASA also induced the expression of cold related genes, *CBF3* and *COR47* under chilling stress (Soliman et al., 2018). In wheat, SA treatment could overcome cold stress induced reduction in photosynthetic rate, biomass, plant height and grain yield by increasing the activities of antioxidant enzymes, osmoprotectant contents, and photochemical efficiency of photosystem II (Wang et al., 2021). SA application on cucumber seedlings led to higher expression and activity of antioxidant enzymes including superoxide dismutase and catalase under chilling stress (Ignatenko et al., 2021). Similarly, in wheat, 100 µM of SA improved the chilling tolerance by increasing the activity of antioxidant enzymes and proline content (Ignatenko et al., 2019). SA treatment led to reduction in cold injury via different metabolic changes, such as increase of disaccharides, synthesis of unsaturated fatty acids, proline, and energy charge (Chen et al., 2020). Aazami and Mahna (2017) reported that increasing the duration of low temperature led to rise in the expression of *CBF4* in grapes following pretreatment with SA (Aazami and Mahna, 2017). Under chilling stress, SA hydroxylase *NahG* transgenic plants showed better growth in comparison to wild type. Growth of SA signaling mutant, *nonexpresser of pr genes 1* (*npr1*), was more than the wild type and but less than SA-deficient genotypes [*NahG* and *enhanced disease susceptibility 5* (*eds5*)] under cold stress. Thus SA leads to growth inhibition during chilling stress via NPR1 dependent and independent pathway (Scott et al., 2004).

Exogenous application of SA in winter wheat caused increase in the ice nucleation activity by affecting the apoplastic proteins thereby enhancing the freezing tolerance (Taşgín et al., 2003). SA could also improve freezing tolerance in spinach and increased the accumulation of proline and ascorbic acid. SA mediated freezing tolerance occurred through NO and H_2O_2 signaling as freezing tolerance was diminished on addition of H_2O_2 or NO scavenger (Shin et al., 2018). Metabolite profiling of combined cold acclimated and SA treated spinach leaves showed higher osmolytes, antioxidants, and SA than cold-acclimated spinach and exhibited better freezing tolerance. SA acclimated treated spinach showed accumulation of trehalose, ascorbic acid, γ-tocopherol, proline, and leucine, and lower mannose and aconitic acid than nonacclimated spinach tissues suggesting of specifically SA-mediated accumulation (Min et al., 2018b). Treatment with 100 μM SA transcriptionally upregulated *WRKY19*, *HEAT SHOCK TRANSCRIPTION FACTOR* (*HSF3*), *MITOCHONDRIAL ALTERNATIVE OXIDASE* (*AOX1a*), and *HEAT SHOCK PROTEIN* (*HSP70*) under freezing stress and conferred increase in antioxidant capacity and higher photochemistry efficiency of photosystem II (Wang et al., 2020c). In addition, cold priming increased the SA biosynthesis and endogenous levels of SA and improved the antioxidant capacity and expression of cold regulated genes thus leading to enhanced freezing tolerance (Wang et al., 2021). Mutant of *SAP AND MIZ1 DOMAIN-CONTAINING LIGASE1* (*SIZ1*) consisting of high endogenous levels of SA showed reduced expression of *CBF3* and its regulon genes, thereby displaying chilling and freezing sensitivity (Fig. 11.1B). Decline in the SA levels in *siz1−2 NahG* conferred upregulation of cold-induced genes and freezing resistance. Similarly, another SA accumulating mutant *accelerated cell death 6* (*acd6*) was freezing sensitive and *acd6 NahG* was resistant to freezing stress (Miura and Ohta, 2010). Hence, SA imparts freezing tolerance by modulating the cold signaling genes (Miura and Ohta, 2010) (Fig. 11.1).

11.4.9 Role of strigolactones in regulating heat stress

SLs are a class of compounds derived from terpenoids and were discovered as (+)-strigol, which has been found to stimulate the seed germination from the parasitic plant striga (Cook et al., 1966; Bürger and Chory, 2020). SLs are important to have a role in several agricultural important processes, such as shoot branching, symbiosis of the arbuscular mycorrhizal, and germination of parasitic plants (Bürger and Chory, 2020). SLs are synthesized from β-carotene by the actions of β-carotene isomerase DWARF 27 (D27). Further steps creates 9-cis-aldehyde by the enzyme CAROTENOID CLEAVAGE DIOXYGENASE 7 (CCD7) and synthesis of carlactone by the enzyme CAROTENOID CLEAVAGE DIOXYGENASE 8 (CCD8) from 9-*cis*-aldehyde (Alder et al., 2012). Studies in *Arabidopsis*, rice and petunia suggest that SLs are transported locally as well as travel to long-distance transport. Through a series of grafting experiments using the *Arabidopsis* mutants of *MORE AXILLARY GROWTH* (*MAX*), *DWARF 27* (*D27*), and *LATERAL BRANCHING OXIDOREDUCTASE* (*LBO*) suggested that many SL intermediates may be mobile signals from roots to shoots (Mashiguchi et al., 2021).

Increasing experimental evidence suggests that SLs and their downstream signaling components play crucial roles in the coping to abiotic stresses. In a recent report, a rac-

GR24 (a synthetic SL analog) application showed enhanced tolerance of lupine seedlings toward heat stress. Priming of seeds with GR-24 showed highest germination rate, increased amount of proline and reduced lipid peroxidation. Also, GR-24 enhanced the activities of the enzymes involved in antioxidant and glyoxalase systems in lupine seedlings (Omoarelojie et al., 2020). In another report, SL (GR-24) promoted the growth elongation of the crown roots in tall fescue under normal as well as heat stress conditions and also alleviated heat-inhibition of root growth. This GR-24 governed root growth elongation was accompanied with increased cell numbers, induced expression of cell cycle-associated genes, and reduced expression of auxin transport-related genes in the crown roots (Hu et al., 2018). Application of GR24 not only promotes root elongation in tall fescue but also promotes the leaf elongation and alleviates the heat inhibition of leaf growth through enhanced expression of cell cycle and decreased expression of auxin transport related genes (Hu et al., 2019).

11.4.10 Role of strigolactones in regulating cold stress

SLs not only regulate plant response to high temperature stress but also affect low temperature stress responses. In Rape seedlings, low temperature inhibits the growth of the root and exogenous SL (GR24) application alleviates this low temperature response. At the same time, GR24 application could also promote the cell viability, improved soluble protein, proline content, antioxidant activity and inhibits the accumulation of ROS (Zhang et al., 2020b). In pea, mutants defective in SL biosynthesis and signaling (*ramosus3* (*rms3*), *rms4*, and *rms5*) normally show greater shoot branching with higher leaf chlorophylls and carotenoids. Exposure of these mutants to dark chilling caused significantly decreased shoot fresh weights. Moreover, chilling induced inhibition of photosynthetic carbon assimilation was more in *Arabidopsis more axillary growth3−9* (*max3−9*), *max4−1*, and *max2−1*, which were compromised in SL biosynthesis or signaling (Cooper et al., 2018). In another report, cold stress significantly induced the expression of SL signaling pathway genes. All these studies indicate a pivotal role of SLs in mediated cold stress response in plants (Prerostova et al., 2021a).

11.5 Conclusions and perspectives

Extreme temperature conditions resulting due to climate change pose a threat for agriculture as it affects the yield and productivity of the crops. Plants are sessile beings and have developed complex mechanisms to adapt to the fluctuating environmental conditions. Exposure to extreme temperatures activates numerous cellular and molecular mechanisms to achieve adequate physiological responses. Phytohormones are key regulators of development as well as stress responses in plants. As discussed, plants employ various phytohormone signaling, redox mechanisms, and stress signaling to ensure survival in critical environmental conditions. In the past decades, large numbers of studies have been done elucidating the involvement of different phytohormones in stress response, but the interaction of different phytohormone signaling machineries with each other during heat, cold, and freezing is complex and less explored. In addition, phytohormones do not exist in uniformity but form a differential spatio-temporal gradient within the plant tissue. For example, auxin creates a spatio-temporal gradient, which is required for proper organ formation. Other than auxin, there are other hormones, which play important role in plant growth and organ formation. During stress responses, plants cease their growth and promote stress survival. However, when plants feel stress removal in its vicinity, they restart to grow. Recent studies showed how levels of different phytohormones are altered in response to heat, cold, and freezing stress, but their spatio-temporal alteration during poststress period to regulate organogenesis still remains an unexplored area of research. The levels of phytohormones and the expression and activity of core signaling factors of different phytohormone signaling pathway show diurnal variation. The temporal interplay of these diverse phytohormones in response to an anticipatory or sudden stress at certain time of the day can be a study of interest to raise chronobiologically smart crops. Also, under natural environment, plants are not only challenged by single exposure of stress. Rather, stresses are combined, multiple, chronic and irregular in nature. Exposure of plants to mild stress leads to a heightened response to future stress which is described by a phenomenon known as stress memory. Like phytohormones, sugars are also vital molecules for organism's growth. Recent reports suggest that sugars are not only important for plant growth but are also important signaling molecules for stress responses. Recently, Sharma et al. (2019, 2021) showed that sugars, such as glucose and its signaling machinery, are an important regulator of thermomemory in plants. However, the role of phytohormones in governing stress memory in plants still remains elusive. Thus future work in these directions will open a gateway for the role of phytohormones, their crosstalk in agricultural use for better survival, and stress tolerance to plants.

Acknowledgments

The research in AL laboratory is supported by a Core Grant from the National Institute of Plant Genome Research. MS duly acknowledges research fellowship from National Institute of Plant Genome Research, Govt of India. HBS and HKB duly acknowledge research fellowships from Department of Biotechnology. The authors acknowledge DBT-eLibrary Consortium (DeLCON) for providing access to e-resources.

Conflict of interest statement

The authors declare that the research was conducted in the absence of any commercial or financial relationships that could be construed as a potential conflict of interest.

Author contributions

MS, HBS, HKB, and AL conceptualized the article. MS, HBS, and HKB wrote the original draft. MS, HBS, and HKB prepared the figures. MS and HKB prepared the table. AL supervised and complemented the writing. All authors reviewed and finalized the article.

References

Aazami, M.A., Mahna, N., 2017. Salicylic acid affects the expression of VvCBF4 gene in grapes subjected to low temperature. J. Genet. Eng. Biotechnol. 15, 257–261. Available from: https://doi.org/10.1016/j.jgeb.2017.01.005.

Achard, P., Gong, F., Cheminant, S., Alioua, M., Hedden, P., Genschika, P., 2008a. The cold-inducible CBF1 factor-dependent signaling pathway modulates the accumulation of the growth-repressing DELLA proteins via its effect on gibberellin metabolism. Plant Cell 20, 2117–2129. Available from: https://doi.org/10.1105/tpc.108.058941.

Achard, P., Renou, J.P., Berthomé, R., Harberd, N.P., Genschik, P., 2008b. Plant DELLAs restrain growth and promote survival of adversity by reducing the levels of reactive oxygen species. Curr. Biol. 18, 656–660. Available from: https://doi.org/10.1016/j.cub.2008.04.034.

Ahammed, G.J., Li, X., Liu, A., Chen, S., 2020. Brassinosteroids in plant tolerance to abiotic stress. J. Plant Growth Regul. 39, 1451–1464. Available from: https://doi.org/10.1007/s00344-020-10098-0.

Alder, A., Jamil, M., Marzorati, M., Bruno, M., Vermathen, M., Bigler, P., et al., 2012. The path from β-carotene to carlactone, a strigolactone-like plant hormone. Science (80-) 335, 1348–1351. Available from: https://doi.org/10.1126/science.1218094.

An, J.P., Wang, X.F., Zhang, X.W., You, C.X., Hao, Y.J., 2021. Apple B-box protein BBX37 jasmonic acid mediated cold tolerance through the JAZ-BBX37-ICE1-CBF pathway and undergoes MIEL1-mediated ubiquitination and degradation. New Phytol. Available from: https://doi.org/10.1111/nph.17050.

Andrés, F., Porri, A., Torti, S., Mateos, J., Romera-Branchat, M., García-Martínez, J.L., et al., 2014. SHORT VEGETATIVE PHASE reduces gibberellin biosynthesis at the Arabidopsis shoot apex to regulate the floral transition. Proc. Natl. Acad. Sci. U. S. A. 111, E2760–E2769. Available from: https://doi.org/10.1073/PNAS.1409567111.

Ashraf, M.A., Rahman, A., 2019. Cold stress response in Arabidopsis thaliana is mediated by GNOM ARF-GEF. Plant J. Available from: https://doi.org/10.1111/tpj.14137.

Aslam, M., Sugita, K., Qin, Y., Rahman, A., 2020. Aux/iaa14 regulates microrna-mediated cold stress response in arabidopsis roots. Int. J. Mol. Sci. Available from: https://doi.org/10.3390/ijms21228441.

Asthir, B., Bhatia, S., 2014. In vivo studies on artificial induction of thermotolerance to detached panicles of wheat (Triticum aestivum L) cultivars under heat stress. J. Food Sci. Technol. 51, 118–123. Available from: https://doi.org/10.1007/s13197-011-0458-1.

Ba, L.J., Kuang, J.F., Chen, J.Y., Lu, W.J., 2016. MaJAZ1 attenuates the MaLBD5-mediated transcriptional activation of jasmonate biosynthesis gene MaAOC2 in regulating cold tolerance of banana fruit. J. Agric. Food Chem. Available from: https://doi.org/10.1021/acs.jafc.5b05005.

Balasubramanian, S., Sureshkumar, S., Lempe, J., Weigel, D., 2006. Potent induction of Arabidopsis thaliana flowering by elevated growth temperature. PLoS Genet. 2, e106. Available from: https://doi.org/10.1371/JOURNAL.PGEN.0020106.

Balfagón, D., Sengupta, S., Gómez-Cadenas, A., Fritschi, F.B., Azad, R.K., Mittler, R., et al., 2019. Jasmonic acid is required for plant acclimation to a combination of high light and heat stress. Plant Physiol. Available from: https://doi.org/10.1104/pp.19.00956.

Bao, G., Zhuo, C., Qian, C., Xiao, T., Guo, Z., Lu, S., 2016. Co-expression of NCED and ALO improves vitamin C level and tolerance to drought and chilling in transgenic tobacco and stylo plants. Plant Biotechnol. J. 14, 206–214. Available from: https://doi.org/10.1111/pbi.12374.

Beard, R.A., Anderson, D.J., Bufford, J.L., Tallman, G., 2012. Heat reduces nitric oxide production required for auxin-mediated gene expression and fate determination in tree tobacco guard cell protoplasts. Plant Physiol. Available from: https://doi.org/10.1104/pp.112.200089.

Belintani, N.G., Guerzoni, J.T.S., Moreira, R.M.P., Vieira, L.G.E., 2012. Improving low-temperature tolerance in sugarcane by expressing the ipt gene under a cold inducible promoter.

Belkhadir, Y., Jaillais, Y., 2015. The molecular circuitry of brassinosteroid signaling. New Phytol. Available from: https://doi.org/10.1111/nph.13269.

Bellstaedt, J., Trenner, J., Lippmann, R., Poeschl, Y., Zhang, X., Friml, J., et al., 2019. A mobile auxin signal connects temperature sensing in cotyledons with growth responses in hypocotyls. Plant Physiol. Available from: https://doi.org/10.1104/pp.18.01377.

Bürger, M., Chory, J., 2020. The many models of strigolactone signaling. Trends Plant Sci. Available from: https://doi.org/10.1016/j.tplants.2019.12.009.

Casal, J.J., Balasubramanian, S., 2019. Thermomorphogenesis. Annu. Rev. Plant Biol. Available from: https://doi.org/10.1146/annurev-arplant-050718-095919.

Chen, M., Guo, H., Chen, S., Li, T., Li, M., Rashid, A., et al., 2019a. Methyl jasmonate promotes phospholipid remodeling and jasmonic acid signaling to alleviate chilling injury in peach fruit. J. Agric. Food Chem. Available from: https://doi.org/10.1021/acs.jafc.9b03853.

Chen, W.J., Wang, X., Yan, S., Huang, X., Yuan, H.M., 2019b. The ICE-like transcription factor HbICE2 is involved in jasmonate-regulated cold tolerance in the rubber tree (Hevea brasiliensis. Plant Cell Rep. Available from: https://doi.org/10.1007/s00299-019-02398-x.

Chen, L., Zhao, X., Wu, J., He, Y., Yang, H., 2020. Metabolic analysis of salicylic acid-induced chilling tolerance of banana using NMR. Food Res. Int. 128. Available from: https://doi.org/10.1016/j.foodres.2019.108796.

Chen, S., Chen, M., Li, Y., Huang, X., Niu, D., Rashid, A., et al., 2021. Adjustments of both phospholipids and sphingolipids contribute to cold tolerance in stony hard peach fruit by continuous ethylene. Postharvest Biol. Technol. 171, 111332. Available from: https://doi.org/10.1016/j.postharvbio.2020.111332.

Cheng, Y., Dai, X., Zhao, Y., 2006. Auxin biosynthesis by the YUCCA flavin monooxygenases controls the formation of floral organs and vascular tissues in Arabidopsis. Genes Dev. Available from: https://doi.org/10.1101/gad.1415106.

Cheng, M.C., Liao, P.M., Kuo, W.W., Lin, T.P., 2013. The arabidopsis ETHYLENE RESPONSE FACTOR1 regulates abiotic stress-responsive gene expression by binding to different cis-acting elements in response to different stress signals. Plant Physiol. 162, 1566–1582. Available from: https://doi.org/10.1104/pp.113.221911.

Choi, H., Oh, E., 2016. PIF4 integrates multiple environmental and hormonal signals for plant growth regulation in Arabidopsis. Mol. Cells 39, 587. Available from: https://doi.org/10.14348/MOLCELLS.2016.0126.

Clarke, S.M., Cristescu, S.M., Miersch, O., Harren, F.J.M., Wasternack, C., Mur, L.A.J., 2009. Jasmonates act with salicylic acid to confer basal thermotolerance in Arabidopsis thaliana. New Phytol. Available from: https://doi.org/10.1111/j.1469-8137.2008.02735.x.

Colebrook, E.H., Thomas, S.G., Phillips, A.L., Hedden, P., 2014. The role of gibberellin signalling in plant responses to abiotic stress. J. Exp. Biol. Available from: https://doi.org/10.1242/jeb.089938.

Cook, C.E., Whichard, L.P., Turner, B., Wall, M.E., Egley, G.H., 1966. Germination of witchweed (striga lutea lour.): isolation and properties of a potent stimulant. Science (80-). Available from: https://doi.org/10.1126/science.154.3753.1189.

Cooper, J.W., Hu, Y., Beyyoudh, L., Yildiz Dasgan, H., Kunert, K., Beveridge, C.A., et al., 2018. Strigolactones positively regulate chilling tolerance in pea and in Arabidopsis. Plant Cell Environ. Available from: https://doi.org/10.1111/pce.13147.

Cortleven, A., Leuendorf, J.E., Frank, M., Pezzetta, D., Bolt, S., Schmülling, T., 2019. Cytokinin action in response to abiotic and biotic stresses in plants. Plant Cell Environ. 42, 998–1018. Available from: https://doi.org/10.1111/pce.13494.

Dang, F.F., Wang, Y.N., Yu, L., Eulgem, T., Lai, Y., Liu, Z.Q., et al., 2013. CaWRKY40, a WRKY protein of pepper, plays an important role in the regulation of tolerance to heat stress and resistance to Ralstonia solanacearum infection. Plant Cell Environ. Available from: https://doi.org/10.1111/pce.12011.

Dat, J.F., Lopez-Delgado, H., Foyer, C.H., Scott, I.M., 2000. Effects of salicylic acid on oxidative stress and thermotolerance in tobacco. J. Plant Physiol. 156, 659–665. Available from: https://doi.org/10.1016/S0176-1617(00)80228-X.

Deal, R.B., Topp, C.N., McKinney, E.C., Meagher, R.B., 2007. Repression of flowering in Arabidopsis requires activation of FLOWERING LOCUS C expression by the histone variant H2A.Z. Plant Cell 19, 74. Available from: https://doi.org/10.1105/TPC.106.048447.

Dhaubhadel, S., Browning, K.S., Gallie, D.R., Krishna, P., 2002. Brassinosteroid functions to protect the translational machinery and heat-shock protein synthesis following thermal stress. Plant J. 29, 681–691. Available from: https://doi.org/10.1046/j.1365-313X.2002.01257.x.

Divi, U.K., Krishna, P., 2010. Overexpression of the brassinosteroid biosynthetic gene AtDWF4 in Arabidopsis seeds overcomes abscisic acid-induced inhibition of germination and increases cold tolerance in transgenic seedlings. J. Plant Growth Regul. 29, 385–393. Available from: https://doi.org/10.1007/s00344-010-9150-3.

Dobrá, J., Černý, M., Štorchová, H., Dobrev, P., Skalák, J., Jedelský, P.L., et al., 2015. The impact of heat stress targeting on the hormonal andtranscriptomic response in Arabidopsis. Plant Sci. 231, 52–61. Available from: https://doi.org/10.1016/j.plantsci.2014.11.005.

Dong, M.A., Bufford, J.L., Oono, Y., Church, K., Dau, M.Q., Michels, K., et al., 2007. Heat suppresses activation of an auxin-responsive promoter in cultured guard cell protoplasts of tree tobacco. Plant Physiol. Available from: https://doi.org/10.1104/pp.107.104646.

Du, H., Liu, H., Xiong, L., 2013. Endogenous auxin and jasmonic acid levels are differentially modulated by abiotic stresses in rice. Front. Plant Sci. Available from: https://doi.org/10.3389/fpls.2013.00397.

Dubas, E., Moravčíková, J., Libantová, J., Matušíková, I., Benková, E., Zur, I., et al., 2014. The influence of heat stress on auxin distribution in transgenic B. napus microspores and microspore-derived embryos. Protoplasma. Available from: https://doi.org/10.1007/s00709-014-0616-1.

Dubois, M., Van den Broeck, L., Inzé, D., 2018. The pivotal role of ethylene in plant growth. Trends Plant Sci. Available from: https://doi.org/10.1016/j.tplants.2018.01.003.

Eremina, M., Unterholzner, S.J., Rathnayake, A.I., Castellanos, M., Khan, M., Kugler, K.G., et al., 2016. Brassinosteroids participate in the control of basal and acquired freezing tolerance of plants. Proc. Natl. Acad. Sci. U. S. A. 113, E5982−E5991. Available from: https://doi.org/10.1073/pnas.1611477113.

Fang, P., Yan, M., Chi, C., Wang, M., Zhou, Y., Zhou, J., et al., 2019. Brassinosteroids act as a positive regulator of photoprotection in response to chilling stress. Plant Physiol. 180, 2061−2076. Available from: https://doi.org/10.1104/pp.19.00088.

Fang, P., Wang, Y., Wang, M., Wang, F., Chi, C., Zhou, Y., et al., 2021. Crosstalk between brassinosteroid and redox signaling contributes to the activation of CBF expression during cold responses in tomato. Antioxidants 10. Available from: https://doi.org/10.3390/antiox10040509.

Feng, B., Zhang, C., Chen, T., Zhang, X., Tao, L., Fu, G., 2018. Salicylic acid reverses pollen abortion of rice caused by heat stress. BMC Plant Biol. 18, 1−16. Available from: https://doi.org/10.1186/s12870-018-1472-5.

Firon, N., Pressman, E., Meir, S., Khoury, R., Altahan, L., 2012. Ethylene is involved in maintaining tomato (Solanum lycopersicum) pollen quality under heat-stress conditions. AoB Plants 2012, 1−9. Available from: https://doi.org/10.1093/aobpla/pls024.

Franklin, K.A., Lee, S.H., Patel, D., Kumar, S.V., Spartz, A.K., Gu, C., et al., 2011. Phytochrome-Interacting Factor 4 (PIF4) regulates auxin biosynthesis at high temperature. Proc. Natl. Acad. Sci. U. S. A. 108, 20231−20235. Available from: https://doi.org/10.1073/pnas.1110682108.

Fu, X., Liu, H., Xu, J., Tang, J., Shang, X., 2014. Primary metabolite mobilization and hormonal regulation during seed dormancy release in Cornus japonica var. chinensis. Scand. J. For. Res. 29, 542−551. Available from: https://doi.org/10.1080/02827581.2014.922608.

Fukaki, H., Fujisawa, H., Tasaka, M., 1996. Gravitropic response of inflorescence stems in arabidopsis thaliana. Plant Physiol. Available from: https://doi.org/10.1104/pp.110.3.933.

Galvão, V.C., Collani, S., Horrer, D., Schmid, M., 2015. Gibberellic acid signaling is required for ambient temperature-mediated induction of flowering in Arabidopsis thaliana. Plant J. 84, 949−962. Available from: https://doi.org/10.1111/tpj.13051.

García, A., Aguado, E., Cebrián, G., Iglesias, J., Romero, J., Martínez, C., et al., 2020. Effect of ethylene-insensitive mutation etr2b on postharvest chilling injury in zucchini fruit. Agriculture 10, 1−12. Available from: https://doi.org/10.3390/agriculture10110532.

Ghani, M.A., Abbas, M.M., Ali, B., Ziaf, K., Azam, M., Anjum, R., et al., 2021. Role of salicylic acid in heat stress tolerance in tri-genomic brassica napus l. Bioagro 33, 13−20. Available from: https://doi.org/10.51372/BIOAGRO331.2.

Goossens, J., Fernández-Calvo, P., Schweizer, F., Goossens, A., 2016. Jasmonates: signal transduction components and their roles in environmental stress responses. Plant Mol. Biol. Available from: https://doi.org/10.1007/s11103-016-0480-9.

Guan, S., Xu, Q., Ma, D., Zhang, W., Xu, Z., Zhao, M., et al., 2019. Transcriptomics profiling in response to cold stress in cultivated rice and weedy rice. Gene. Available from: https://doi.org/10.1016/j.gene.2018.10.066.

Hannah, M.A., Heyer, A.G., Hincha, D.K., 2005. A global survey of gene regulation during cold acclimation in Arabidopsis thaliana. PLoS Genet. Available from: https://doi.org/10.1371/journal.pgen.0010026.

Heidari, P., Entazari, M., Ebrahimi, A., Ahmadizadeh, M., Vannozzi, A., Palumbo, F., et al., 2021. Exogenous ebr ameliorates endogenous hormone contents in tomato species under low-temperature stress. Horticulturae 7. Available from: https://doi.org/10.3390/horticulturae7040084.

Higashitani, A., 2013. High temperature injury and auxin biosynthesis in microsporogenesis. Front. Plant Sci. Available from: https://doi.org/10.3389/fpls.2013.00047.

Hu, Q., Zhang, S., Huang, B., 2018. Strigolactones and interaction with auxin regulating root elongation in tall fescue under different temperature regimes. Plant Sci. Available from: https://doi.org/10.1016/j.plantsci.2018.03.008.

Hu, Y., Jiang, L., Wang, F., Yu, D., 2013. Jasmonate regulates the INDUCER OF CBF expression-C-repeat binding factor/dre binding factor1 Cascade and freezing tolerance in Arabidopsis. Plant Cell. Available from: https://doi.org/10.1105/tpc.113.112631.

Hu, Q., Zhang, S., Huang, B., 2019. Strigolactones promote leaf elongation in tall fescue through upregulation of cell cycle genes and downregulation of auxin transport genes in tall fescue under different temperature regimes. Int. J. Mol. Sci. Available from: https://doi.org/10.3390/ijms20081836.

Hu, Z., Huang, X., Amombo, E., Liu, A., Fan, J., Bi, A., et al., 2020. The ethylene responsive factor CdERF1 from bermudagrass (Cynodon dactylon) positively regulates cold tolerance. Plant Sci. 294, 110432. Available from: https://doi.org/10.1016/j.plantsci.2020.110432.

Huang, J., Zhao, X., Bürger, M., Wang, Y., Chory, J., 2021. Two interacting ethylene response factors regulate heat stress response. Plant Cell 33, 338−357. Available from: https://doi.org/10.1093/plcell/koaa026.

Hussain, M., Khan, T.A., Yusuf, M., Fariduddin, Q., 2019. Silicon-mediated role of 24-epibrassinolide in wheat under high-temperature stress. Environ. Sci. Pollut. Res. 26, 17163−17172. Available from: https://doi.org/10.1007/s11356-019-04938-0.

Ibañez, C., Delker, C., Martinez, C., Bürstenbinder, K., Janitza, P., Lippmann, R., et al., 2018. Brassinosteroids dominate hormonal regulation of plant thermomorphogenesis via BZR1. Curr. Biol. 28, 303−310.e3. Available from: https://doi.org/10.1016/j.cub.2017.11.077.

Ignatenko, A., Talanova, V., Repkina, N., Titov, A., 2019. Exogenous salicylic acid treatment induces cold tolerance in wheat through promotion of antioxidant enzyme activity and proline accumulation. Acta Physiol. Plant 41, 1−10. Available from: https://doi.org/10.1007/s11738-019-2872-3.

Ignatenko, A.A., Talanova, V.V., Repkina, N.S., Titov, A.F., 2021. Effect of salicylic acid on antioxidant enzymes and cold tolerance of cucumber plants. Russ. J. Plant Physiol. 68, 491−498. Available from: https://doi.org/10.1134/S1021443721020059.

Illgen, S., Zintl, S., Zuther, E., Hincha, D.K., Schmülling, T., 2020. Characterisation of the ERF102 to ERF105 genes of Arabidopsis thaliana and their role in the response to cold stress. Plant Mol. Biol. 103, 303−320. Available from: https://doi.org/10.1007/s11103-020-00993-1.

Jain, M., Khurana, J.P., 2009. Transcript profiling reveals diverse roles of auxin-responsive genes during reproductive development and abiotic stress in rice. FEBS J. Available from: https://doi.org/10.1111/j.1742-4658.2009.07033.x.

Jameson, P.E., Song, J., 2016. Cytokinin: a key driver of seed yield. J. Exp. Bot. 67, 593−606. Available from: https://doi.org/10.1093/JXB/ERV461.

Jang, G., Yoon, Y., Choi, Y.D., 2020. Crosstalk with jasmonic acid integrates multiple responses in plant development. Int. J. Mol. Sci. Available from: https://doi.org/10.3390/ijms21010305.

Jegadeesan, S., Chaturvedi, P., Ghatak, A., Pressman, E., Meir, S., Faigenboim, A., et al., 2018. Proteomics of heat-stress and ethylene-mediated thermotolerance mechanisms in tomato pollen grains. Front. Plant Sci. 871, 1−20. Available from: https://doi.org/10.3389/fpls.2018.01558.

Jeon, J., Kim, J., 2013. Arabidopsis response regulator1 and Arabidopsis histidine phosphotransfer protein2 (AHP2), AHP3, and AHP5 function in cold signaling. Plant Physiol. 161, 408−424. Available from: https://doi.org/10.1104/pp.112.207621.

Jeon, J., Cho, C., Lee, M.R., Van Binh, N., Kim, J., 2016. CYTOKININ RESPONSE FACTOR2 (CRF2) and CRF3 regulate lateral root development in response to cold stress in arabidopsis. Plant Cell 28, 1828−1843. Available from: https://doi.org/10.1105/tpc.15.00909.

Kagale, S., Divi, U.K., Krochko, J.E., Keller, W.A., Krishna, P., 2007. Brassinosteroid confers tolerance in Arabidopsis thaliana and Brassica napus to a range of abiotic stresses. Planta 225, 353−364. Available from: https://doi.org/10.1007/s00425-006-0361-6.

Karunadasa, S., Kurepa, J., Smalle, J.A., 2022. Gain-of-function of the cytokinin response activator ARR1 increases heat shock tolerance in Arabidopsis thaliana. Plant Signal. Behav. 31 17 (1), 2073108. Available from: https://doi.org/10.1080/15592324.2022.2073108.

Kaur, H., Ozga, J.A., Reinecke, D.M., 2021. Balancing of hormonal biosynthesis and catabolism pathways, a strategy to ameliorate the negative effects of heat stress on reproductive growth. Plant Cell Environ. Available from: https://doi.org/10.1111/pce.13820.

Kazan, K., 2013. Auxin and the integration of environmental signals into plant root development. Ann. Bot. Available from: https://doi.org/10.1093/aob/mct229.

Kazan, K., 2015. Diverse roles of jasmonates and ethylene in abiotic stress tolerance. Trends Plant Sci. Available from: https://doi.org/10.1016/j.tplants.2015.02.001.

Khan, A., Bilal, S., Khan, A.L., Imran, M., Shahzad, R., Al-Harrasi, A., et al., 2020. Silicon and gibberellins: synergistic function in harnessing aba signaling and heat stress tolerance in date palm (Phoenix dactylifera L.). Plants 9. Available from: https://doi.org/10.3390/plants9050620.

Kieber, J.J., Schaller, G.E., 2018. Cytokinin signaling in plant development. Development. Available from: https://doi.org/10.1242/dev.149344.

Kim, S.Y., Kim, B.H., Lim, C.J., Lim, C.O., Nam, K.H., 2010. Constitutive activation of stress-inducible genes in a brassinosteroid-insensitive 1 (bri1) mutant results in higher tolerance to cold. Physiol. Plant. 138, 191–204. Available from: https://doi.org/10.1111/j.1399-3054.2009.01304.x.

Ko, C.-B., Young Min, W., Lee, D.J., Lee, M.-C., Kim, C., 2007. Enhanced tolerance to heat stress in transgenic plants expressing the GASA4 gene. Plant Physiol. Biochem. 45, 722–728. Available from: https://doi.org/10.1016/j.plaphy.2007.07.010.

Koini, M.A., Alvey, L., Allen, T., Tilley, C.A., Harberd, N.P., Whitelam, G.C., et al., 2009. High temperature-mediated adaptations in plant architecture require the bHLH transcription factor PIF4. Curr. Biol. 19, 408–413. Available from: https://doi.org/10.1016/j.cub.2009.01.046.

Kosová, K., Prášil, I.T., Vítámvás, P., Dobrev, P., Motyka, V., Floková, K., et al., 2012. Complex phytohormone responses during the cold acclimation of two wheat cultivars differing in cold tolerance, winter Samanta and spring Sandra. J. Plant Physiol. Available from: https://doi.org/10.1016/j.jplph.2011.12.013.

Kothari, A., Lachowiec, J., 2021. Roles of brassinosteroids in mitigating heat stress damage in cereal crops. Int. J. Mol. Sci. Available from: https://doi.org/10.3390/ijms22052706.

Kumar, S.V., Lucyshyn, D., Jaeger, K.E., Alós, E., Alvey, E., Harberd, N.P., et al., 2012. Transcription factor PIF4 controls the thermosensory activation of flowering. Nature. Available from: https://doi.org/10.1038/nature10928.

Lämke, J., Brzezinka, K., Altmann, S., Bäurle, I., 2016. A hit-and-run heat shock factor governs sustained histone methylation and transcriptional stress memory. EMBO J. 35, 162–175. Available from: https://doi.org/10.15252/embj.

Larkindale, J., Knight, M.R., 2002. Protection against heat stress-induced oxidative damage in Arabidopsis involves calcium, abscisic acid, ethylene, and salicylic acid. Plant Physiol. 128, 682–695. Available from: https://doi.org/10.1104/pp.010320.

Larkindale, J., Huang, B., 2004. Thermotolerance and antioxidant systems in Agrostis stolonifera: involvement of salicylic acid, abscisic acid, calcium, hydrogen peroxide, and ethylene. J. Plant Physiol. 161, 405–413. Available from: https://doi.org/10.1078/0176-1617-01239.

Lee, B.H., Henderson, D.A., Zhu, J.K., 2005. The Arabidopsis cold-responsive transcriptome and its regulation by ICE1. Plant Cell. Available from: https://doi.org/10.1105/tpc.105.035568.

Lenka, S.K., Muthusamy, S.K., Chinnusamy, V., Bansal, K.C., 2018. Ectopic expression of rice PYL3 enhances cold and drought tolerance in Arabidopsis thaliana. Mol. Biotechnol. 60, 350–361. Available from: https://doi.org/10.1007/s12033-018-0076-5.

Li, H., Liu, S.S., Yi, C.Y., Wang, F., Zhou, J., Xia, X.J., et al., 2014. Hydrogen peroxide mediates abscisic acid-induced HSP70 accumulation and heat tolerance in grafted cucumber plants. Plant Cell Environ. 37, 2768–2780. Available from: https://doi.org/10.1111/pce.12360.

Li, H., Ye, K., Shi, Y., Cheng, J., Zhang, X., Yang, S., 2017. BZR1 positively regulates freezing tolerance via CBF-dependent and CBF-independent pathways in Arabidopsis. Mol. Plant 10, 545–559. Available from: https://doi.org/10.1016/j.molp.2017.01.004.

Li, B., Gao, K., Ren, H., Tang, W., 2018. Molecular mechanisms governing plant responses to high temperatures. J. Integr. Plant Biol. 60, 757–779. Available from: https://doi.org/10.1111/JIPB.12701.

Li, G., Zhang, C., Zhang, G., Fu, W., Feng, B., Chen, T., et al., 2020a. Abscisic acid negatively modulates heat tolerance in rolled leaf rice by increasing leaf temperature and regulating energy homeostasis. Rice 13. Available from: https://doi.org/10.1186/s12284-020-00379-3.

Li, M., Jannasch, A.H., Jiang, Y., 2020b. Growth and hormone alterations in response to heat stress in perennial ryegrass accessions differing in heat tolerance. J. Plant Growth Regul. 39, 1022–1029. Available from: https://doi.org/10.1007/s00344-019-10043-w.

Li, P., Tian, J., Guo, C., Luo, S., Li, J., 2021. Interaction of gibberellin and other hormones in almond anthers: phenotypic and physiological changes and transcriptomic reprogramming. Hortic. Res. 8, 94. Available from: https://doi.org/10.1038/s41438-021-00527-w.

Liu, X., Huang, B., 2018. Cytokinin effects on creeping bentgrass response to heat stress. Crop Sci. 42, 466. Available from: https://doi.org/10.2135/cropsci2002.4660.

Liu, Q., Galli, M., Liu, X., Federici, S., Buck, A., Cody, J., et al., 2019. NEEDLE1 encodes a mitochondria localized ATP-dependent metalloprotease required for thermotolerant maize growth. Proc. Natl. Acad. Sci. U. S. A. Available from: https://doi.org/10.1073/pnas.1907071116.

Liu, Y., Zhang, M., Meng, Z., Wang, B., Chen, M., 2020. Research progress on the roles of cytokinin in plant response to stress. Int. J. Mol. Sci. 21, 1–18. Available from: https://doi.org/10.3390/ijms21186574.

Lu, X., Liu, W., Xiang, C., Li, X., Wang, Q., Wang, T., et al., 2020. Genome-wide characterization of gras family and their potential roles in cold tolerance of cucumber (Cucumis sativus L.). Int. J. Mol. Sci. 21, 1–14. Available from: https://doi.org/10.3390/ijms21113857.

Ma, Q.-H., 2008. Genetic engineering of cytokinins and their application to agriculture. Crit. Rev. Biotechnol. 28, 213–232. Available from: https://doi.org/10.1080/07388550802262205.

Majláth, I., Szalai, G., Soós, V., Sebestyén, E., Balázs, E., Vanková, R., et al., 2012. Effect of light on the gene expression and hormonal status of winter and spring wheat plants during cold hardening. Physiol. Plant. Available from: https://doi.org/10.1111/j.1399-3054.2012.01579.x.

Martínez, C., Espinosa-Ruíz, A., Lucas, M., Bernardo-García, S., Franco-Zorrilla, J.M., Prat, S., 2018. PIF 4-induced BR synthesis is critical to diurnal and thermomorphogenic growth. EMBO J. 37, 1–15. Available from: https://doi.org/10.15252/embj.201899552.

Mashiguchi, K., Seto, Y., Yamaguchi, S., 2021. Strigolactone biosynthesis, transport and perception. Plant J. Available from: https://doi.org/10.1111/tpj.15059.

Min, D., Li, F., Zhang, X., Cui, X., Shu, P., Dong, L., et al., 2018a. SlMYC2 involved in methyl jasmonate-induced tomato fruit chilling tolerance. J. Agric. Food Chem. Available from: https://doi.org/10.1021/acs.jafc.8b00299.

Min, K., Showman, L., Perera, A., Arora, R., 2018b. Salicylic acid-induced freezing tolerance in spinach (Spinacia oleracea L.) leaves explored through metabolite profiling. Environ. Exp. Bot. 156, 214–227. Available from: https://doi.org/10.1016/j.envexpbot.2018.09.011.

Min, D., Zhou, J., Li, J., Ai, W., Li, Z., Zhang, X., et al., 2021. SlMYC2 targeted regulation of polyamines biosynthesis contributes to methyl jasmonate-induced chilling tolerance in tomato fruit. Postharvest Biol. Technol. Available from: https://doi.org/10.1016/j.postharvbio.2020.111443.

Ming, R., Zhang, Y., Wang, Y., Khan, M., Dahro, B., Liu, J.H., 2021. The JA-responsive MYC2-BADH-like transcriptional regulatory module in Poncirus trifoliata contributes to cold tolerance by modulation of glycine betaine biosynthesis. New Phytol. Available from: https://doi.org/10.1111/nph.17063.

Mishra, B.S., Singh, M., Aggrawal, P., Laxmi, A., 2009. Glucose and auxin signaling interaction in controlling arabidopsis thaliana seedlings root growth and development. PLoS One 4. Available from: https://doi.org/10.1371/journal.pone.0004502.

Mitchell, J.W., Mandava, N., Worley, J.F., Plimmer, J.R., Smith, M.V., 1970. Brassins-a new family of plant hormones from rape pollen. Nature 225, 1065–1066. Available from: https://doi.org/10.1038/2251065a0.

Miura, K., Ohta, M., 2010. SIZ1, a small ubiquitin-related modifier ligase, controls cold signaling through regulation of salicylic acid accumulation. J. Plant Physiol. 167, 555–560. Available from: https://doi.org/10.1016/j.jplph.2009.11.003.

Nakashima, K., Yamaguchi-Shinozaki, K., Shinozaki, K., 2014. The transcriptional regulatory network in the drought response and its crosstalk in abiotic stress responses including drought, cold, and heat. Front. Plant Sci. 5, 170. Available from: https://doi.org/10.3389/fpls.2014.00170.

Ogweno, J.O., Song, X.S., Shi, K., Hu, W.H., Mao, W.H., Zhou, Y.H., et al., 2008. Brassinosteroids alleviate heat-induced inhibition of photosynthesis by increasing carboxylation efficiency and enhancing antioxidant systems in Lycopersicon esculentum. J. Plant Growth Regul. 27, 49–57. Available from: https://doi.org/10.1007/s00344-007-9030-7.

Ohama, N., Sato, H., Shinozaki, K., Yamaguchi-Shinozaki, K., 2017. Transcriptional regulatory network of plant heat stress response. Trends Plant Sci. Available from: https://doi.org/10.1016/j.tplants.2016.08.015.

Omoarelojie, L.O., Kulkarni, M.G., Finnie, J.F., Pospíšil, T., Strnad, M., Van Staden, J., 2020. Synthetic strigolactone (rac-GR24) alleviates the adverse effects of heat stress on seed germination and photosystem II function in lupine seedlings. Plant Physiol. Biochem. Available from: https://doi.org/10.1016/j.plaphy.2020.07.043.

Ozga, J.A., Kaur, H., Savada, R.P., Reinecke, D.M., 2017. Hormonal regulation of reproductive growth under normal and heat-stress conditions in legume and other model crop species. J. Exp. Bot. Available from: https://doi.org/10.1093/jxb/erw464.

Park, Y.J., Lee, H.J., Gil, K.E., Kim, J.Y., Lee, J.H., Lee, H., et al., 2019. Developmental programming of thermonastic leaf movement. Plant Physiol. 180, 1185–1197. Available from: https://doi.org/10.1104/pp.19.00139.

Petrášek, J., Friml, J., 2009. Auxin transport routes in plant development. Development. Available from: https://doi.org/10.1242/dev.030353.

Planas-Riverola, A., Gupta, A., Betegoń-Putze, I., Bosch, N., Ibanęs, M., Cano-Delgado, A.I., 2019. Brassinosteroid signaling in plant development and adaptation to stress. Development. Available from: https://doi.org/10.1242/dev.151894.

Poór, P., Nawaz, K., Gupta, R., Ashfaque, F., Khan, M.I.R., 2021. Ethylene involvement in the regulation of heat stress tolerance in plants. Plant Cell Rep. Available from: https://doi.org/10.1007/s00299-021-02675-8.

Prerostova, S., Dobrev, P.I., Kramna, B., Gaudinova, A., Knirsch, V., Spichal, L., et al., 2020a. Heat acclimation and inhibition of cytokinin degradation positively affect heat stress tolerance of Arabidopsis. Front. Plant Sci. Available from: https://doi.org/10.3389/fpls.2020.00087.

Prerostova, S., Dobrev, P.I., Kramna, B., Gaudinova, A., Knirsch, V., Spichal, L., et al., 2020b. Heat acclimation and inhibition of cytokinin degradation positively affect heat stress tolerance of Arabidopsis. Front. Plant Sci. 11, 87. Available from: https://doi.org/10.3389/fpls.2020.00087.

Prerostova, S., Dobrev, P.I., Kramna, B., Gaudinova, A., Knirsch, V., Spichal, L., et al., 2020c. Heat acclimation and inhibition of cytokinin degradation positively affect heat stress tolerance of Arabidopsis. Front. Plant Sci. 11, 1–14. Available from: https://doi.org/10.3389/fpls.2020.00087.

Prerostova, S., Dobrev, P.I., Knirsch, V., Jarosova, J., Gaudinova, A., Zupkova, B., et al., 2021a. Light quality and intensity modulate cold acclimation in arabidopsis. Int. J. Mol. Sci. Available from: https://doi.org/10.3390/ijms22052736.

Prerostova, S., Zupkova, B., Petrik, I., Simura, J., Nasinec, I., Kopecky, D., et al., 2021b. Hormonal responses associated with acclimation to freezing stress in Lolium perenne. Environ. Exp. Bot. 182, 104295. Available from: https://doi.org/10.1016/j.envexpbot.2020.104295.

Qu, T., Liu, R., Wang, W., An, L., Chen, T., Liu, G., et al., 2011. Brassinosteroids regulate pectin methylesterase activity and AtPME41 expression in Arabidopsis under chilling stress. Cryobiology 63, 111–117. Available from: https://doi.org/10.1016/j.cryobiol.2011.07.003.

Quint, M., Delker, C., Franklin, K.A., Wigge, P.A., Halliday, K.J., Van Zanten, M., 2016. Molecular and genetic control of plant thermomorphogenesis. Nat. Plants 2. Available from: https://doi.org/10.1038/nplants.2015.190.

Quintero-Calderón, E.H., Sánchez-Reinoso, A.D., Chá'Vez-Arias, C.C., Garces-Varon, G., Restrepo-Díaz, H., 2021. Rice seedlings showed a higher heat tolerance through the foliar application of biostimulants. Not. Bot. Horti Agrobot. Cluj Napoca 49, 1–20. Available from: https://doi.org/10.15835/nbha49112120.

Reed, J.W., Wu, M.F., Reeves, P.H., Hodgens, C., Yadav, V., Hayes, S., et al., 2018. Three auxin response factors promote hypocotyl elongation1,2[open]. Plant Physiol. Available from: https://doi.org/10.1104/PP.18.00718.

Robison, J.D., Yamasaki, Y., Randall, S.K., 2019. The ethylene signaling pathway negatively impacts CBF/DREB-regulated cold response in soybean (Glycine max. Front. Plant Sci. 10, 1–18. Available from: https://doi.org/10.3389/fpls.2019.00121.

Rubio, S., Noriega, X., Pérez, F.J., 2019. Abscisic acid (ABA) and low temperatures synergistically increase the expression of CBF/DREB1 transcription factors and cold-hardiness in grapevine dormant buds. Ann. Bot. 123, 681–689. Available from: https://doi.org/10.1093/aob/mcy201.

Sakata, T., Oshino, T., Miura, S., Tomabechi, M., Tsunaga, Y., Higashitani, N., et al., 2010. Auxins reverse plant male sterility caused by high temperatures. Proc. Natl. Acad. Sci. U. S. A. Available from: https://doi.org/10.1073/pnas.1000869107.

Salehin, M., Bagchi, R., Estelle, M., 2015. ScfTIR1/AFB-based auxin perception: mechanism and role in plant growth and development. Plant Cell. Available from: https://doi.org/10.1105/tpc.114.133744.

Sayyari, M., Babalar, M., Kalantari, S., Martínez-Romero, D., Guillén, F., Serrano, M., et al., 2011. Vapour treatments with methyl salicylate or methyl jasmonate alleviated chilling injury and enhanced antioxidant potential during postharvest storage of pomegranates. Food Chem. Available from: https://doi.org/10.1016/j.foodchem.2010.07.036.

Scott, I.M., Clarke, S.M., Wood, J.E., Mur, L.A.J., 2004. Salicylate accumulation inhibits growth at chilling temperature in Arabidopsis. Plant Physiol. 135, 1040–1049. Available from: https://doi.org/10.1104/pp.104.041293.

Seki, M., Ishida, J., Narusaka, M., Fujita, M., Nanjo, T., Umezawa, T., et al., 2002. Monitoring the expression pattern of around 7,000 Arabidopsis genes under ABA treatments using a full-length cDNA microarray. Funct. Integr. Genomics 2, 282–291. Available from: https://doi.org/10.1007/s10142-002-0070-6.

Setsungnern, A., Muñoz, P., Pérez-Llorca, M., Müller, M., Thiravetyan, P., Munné-Bosch, S., 2020. A defect in BRI1-EMS-SUPPRESSOR 1 (bes1)-mediated brassinosteroid signaling increases photoinhibition and photo-oxidative stress during heat stress in Arabidopsis. Plant Sci. 296. Available from: https://doi.org/10.1016/j.plantsci.2020.110470.

Sezgin Muslu, A., Kadıoğlu, A., 2021. Role of abscisic acid, osmolytes and heat shock factors in high temperature thermotolerance of Heliotropium thermophilum. Physiol. Mol. Biol. Plants 27, 861–871. Available from: https://doi.org/10.1007/s12298-021-00975-7.

Shah Jahan, M., Wang, Y., Shu, S., Zhong, M., Chen, Z., Wu, J., et al., 2019. Exogenous salicylic acid increases the heat tolerance in Tomato (Solanum lycopersicum L) by enhancing photosynthesis efficiency and improving antioxidant defense system through scavenging of reactive oxygen species. Sci. Hortic. (Amsterdam) 247, 421–429. Available from: https://doi.org/10.1016/j.scienta.2018.12.047.

Sharma, M., Laxmi, A., 2016. Jasmonates: emerging players in controlling temperature stress tolerance. Front. Plant Sci. Available from: https://doi.org/10.3389/fpls.2015.01129.

Sharma, M., Banday, Z.Z., Shukla, B.N., Laxmi, A., 2019. Glucose-regulated hlp1 acts as a key molecule in governing thermomemory. Plant Physiol. 180. Available from: https://doi.org/10.1104/pp.18.01371.

Sharma, M., Jamsheer, M.K., Narayan Shukla, B., Sharma, M., Awasthi, P., Kumar Mahtha, S., et al., 2020. TOR coordinates with transcriptional and chromatin machinery to regulate 1 thermotolerance and thermomemory 2. bioRxiv.

Sharma, E., Borah, P., Kaur, A., Bhatnagar, A., Mohapatra, T., Kapoor, S., et al., 2021. A comprehensive transcriptome analysis of contrasting rice cultivars highlights the role of auxin and ABA responsive genes in heat stress response. Genomics . Available from: https://doi.org/10.1016/j.ygeno.2021.03.007.

Shekhawat, K., Saad, M.M., Sheikh, A., Mariappan, K., Al-Mahmoudi, H., Abdulhakim, F., et al., 2021. Root endophyte induced plant thermotolerance by constitutive chromatin modification at heat stress memory gene loci. EMBO Rep. 22, 1–15. Available from: https://doi.org/10.15252/embr.202051049.

Shi, Q., Bao, Z., Zhu, Z., Ying, Q., Qian, Q., 2006. Effects of different treatments of salicylic acid on heat tolerance, chlorophyll fluorescence, and antioxidant enzyme activity in seedlings of Cucumis sativa L. Plant Growth Regul. 48, 127–135. Available from: https://doi.org/10.1007/s10725-005-5482-6.

Shi, Y., Tian, S., Hou, L., Huang, X., Zhang, X., Guo, H., et al., 2012. Ethylene signaling negatively regulates freezing tolerance by repressing expression of CBF and type-A ARR genes in Arabidopsis. Plant Cell 24, 2578–2595. Available from: https://doi.org/10.1105/tpc.112.098640.

Shi, X., Gupta, S., Rashotte, A.M., 2014. Characterization of two tomato AP2/ERF genes, SlCRF1 and SlCRF2 in hormone and stress responses. Plant Cell Rep. 33, 35–45. Available from: https://doi.org/10.1007/s00299-013-1510-6.

Shibasaki, K., Uemura, M., Tsurumi, S., Rahman, A., 2009. Auxin response in arabidopsis under cold stress: underlying molecular mechanisms. Plant Cell . Available from: https://doi.org/10.1105/tpc.109.069906.

Shin, H., Min, K., Arora, R., 2018. Exogenous salicylic acid improves freezing tolerance of spinach (Spinacia oleracea L.) leaves. Cryobiology 81, 192–200. Available from: https://doi.org/10.1016/j.cryobiol.2017.10.006.

Skalák, J., Cerný, M., Jedelský, P., Dobrá, J., Ge, E., Novák, J., et al., 2016. Stimulation of ipt overexpression as a tool to elucidate the role of cytokinins in high temperature responses of Arabidopsis thaliana. J. Exp. Bot. 67, 2861–2873. Available from: https://doi.org/10.1093/jxb/erw129.

Smith, S.M., 2014. Q&A: what are strigolactones and why are they important to plants and soil microbes? BMC Biol. Available from: https://doi.org/10.1186/1741-7007-12-19.

Snyman, M., Cronjé, M.J., 2008. Modulation of heat shock factors accompanies salicylic acid-mediated potentiation of Hsp70 in tomato seedlings. J. Exp. Bot. 59, 2125–2132. Available from: https://doi.org/10.1093/jxb/ern075.

Soliman, M.H., Alayafi, A.A.M., El Kelish, A.A., Abu-Elsaoud, A.M., 2018. Acetylsalicylic acid enhance tolerance of Phaseolus vulgaris L. to chilling stress, improving photosynthesis, antioxidants and expression of cold stress responsive genes. Bot. Stud. 59, 1–17. Available from: https://doi.org/10.1186/s40529-018-0222-1.

Song, Z., Fan, N., Jiao, G., Liu, M., Wang, X., Jia, H., 2019. Overexpression of OsPT8 increases auxin content and enhances tolerance to high-temperature stress in nicotiana tabacum. Genes (Basel). Available from: https://doi.org/10.3390/genes10100809.

Stavang, J.A., Gallego-Bartolomé, J., Gómez, M.D., Yoshida, S., Asami, T., Olsen, J.E., et al., 2009. Hormonal regulation of temperature-induced growth in Arabidopsis. Plant J. Available from: https://doi.org/10.1111/j.1365-313X.2009.03983.x.

Sun, J., Qi, L., Li, Y., Chu, J., Li, C., 2012. Pif4-mediated activation of yucca8 expression integrates temperature into the auxin pathway in regulating arabidopsis hypocotyl growth. PLoS Genet. 8. Available from: https://doi.org/10.1371/journal.pgen.1002594.

Sun, X., Zhao, T., Gan, S., Ren, X., Fang, L., Karungo, S.K., et al., 2016. Ethylene positively regulates cold tolerance in grapevine by modulating the expression of ETHYLENE RESPONSE FACTOR 057. Sci. Rep. 6, 1–14. Available from: https://doi.org/10.1038/srep24066.

Suzuki, N., Bassil, E., Hamilton, J.S., Inupakutika, M.A., Zandalinas, S.I., Tripathy, D., et al., 2016. ABA is required for plant acclimation to a combination of salt and heat stress. PLoS One 11. Available from: https://doi.org/10.1371/journal.pone.0147625.

SV, K., PA, W., 2010. H2A.Z-containing nucleosomes mediate the thermosensory response in Arabidopsis. Cell 140, 136–147. Available from: https://doi.org/10.1016/J.CELL.2009.11.006.

Taşgín, E., Atící, Ö., Nalbantoğlu, B., 2003. Effects of salicylic acid and cold on freezing tolerance in winter wheat leaves. Plant Growth Regul. 41, 231–236. Available from: https://doi.org/10.1023/B:GROW.0000007504.41476.c2.

Tian, X., Wang, Z., Li, X., Lv, T., Liu, H., Wang, L., et al., 2015. Characterization and functional analysis of pyrabactin resistance-like abscisic acid receptor family in rice. Rice 8, 1–13. Available from: https://doi.org/10.1186/s12284-015-0061-6.

Trapero-Mozos, A., Ducreux, L.J.M., Bita, C.E., Morris, W., Wiese, C., Morris, J.A., et al., 2018. A reversible light- and genotype-dependent acquired thermotolerance response protects the potato plant from damage due to excessive temperature. Planta . Available from: https://doi.org/10.1007/s00425-018-2874-1.

Tsai, W.A., Weng, S.H., Chen, M.C., Lin, J.S., Tsai, W.S., 2019. Priming of plant resistance to heat stress and tomato yellow leaf curl Thailand virus with plant-derived materials. Front. Plant Sci. Available from: https://doi.org/10.3389/fpls.2019.00906.

Tuttle, K.M., Martinez, S.A., Schramm, E.C., Takebayashi, Y., Seo, M., Steber, C.M., 2015. Grain dormancy loss is associated with changes in ABA and GA sensitivity and hormone accumulation in bread wheat, Triticum aestivum (L.). Seed Sci. Res. 25, 179–193. Available from: https://doi.org/10.1017/S0960258515000057.

Verma, V., Ravindran, P., Kumar, P.P., 2016. Plant hormone-mediated regulation of stress responses. BMC Plant Biol. Available from: https://doi.org/10.1186/s12870-016-0771-y.

Verma, R.K., Santosh Kumar, V.V., Yadav, S.K., Pushkar, S., Rao, M.V., Chinnusamy, V., 2019. Overexpression of ABA receptor PYL10 gene confers drought and cold tolerance to indica rice. Front. Plant Sci. 10. Available from: https://doi.org/10.3389/fpls.2019.01488.

Vincent, C., Mesa, T., Munné-Bosch, S., 2020. Hormonal interplay in the regulation of fruit ripening and cold acclimation in avocados. J. Plant Physiol. Available from: https://doi.org/10.1016/j.jplph.2020.153225.

Vishwakarma, K., Upadhyay, N., Kumar, N., Yadav, G., Singh, J., Mishra, R.K., et al., 2017. Abscisic acid signaling and abiotic stress tolerance in plants: a review on current knowledge and future prospects. Front. Plant Sci. Available from: https://doi.org/10.3389/fpls.2017.00161.

Wang, F., Guo, Z., Li, H., Wang, M., Onac, E., Zhou, J., et al., 2016a. Phytochrome a and b function antagonistically to regulate cold tolerance via abscisic acid-dependent jasmonate signaling1. Plant Physiol. Available from: https://doi.org/10.1104/pp.15.01171.

Wang, R., Zhang, Y., Kieffer, M., Yu, H., Kepinski, S., Estelle, M., 2016b. HSP90 regulates temperature-dependent seedling growth in Arabidopsis by stabilizing the auxin co-receptor F-box protein TIR1. Nat. Commun. Available from: https://doi.org/10.1038/ncomms10269.

Wang, G.Z., Ma, C.J., Luo, Y., Zhou, S.S., Zhou, Y., Ma, X.L., et al., 2018. Proteome and transcriptome reveal involvement of heat shock proteins and indoleacetic acid metabolism process in lentinula edodes thermotolerance. Cell. Physiol. Biochem. Available from: https://doi.org/10.1159/000494784.

Wang, H.Q., Liu, P., Zhang, J.W., Zhao, B., Ren, B.Z., 2020a. Endogenous hormones inhibit differentiation of young ears in maize (Zea mays L.) under heat stress. Front. Plant Sci. Available from: https://doi.org/10.3389/fpls.2020.533046.

Wang, J., Song, L., Gong, X., Xu, J., Li, M., 2020b. Functions of jasmonic acid in plant regulation and response to abiotic stress. Int. J. Mol. Sci. Available from: https://doi.org/10.3390/ijms21041446.

Wang, W., Wang, X., Zhang, J., Huang, M., Cai, J., Zhou, Q., et al., 2020c. Salicylic acid and cold priming induce late-spring freezing tolerance by maintaining cellular redox homeostasis and protecting photosynthetic apparatus in wheat. Plant Growth Regul. 90, 109−121. Available from: https://doi.org/10.1007/s10725-019-00553-8.

Wang, W., Wang, X., Huang, M., Cai, J., Zhou, Q., Dai, T., et al., 2021. Alleviation of field low-temperature stress in winter wheat by exogenous application of salicylic acid. J. Plant Growth Regul. 40, 811−823. Available from: https://doi.org/10.1007/s00344-020-10144-x.

Wasternack, C., Forner, S., Strnad, M., Hause, B., 2013. Jasmonates in flower and seed development. Biochimie . Available from: https://doi.org/10.1016/j.biochi.2012.06.005.

Wu, Y.S., Yang, C.Y., 2019. Ethylene-mediated signaling confers thermotolerance and regulates transcript levels of heat shock factors in rice seedlings under heat stress. Bot. Stud. 60. Available from: https://doi.org/10.1186/s40529-019-0272-z.

Wu, C., Cui, K., Wang, W., Li, Q., Fahad, S., Hu, Q., et al., 2016. Heat-induced phytohormone changes are associated with disrupted early reproductive development and reduced yield in rice. Sci. Rep. 6. Available from: https://doi.org/10.1038/srep34978.

Wu, C., Cui, K., Wang, W., Li, Q., Fahad, S., Hu, Q., et al., 2017. Heat-induced cytokinin transportation and degradation are associated with reduced panicle cytokinin expression and fewer spikelets per panicle in rice. Front. Plant Sci. 8. Available from: https://doi.org/10.3389/fpls.2017.00371.

Wyatt, S.E., Rashotte, A.M., Shipp, M.J., Robertson, D., Muday, G.K., 2002. Mutations in the gravity persistence signal loci in Arabidopsis disrupt the perception and/or signal transduction of gravitropic stimuli. Plant Physiol. Available from: https://doi.org/10.1104/pp.102.010579.

Xia, J., Zhao, H., Liu, W., Li, L., He, Y., 2009. Role of cytokinin and salicylic acid in plant growth at low temperatures. Plant Growth Regul. 57, 211−221. Available from: https://doi.org/10.1007/S10725-008-9338-8.

Xia, X.J., Fang, P.P., Guo, X., Qian, X.J., Zhou, J., Shi, K., et al., 2018. Brassinosteroid-mediated apoplastic H$_2$O$_2$-glutaredoxin 12/14 cascade regulates antioxidant capacity in response to chilling in tomato. Plant Cell Environ. 41, 1052−1064. Available from: https://doi.org/10.1111/pce.13052.

Xu, Y.H., Liao, Y.C., Zhang, Z., Liu, J., Sun, P.W., Gao, Z.H., et al., 2016. Jasmonic acid is a crucial signal transducer in heat shock induced sesquiterpene formation in Aquilaria sinensis. Sci. Rep. Available from: https://doi.org/10.1038/srep21843.

Xu, C., Xia, Z., Huang, Z., Xia, C., Huang, J., 2019. Understanding the physiological and transcriptional mechanism of reproductive stage soybean in response to heat stress. Crop Breed. Genet. Genomics. Available from: https://doi.org/10.20900/cbgg20200004.

Yang, Y.-X., Ahammed, G., Wu, C., Fan, S., Zhou, Y.-H., 2015. Crosstalk among jasmonate, salicylate and ethylene signaling pathways in plant disease and immune responses. Curr. Protein Pept. Sci. Available from: https://doi.org/10.2174/1389203716666150330141638.

Yang, D., Li, Y., Shi, Y., Cui, Z., Luo, Y., Zheng, M., et al., 2016. Exogenous cytokinins increase grain yield of winter wheat cultivars by improving stay-green characteristics under heat stress. PLoS One 11. Available from: https://doi.org/10.1371/journal.pone.0155437.

Yang, G.Y., Zhang, W.H., Sun, Y.D., Zhang, T.T., Hu, D., Zhai, M.Z., 2017. Two novel WRKY genes from Juglans regia, JrWRKY6 and JrWRKY53, are involved in abscisic acid-dependent stress responses. Biol. Plant. 61, 611−621. Available from: https://doi.org/10.1007/s10535-017-0723-x.

Yang, J., Duan, G., Li, C., Liu, L., Han, G., Zhang, Y., et al., 2019. The crosstalks between jasmonic acid and other plant hormone signaling highlight the involvement of jasmonic acid as a core component in plant response to biotic and abiotic stresses. Front. Plant Sci. Available from: https://doi.org/10.3389/fpls.2019.01349.

Ye, K., Li, H., Ding, Y., Shi, Y., Song, C., Gong, Z., et al., 2019. BRASSINOSTEROID-INSENSITIVE2 negatively regulates the stability of transcription factor ICE1 in response to cold stress in arabidopsis. Plant Cell 31, 2682−2696. Available from: https://doi.org/10.1105/tpc.19.00058.

Zandalinas, S.I., Rivero, R.M., Martínez, V., Gómez-Cadenas, A., Arbona, V., 2016. Tolerance of citrus plants to the combination of high temperatures and drought is associated to the increase in transpiration modulated by a reduction in abscisic acid levels. BMC Plant Biol. 16, 1−16. Available from: https://doi.org/10.1186/s12870-016-0791-7.

Zazímalová, E., Murphy, A.S., Yang, H., Hoyerová, K., Hosek, P., 2010. Auxin transporters−why so many? Cold Spring Harb. Perspect. Biol. Available from: https://doi.org/10.1101/cshperspect.a001552.

Zhang, Z., Huang, R., 2010. Enhanced tolerance to freezing in tobacco and tomato overexpressing transcription factor TERF2/LeERF2 is modulated by ethylene biosynthesis. Plant Mol. Biol. 73, 241–249. Available from: https://doi.org/10.1007/s11103-010-9609-4.

Zhang, S., Wang, X., 2011. Overexpression of GASA5 increases the sensitivity of Arabidopsis to heat stress. J. Plant Physiol. 168, 2093–2101. Available from: https://doi.org/10.1016/j.jplph.2011.06.010.

Zhang, L., Zhang, L., Xia, C., Zhao, G., Liu, J., Jia, J., et al., 2015. A novel wheat bZIP transcription factor, TabZIP60, confers multiple abiotic stress tolerances in transgenic Arabidopsis. Physiol. Plant. 153, 538–554. Available from: https://doi.org/10.1111/ppl.12261.

Zhang, L., Liu, X., Gaikwad, K., Kou, X., Wang, F., Tian, X., et al., 2017. Mutations in eIF5b confer thermosensitive and pleiotropic phenotypes via translation defects in Arabidopsis thaliana. Plant Cell. Available from: https://doi.org/10.1105/tpc.16.00808.

Zhang, C., Li, G., Chen, T., Feng, B., Fu, W., Yan, J., et al., 2018. Heat stress induces spikelet sterility in rice at anthesis through inhibition of pollen tube elongation interfering with auxin homeostasis in pollinated pistils. Rice . Available from: https://doi.org/10.1186/s12284-018-0206-5.

Zhang, Q., Kong, X., Yu, Q., Ding, Y., Li, X., Yang, Y., 2019. Responses of PYR/PYL/RCAR ABA receptors to contrasting stresses, heat and cold in Arabidopsis. Plant Signal. Behav. 14. Available from: https://doi.org/10.1080/15592324.2019.1670596.

Zhang, X.W., Liu, F.J., Zhai, J., Li, F.D., Bi, H.G., Ai, X.Z., 2020a. Auxin acts as a downstream signaling molecule involved in hydrogen sulfide-induced chilling tolerance in cucumber. Planta. Available from: https://doi.org/10.1007/s00425-020-03362-w.

Zhang, X., Zhang, L., Sun, Y., Zheng, S., Wang, J., Zhang, T., 2020b. Hydrogen peroxide is involved in strigolactone induced low temperature stress tolerance in rape seedlings (Brassica rapa L.). Plant Physiol. Biochem. Available from: https://doi.org/10.1016/j.plaphy.2020.11.006.

Zhao, Y., 2010. Auxin biosynthesis and its role in plant development. Annu. Rev. Plant Biol. Available from: https://doi.org/10.1146/annurev-arplant-042809-112308.

Zhao, D., Shen, L., Fan, B., Yu, M., Zheng, Y., Lv, S., et al., 2009. Ethylene and cold participate in the regulation of LeCBF1 gene expression in postharvest tomato fruits. FEBS Lett. 583, 3329–3334. Available from: https://doi.org/10.1016/j.febslet.2009.09.029.

Zhao, M.L., Wang, J.N., Shan, W., Fan, J.G., Kuang, J.F., Wu, K.Q., et al., 2013. Induction of jasmonate signalling regulators MaMYC2s and their physical interactions with MaICE1 in methyl jasmonate-induced chilling tolerance in banana fruit. Plant Cell Environ. Available from: https://doi.org/10.1111/j.1365-3040.2012.02551.x.

Zhao, M., Liu, W., Xia, X., Wang, T., Zhang, W.H., 2014. Cold acclimation-induced freezing tolerance of Medicago truncatula seedlings is negatively regulated by ethylene. Physiol. Plant. 152, 115–129. Available from: https://doi.org/10.1111/ppl.12161.

Zhao, J., Zhang, S., Yang, T., Zeng, Z., Huang, Z., Liu, Q., et al., 2015. Global transcriptional profiling of a cold-tolerant rice variety under moderate cold stress reveals different cold stress response mechanisms. Physiol. Plant . Available from: https://doi.org/10.1111/ppl.12291.

Zhao, C., Liu, B., Piao, S., Wang, X., Lobell, D.B., Huang, Y., et al., 2017. Temperature increase reduces global yields of major crops in four independent estimates. Proc. Natl. Acad. Sci. U. S. A. Available from: https://doi.org/10.1073/pnas.1701762114.

Zwack, P.J., Compton, M.A., Adams, C.I., Rashotte, A.M., 2016. Cytokinin response factor 4 (CRF4) is induced by cold and involved in freezing tolerance. Plant Cell Rep. 35, 573–584. Available from: https://doi.org/10.1007/s00299-015-1904-8.

Drought-induced plant miRNAome and phytohormone signaling cross-talk

Bhuvnesh Kapoor[1],*, Pankaj Kumar[2],*, Rajnish Sharma[2] and Mohammad Irfan[3]

[1]University Institute of Biotechnology (UIBT), Chandigarh University, Mohali, Punjab, India
[2]Department of Biotechnology, Dr. Yashwant Singh Parmar University of Horticulture and Forestry, Solan, Himachal Pradesh, India [3]Plant Biology Section, School of Integrative Plant Science, Cornell University, Ithaca, NY, United States

12.1 Introduction

microRNAs (miRNAs) are evolutionary conserved short endogenous RNAs with a length of 20–24 nucleotides that assist in the expression regulation at the transcriptional and posttranscriptional stages (Ferdous et al., 2015; Shriram et al., 2016; Bhogireddy et al., 2021; Gelaw and Sanan-Mishra, 2021). A nuclease Dicer-like protein is known to cleave plant miRNA that leads to downregulation or silencing of the target genes, mostly transcription factors (TFs). A large number of miRNAs are recently found to be either upregulated or downregulated under the effect of drought stress (Gelaw and Sanan-Mishra, 2021). Among the several abiotic factors that affect plant performance and yield, drought stress is one of the most well-known global issues. As a result, it is critical in the current situation to place a greater priority on the development of new drought-tolerant crops (Kapoor et al., 2021; Sharma et al., 2021). In nature, plants have devised many strategies to deal with adverse conditions, such as drought avoidance, by arranging for the most vulnerable periods of development when the stress is at its lowest levels. Meanwhile, other approaches involve maintaining a high tissue water potential combined with improved

---

* Authors contributed equally and share the first authorship.

water acquisition via a deep root system with reduced water loss via transpiration control (Ferdous et al., 2015; Shriram et al., 2016). Understanding how agricultural plants adapt to harsh environmental conditions and perform better to maintain yield is essential. Furthermore, in nature, plants are rarely subjected to singular stress at a period. Rather, they are subjected to a variety of stressors, each of which prompts a distinct response based on the developmental stage, the duration, and stress types, and also the intensity of the various stresses. Plants have to precisely deploy valuable resources to cope with stresses and prevent themselves from additional stressors havocs that may cooccur when stressors reoccurrence. To do so, a complete remodeling of transcriptome, proteome, and metabolome is ensured by plant stress response pathways (Sharma et al., 2021). Under drought stress, epigenetic changes also influence the expression of certain genes from the chromatic remodeling sites at the genomic level. Alternative splicing of the resulting transcripts might result in the production of stress-relevant messenger RNA (mRNA) isoforms (Gelaw and Sanan-Mishra, 2021). Furthermore, through nuclease-assisted cleavage and promoter methylation, long noncoding RNA (lncRNA) also influences the function of pre-processed transcripts (Bhogireddy et al., 2021).

Plants respond to a variety of biotic and abiotic stressors, each with its unique mechanism of action (Khan et al., 2021a,b; Nazir et al., 2021, 2022). All these responses somehow show some points of convergence and others also exhibits unique stress-specific signaling pathway. Plant perceives different stresses that are further relayed by downstream secondary messengers to effector molecules (Kapoor et al., 2021). In case of drought and salt stresses, plants produced osmotic imbalances that are further amplified by the combined action of ABA-dependent gene expression of key stress-responsive genes(*LEA*, chaperons, *HSPs*, and protein inhibitors) as well as ROS, MAPKs, calcium signaling, and TFs (Kapoor et al., 2021). Furthermore, spatiotemporal regulation thought these elements is coordinated by binding of effectors gene to specific TFs. Heat shock-responsive elements (*HSRE*), ethylene (ET)-responsive elements (*ETRE*), ABA-responsive elements (*ARE*), gibberellic acid-responsive elements (*GARE*), *MYBs*, and *MYC*-binding elements are some of the TFs that have been identified and characterized in numerous crops so far (Shriram et al., 2016; Bhogireddy et al., 2021; Gelaw and Sanan-Mishra, 2021). On the other hand, genome-wide sequencing projects are making discoveries of new miRNA species with increasing depths of sequence information. These large-scale sequencing data have been able to show the novel targets of plant mRNAome that further enhance our understanding of their molecular and physiological actions. For example, in apple, genome-wide sequencing identified drought-induced upregulation of *miR156*, *miR395*, *mi408a*, *miR5225*, and *miRn244* levels, while subsequent target analysis indicated that *miR156* and *miR244* have a significant role in osmotic stress tolerance by modulating the expression level of *SPL-6a*, *SPL 9*, *SPL 12*, *SPL 13a*, *SPL2*, and zinc finger TFs (Niu et al., 2019). At the core of these signaling responses are phytohormones that further facilitate the action of sucrose nonfermenting related kinases (*SnRKs*) and abscisic acid insensitive (*ABI*) proteins while coordinating with ROS, sugar, and calcium signaling.

Phytohormones, including ABA, AUX, GA, CK, BR, JA, SA, and SLs, play a central role in a highly coordinated signaling cross-talk mechanism (Khan et al., 2020a,b; Nazir et al., 2019a,b). This signaling cross-talk among phytohormones enabled plants to appropriately coordinate the developmental stages from seed germination to senescence as well as

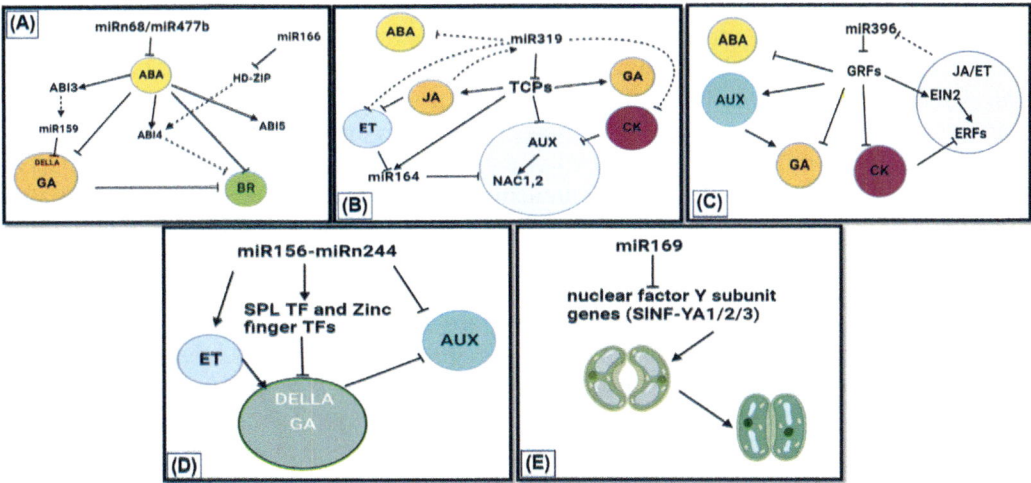

FIGURE 12.1 Overview of miRNA-mediated phytohormone cross-talk and associated functions. Images (A–E) summarize the molecular cross-talk between the phytohormone and specific mRNAs under drought stress and some developmental phases. Solid arrows, positive regulation; blocked arrows, negative regulation; double-headed arrow, mutual regulation; dotted arrow, involving other factors in between. *ARF*, Auxin-responsive factor; *ABA*, abscisic acid; *AUX*, auxin; *SA*, salicylic acid; *CK*, cytokinin; *GA*, gibberellin; *ET*, ethylene; *JA*, jasmonic acid; *BR*, brassinosteroids; *GRF*, growth regulating factor; *SPL*, squamosa promoter binding-like; *TCP*, teosinte branched1/cincinnata/proliferating cell factor.

response to exogenous and endogenous cues (biotic and abiotic factors) (Sharma et al., 2021; Kapoor et al., 2021). Meanwhile, ABA has been proved to be a central phytohormone that enables signaling cross-talk during various abiotic stresses. The active plant miRNAome represents *miRNAs 156, 159, 160/161, 162, 164, 165/166, 167, 168, 169, 170/171, 172, 319, 390, 393, 395, 396, 397, 398, 399,* and *400*, across various plant species that take part in an adaptive response to particular growth and developmental stage as well induced stresses (Ferdous et al., 2015; Gelaw and Sanan-Mishra, 2021). The overview of miRNA-mediated phytohormone cross-talk and associated functions are represented in Fig. 12.1.

12.2 Drought-responsive miRNAome

The evolutionary adaptations enabled plants to develop multilayered perception and signaling response systems for all kinds of environmental stresses. Meanwhile, drought is one the most often encountered stress that affects plants through the expression of various stress-responsive genes and metabolites, such as ABA-dependent LEA, RAB (response to abscisic acid), dehydrin, proline, cold regulated (COR), Rubisco (5-biphosphate carboxylase-oxygenase), glutathione S-transferase, helicase, vacuolar acid invertase, Ca^{2+} signaling, H_2S, and overall sugar metabolism. Understandably, the regulation of these wide classes of genes and metabolites has been posttranscriptionally regulated by various classes of miRNAs and their TF targets as suggested by various studies (Zhang et al., 2011; Kouhi et al., 2020). Collectively, drought stress-responsive miRNAs have been identified in a number of plant species, including *Arabidopsis* (Liu et al., 2008; Shen et al., 2013; Song et al., 2013), apple (Niu et al., 2019), rice (Zhao et al., 2007; Zhou et al., 2010b), tomato (Liu et al., 2017), Safflower (Kouhi et al., 2020), cassava (Li et al., 2017), and *Brassica juncea* (Bhatia et al., 2020). Genome-wide sequencing technology has facilitated the identification of drought-induced miRNAs that are further used to track the target genes as well as TFs. For example, in apple, genome-wide sequencing identified drought-induced upregulation of *miR156, miR395, mi408a, miR5225,* and *miRn244* levels, while subsequent target analysis indicated that *miR156* and *miRn244* involved in the osmotic stress tolerance by altering the gene expression of *SPL-6a, 9, 12,* and *13a, SPL2* and zinc finger TFs (Niu et al., 2019). The upregulation and downregulation of miRNAs are involved in different developmental stages that further showed the linkage with the regulation of drought tolerance (Bai et al., 2017). Furthermore, miRNAome studies provided a vantage point to reveal the target genes involved in different plant response signaling pathways. Zhang et al. (2011) studied the constitutive overexpression of *Sly-miR169c* in transgenic tomatoes that led to the reduced leaf water loss by limiting the stomatal openings and rate of transpiration hence enhancing drought tolerance as compared to the control plants. However, it has also been shown that the levels of miRNA expression during drought stress response were species-dependent (Ferdous et al., 2015). For example, in drought-stressed rice, genome-wide profiling, and miRNA analysis revealed a total of 30 miRNAs, 11 of which were found to be downregulated (*miR170, miR172, miR397, miR408, miR529, miR896, miR1030, miR1035, miR1050, miR1088,* and *miR1126*), eight of which were upregulated (*miR395, miR474, miR845, miR851, miR901, miR903,* and *miR1125*). Although nine miRNA from drought-stressed *Arabidopsis* displayed distinct expression patterns (*miR156, miR168, miR170, miR171, miR172, miR319, miR396, miR397,* and *miR408*) (Zhuo et al., 2010ab). Meanwhile, drought has been reported to increase the expression of *miR169* in some plants, that is, tomato and *Arabidopsis* (Yu et al., 2019; Rao et al., 2020), whereas it is downregulated in maize (Luan et al., 2015) under similar conditions. A different miRNA expression within species was also reported in wheat, where *miR160a, miR164b, miR166h, miR169d,* and *miR444d.3* were found downregulated in drought-resistant "Hanxuan10" genotype as compared to the drought-susceptible "Zhengyin1" after drought stress (Ma et al., 2015). However, these subtle differences in miRNA expressions may reflect the degree of drought stress affecting normal plant physiology and development that varies in an inter-species as well as intraspecies fashion.

The miRNAome represents a dynamic regulatory system that operates in a spatiotemporal manner and allows plants to cope up with real-time challenges. Furthermore, differential expression of miRNAs was also found between the different plant tissues as well as developmental stages. Under drought stress, the overexpression of 40 miRNAs was reported along with the downregulation of 37 miRNAs representing 23 families in rice shoots, whereas drought causes the upregulation of 65 miRNAs and the downregulation of 20 miRNAs from 46 families in rice roots (Zhang et al., 2017). Drought-induced up-or downregulation of miRNA might be useful for engineering drought tolerance in plants, as miRNA targets are likely to contain genes that contribute both positively and negatively to tolerance. When miRNAs are upregulated, their targets are downregulated under the same conditions, and vice versa. Overexpression of the target gene or silencing of the corresponding miRNA could be used to increase the accumulation of target genes that contribute to drought tolerance. Under adverse conditions, such as drought and high temperatures, plants can accumulate a substantial quantity of ROS, which can seriously hinder plant growth and development while the antioxidant system can scavenge most of the ROS, which reduces oxidative damage. The number of antioxidant enzymes, as well as patterns of miRNAs and associated target genes related to antioxidant systems, have been disclosed in *Zanthoxylum bungeanum* during drought stress (Fei et al., 2020). The findings from this study revealed that during drought stress, APX, POD, CAT, MDA, and proline genes were responded positively while SOD and associated genes were negatively regulated revealing their supportive role. Furthermore, miRNAs and their target genes have a negative correlation, suggesting that miRNAs can limit the expression of associated genes and are significant regulators in the antioxidant system (Fei et al., 2020). Drought-responsive miRNAome and their target genes in different plant species are represented in Table 12.1.

12.3 Drought-induced miRNAome as mediator of phytohormonal signaling cross-talk

Phytohormones are central signaling molecules that are involved in the regulation of almost all biological activities of plant growth and development. Phytohormones including ABA, AUX, GA, CK, BR, JA, SA, and SLs have been shown to play a key role in a highly coordinated signaling cross-talk mechanism (Kapoor et al., 2021; Sharma et al., 2021). This signaling cross-talk among phytohormones enabled plants to appropriately coordinate the developmental stages from seed germination to senescence as well as response to exogenous and endogenous cues (biotic and abiotic factors). Meanwhile, ABA has been proved to be a central phytohormone that enables signaling cross-talk during various abiotic stresses, such as drought, salinity, cold, and heavy metal. *ABSCISIC ACID-INSENSITIVE* proteins, *ABI1* and *ABI2* homologs of type II phosphatases, are known to negatively control the ABA response in the ABA signaling pathway; however, *ABI 3, 4,* and *5 TFs* are positive regulators of the ABA response. Given that *ABI* factors are essential for signaling cross-talk, it is not unexpected that they are prime miRNA targets. This inactivation of protein phosphatases further controls the activation of ABA-dependent gene expression and ion channels as a result of activated *SNF1-type* kinases (SnRKs) activity, (Raghavendra et al., 2010). *SnRK2* are ABA signaling regulators that affect the plant's adaptive response to drought-induced osmotic stress. Meanwhile, angiosperms have acquired ABA-insensitive *SnRK2s* that control miRNA degradation and enhance plant development in response to osmotic stressors. Furthermore, during osmotic stress, B4 Raf-like MAP kinase kinase kinases (MAPKKKs) phosphorylate and activate subclass I *SnRK2s*, indicating that these MAPKKKs are upstream factors of subclass I *SnRK2* and are triggered immediately by osmotic stress (Soma et al., 2020).

Over the years, miRNAs have been associated with stress-induced signaling cross-talk among phytohormones due to their trans-regulation capabilities and critical developmental functions. For example, small RNA and degradome sequencing investigations have recently revealed that numerous ABA-responsive miRNAs, such as *Peu-miR-n68* and *Peu-miR477b*, act as mediators of ABA-BR and ABA-GA cross-talk signaling (Lian et al., 2018; Sha and Wang, 2020).

TABLE 12.1   Drought-responsive miRNAome and their target genes in different plant species.

miRNA	Targets	Crops/species	References
miR396 miR167 miR168 miR157 miR408 miR171 miR393 mIR394a/b	AFB2, ABI3, ABI4, ABI5, ABF3, ARF8, ABF4, TIR1, AFB3, GRF1, GRF2, GRF3, CSD1, CSD2	Arabidopsis thaliana	Liu et al. (2008) Song et al. (2013)
lncRNAs TCONS_00051908 TCONS_00088973 TCONS_00081575 TCONS_00047156	miR156, miR159, miR172, miR319, miR399, HSPs, MYBs	Brassica juncea	Bhatia et al. (2020)
miR156 miR162 miR164 miR166 miR172 miR398 miR408	SPL, NAC, HD-ZIP, AP2/ERF, CSD1, CUP	Carthamus tinctorius	Kouhi et al. (2020)
miR160 miR167	ARF6, ARF8, ARF10, ARF16 and ARF17	Dimocarpus longan	Lin et al. (2015)
miR5071 miR393 miR156 miR168 miR166 miR167 miR894	OsMLA10-like gene, auxin receptors TIR1/AFBs	Hordeum vulgare	Bai et al. (2017)
miR156 miR395 mi408a miR5225 miRn244	RNA pol II C-terminal domain phosphatase, WRKY33, receptor kinase, SPL2, SPL4, SPL9, SPL12	Malus spp.	Niu et al. (2019)
lincRNA101 lincRNA 391 lincRNA 356	NAC and NUCLEAR FACTOR Y (NF-Y)	Manihot esculenta	Li et al. (2017)
miR169g miR528−5p miR827 miR167d	Laccase-encoding gene (LOC_Os01g62600.1), OsSPX-MFS1, OsSPX-MFS2, OsARF12	Oryza sativa	Zhao et al. (2017)
Peu-miR-n68 miR394−5p miR-n30 miR-n68 miR530b Peu-miR477b	RGL1, BAK1, HD-ZIP-III, FBX6, bHLH, SAP1	Populus euphratica	Lian et al. (2018)

(Continued)

TABLE 12.1 (Continued)

miRNA	Targets	Crops/species	References
mireap-m0020−5p mireap-m0104−3p mireap-m0086−5p mireap-m0043−3p mireap-m0049−5p mireap-m0017−3 mireap-m0043−3p mireap-m0020−5p mireap-m0067−5p	ALDH, CaBP, carboxypeptidase, cytochrome P450, disease resistance protein RPM1, POD, SPS, HS 70 kDa protein, HSP20 family protein and xyloglucan endotransglucosylase/hydrolase, MYB, TGA transcription factor, bZIP transcription factor	Saccharum officinarum	Selvi et al. (2021)
miR156a miR164a-5p slymiR171d Sly319c-3p	SPL, AP2-like, NAC, MYB, GRAS, TCP, ARFs	Solanum lycopersicum	Liu et al. (2017) and Bouzroud et al. (2018)
Sto-miR398b Sto-miR5225 Sto-miR482d	AUX1, CRE1, APF, ETR	Sophora tonkinensis	Liang et al. (2021)
miR156k miR444c.1 miR159a miR160a miR166h	SBP TF, MADS-box TF, MYB3, ARFs, NAC, Class III HD-ZIP protein 4 (HD-ZIP4 III)	Triticum aestivum	Ma et al. (2015)
miR396a-5p ath-miR167a3p ath-miR834 ath-miR169b-3p ath-miR447−3p ath-miR773−3p ath-miR397b ath-miR859 ath-miR5632−5p ath-miR1888a ath-miR5638a ath-miR398a-3p ath-miR3434−3p	SOD1, SOD2, PRX2E, CAT, APX3, GPX1, P5CS, JAR1, ABI, MAPK1, PDI52, RBOHC, NRX1, TCTP	Zanthoxylum bungeanum	Fei et al. (2020)

ABA, abscisic acid; ALDH, aldehyde dehydrogenase; AP2, APPETALA2-like; APX, ascorbate peroxidase; ARE, ABA-responsive elements; ARF, Auxin-responsive factor; AUX, auxin; BR, brassinosteroids; CaBP, calmodulin-binding protein; CAT, catalase; CK, cytokinin; CSD1, superoxide dismutase [Cu-Zn] 1; CUP, cupredoxin; ERF, ethylene responsive factor; ET, ethylene; GA, gibberellin; GARE, gibberellic acid-responsive elements; GRF, growth regulating factor; HSP, heat shock protein; JA, jasmonic acid; JAR1, jasmonic acid-amido synthetase1; LEA, late embryogenesis proteins; MADS-box TF, MIKC-type MADS-box transcription factor; MOD, malondialdehyde; MPKs, MAPK phosphorylated kinases; NO, nitrous oxide; NRX1, probable nucleoredoxin 1; P5CS, Delta-1-pyrroline-5-carboxylate synthase; PDI52, protein disulfide-isomerase 5−2 isoform X1; PEG, polyethylene glycol; POD, peroxidase; PRX2E, peroxiredoxin-2E; RBOHC, respiratory burst oxidase homolog protein C; ROS, reactive oxygen species; SA, salicylic acid; SOD, superoxide dismutase; SPL, squamosa promoter binding-like; SPL, squamosa promoter binding-like; SPS, sucrose-phosphate synthase; TCTP, translationally controlled tumor protein homolog; TF, transcription factor.

Another study discovered miRNA targeting genes, such as *AUX1*, *CRE1*, *APF*, and *ETR*, which play critical roles in the AUX, CK, ABA, and ET signal transduction pathways (Liang et al., 2021). Auxin has a role in root modeling and organ polarity, which both vary in response to drought stress. In the context of root architecture modeling, the antagonistic relationship between AUX and ABA has been systematically investigated (Nibau et al., 2008; Fukaki and Tasaka, 2009; Jeon et al., 2017). Auxin-promote cell multiplication and lateral root meristem growth while during drought stress ABA suppress these functions by tissue-specific expression of the *LATERAL ORGAN BOUNDARIES (LOB) DOMAIN/ASYMMETRIC LEAVES2-LIKE (LBD/ASL)* gene and TF *MYB86* (Jeon et al., 2017; Li et al., 2021). Furthermore, *miR164, 167, 160, 393,* and *394,* which govern levels of free AUX and hence lateral root growth, are similarly responsive to ABA treatment through their impact on AUX-responsive factors (*ARFs*). The repression of TFs, such as *ARF6, ARF8, ARF10, ARF16,* and *ARF17,* by *miR160* and *miR167* over-expression promotes germination as well as postgermination stages in *Arabidopsis* and *Dimocarpus longan* (longan) (Liu et al., 2008; Lin et al., 2015). Consequently, the *miR160/miR167/ ARFs* network is a powerful regulatory junction through which AUX may affect both the ABA and JA signaling pathways. Postembryonic development, root initiation, and reproductive organ maturation, among other developmental processes, appear to be regulated in this manner. Similarly, in tomatoes, drought significantly increased the expression of *SlARF2A, SlARF2B, SlARF4, SlARF7A, SlARF8A,* and *SlARF9A* while *SlARF3, SlARF10A,* and *SlARF19* were repressed (Bouzroud et al., 2018). Furthermore, *miR156* and *miR172* exhibit the capacity to downregulate TF families, such as *SQUAMOSA PROMOTER BINDING-LIKE (SPL)* and *APPETALA2-like (AP2-like),* respectively. In another miRNome study, differential expression analysis revealed 11 and 34 miRNAs unique to drought-tolerant and drought-sensitive maize lines, respectively. Meanwhile, target analysis revealed that the tolerant line's *miR164-MYB* and *miR164-NAC* modules modulated the stress response in an ABA (abscisic acid)-dependent way, whereas the sensitive line's *miR156-SPL* and *miR160-ARF* modules inhibited metabolism in drought-exposed leaves. In another study, an *miR167a* was discovered to target the evolutionary conserved IAA-Ala Resistant3 (*IAR3*) gene, which hydrolyzes an inactive version of auxin (IAA-alanine) and releases bioactive AUX (IAA), a key phytohormone for root growth and architecture in *Arabidopsis thaliana* (Kinoshita et al., 2012). miR396, *GROWTH REGULATING FACTORS* (GRFs), and *GRF-INTERACTING FACTORS* (GIFs) have been identified as part of a regulatory module that regulate the growth of numerous tissues and organs in several species (Liebsch and Palatnik, 2020). Recently, this module has been found associated with key genes of phytohormone regulation in plants. *Hyo-*miRNA396b was found downregulated under the PEG (15%) induced drought stress and its expression was also fluctuated by ABA treatment (0.38 mM) in *Hylocereus polyrhizus* (Li et al., 2019). Furthermore, more than 60 genes were enlisted by Hewezi and Baum (2012) to be involved in all six phytohormone signal transduction in *A. thaliana,* including several genes coding for ET and ARF as well as genes directly involved in the manufacturing of bioactive GA, CK, and ABA. Meanwhile, it is speculated that miR396-GRF mediated regulation of ABA-responsive gene and *ETHYLENE RESPONSE FACTOR* (ERF) genes expression may be evolutionary conserved in plants (Müller and Munné-Bosch, 2015). Such as overexpression of *SlERF5* (*AP2/ERF* TF family) in transgenic tomato plants exhibiting increased drought tolerance and relative water content as compared to wild type (Pan et al., 2012). The active participation of the *miR396-GRF* complex in root stress adaptations represents its potential role in cross-talk signaling between CK, ABA, and ET. In both *cis* and *trans* modalities, mRNAs have been shown to collaborate closely with lncRNAs. Recently, Qiu et al. (2020) reported that PEG (polyethylene glycol) induced drought stress in Tibetan wild barley geno-types namely XZ5 as drought-tolerant and XZ54 as drought-sensitive, and also in one drought-tolerant cultivar Tadmor. A total of 23,651 new lncRNAs have been discovered. lncRNA target prediction found 1279 potential lncRNA—mRNA pairs, 503 *cis*-acting and 776 *trans*-acting. As per pathway enrichment analysis, the targets were enriched in molecular processes related to plant hormone signal transduction, diterpenoid biosynthesis, and ascorbate/aldarate metabolism. The functional relevance of lncRNA—mRNA interactions was discovered to modulate kinase-mediated signal transduction. *SMG1* (serine/threonine-protein kinase) is thought to be important for drought tolerance in XZ5 (Qiu et al., 2020). The processing of pre-RNA by the spliceosome provides RNA availability for cellular localization and translation. RNA-binding proteins that affect posttranscriptional regulation typically make up this complex. The mRNA interactome of *A. thaliana* under drought stress revealed 44 proteins linked with spliceosomes and 32 proteins related to stress granules (Marondedze et al., 2020). Splicing factor discovery reflects direct and/or indirect stress-induced splicing events that have a direct impact on transcriptome and proteome modifications. A new class of mobile mRNAs that respond to drought stress was found recently, that is, *PbDRM* (*Pyrus betulaefolia DROUGHT-RESPONSIVE MOBILE GENE*). The virus-induced gene silencing of this PbDRM complex leads to the increased sensitivity of *P. betulaefolia* to drought stress. It was also determined that *PbDRM* mRNA functions as a phloem-mobile signal in pear during drought stress (Hao et al., 2020).

12.4 Drought stress—induced miRNAome and epigenetic regulation

Genes that are activated at spatiotemporal levels account for the majority of epigenetic control of gene expression. Although under drought stress, epigenetic regulation and phytohormone cross-talk are well described to modify the chromatin structure, miRNA-assisted regulation of chromatin are nascent at the current levels of understanding. While miRNAs primarily exert posttranscriptional gene silencing and do not induce direct epigenetic alterations, they may have an indirect impact on the epigenetic landscape by regulating transcript levels for certain chromatin or epigenetic modifiers proteins (Luján-Soto and Dinkova, 2021). The *SWR1* chromatin remodeling complex (*SWR1-C*) is known to produce variant nucleosomes by switching the histone H2A-H2B dimer with the H2A.Z-H2B dimer. In *Arabidopsis*, mutants of *SWR1-C* components lowered the transcript level of *miR156* and *miR164*, resulting in morphological modifications, implying that *SWR1-C* promotes fine regulation of plant growth by establishing a transcriptional equilibrium between target mRNAs and miRNAs (Choi et al., 2016). In an *Arabidopsis* embryogenic culture, extensive TF gene modulation is followed by differential expression of many miRNAs, implying a role for miRNA control in the embryogenic transition of plant somatic cells (Wójcik and Gaj, 2016). Furthermore, *miR165, miR166, miR167*, and *miR390* were reported in major somatic embryogenesis transcriptomes (Wójcikowska et al., 2020). The complementary sites of *Arabidopsis PHABULOSA* (*PHB*) and *PHAVOLUTA* (*PHV*) mRNA genes controlling abaxial and adaxial leaf fates have been altered in the dominant mutants of *PHB* and *PHV* TF gene. These findings imply a process where the miRNA interactions with processed *PHB mRNA* alter the chromatin structure of the PHB template DNA in differentiated cells (Bao et al., 2004). Several abiotic stressors, particularly in resistant cultivars, produce *TCP19* (Class I TCP TF). A rice*OsTCP19* interacts directly with *ULT1* (trithorax group), a component that recruits *ATX1* (*Arabidopsis* homolog of trithorax1), a histone methyltransferase, and suppresses the Polycomb group (*PcG*) of gene repression complexes in *Arabidopsis*. As a result, it plays a role in chromatin alteration during times of stress (Mukhopadhyay and Tyagi, 2015). Many effector TFs (*MYBs, NAC, GRAS, AP2/ERF, SPLs, HD-ZIPs, TCP,* and *NFYA*) are regulated by miRNA in response to stress, and most of these are common to drought, salt, and heavy metal stressors. Moreover, these TFs are involved in metabolic processes, such as cell-wall modeling and cellulose synthesis, hormonal signal transmission, sugar and starch metabolism and breakdown, and plant growth and development.

12.5 Conclusion

Complex regulatory networks are involved in phytohormonal cross-talk signaling that eventually enables plants to balance and coordinate the developmental stages from seed germination to senescence as well as response to exogenous and endogenous cues (biotic and abiotic factors). Meanwhile, miRNAs are implicated in this cross-talk, as evidenced by the research cited in this chapter. Advanced high-throughput sequencing platforms at genomic levels have facilitated the identification of drought-induced miRNAs that are further used to track the target genes as well as TFs. The variety of hormone-regulated miRNAs and miRNA targets implicated in hormone signaling continues to rise as more research employs short RNAs and mRNA degradome libraries from various cited plant species. Additional phytohormone routes (e.g., BRs, SA, plant peptide hormones, polyamines, NO, and SLs) have recently been discovered that can be controlled by miRNA. The miRNA/hormone network becomes increasingly complicated as research develops, necessitating more precise investigations, to unravel the genetic mechanisms underlying hormonal interaction in the spatiotemporal context of a biological event, for example, loss of miRNA function. While miRNAs generally suppress genes after transcription and do not cause direct epigenetic changes, they can influence the epigenetic landscape indirectly by modulating transcript levels for specific chromatin or epigenetic modifier proteins. Ultimately, miRNAome represents a dynamic regulatory system that operates in a spatiotemporal manner and allows plants to cope up with real-time challenges. Furthermore, differential expression of miRNAs was also found between the different plant tissues as well as developmental stages.

References

Bai, B., Peviani, A., van der Horst, S., Gamm, M., Snel, B., Bentsink, L., Hanson, J., 2017. Extensive translational regulation during seed germination revealed by polysomal profiling. New Phytol. 214 (1), 233–244.

Bao, N., Lye, K.W., Barton, M.K., 2004. MicroRNA binding sites in *Arabidopsis* class III HD-ZIP mRNAs are required for methylation of the template chromosome. Dev. Cell 7 (5), 653–662.

Bhatia, G., Singh, A., Verma, D., Sharma, S., Singh, K., 2020. Genome-wide investigation of regulatory roles of lncRNAs in response to heat and drought stress in *Brassica juncea* (Indian mustard). Environ. Exp. Bot. 171, 103922.

Bhogireddy, S., Mangrauthia, S.K., Kumar, R., Pandey, A.K., Singh, S., Jain, A., et al., 2021. Regulatory non-coding RNAs: a new frontier in regulation of plant biology. Funct. Integr. Genomics 20, 1–8.

Bouzroud, S., Gouiaa, S., Hu, N., Bernadac, A., Mila, I., Bendaou, N., et al., 2018. Auxin response factors (ARFs) are potential mediators of auxin action in tomato response to biotic and abiotic stress (*Solanum lycopersicum*). PLoS One 13 (2), e0193517.

Choi, K., Kim, J., Müller, S.Y., Oh, M., Underwood, C., Henderson, I., et al., 2016. Regulation of microRNA-mediated developmental changes by the SWR1 chromatin remodeling complex. Plant Physiol. 171 (2), 1128–1143.

Fei, X., Li, J., Kong, L., Hu, H., Tian, J., Liu, Y., et al., 2020. miRNAs and their target genes regulate the antioxidant system of *Zanthoxylum bungeanum* under drought stress. Plant Physiol. Biochem. 150, 196–203.

Ferdous, J., Hussain, S.S., Shi, B.J., 2015. Role of micro RNAs in plant drought tolerance. Plant Biotechnol. J. 13 (3), 293–305.

Fukaki, H., Tasaka, M., 2009. Hormone interactions during lateral root formation. Plant Mol. Biol. 69 (4), 437–449.

Gelaw, T.A., Sanan-Mishra, N., 2021. Non-coding RNAs in response to drought stress. Int. J. Mol. Sci. 22 (22), 12519.

Hao, L., Zhang, Y., Wang, S., Zhang, W., Wang, S., Xu, C., et al., 2020. A constitutive and drought-responsive mRNA undergoes long-distance transport in pear (*Pyrus betulaefolia*) phloem. Plant Sci. 293, 110419.

Hewezi, T., Baum, T.J., 2012. Complex feedback regulations govern the expression of miRNA396 and its GRF target genes. Plant Signal. Behav. 7 (7), 749–751.

Jeon, E., Young Kang, N., Cho, C., Joon Seo, P., Chung Suh, M., Kim, J., 2017. LBD14/ASL17 positively regulates lateral root formation and is involved in ABA response for root architecture in *Arabidopsis*. Plant Cell Physiol. 58 (12), 2190–2201.

Kapoor, B., Kumar, P., Sharma, R., Kumar, A., 2021. Regulatory interactions in phytohormone stress signaling implying plants resistance and resilience mechanisms. J. Plant Biochem. Biotechnol. 10, 1–6.

Khan, M.I.R., Jahan, B., AlAjmi, M.F., Rehman, M.T., Khan, N.A., 2020a. Ethephon mitigates nickel stress by modulating antioxidant system, glyoxalase system and proline metabolism in Indian mustard. Physiol. Mol. Biol. Plants 26 (6), 1201–1213.

Khan, M.I.R., Trivellini, A., Chhillar, H., Chopra, P., Ferrante, A., Khan, N.A., et al., 2020b. The significance and functions of ethylene in flooding stress tolerance in plants. Environ. Exp. Bot. 179, 104188.

Khan, M.I.R., Ashfaque, F., Chhillar, H., Irfan, M., Khan, N.A., 2021a. The intricacy of silicon, plant growth regulators and other signaling molecules for abiotic stress tolerance: An entrancing crosstalk between stress alleviators. Plant Physiol. Biochem. 162, 36–47.

Khan, M.I.R., Jahan, B., AlAjmi, M.F., Rehman, M.T., Iqbal, N., Irfan, M., et al., 2021b. Crosstalk of plant growth regulators protects photosynthetic performance from arsenic damage by modulating defense systems in rice. Ecotoxicol. Environ. Saf. 222, 112535.

Kinoshita, N., Wang, H., Kasahara, H., Liu, J., MacPherson, C., Machida, Y., et al., 2012. IAA-Ala Resistant3, an evolutionarily conserved target of miR167, mediates *Arabidopsis* root architecture changes during high osmotic stress. Plant Cell 24 (9), 3590–3602.

Kouhi, F., Sorkheh, K., Ercisli, S., 2020. MicroRNA expression patterns unveil differential expression of conserved miRNAs and target genes against abiotic stress in safflower. PLoS One 15 (2), e0228850.

Li, S., Yu, X., Lei, N., Cheng, Z., Zhao, P., He, Y., et al., 2017. Genome-wide identification and functional prediction of cold and/or drought-responsive lncRNAs in cassava. Sci. Rep. 7 (1), 1–7.

Li, A.L., Wen, Z., Yang, K., Wen, X.P., 2019. Conserved miR396b-GRF regulation is involved in abiotic stress responses in pitaya (*Hylocereu spolyrhizus*). Int. J. Mol. Sci. 20 (10), 2501.

Li, H., Testerink, C., Zhang, Y., 2021. How roots and shoots communicate through stressful times. Trends Plant Sci. 26 (9), 940–952.

Lian, C., Yao, K., Duan, H., Li, Q., Liu, C., Yin, W., Xia, X., 2018. Exploration of ABA responsive miRNAs reveals a new hormone signaling crosstalk pathway regulating root growth of Populus euphratica. Int. J. Mol. Sci. 19 (5), 1481.

Liang, Y., Wei, K., Wei, F., Qin, S., Deng, C., Lin, Y., et al., 2021. Integrated transcriptome and small RNA sequencing analyses reveal a drought stress response network in Sophora tonkinensis. BMC Plant Biol. 21 (1), 1–20.

Liebsch, D., Palatnik, J.F., 2020. MicroRNA miR396, GRF transcription factors and GIF co-regulators: a conserved plant growth regulatory module with potential for breeding and biotechnology. Curr. Opin. Plant Biol. 53, 31–42.

Lin, Y., Lai, Z., Lin, L., Lai, R., Tian, Q., Ye, W., et al., 2015. Endogenous target mimics, microRNA167, and its targets ARF6 and ARF8 during somatic embryo development in *Dimocarpus longan* Lour. Mol. Breed. 35 (12), 1–5.

Liu, H.H., Tian, X., Li, Y.J., Wu, C.A., Zheng, C.C., 2008. Microarray-based analysis of stress-regulated microRNAs in *Arabidopsis thaliana*. RNA 14 (5), 836–843.

Liu, M., Yu, H., Zhao, G., Huang, Q., Lu, Y., Ouyang, B., 2017. Profiling of drought-responsive microRNA and mRNA in tomato using high-throughput sequencing. BMC Genomics 18 (1), 1–8.

Luan, M., Xu, M., Lu, Y., Zhang, L., Fan, Y., Wang, L., 2015. Expression of zma-miR169 miRNAs and their target ZmNF-YA genes in response to abiotic stress in maize leaves. Gene 555 (2), 178–185.

Luján-Soto, E., Dinkova, T.D., 2021. Time to wake up: epigenetic and small-RNA-mediated regulation during seed germination. Plants 10, 236.

Ma, X., Xin, Z., Wang, Z., Yang, Q., Guo, S., Guo, X., et al., 2015. Identification and comparative analysis of differentially expressed miRNAs in leaves of two wheat (*Triticum aestivum* L.) genotypes during dehydration stress. BMC Plant Biol. 15 (1), 1–5.

Marondedze, C., Thomas, L., Lilley, K.S., Gehring, C., 2020. Drought stress causes specific changes to the spliceosome and stress granule components. Front. Mol. Biosci. 6, 163.

Mukhopadhyay, P., Tyagi, A.K., 2015. OsTCP19 influences developmental and abiotic stress signaling by modulating ABI4-mediated pathways. Sci. Rep. 5 (1), 1–2.

Müller, M., Munné-Bosch, S., 2015. Ethylene response factors: a key regulatory hub in hormone and stress signaling. Plant Physiol. 169 (1), 32–41.

Nazir, F., Hussain, A., Fariduddin, Q., 2019a. Hydrogen peroxide modulate photosynthesis and antioxidant system in tomato (*Solanum lycopersicum* L.) plants under copper stress. Chemosphere 230, 544–558.

Nazir, F., Hussain, A., Fariduddin, Q., 2019b. Interactive role of epibrassinolide and hydrogen peroxide in regulating stomatal physiology, root morphology, photosynthetic and growth traits in *Solanum lycopersicum* L. under nickel stress. Environ. Exp. Bot. 162, 479–495.

Nazir, F., Hussain, A., Fariduddin, Q., Tanveer, A.K., 2021. Brassinosteroid and hydrogen peroxide improve photosynthetic efficiency and maintain chloroplast ultrastructure, stomatal movement, root morphology, cell viability and reduce Cu-triggered oxidative burst in tomato. Ecotoxicol. Environ. Saf. 207, 111081.

Nazir, F., Qazi, F., Khan, M.T.A., 2022. Interaction between brassinosteroids and hydrogen peroxide networking signal molecules in plants. In: Brassinosteroids Signalling, Springer, Singapore, pp. 59–79.

Nibau, C., Gibbs, D.J., Coates, J.C., 2008. Branching out in new directions: the control of root architecture by lateral root formation. New Phytol. 179 (3), 595–614.

Niu, C., Li, H., Jiang, L., Yan, M., Li, C., Geng, D., et al., 2019. Genome-wide identification of drought-responsive microRNAs in two sets of *Malus* from interspecific hybrid progenies. Horticult. Res. 6, 75. Available from: https://doi.org/10.1038/s41438-019-0157-z.

Pan, Y., Seymour, G.B., Lu, C., Hu, Z., Chen, X., Chen, G., 2012. An ethylene response factor (ERF5) promoting adaptation to drought and salt tolerance in tomato. Plant Cell Rep. 31 (2), 349–360.

Qiu, C.W., Liu, L., Feng, X., Hao, P.F., He, X., Cao, F., et al., 2020. Genome-wide identification and characterization of drought stress-responsive microRNAs in Tibetan wild barley. Int. J. Mol. Sci. 21 (8), 2795.

Raghavendra, A.S., Gonugunta, V.K., Christmann, A., Grill, E., 2010. ABA perception and signalling. Trends in plant science 15(7), 395–401.

Rao, S., Balyan, S., Jha, S., Mathur, S., 2020. Novel insights into expansion and functional diversification of MIR169 family in tomato. Planta. 251 (2), 1–7.

Selvi, A., Devi, K., Manimekalai, R., Prathima, P.T., Valiyaparambth, R., Lakshmi, K., 2021. High-throughput miRNA deep sequencing in response to drought stress in sugarcane. 3 Biotechnol. 11 (7), 1–8.

Sha, Y., Wang, J.-W., 2020. The cross-talk between microRNAs and gibberellin signaling in plants. Plant Cell Physiol. 61, 1880–1890.

Sharma, M., Irfan, M., Kumar, A., Kumar, P., Datta, A., 2021. Recent insights into plant circadian clock response against abiotic stress. J. Plant Growth Regul. 23, 1–4.

Shen, J., Xing, T., Yuan, H., Liu, Z., Jin, Z., Zhang, L., et al., 2013. Hydrogen sulfide improves drought tolerance in *Arabidopsis thaliana* by microRNA expressions. PLoS One 8 (10), e77047.

Shriram, V., Kumar, V., Devarumath, R.M., Khare, T.S., Wani, S.H., 2016. MicroRNAs as potential targets for abiotic stress tolerance in plants. Front. Plant Sci. 14 (7), 817.

Soma, F., Takahashi, F., Suzuki, T., Shinozaki, K., Yamaguchi-Shinozaki, K., 2020. Plant Raf-like kinases regulate the mRNA population upstream of ABA-unresponsive SnRK2 kinases under drought stress. Nat. Commun. 11 (1), 1–2.

Song, J.B., Gao, S., Sun, D., Li, H., Shu, X.X., Yang, Z.M., 2013. miR394 and LCR are involved in Arabidopsis salt and drought stress responses in an abscisic acid-dependent manner. BMC Plant Biol. 13 (1), 1–6.

Wójcik, A.M., Gaj, M.D., 2016. miR393 contributes to the embryogenic transition induced in vitro in *Arabidopsis* via the modification of the tissue sensitivity to auxin treatment. Planta 244 (1), 231–243.

Wójcikowska, B., Wójcik, A.M., Gaj, M.D., 2020. Epigenetic regulation of auxin-induced somatic embryogenesis in plants. Int. J. Mol. Sci. 21 (7), 2307.

Yu, Y., Ni, Z., Wang, Y., Wan, H., Hu, Z., Jiang, Q., et al., 2019. Overexpression of soybean miR169c confers increased drought stress sensitivity in transgenic *Arabidopsis thaliana*. Plant Sci. 285, 68–78.

Zhang, X., Zou, Z., Gong, P., Zhang, J., Ziaf, K., Li, H., et al., 2011. Over-expression of microRNA169 confers enhanced drought tolerance to tomato. Biotechnol. Lett. 33 (2), 403–409.

Zhao, B., Liang, R., Ge, L., Li, W., Xiao, H., Lin, H., et al., 2007. Identification of drought-induced microRNAs in rice. Biochem. Biophys. Res. Commun. 354 (2), 585–590.

Zhao, Y., Wen, H., Teotia, S., Du, Y., Zhang, J., Li, J., et al., 2017. Suppression of microRNA159 impacts multiple agronomic traits in rice (Oryza sativa L.). BMC plant biology 17(1), 1–13.

Zhang, J.W., Long, Y., Xue, M.D., Xiao, X.G., Pei, X.W., 2017. Identification of microRNAs in response to drought in common wild rice (*Oryza rufipogon* Griff.) shoots and roots. PLoS One 12 (1), e0170330.

Zhuo, M., Gu, L., Li, P., Song, X., Wei, L., Chen, Z., Cao, X., 2010a. Degradome sequencing reveals endogenous small RNA targets in rice (*Oryza sativa* L. ssp. indica). Front. Biol. 5 (1), 67–90.

Zhou, L., Liu, Y., Liu, Z., Kong, D., Duan, M., Luo, L., 2010b. Genome-wide identification and analysis of drought-responsive microRNAs in *Oryza sativa*. J. Exp. Bot. 61 (15), 4157–4168.

Interaction between the key defense-related phytohormones and polyamines in crops

Ágnes Szepesi and Péter Poór

Department of Plant Biology, Faculty of Science and Informatics, University of Szeged, Szeged, Hungary

13.1 Introduction

Polyamines (PAs) are essential polycationic molecules with many amino groups, possessing regulatory functions not only in plant growth and development but also in biotic and abiotic stress responses (Gill and Tuteja, 2010; Tiburcio et al., 2014). Under their biosynthesis and catabolism, their mechanism provides PAs in different forms: free, conjugated, or bounded in plants (Minocha et al., 2014; Shi and Chan, 2014; Alcázar et al., 2020). These complex features give a multifunctional role to PAs in the development and also in the engineered enhancement of stress tolerance; however, our knowledge is limited about PAs interactions with defense-related phytohormones (Hussain et al., 2011; Pál et al., 2015; Khajuria et al., 2018). The picture could be more complex as there is also crosstalk between oxidative stress responses thus inducing adaptation responses (Pottosin et al., 2014). As a consequence, PAs exhibit different roles under both biotic and abiotic stress (Seifi and Shelp, 2019). Several studies have suggested that PAs could be prominent plant growth regulators in the service of sustainable agriculture inducing abiotic stress tolerance in plants (Szepesi, 2019).

Plant responses to diverse environmental conditions are regulated by various phytohormones (Cramer et al., 2011; Peleg and Blumwald, 2011). Traditionally, salicylic acid (SA), jasmonates (JAs), ethylene (ET), and abscisic acid are considered the main defense-related phytohormones, whereas auxins (IAAs), gibberellins (GAs), cytokinins (CKs), and brassinosteroids (BRs) are associated with plant growth and developmental processes. At the same time, phytohormones are in a close relationship and all of them regulate the stress responses or developmental processes of plants in a complex manner (Kazan, 2015; Xia et al., 2015).

In this chapter, we highlighted the current state of knowledge about the effects of defense-related phytohormones, such as SA, JA, and ET on PA metabolism in different plant species, genotypes, or organs, while the effect of exogenous PA application was also discussed on the biosynthesis and signaling of these hormones. Moreover, the crosstalk was also reviewed between phytohormones on PA metabolism. This knowledge can help to increase our ability to improve stress resistance by modulating PA metabolism and phytohormone-mediated defense signaling in crops in the future under the changing environment.

13.2 The main steps of polyamine metabolism

PAs could be synthesized from amino acids, L-arginine, or L-ornithine to producing the diamine putrescine (Put), which can be a substrate for biosynthesis of higher Pas, such as triamine spermidine (Spd) and tetramine spermine (Spm) (Fig. 13.1). However, it is a species-dependent manner from which pathway could be found, for example, in *Arabidopsis thaliana*, ornithine decarboxylase (ODC) is missing (Urano et al., 2003). These PAs are under the most intensive examination, but other Pas, such as cadaverine (Cad) or thermospermine (T-Spm), have been also deciphered to have an important role in the regulation of plant development

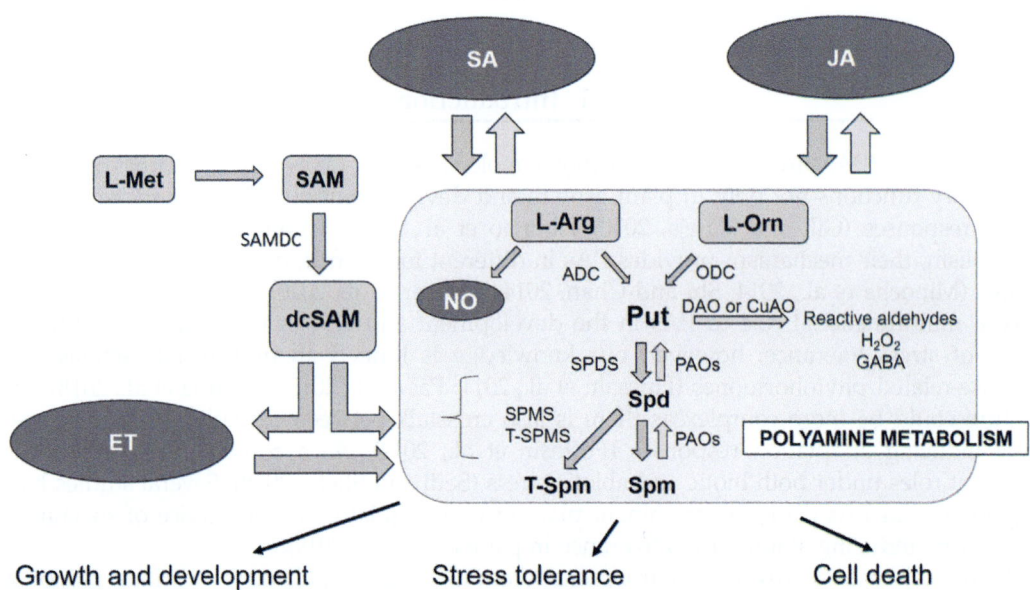

FIGURE 13.1 Schematic diagram of the possible interaction between the key defense-related phytohormones and PA metabolism in crops. *ADC*, arginine decarboxylase; *DAO or CuAO*, diamine oxidase or copper amine oxidase; *dcSAM*, decarboxylated SAM; *ET*, ethylene; *GABA*, Ψ amino butyric acid; *H₂O₂*, hydrogen peroxide; *JA*, jasmonic acid; *L-Arg*, L-arginine; *L-Met*, L-methionine; *L-Orn*, L-ornithine; *NO*, nitric oxide; *ODC*, ornithine decarboxylase; *PAOs*, polyamine oxidases; *Put*, putrescine; *SA*, salicylic acid; *SAMDC*, S-adenosylmethionine decarboxylase; *SAM*, S-adenosyl methionine; *Spd*, spermidine; *SPDS*, spermidine synthase; *Spm*, spermine; *SPMS*, spermine synthase; *T-SPMS*, T-SPM synthase.

and stress responses (Jancewicz et al., 2016). Several reviews describe the precise mechanism of biosynthetic pathways of PAs and suggest different modes of studying these processes, including the use of pharmacological inhibitors or transgenic modifications (Alcázar et al., 2010; Cheng et al., 2018). One of the most important interactions is the ET-PA biosynthetic pathway connecting by decarboxylation of S-adenosylmethionine (SAM). As decarboxylated SAM generated by S-adenosylmethionine decarboxylase (SAMDC) is essential to generate Spd, some studies suggest that the biosynthetic pathways of PAs and ET are competitive (Majumdar et al., 2017). The enzyme, Spd synthase (SPDS), is involved in the synthesis of Spd, whereas Spm synthase (SPMS) catalyzes the Spm formation. Moreover, there is evidence for the involvement of SPDS in some senescence-related processes in plants as overexpression of this enzyme from yeast altered the ripening, senescence, and decay of tomato fruits (Nambeesan et al., 2010).

Degradation of PAs has also a critical function in plant development and defense as it could stabilize the balance of PA-related responses (Cona et al., 2006; Moschou et al., 2008). PA catabolic events are mediated by two different types of enzymes: diamine oxidases or copper-containing amine oxidases (DAOs or CuAOs) and FAD-containing polyamine oxidases (PAOs). During catabolic processes, reactive aldehydes and hydrogen peroxide (H_2O_2) can be synthesized as by-products influencing the stress responses of plants (Yu et al., 2019). For example, PA catabolism was demonstrated that could be responsible for inducing leaf senescence (Sobieszczuk-Nowicka, 2017). Many data suggest that some PAOs are able to convert higher PAs into Spd or Put (back conversion), while others only take part in terminal catabolism. Interestingly, during catabolism of PAs, gamma-amino-butyric acid (GABA) also could be synthesized, which is involved in different plant abiotic stress responses (Gupta et al., 2016). Moreover, (Takács et al., 2017) provided evidence that PAO was responsible for nitric oxide (NO) production, dependent on studied plant organs, together with H_2O_2 in tomatoes exposed to salt stress, suggesting that even NO synthesis can be triggered because of the catabolism of PAs. At the same time, CuAOs and PAOs show different roles in plant tissue differentiation and organ development (Tavladoraki et al., 2016; Kaszler et al., 2021).

Recently, some reports revealed the GABA-induced responses in plant stress responses (Li et al., 2021). Kim et al. (2013) showed in pepper that arginine decarboxylase (ADC) is essential for proper PAs and GABA signal pathways in cell death processes and defense. GABA could act as a priming agent protecting plants against *Botrytis cinerea* via ROS production (Janse van Rensburg and Van den Ende, 2020) showing that GABA can be a useful molecule to induce biotic stress tolerance in agriculture. Exogenously added GABA could rapidly metabolize to succinic acid feeding the TCA cycle (Hijaz and Killiny, 2019). Inhibition of GABA transaminases was reported to enhance GABA accumulation and resulted in the development of dwarf and infertile mutant tomato plants, demonstrating the importance of GABA homeostasis in plant development (Koike et al., 2013). GABA took part in inducing heat and drought tolerance in creeping bentgrass (*Agrostis stolonifera*) by altering the transcript levels of stress-responsive genes and transcriptional factors (Li et al., 2018). GABA also altered endogenous PAs and organic metabolites in creeping bentgrass (Li et al., 2020). A connection between phytohormones and PAs was revealed in regulating plant stress responses via altered GABA pathway (Podlešáková et al., 2019). It can be concluded that GABA and related amino acids could take part in plant immune responses (Tarkowski et al., 2020).

13.3 Salicylic acid and polyamines

Defense mechanisms of plants are regulated by several phytohormones, such as SA (Khan et al., 2015; Vlot et al., 2009). It is well known that this phenolic compound plays a significant role in the regulation of defense signaling under various abiotic and biotic stress effects in plants (Hayat et al., 2010; Rivas-San Vicente and Plasencia, 2011). Moreover, SA is crucial in the induction of both local and systemic acquired resistance after pathogen infection (Klessig et al., 2018). There is a close interaction between ROS metabolism and SA. A lethal increase in the levels of SA upon various stressors induces the rapid accumulation of ROS, leading to oxidized proteins and lipids, and finally results in local cell death (Poór, 2020). At the same time, a lower concentration of endogenous SA regulates various components of defense responses of plants, such as the enzymatic and nonenzymatic antioxidants, several antimicrobial components of plants and pathogenesis-related (PR) proteins (Maruri-López et al., 2019; Tripathi et al., 2019). Thus the interaction between SA and PAs could be crucial in the aspect of the output of stress-induced defense reaction of plants.

It was found that exogenous treatments with SA induced the accumulation of several PAs in various organs of plants. SA increased the levels of Put, Spd, and Spm in maize (*Zea mays*) seedlings (Szalai et al., 2016), peach (*Prunus persica*) fruits (Cao et al., 2010), potato (*Solanum tuberosum*) shoots (Ma et al., 2018), "Qingnai" plums (*Prunus salicina*) fruits (Luo et al., 2011), and tomato (*Solanum lycopersicum*) roots and leaves (Szepesi et al., 2009; Gharbi et al., 2016; Takács et al., 2016, 2021) by concentration- and time-dependent manner. We can conclude that the accumulation of PAs induced by SA is dependent on plant genotypes and plant organs (Szalai et al., 2017). In addition, Otálora et al. (2020) observed that the increase in PA contents induced by SA is highly dependent on the temperature, PA levels were elevated by higher temperature under SA treatment. Takács et al. (2016) found that this process is dependent on the presence of light or other phytohormones, such as ET (Takács et al., 2021). Other authors analyzed the expression of various biosynthetic genes of PAs after SA treatments. First, Zhang et al. (2011) measured that SA increased the gene expression and activity of ADC and ODC, as well as the accumulation of Put, Spd and Spm in fruits of tomato. Later, Mo et al. (2015a) reported that SA elevated the expression levels of *AtADC2*, *AtSAMDC1/2*, *AtSPDS1/2*, *AtSPMS*, *AtPAO2*, and *AtPAO3*, indicating that PA biosynthesis was entirely induced by SA. In addition, GhACL5, encoding thermospermine (T-Spm) synthase from cotton, was induced by SA (Mo et al., 2015b). Later, they found that expression of GhSPMS and GhSAMDC was also induced by SA in *Arabidopsis* within 12 hours (Mo et al., 2016).

Not only SA elevated PA levels but also several PAs influenced SA contents and signaling. Put induced SA accumulation in the leaves of *Arabidopsis* (Liu et al., 2020), roots of maize (Szalai et al., 2017), and leaves of wheat plants (Pál et al., 2019). Exogenously applied Spm via the root system also elevated SA contents in the leaves but not in the roots of wheat (Pál et al., 2019) and maize (Szalai et al., 2017). It increased PR-1, PR-2, PR-3, and PR-5 protein accumulation and the activity of SA-induced protein kinase (SIPK) in tobacco (*Nicotiana tabacum*) detached leaves (Yamakawa et al., 1998; Takahashi et al., 2003).

In addition, it was observed that the overexpression of *S-adenosylmethionine decarboxylase* (*GhSAMDC*) from cotton resulted in higher SA levels in *Arabidopsis* (Mo et al., 2016), but the accumulation of SA declined in *CaADC1*-silenced *Nicotiana benthamiana* leaves compared with control leaves under *Xanthomonas campestris* infection (Kim et al., 2013).

Thus we can conclude that SA and Put, as well as Spm can increase each other's levels in various plant genotypes and organs by concentration- and time-dependent manner.

13.4 Jasmonic acid and polyamines

Besides SA, jasmonic acid (JA) also plays a regulatory role under plant—pathogen interactions and in resistance processes upon various abiotic stress (Van der Ent et al., 2009; Santino et al., 2013; Wasternack, 2015). JA induces the production of several antifungal proteins, such as defensin, as well as protease inhibitors within hours after the infection or wounding, as well as increases the activity of other defense-related enzymes (Hu et al., 2009; Wasternack and Hause, 2013). However, the interaction between JA and PAs has not been known in full detail.

Methyl jasmonate (MeJA) increased the content of Put in barley (*Hordeum vulgare*) leaves (Walters et al., 2002), buckwheat (*Fagopyrum esculentum*) cotyledons (Horbowicz et al., 2015), Egyptian henbane (*Hyoscyamus muticus*) root culture (Biondi et al., 2000), loquat (*Eriobotrya japonica*) fruit (Cao et al., 2014), potato (*S. tuberosum*) root (Mader, 1999), tobacco (*N. tabacum*) cells (Imanishi et al., 1998), and tomato fruit (Min et al., 2020). Similar to Put, Spm was also elevated by MeJA in barley leaves (Walters et al., 2002), Egyptian henbane root culture (Biondi et al., 2000), lima bean (*Phaseolus lunatus*) leaves (Ozawa et al., 2009), loquat fruit (Cao et al., 2014), and tomato fruits (Min et al., 2020). Spd was also enhanced by MeJA in Egyptian henbane root culture (Biondi et al., 2000), loquat fruit (Cao et al., 2014), potato root (Mader, 1999), and tomato fruit (Min et al., 2020). MeJA analog *n*-propyl dihydrojasmonate elevated also Put, Spd, and Spm contents in the epicarp and mesocarp of peach (*P. persica*) fruits after 21 days (Ziosi et al., 2009). These and other results (Mader, 1999) confirmed the organ-specific effects of JA on the elevation of PA contents. In addition, Min et al. (2020) confirmed the time-dependent effects of JA on the biosynthesis of various PAs. They observed that MeJA increased the activity of arginase, ADC, and ornithine aminotransferase (OAT) after 6, 9, and 12 days, and expression of *SlARG2*, *SlADC*, ornithine decarboxylase (*SlODC*), and *SlOAT* already after 6 days of the treatment. However, increased accumulation of Put and Spd was observed after only 3 days, and Spm was increased after 9 days (Min et al., 2020). Others also found the rapid expression of ODC, SAMDC, SAMS, and PMT by MeJA after 6 hours, before the detection of the elevated Put content after 24 hours (Imanishi et al., 1998). At the same time, the activation of catabolic enzymes was also detected. MeJA elevated the activity of DAO in barley leaves (Walters et al., 2002) and the expression of *OsPAO2* and *OsPAO6* in rice (*Oryza sativa*) seedlings after 6 hours (Sagor et al., 2021).

Close interaction between JA and PAs was also confirmed by exogenous Spm treatments. It was found that Spm application increased JA levels in apple (*Malus sylvestris*) fruits (Yoshikawa et al., 2007) and mangosteen (*Garcinia mangostana*) fruits (Kondo et al., 2004) within days which was beneficial in cold stress resistance. However, exogenous Spm treatment did not change basically JA levels in the leaf but increased it under osmotic stress as compared to the polyethylene glycol (9%)-treated soybeans (*Glycine max*) plants

(Radhakrishnan and Lee, 2013). At the same time, the effects of exogenous Spd and Spm on JA levels have not been investigated.

It can be concluded that JA and PA can also regulate each other's levels similarly to SA.

13.5 Ethylene and polyamines

Not only SA and JA play role in the regulation of plant responses against various stressors, but the gaseous ET also takes part in the fine regulation of defense mechanisms in plants (Khan et al., 2019, 2020a, 2020b, 2021b; Koornneef and Pieterse, 2008; Poór et al., 2021a). It is well known while SA mainly controls plant responses against bio-trophic pathogens, JA and ET are basically involved in defense processes against insects or necrotrophic pathogens that kill cells (Glazebrook, 2005). However, the role of ET in stress responses is rather contradictory since it can promote disease or stress resistance but it can also induce cell death of plants in a concentration-dependent manner, mostly by regulating the metabolism of ROS (van Loon et al., 2006; Poór et al., 2013; Borbély et al., 2019). ET is synthesized from SAM by the conversion of SAM to 1-aminocyclopropane-1-carboxylic acid (ACC) and then the oxidation of ACC to ET is catalyzed by ACC synthase (ACS) and ACC oxidase (ACO), respectively (Sun et al., 2017). While there is a common precursor for both ET and PA biosynthesis, and the relationship between ET and PAs is significantly examined but the currently available data is contradictory, especially in crops (Pál and Janda, 2017).

The effects of ET can be detected by the exogenous application of ET precursor ACC, however, it has ET-independent signaling effects (Van de Poel, 2020), respectively. In addition, ethephon as ET-releasing compound, as well as ET gas are also used to detect its effects on crops, especially in fruits. Thus scientific results could be basically different depending not only on the duration of the application or the investigated plant species and organs but also on ET treatments/chemicals. First, it was found that treatments with ACC resulted in significantly higher Put levels in barley root (Tamai et al., 1999) and spikelets, grains, and panicles of rice (Wang et al., 2012; Chen et al., 2013; Zhang et al., 2017). Ethephon treatment also increased Put in olive trees (*Olea europaea*) fruits (Gil-Amado and Gomez-Jimenez, 2012), spikelets and grains of rice (Wang et al., 2012; Chen et al., 2013), and wheat germinating grains (Yang et al., 2014). ET gas also stimulated the accumulation of Put within 24 hours in *Arabidopsis* by concentration-dependent manner (Rakitin et al., 2009) and promoted the accumulation of Cad in common ice plant (*Mesembryanthemum crystallinum*) leaves (Shevyakova et al., 2001), respectively. At the same time, ET gas did not change Put level in potato (*S. tuberosum*) tuber after 33 weeks (Jeong et al., 2002). The action of ET can be blocked by using the ET biosynthesis inhibitor aminoethoxyvinylglycine (AVG) or CoCl$_2$. The ET receptor can be deactivated using 1-methylcyclopropene (1-MCP) or silver ions as by silver thiosulfate (STS) or AgNO$_3$. Interestingly, the application of AVG also increased Put levels in seedlings of barley after 3 days (Locke et al., 2000) and in in vitro cotyledonary explants of cucumber after 18 days (Zhu and Chen, 2005). Surprisingly, treatments with ACC, AVG, and STS did not change total PA levels in pepper (*Capsicum annuum*) in vitro plants during 30 days (Batista et al., 2013). At the same time, the application of AVG resulted in higher Put levels after 1 day, and then, it decreased in tissue culture of tobacco (Torrigiani et al., 2003) showing the rapid time-dependent effects of the ET biosynthesis inhibitor. AVG decreased Put content in somatic embryos of *Medicago sativa* (Huang et al., 2001), spikelets, grains, and panicles of rice (Wang et al., 2012; Chen et al., 2013; Zhang et al., 2017), and grains of wheat (Yang et al., 2014). Not only the inhibition of biosynthesis but also the application of the ET action inhibitor 1-MCP resulted in lower Put levels in apple fruits after 46 weeks (Deyman et al., 2014). At the same time, 1-MCP treatments did not change or increased Put levels in apple fruits after 15 days (Pang et al., 2006), nor in nectarine (*P. persica*) fruit (Bregoli et al., 2002), pepper fruits (Cao et al., 2012), and tomato fruits (Tassoni et al., 2006). In addition, slightly higher Put content was measured in ET receptor mutant *Never ripe* tomato leaves, respectively (Takács et al., 2021). These results suggest that there are different effects of the application of ET biosynthesis inhibitors or the inhibition of ET action on PA synthesis in crops. Furthermore, the level of Put is highly depending on its biosynthesis mediated by ODC (indirectly arginase) and ADC and the activity of the catabolic enzyme DAO. AVG enhanced ADC and ODC but reduced DAO activity after 7 days in tobacco tissue culture (Torrigiani et al., 2003). Application of 1-MCP increased the expression of ODC but suppressed ADC in the first days in tomato fruits (Tassoni et al., 2006). The expression of tomato ADC and ODC, as well as DAO activity, was not different in *N. ripe* as compared to wild-type tomato leaves (Takács et al., 2021).

In contrast to Put, levels of Spd and Spm can change differently upon modification of ET biosynthesis or action. ACC exposure did not change Spd and Spm contents in barley roots (Tamai et al., 1999) and in pepper in vitro plants (Batista et al., 2013), but it reduced their levels in spikelets, grains, and panicles of rice (Wang et al., 2012; Chen et al., 2013; Zhang et al., 2017). Ethephon treatment also decreased Spd and Spm contents in olive fruits (Gil-Amado and Gomez-Jimenez, 2012), spikelets and grains of rice (Wang et al., 2012; Chen et al., 2013), and wheat germinating grains (Yang et al., 2014). Ethephone also increased the expression of thermospermine (T-Spm) synthase in *Arabidopsis* (Mo et al., 2015b). Treatments with AVG increased Spm and Spd contents in barley seedlings (Locke et al., 2000), spikelets, grains, and panicles of rice (Wang et al., 2012; Chen et al., 2013; Zhang et al., 2017), tobacco tissue culture (Torrigiani et al., 2003), and wheat germinating grains (Yang et al., 2014). In addition, content of Spd increased in *M. sativa* somatic embryo (Huang et al., 2001) and Spm was accumulated in cucumber in vitro explants (Zhu and Chen, 2005) upon AVG. Treatment with 1-MCP, also increased Spd and Spm contents in the first 15 days in apple fruit (Pang et al., 2006), pepper fruit (Cao et al., 2012), and tomato fruits (Tassoni et al., 2006). Levels of Spd increased in nectarine fruit (Bregoli et al., 2002) after 1-MCP exposure. Furthermore, treatment with $CoCl_2$ increased Spd, Spm, and Cad contents in olive fruits (Gil-Amado and Gomez-Jimenez, 2012). At the same time, Spd and Spm levels were basically lower in *N. ripe* tomato leaves (Takács et al., 2021). These results suggest that there are different effects of the inhibition of ET action in fruits and leaves. AVG by concentration-dependent manner increased the activity of SAMDC in *M. sativa* somatic embryos (Huang et al., 2001), peach fruits (Bregoli et al., 2002), and tobacco tissue culture (Torrigiani et al., 2003) according to the higher Spd and Spm levels. 1-MCP also promoted the expression of *SAMDC*, *Spd synthase*, and *Spm synthase* in tomato fruits in time-dependent manner (Tassoni et al., 2006).

Inhibition of ET biosynthesis or its action can also modulate the levels of PAs but also PAs can influence ET biosynthesis and emission. Put, Spd, and Spm elevated the expression of ET biosynthesis genes, *ACS1* and *ACO1* in Chinese pink (*Dianthus chinensis*) in vitro shoots (Sreelekshmi and Siril, 2020). Exogenous application of Put significantly elevated ET production in the grains during the grain filling stage in wheat (Liu et al., 2016) and in rice leaves (Quinet et al., 2010). In contrast to this, Put significantly reduced ET production in common chicory (*Cichorium intybus*) shoot culture (Bais et al., 2000), peach fruit (Bregoli et al., 2002), and wheat leaf (Hassanein et al., 2013). At the same time, Put did not change ET levels in pepper in vitro plants (Batista et al., 2013), panicles and spikelets of rice (Wang et al., 2012; Zhang et al., 2017), and strawberry (*Fragaria ananassa*) fruits (Guo et al., 2018). Overexpression of *ADC* and mouse *ODC* resulted in higher Put but reduced ET levels in tomato fruits (Pandey et al., 2015; Gupta et al., 2019).

Spd treatments decreased ET production in kiwi (*Actinidia deliciosa*) fruits (Jhalegar et al., 2012), peach fruit (Bregoli et al., 2002), pepper in vitro plants (Batista et al., 2013), panicles and spikelets of rice (Wang et al., 2012; Zhang et al., 2017), and grains of wheat (Liu et al., 2013, 2016; Yang et al., 2014), but it increased ACC content in white spruce (*Picea glauca*) somatic embryo (Meskaoui and Trembaly, 2009).

Application of Spm showed (resulted in) more complicate results. Spm treatments reduced ET production in kiwi (*A. deliciosa*) fruits (Jhalegar et al., 2012), peach fruit (Bregoli et al., 2002), pepper in vitro plants (Batista et al., 2013), spikelets of rice (Wang et al., 2012), soybean (*G. max*) pods (Song et al., 2018), and wheat grains (Liu et al., 2013, 2016), similarly to Spd. However, Spm treatments elevated ET and ACC contents and ACO activity in chick-pea (*Cicer arietinum*) germinating seeds (Gómez-Jiménez et al., 2001), hibiscus (*Hibiscus syriacus*) flowers (Seo et al., 2009), maize (*Z. mays*) leaves (Talaat and Shawky, 2016), and strawberry fruits (Guo et al., 2018).

Experiments with inhibitors of PA biosynthesis, such as methylglyoxal-bis (guanylhydrazone) (MGBG), or by using transgenic plants can help to understand the very complicated PAs–ET relationships in crops. Inhibition of SAMDC by MGBG increased ET emission in cucumber in vitro plants (Zhu and Chen, 2005), hibiscus flowers (Seo et al., 2009), panicles and spikelets of rice (Wang et al., 2012; Zhang et al., 2017), strawberry fruits (Guo et al., 2018), and wheat grains (Yang et al., 2014) but did not change ET production in pepper (Batista et al., 2013). Interestingly, overexpression of *MfSAMS1* did not change Put, Spd, and Spm levels and the ET production in tobacco (Guo et al., 2014). At the same time, overexpression of *MfERF1* resulted in a higher transcript levels of *SAMS*, *SAMDC1*, *SAMDC2*, *SPDS1*, *SPDS2*, and *SPMS*, as well as *PAO* as compared to control tobacco leaves (Zhuo et al., 2018), suggesting the regulatory role of ET on PA biosynthesis. Overexpression of *human-SAMDC* resulted in reduced ET levels but higher Put, Spd, and Spm contents in tomato fruits (Madhulatha et al., 2014). At the same time, ET emission decreased in both *FaSAMDC* RNAi and OE fruits of strawberries (Guo et al., 2018). Overexpression of yeast *spermidine synthase* (*ySpdSyn*) in tomato also resulted in changes in the rate of ET production (Nambeesan et al., 2010). ACC and ET contents increased significantly in RNAi lines and decreased in OE lines of germinating seeds of rice suggesting that *OsSPMS1* affects ET synthesis (Tao et al., 2018). It can be concluded that the ET-PAs relationship is very complicated that depends on various endogenous and exogenous factors, such as developmental stages or presence of various stressors.

13.6 Conclusions

Despite the fact that the interaction between phytohormones and PAs, especially the role of ET on Pas, has been intensively studied over the last 30 years, there are many gaps in our knowledge regarding to the regulation of SA, JA, and ET on PA metabolism of crops. In this chapter, we highlighted the current state of the effects of defense-related phytohormones on PA levels in different plant species, genotypes, or organs, while the effect of exogenous PA application was also discussed on phytohormone biosynthesis and signaling. Moreover, the crosstalk was reviewed between phytohormones on PA metabolism.

Based on the revised works, we can conclude that all defense-related phytohormones influenced PA metabolism and exogenous treatments with various PAs significantly changed the accumulation of the investigated hormones in various plant organs (fruit, leaf, and root), but there can be significant differences between certain plant species and organs. All of the investigated phytohormones, SA, JA, and ET increased the levels of Put. Spd and Spm contents were elevated by SA but mostly reduced by ET precursor ACC or ethephon. Fluctuation of PA metabolism can be dependent on the dose and duration of hormone treatments, such as it was detected in the case of JA. Only a few authors investigated the impact of the different concentrations of phytohormones on the selected plant species and organs. Moreover, PA metabolism shows distinct patterns of the biosynthesis, conversion, and degradation, which can be altered by the defense-related hormones in a time-dependent manner. The interactions between the hormones in PA metabolisms such as in the case of SA and ET should be investigated. Unfortunately, most of the authors determined PA and/or hormone levels only at one time-point after the treatments, although the rise may vary from some hours to a few days or weeks, especially in the case of fruits. Some authors detected the effects of hormone biosynthesis inhibitors and inhibitors of receptors on PA metabolism in the case of ET, therefore influencing the stress tolerance mechanisms in the investigated plant species and organs, mostly in fruits. These data contributed to the better understanding of the effects of ET on PA metabolism, respectively. In the case of ET, many transgenic plants also served to increase the scientific knowledge of this interaction. Not only hormones altered PA levels, but also several PAs influenced hormone contents and signaling. Put induced SA accumulation while Spm treatment resulted in higher SA and JA levels as well. In contrast, Put and Spd significantly reduced ET production in many plant species but it was found that Spm could both induce and reduce ET emission. This complicated relationship needs further research.

In the future, the accurate investigation of the role of key enzymes in PA metabolisms, such as arginase, ADC, ODC, SPDS, SPMS, SAMDC, PAO, and DAO, would provide new insights into the complex signaling pathways under different hormonal conditions. It would be interesting to examine the combined role of phytohormones with PAs in the plant responses to different stressors, such as drought and pathogens. Understanding the mechanism that can regulate hormone levels upon exogenous PA treatments at the cellular, tissue, organ, or whole plant levels have a great importance in plant biology and agriculture. Precise investigation of the effects of exogenous PA treatments on fruits could be used to increase the yield and stress tolerance under today's changing environment.

Acknowledgments

We apologize to those colleagues whose work was not reviewed here. This work was supported by the grants from the National Research, Development and Innovation Office of Hungary—NKFIH (Grant no. NKFIH FK 138867 and FK 129061) and by the UNKP-21−5 New National Excellence Program of the Ministry of Human Capacities. Péter Poór was supported by the János Bolyai Research Scholarship of the Hungarian Academy of Sciences.

Conflicts of interest

The authors declare that the research was conducted in the absence of any commercial or financial relationships that could be construed as a potential conflict of interest.

Author contributions

Authors contributed equally to this work.

References

Alcázar, R., Altabella, T., Marco, F., Bortolotti, C., Reymond, M., Koncz, C., et al., 2010. Polyamines: molecules with regulatory functions in plant abiotic stress tolerance. Planta 231 (6), 1237−1249.

Alcázar, R., Bueno, M., Tiburcio, A.F., 2020. Polyamines: small amines with large effects on plant abiotic stress tolerance. Cells 9 (11), 2373.

Bais, H.P., Sudha, G.S., Ravishankar, G.A., 2000. Putrescine and silver nitrate influences shoot multiplication, in vitro flowering and endogenous titers of polyamines in Cichorium intybus L. cv. Lucknow local. J. Plant Growth Regul. 19 (2), 238−248.

Batista, D.S., Dias, L.L.C., Macedo, A.F., do Rêgo, M.M., do Rêgo, E.R., Floh, E.I.S., et al., 2013. Suppression of ethylene levels promotes morphogenesis in pepper (Capsicum annuum L. Vitro Cell Dev. Biol. Plant 49 (6), 759−764.

Biondi, S., Fornale, S., Oksman-Caldentey, K.M., Eeva, M., Agostani, S., Bagni, N., 2000. Jasmonates induce over-accumulation of methylputrescine and conjugated polyamines in Hyoscyamus muticus L. root cultures. Plant Cell Rep. 19 (7), 691−697.

Borbély, P., Bajkán, S., Poór, P., Tari, I., 2019. Exogenous 1-Aminocyclopropane-1-carboxylic acid controls photosynthetic activity, accumulation of reactive oxygen or nitrogen species and macroelement content in tomato in long-term experiments. J. Plant Growth Regul. 38 (3), 1110−1126.

Bregoli, A.M., Scaramagli, S., Costa, G., Sabatini, E., Ziosi, V., Biondi, S., et al., 2002. Peach (Prunus persica) fruit ripening: aminoethoxyvinylglycine (AVG) and exogenous polyamines affect ethylene emission and flesh firmness. Physiol. Plant. 114 (3), 472−481.

Cao, S., Hu, Z., Zheng, Y., Lu, B., 2010. Synergistic effect of heat treatment and salicylic acid on alleviating internal browning in cold-stored peach fruit. Postharvest Biol. Technol. 58 (2), 93−97.

Cao, S., Yang, Z., Zheng, Y., 2012. Effect of 1-methylcyclopene on senescence and quality maintenance of green bell pepper fruit during storage at 20 C. Postharvest. Biol. Technol. 70, 1−6.

Cao, S., Cai, Y., Yang, Z., Joyce, D.C., Zheng, Y., 2014. Effect of MeJA treatment on polyamine, energy status and anthracnose rot of loquat fruit. Food Chem. 145, 86−89.

Chen, T., Xu, Y., Wang, J., Wang, Z., Yang, J., Zhang, J., 2013. Polyamines and ethylene interact in rice grains in response to soil drying during grain filling. J. Exp. Bot. 64 (8), 2523−2538.

Cheng, B., Li, Z., Liang, L., Cao, Y., Zeng, W., Zhang, X., et al., 2018. The γ-aminobutyric acid (GABA) alleviates salt stress damage during seeds germination of white clover associated with Na+/K+ transportation, dehydrins accumulation, and stress-related genes expression in white clover. Int. J. Mol. Sci. 19 (9), 2520.

Cona, A., Rea, G., Angelini, R., Federico, R., Tavladoraki, P., 2006. Functions of amine oxidases in plant development and defence. Trends Plant Sci. 11 (2), 80–88.

Cramer, G.R., Urano, K., Delrot, S., Pezzotti, M., Shinozaki, K., 2011. Effects of abiotic stress on plants: a systems biology perspective. BMC Plant Biol. 11 (1), 1–14.

Deyman, K.L., Brikis, C.J., Bozzo, G.G., Shelp, B.J., 2014. Impact of 1-methylcyclopropene and controlled atmosphere storage on polyamine and 4-aminobutyrate levels in "Empire" apple fruit. Front. Plant Sci. 5, 144.

Gharbi, E., Martínez, J.P., Benahmed, H., Fauconnier, M.L., Lutts, S., Quinet, M., 2016. Salicylic acid differently impacts ethylene and polyamine synthesis in the glycophyte Solanum lycopersicum and the wild-related halophyte Solanum chilense exposed to mild salt stress. Physiol. Plant. 158 (2), 152–167.

Gil-Amado, J.A., Gomez-Jimenez, M.C., 2012. Regulation of polyamine metabolism and biosynthetic gene expression during olive mature-fruit abscission. Planta 235 (6), 1221–1237.

Gill, S.S., Tuteja, N., 2010. Polyamines and abiotic stress tolerance in plants. Plant Signal. Behav. 5 (1), 26–33.

Glazebrook, J., 2005. Contrasting mechanisms of defense against biotrophic and necrotrophic pathogens. Annu. Rev. Phytopathol. 43, 205–227.

Gómez-Jiménez, M.D.C., García-Olivares, E., Matilla, A.J., 2001. 1-Aminocyclopropane-1-carboxylate oxidase from embryonic axes of germinating chick-pea (Cicer arietinum L.) seeds: cellular immunolocalization and alterations in its expression by indole-3-acetic acid, abscisic acid and spermine. Seed Sci. Res. 11 (3), 243–253.

Guo, Z., Tan, J., Zhuo, C., Wang, C., Xiang, B., Wang, Z., 2014. Abscisic acid, H2O2 and nitric oxide interactions mediated cold-induced S-adenosylmethionine synthetase in Medicago sativa subsp. falcata that confers cold tolerance through up-regulating polyamine oxidation. Plant Biotechnol. J. 12 (5), 601–612.

Guo, J., Wang, S., Yu, X., Dong, R., Li, Y., Mei, X., et al., 2018. Polyamines regulate strawberry fruit ripening by abscisic acid, auxin, and ethylene. Plant. Physiol. 177 (1), 339–351.

Gupta, K., Sengupta, A., Chakraborty, M., Gupta, B., 2016. Hydrogen peroxide and polyamines act as double edged swords in plant abiotic stress responses. Front. Plant Sci. 7, 1343.

Gupta, A., Pandey, R., Sinha, R., Chowdhary, A., Pal, R.K., Rajam, M.V., 2019. Improvement of post-harvest fruit characteristics in tomato by fruit-specific over-expression of oat arginine decarboxylase gene. Plant Growth Regul. 88 (1), 61–71.

Hassanein, R.A., El-Khawas, S.A., Ibrahim, S.K., El-Bassiouny, H.M., Mostafa, H.A., Abdel-Monem, A.A., 2013. Improving the thermo tolerance of wheat plant by foliar application of arginine or putrescine. Pak. J. Bot. 45 (1), 111–118.

Hayat, Q., Hayat, S., Irfan, M., Ahmad, A., 2010. Effect of exogenous salicylic acid under changing environment: a review. Environ. Exp. Bot. 68 (1), 14–25.

Hijaz, F., Killiny, N., 2019. Exogenous GABA is quickly metabolized to succinic acid and fed into the plant TCA cycle. Plant Signal. Behav. 14 (3), e1573096.

Horbowicz, M., Wiczkowski, W., Szawara-Nowak, D., Sawicki, T., Kosson, R., Sytykiewicz, H., 2015. The level of flavonoids and amines in de-etiolated and methyl jasmonate treated seedling of common buckwheat. Phytochem. Lett. 13, 15–19.

Hu, X., Wansha, L., Chen, Q., Yang, Y., 2009. Early signals transduction linking the synthesis of jasmonic acid in plant. Plant Signal. Behav. 4 (8), 696–697.

Huang, X.L., Li, X.J., Li, Y., Huang, L.Z., 2001. The effect of AOA on ethylene and polyamine metabolism during early phases of somatic embryogenesis in Medicago sativa. Physiol. Plant. 113 (3), 424–429.

Hussain, S.S., Ali, M., Ahmad, M., Siddique, K.H., 2011. Polyamines: natural and engineered abiotic and biotic stress tolerance in plants. Biotechnol. Adv. 29 (3), 300–311.

Imanishi, S., Hashizume, K., Nakakita, M., Kojima, H., Matsubayashi, Y., Hashimoto, T., et al., 1998. Differential induction by methyl jasmonate of genes encoding ornithine decarboxylase and other enzymes involved in nicotine biosynthesis in tobacco cell cultures. Plant Mol. Biol. 38 (6), 1101–1111.

Jancewicz, A.L., Gibbs, N.M., Masson, P.H., 2016. Cadaverine's functional role in plant development and environmental response. Front. Plant Sci. 7, 870.

Janse van Rensburg, H.C., Van den Ende, W., 2020. Priming with γ-aminobutyric acid against Botrytis cinerea reshuffles metabolism and reactive oxygen species: dissecting signalling and metabolism. Antioxidants 9 (12), 1174.

Jeong, J.C., Prange, R.K., Daniels-Lake, B.J., 2002. Long-term exposure to ethylene affects polyamine levels and sprout development in 'Russet Burbank' and 'Shepody' potatoes. J. Am. Soc. Hortic. Sci. 127 (1), 122–126.

Jhalegar, M.J., Sharma, R.R., Pal, R.K., Rana, V., 2012. Effect of postharvest treatments with polyamines on physiological and biochemical attributes of kiwifruit (Actinidia deliciosa) cv. Allison. Fruits 67 (1), 13−22.

Kaszler, N., Benkó, P., Bernula, D., Szepesi, Á., Fehér, A., Gémes, K., 2021. Polyamine metabolism is involved in the direct regeneration of shoots from arabidopsis lateral root primordia. Plants 10 (2), 305.

Kazan, K., 2015. Diverse roles of jasmonates and ethylene in abiotic stress tolerance. Trends Plant Sci. 20 (4), 219−229.

Khajuria, A., Sharma, N., Bhardwaj, R., Ohri, P., 2018. Emerging role of polyamines in plant stress tolerance. Curr. Protein Peptide Sci. 19 (11), 1114−1123.

Khan, M.I.R., Fatma, M., Per, T.S., Anjum, N.A., Khan, N.A., 2015. Salicylic acid-induced abiotic stress tolerance and underlying mechanisms in plants. Front. in plant sci. 6, 462.

Khan, M.I.R., Fatma, M., Per, T.S., Anjum, N.A., Khan, N.A., 2021b. Salicylic acid-induced abiotic stress tolerance and underlying mechanisms in plants. Front. Plant Sci. 6, 462.

Khan, M.I.R., Jahan, B., Alajmi, M.F., Rehman, M.T., Khan, N.A., 2019. Exogenously-sourced ethylene modulates defense mechanisms and promotes tolerance to zinc stress in mustard (Brassica juncea L.). Plants 8 (12), 540.

Khan, M.I.R., Jahan, B., AlAjmi, M.F., Rehman, M.T., Khan, N.A., 2020a. Ethephon mitigates nickel stress by modulating antioxidant system, glyoxalase system and prolinemetabolism in Indian mustard. Physiol. Mol. Biol. Plants 26 (6), 1201−1213.

Khan, M.I.R., Trivellini, A., Chhillar, H., Chopra, P., Ferrante, A., Khan, N.A., et al., 2020b. The significance and functions of ethylene in flooding stress tolerance in plants. Environ. Exp. Bot. 179, 104188.

Kim, N.H., Kim, B.S., Hwang, B.K., 2013. Pepper arginine decarboxylase is required for polyamine and γ-aminobutyric acid signaling in cell death and defense response. Plant Physiol. 162 (4), 2067−2083.

Klessig, D.F., Choi, H.W., Dempsey, D.M.A., 2018. Systemic acquired resistance and salicylic acid: past, present, and future. Mol. Plant Microbe Interact. 31 (9), 871−888.

Koike, S., Matsukura, C., Takayama, M., Asamizu, E., Ezura, H., 2013. Suppression of γ-aminobutyric acid (GABA) transaminases induces prominent GABA accumulation, dwarfism and infertility in the tomato (Solanum lycopersicum L.). Plant Cell Physiol. 54 (5), 793−807.

Kondo, S., Jitratham, A., Kittikorn, M., Kanlayanarat, S., 2004. Relationships between jasmonates and chilling injury in mangosteens are affected by spermine. HortScience 39 (6), 1346−1348.

Koornneef, A., Pieterse, C.M., 2008. Cross talk in defense signaling. Plant Physiol. 146 (3), 839−844.

Li, L., Dou, N., Zhang, H., Wu, C., 2021. The versatile GABA in plants. Plant Signal. Behav. 16 (3), 1862565.

Li, Z., Cheng, B., Peng, Y., Zhang, Y., 2020. Adaptability to abiotic stress regulated by γ-aminobutyric acid in relation to alterations of endogenous polyamines and organic metabolites in creeping bentgrass. Plant Physiol. Biochem. 157, 185−194.

Li, Z., Peng, Y., Huang, B., 2018. Alteration of transcripts of stress-protective genes and transcriptional factors by γ-aminobutyric acid (GABA) associated with improved heat and drought tolerance in creeping bentgrass (Agrostis stolonifera). Int. J. Mol. Sci. 19 (6), 1623.

Liu, Y., Gu, D., Wu, W., Wen, X., Liao, Y., 2013. The relationship between polyamines and hormones in the regulation of wheat grain filling. PLoS One 8 (10), e78196.

Liu, Y., Liang, H., Lv, X., Liu, D., Wen, X., Liao, Y., 2016. Effect of polyamines on the grain filling of wheat under drought stress. Plant. Physiol. Biochem. 100, 113−129.

Liu, C., Atanasov, K.E., Arafaty, N., Murillo, E., Tiburcio, A.F., Zeier, J., et al., 2020. Putrescine elicits ROS-dependent activation of the salicylic acid pathway in Arabidopsis thaliana. Plant Cell Environ. 43 (11), 2755−2768.

Locke, J.M., Bryce, J.H., Morris, P.C., 2000. Contrasting effects of ethylene perception and biosynthesis inhibitors on germination and seedling growth of barley (Hordeum vulgare L.). J. Exp. Bot. 51 (352), 1843−1849.

Luo, Z., Chen, C., Xie, J., 2011. Effect of salicylic acid treatment on alleviating postharvest chilling injury of 'Qingnai'plum fruit. Postharvest. Biol. Technol. 62 (2), 115−120.

Ma, Z., Ji, J., Zhu, X., Yi, L., Li, Q., Wang, G., et al., 2018. Salicylic acid mitigates hyperhydricity in newly developed potato shoots through reduced oxidation. Botany 96 (11), 793−803.

Mader, J.C., 1999. Effects of jasmonic acid, silver nitrate and L AOPP on the distribution of free and conjugated polyamines in roots and shoots of Solanum tuberosum in vitro. J. Plant Physiol. 154 (1), 79−88.

Madhulatha, P., Gupta, A., Gupta, S., Kumar, A., Pal, R.K., Rajam, M.V., 2014. Fruit-specific over-expression of human S-adenosylmethionine decarboxylase gene results in polyamine accumulation and affects diverse aspects of tomato fruit development and quality. J. Plant Biochem. Biotechnol. 23 (2), 151−160.

Majumdar, R., Shao, L., Turlapati, S.A., Minocha, S.C., 2017. Polyamines in the life of Arabidopsis: profiling the expression of S-adenosylmethionine decarboxylase (SAMDC) gene family during its life cycle. BMC Plant Biol. 17 (1), 1−20.

Maruri-López, I., Aviles-Baltazar, N.Y., Buchala, A., Serrano, M., 2019. Intra and extracellular journey of the phytohormone salicylic acid. Front. Plant Sci. 10, 423.

Meskaoui, A.E., Trembaly, F.M., 2009. Effects of exogenous polyamines and inhibitors of polyamine biosynthesis on endogenous free polyamine contents and the maturation of white spruce somatic embryos. Afr. J. Biotechnol. 8, 6807−6816.

Min, D., Ai, W., Zhou, J., Li, J., Zhang, X., Li, Z., et al., 2020. SlARG2 contributes to MeJA-induced defense responses to Botrytis cinerea in tomato fruit. Pest Manag. Sci. 76 (9), 3292−3301.

Minocha, R., Majumdar, R., Minocha, S.C., 2014. Polyamines and abiotic stress in plants: a complex relationship1. Front. Plant Sci. 5, 175.

Mo, H., Wang, X., Zhang, Y., Zhang, G., Zhang, J., Ma, Z., 2015a. Cotton polyamine oxidase is required for spermine and camalexin signalling in the defence response to Verticillium dahliae. Plant J. 83 (6), 962−975.

Mo, H., Wang, X., Zhang, Y., Yang, J., Ma, Z., 2015b. Cotton ACAULIS5 is involved in stem elongation and the plant defense response to *Verticillium dahliae* through thermospermine alteration. Plant Cell Rep. 34 (11), 1975−1985.

Mo, H.J., Sun, Y.X., Zhu, X.L., Wang, X.F., Zhang, Y., Yang, J., et al., 2016. Cotton S-adenosylmethionine decarboxylase-mediated spermine biosynthesis is required for salicylic acid-and leucine-correlated signaling in the defense response to *Verticillium dahliae*. Planta 243 (4), 1023−1039.

Moschou, P.N., Paschalidis, K.A., Delis, I.D., Andriopoulou, A.H., Lagiotis, G.D., Yakoumakis, D.I., et al., 2008. Spermidine exodus and oxidation in the apoplast induced by abiotic stress is responsible for H_2O_2 signatures that direct tolerance responses in tobacco. Plant Cell 20 (6), 1708−1724.

Nambeesan, S., Datsenka, T., Ferruzzi, M.G., Malladi, A., Mattoo, A.K., Handa, A.K., 2010. Overexpression of yeast spermidine synthase impacts ripening, senescence and decay symptoms in tomato. Plant J. 63 (5), 836−847.

Otálora, G., Piñero, M.C., Collado-González, J., López-Marín, J., Del Amor, F.M., 2020. Exogenous salicylic acid modulates the response to combined salinity-temperature stress in pepper plants (*Capsicum annuum* l. var. tamarin). Plants 9 (12), 1790.

Ozawa, R., Bertea, C.M., Foti, M., Narayana, R., Arimura, G.I., Muroi, A., et al., 2009. Exogenous polyamines elicit herbivore-induced volatiles in lima bean leaves: involvement of calcium, H_2O_2 and Jasmonic acid. Plant Cell Physiol. 50 (12), 2183−2199.

Pál, M., Szalai, G., Janda, T., 2015. Speculation: polyamines are important in abiotic stress signaling. Plant Sci. 237, 16−23.

Pál, M., Janda, T., 2017. Role of polyamine metabolism in plant pathogen interactions. J. Plant Sci. Phytopathol. 1, 095−0100.

Pál, M., Ivanovska, B., Oláh, T., Tajti, J., Hamow, K.Á., Szalai, G., et al., 2019. Role of polyamines in plant growth regulation of Rht wheat mutants. Plant Physiol. Biochem. 137, 189−202.

Pandey, R., Gupta, A., Chowdhary, A., Pal, R.K., Rajam, M.V., 2015. Over-expression of mouse ornithine decarboxylase gene under the control of fruit-specific promoter enhances fruit quality in tomato. Plant Mol. Biol. 87 (3), 249−260.

Pang, X.M., Nada, K., Liu, J.H., Kitashiba, H., Honda, C., Yamashita, H., et al., 2006. Interrelationship between polyamine and ethylene in 1-methylcyclopropene treated apple fruits after harvest. Physiol. Plant. 128 (2), 351−359.

Peleg, Z., Blumwald, E., 2011. Hormone balance and abiotic stress tolerance in crop plants. Curr. Opin. Plant Biol. 14 (3), 290−295.

Podlešáková, K., Ugena, L., Spíchal, L., Doležal, K., De Diego, N., 2019. Phytohormones and polyamines regulate plant stress responses by altering GABA pathway. N. Biotechnol. 48, 53−65.

Poór, P., Kovács, J., Szopkó, D., Tari, I., 2013. Ethylene signaling in salt stress-and salicylic acid-induced programmed cell death in tomato suspension cells. Protoplasma 250 (1), 273−284.

Poór, P., 2020. Effects of salicylic acid on the metabolism of mitochondrial reactive oxygen species in plants. Biomolecules 10 (2), 341.

Poór, P., Nawaz, K., Gupta, R., Ashfaque, F., Khan, M.I.R., 2021. Ethylene involvement in the regulation of heat stress tolerance in plants. Plant Cell Rep 1−24.

Pottosin, I., Velarde-Buendía, A.M., Bose, J., Zepeda-Jazo, I., Shabala, S., Dobrovinskaya, O., 2014. Cross-talk between reactive oxygen species and polyamines in regulation of ion transport across the plasma membrane: implications for plant adaptive responses. J. Exp. Bot. 65 (5), 1271−1283.

Quinet, M., Ndayiragije, A., Lefevre, I., Lambillotte, B., Dupont-Gillain, C.C., Lutts, S., 2010. Putrescine differently influences the effect of salt stress on polyamine metabolism and ethylene synthesis in rice cultivars differing in salt resistance. J. Exp. Bot. 61 (10), 2719–2733.

Radhakrishnan, R., Lee, I.J., 2013. Spermine promotes acclimation to osmotic stress by modifying antioxidant, abscisic acid, and jasmonic acid signals in soybean. J. Plant Growth Regul. 32 (1), 22–30.

Rakitin, V.Y., Prudnikova, O.N., Rakitina, T.Y., Karyagin, V.V., Vlasov, P.V., Novikova, G.V., et al., 2009. Interaction between ethylene and ABA in the regulation of polyamine level in Arabidopsis thaliana during UV-B stress. Russ. J. Plant Physiol. 56 (2), 147–153.

Rivas-San Vicente, M., Plasencia, J., 2011. Salicylic acid beyond defence: its role in plant growth and development. J. Exp. Bot. 62 (10), 3321–3338.

Sagor, G.H.M., Inoue, M., Kusano, T., Berberich, T., 2021. Expression profile of seven polyamine oxidase genes in rice (Oryza sativa) in response to abiotic stresses, phytohormones and polyamines. Physiol. Mol. Biol. Plants 1–7.

Seifi, H.S., Shelp, B.J., 2019. Spermine differentially refines plant defense responses against biotic and abiotic stresses. Front. Plant Sci. 10, 117.

Shevyakova, N.I., Rakitin, V.Y., Duong, D.B., Sadomov, N.G., Kuznetsov, V.V., 2001. Heat shock-induced cadaverine accumulation and translocation throughout the plant. Plant Sci. 161 (6), 1125–1133.

Shi, H., Chan, Z., 2014. Improvement of plant abiotic stress tolerance through modulation of the polyamine pathway. J. Integr. Plant Biol. 56 (2), 114–121.

Sobieszczuk-Nowicka, E., 2017. Polyamine catabolism adds fuel to leaf senescence. Amino Acids 49 (1), 49–56.

Sreelekshmi, R., Siril, E.A., 2020. Influence of polyamines on hyperhydricity reversion and its associated mechanism during micropropagation of China pink (Dianthus chinensis L.). Physiol. Mol. Biol. Plants 26 (10), 2035–2045.

Santino, A., Taurino, M., De Domenico, S., Bonsegna, S., Poltronieri, P., Pastor, V., et al., 2013. Jasmonate signaling in plant development and defense response to multiple (a) biotic stresses. Plant Cell Rep. 32 (7), 1085–1098.

Seo, S.G., Kang, S.W., Shim, I.S., Kim, W., Fujihara, S., 2009. Effects of various chemical agents and early ethylene production on floral senescence of Hibiscus syriacus L. Plant Growth Regul. 57 (3), 251–258.

Song, J., Wu, G., Liu, C., Li, D., 2018. Effect of exogenous spermine on chilling injury and antioxidant defense system of immature vegetable soybean during cold storage. J. Food Sci. Technol. 55 (10), 4297–4303.

Sun, X., Li, Y., He, W., Ji, C., Xia, P., Wang, Y., et al., 2017. Pyrazinamide and derivatives block ethylene biosynthesis by inhibiting ACC oxidase. Nat. Commun. 8 (1), 1–14.

Szalai, G., Pál, M., Árendás, T., Janda, T., 2016. Priming seed with salicylic acid increases grain yield and modifies polyamine levels in maize. Cereal Res. Commun. 44 (4), 537–548.

Szalai, G., Janda, K., Darkó, É., Janda, T., Peeva, V., Pál, M., 2017. Comparative analysis of polyamine metabolism in wheat and maize plants. Plant Physiol. Biochem. 112, 239–250.

Szepesi, Á., Csiszár, J., Gémes, K., Horváth, E., Horváth, F., Simon, M.L., et al., 2009. Salicylic acid improves acclimation to salt stress by stimulating abscisic aldehyde oxidase activity and abscisic acid accumulation, and increases Na+ content in leaves without toxicity symptoms in Solanum lycopersicum L. J. Plant Physiol. 166 (9), 914–925.

Szepesi, Á., 2019. Molecular Mechanisms of Polyamines-Induced Abiotic Stress Tolerance in Plants. In: Approaches for Enhancing Abiotic Stress Tolerance in Plants, CRC Press, pp. 387–404.

Takács, Z., Poór, P., Szepesi, Á., Tari, I., 2017. In vivo inhibition of polyamine oxidase by a spermine analogue MDL-72527, in tomato exposed to sublethal and lethal salt stress. Fun. Plant Biol. 44 (5), 480–492.

Takács, Z., Poór, P., Tari, I., 2016. Comparison of polyamine metabolism in tomato plants exposed to different concentrations of salicylic acid under light or dark conditions. Plant Physiol. Biochem. 108, 266–278.

Takács, Z., Poór, P., Tari, I., 2021. Interaction between polyamines and ethylene in the response to salicylic acid under normal photoperiod and prolonged darkness. Plant Physiol. Biochem. 167, 470–480.

Takahashi, Y., Berberich, T., Miyazaki, A., Seo, S., Ohashi, Y., Kusano, T., 2003. Spermine signalling in tobacco: activation of mitogen-activated protein kinases by spermine is mediated through mitochondrial dysfunction. Plant J. 36 (6), 820–829.

Talaat, N.B., Shawky, B.T., 2016. Dual application of 24-epibrassinolide and spermine confers drought stress tolerance in maize (Zea mays L.) by modulating polyamine and protein metabolism. J. Plant Growth Regul. 35 (2), 518–533.

Tamai, T., Inoue, M., Sugimoto, T., Sueyoshi, K., Shiraishi, N., Oji, Y., 1999. Ethylene-induced putrescine accumulation modulates K + partitioning between roots and shoots in barley seedlings. Physiol. Plant. 106 (3), 296−301.

Tao, Y., Wang, J., Miao, J., Chen, J., Wu, S., Zhu, J., et al., 2018. The spermine synthase OsSPMS1 regulates seed germination, grain size, and yield. Plant Physiol. 178 (4), 1522−1536.

Tarkowski, P., Signorelli, S., Höfte, M., 2020. γ-Aminobutyric acid and related amino acids in plant immune responses: emerging mechanisms of action. Plant Cell Environ. 43 (5), 1103−1116.

Tassoni, A., Watkins, C.B., Davies, P.J., 2006. Inhibition of the ethylene response by 1-MCP in tomato suggests that polyamines are not involved in delaying ripening, but may moderate the rate of ripening or over-ripening. J. Exp. Bot. 57 (12), 3313−3325.

Tavladoraki, P., Cona, A., Angelini, R., 2016. Copper-containing amine oxidases and FAD-dependent polyamine oxidases are key players in plant tissue differentiation and organ development. Front. plant. Sci. 7, 824.

Tiburcio, A.F., Altabella, T., Bitrián, M., Alcázar, R., 2014. The roles of polyamines during the lifespan of plants: from development to stress. Planta 240 (1), 1−18.

Torrigiani, P., Scaramagli, S., Castiglione, S., Altamura, M.M., Biondi, S., 2003. Downregulation of ethylene production and biosynthetic gene expression is associated to changes in putrescine metabolism in shoot-forming tobacco thin layers. Plant. Sci. 164 (6), 1087−1094.

Tripathi, D., Raikhy, G., Kumar, D., 2019. Chemical elicitors of systemic acquired resistance—Salicylic acid and its functional analogs. Curr. Plant Biol. 17, 48−59.

Urano, K., Yoshiba, Y., Nanjo, T., Igarashi, Y., Seki, M., Sekiguchi, F., et al., 2003. Characterization of Arabidopsis genes involved in biosynthesis of polyamines in abiotic stress responses and developmental stages. Plant Cell Environ. 26 (11), 1917−1926.

Van der Ent, S., Van Wees, S.C., Pieterse, C.M., 2009. Jasmonate signaling in plant interactions with resistance-inducing beneficial microbes. Phytochemistry 70 (13−14), 1581−1588.

Van de Poel, B., 2020. Ethylene's fraternal twin steals the spotlight. Nat. Plants 6 (11), 1309−1310.

van Loon, L.C., Geraats, B.P., Linthorst, H.J., 2006. Ethylene as a modulator of disease resistance in plants. Trends Plant Sci. 11 (4), 184−191.

Vlot, A.C., Dempsey, D.M.A., Klessig, D.F., 2009. Salicylic acid, a multifaceted hormone to combat disease. Annu. Rev. Phytopathol. 47, 177−206.

Walters, D., Cowley, T., Mitchell, A., 2002. Methyl jasmonate alters polyamine metabolism and induces systemic protection against powdery mildew infection in barley seedlings. J. Exp. Bot. 53 (369), 747−756.

Wang, Z., Xu, Y., Wang, J., Yang, J., Zhang, J., 2012. Polyamine and ethylene interactions in grain filling of superior and inferior spikelets of rice. Plant Growth Regul. 66 (3), 215−228.

Wasternack, C., Hause, B., 2013. Jasmonates: biosynthesis, perception, signal transduction and action in plant stress response, growth and development. An update to the 2007 review in Annals of Botany. Ann. Bot. 111 (6), 1021−1058.

Wasternack, C., 2015. How jasmonates earned their laurels: past and present. J. Plant Growth Regul. 34 (4), 761−794.

Xia, X.J., Zhou, Y.H., Shi, K., Zhou, J., Foyer, C.H., Yu, J.Q., 2015. Interplay between reactive oxygen species and hormones in the control of plant development and stress tolerance. J. Exp. Bot. 66 (10), 2839−2856.

Yamakawa, H., Kamada, H., Satoh, M., Ohashi, Y., 1998. Spermine is a salicylate-independent endogenous inducer for both tobacco acidic pathogenesis-related proteins and resistance against tobacco mosaic virus infection. Plant Physiol. 118 (4), 1213−1222.

Yang, W., Yin, Y., Li, Y., Cai, T., Ni, Y., Peng, D., et al., 2014. Interactions between polyamines and ethylene during grain filling in wheat grown under water deficit conditions. Plant Growth Regul. 72 (2), 189−201.

Yoshikawa, H., Honda, C., Kondo, S., 2007. Effect of low-temperature stress on abscisic acid, jasmonates, and polyamines in apples. Plant Growth Regul. 52 (3), 199−206.

Yu, Z., Jia, D., Liu, T., 2019. Polyamine oxidases play various roles in plant development and abiotic stress tolerance. Plants 8 (6), 184.

Zhang, X., Shen, L., Li, F., Meng, D., Sheng, J., 2011. Methyl salicylate-induced arginine catabolism is associated with up-regulation of polyamine and nitric oxide levels and improves chilling tolerance in cherry tomato fruit. J. Agric. Food Chem. 59 (17), 9351−9357.

Zhang, W., Chen, Y., Wang, Z., Yang, J., 2017. Polyamines and ethylene in rice young panicles in response to soil drought during panicle differentiation. Plant Growth Regul. 82 (3), 491−503.

Zhu, C., Chen, Z., 2005. Role of polyamines in adventitious shoot morphogenesis from cotyledons of cucumber in vitro. Plant Cell Tissue Organ. Cult. 81 (1), 45−53.

Zhuo, C., Liang, L., Zhao, Y., Guo, Z., Lu, S., 2018. A cold responsive ethylene responsive factor from Medicago falcata confers cold tolerance by up-regulation of polyamine turnover, antioxidant protection, and proline accumulation. Plant Cell Environ. 41 (9), 2021−2032.

Ziosi, V., Bregoli, A.M., Fregola, F., Costa, G., Torrigiani, P., 2009. Jasmonate-induced ripening delay is associated with up-regulation of polyamine levels in peach fruit. J. Plant Physiol. 166 (9), 938−946.

14

Interplay between phytohormones and hydrogen sulfide regulates plant growth and development under normal and abiotic stresses

Harmanjit Kaur[1], Tashima[2], Sandeep Singh[3] and Sofi J. Hussain[4]

[1]Post Graduate Department of Botany, Government College for Girls, Ludhiana, Punjab, India
[2]Department of Botany, Akal University, Bathinda, Punjab, India [3]Department of Botany, Kanya Maha Vidyalaya, Jalandhar, Punjab, India [4]Department of Botany, Central University of Kashmir, Ganderbal, Jammu and Kashmir, India

14.1 Introduction

Phytohormones or plant hormones occur naturally as small organic molecules which impact physiological processes in plants at extremely low concentrations (Davies, 2004; Khan et al., 2019, 2020a,b; Nazir et al., 2019a,b, 2021). From the first finding of auxin as plant hormone (Went, 1935) to the latest new discovery of strigolactones (SL) (Gomez-Roldan et al., 2008), nine different types of plant hormones, auxins, cytokinins (CK), gibberellins (GA), abscisic acid (ABA), ethylene (ETH), brassinosteroids (BR), salicylates (SA), jasmonates (JA), and SL, have been recognized until now. The former five (auxin, CK, GA, ABA, and ETH) are also known as the "traditional" plant hormones, whereas the latter four have been recently added to the increasing phytohormonal lineage (Fleet and Williams, 2011). These hormones may function either adjacent to or at distant from the site of production to modulate responses to environmental stimulus or hereditarily programmed developmental alterations (Davies, 2004). Plant hormones therefore play a crucial part in regulating plant response to several abiotic stresses, through which the plant may endeavor to escape from the stressful surroundings by decreasing its growth, with an intention to direct its resources

Plant Hormones in Crop Improvement
DOI: https://doi.org/10.1016/B978-0-323-91886-2.00015-X

toward combating the stress (Skirycz and Inzé, 2010). As a matter of fact, stress signal initiates the signal transduction pathways in plants, with plant hormones functioning as the bottom-line transducers (Harrison, 2012).

Hydrogen sulfide (H_2S) is a colorless gas having a distinctive foul smell. Unlike from its function as a phytotoxin at greater levels, H_2S at lesser concentrations play an important role in various stages of plant life cycle, for example, growth, enlargement, and physiological processes comprising seed germination, root morphogenesis, photosynthesis, flower senescence, and abiotic and biotic stress responses (Chen et al., 2011; Zhang et al., 2010, 2011; Jin et al., 2013; Fu et al., 2018; Luo et al., 2020; Deng et al., 2020). Thus, it permits plant viability and adaptability and positively influences several vital aspects of plant growth (Zhang et al., 2021; Zhou et al., 2021). Increasing evidence has indicated the signaling function of H_2S in plants, with similar significance as that of other signaling molecules, like nitric oxide (NO) and hydrogen peroxide (H_2O_2) (Calderwood and Kopriva, 2014; Aroca et al., 2018; Aroca et al., 2020). Various researchers have reported that H_2S acts as a critical messenger in plant resistance signaling against different abiotic stresses at low physiological levels (Jin et al., 2011; Li et al., 2013a,b; Chen et al., 2013).

In plants, H_2S is synthesized in the chloroplast via photosynthetic sulfate assimilation pathway (Gotor et al., 2019). It has been revealed that plants can energetically produce endogenous H_2S under nonstressed, particularly biotic or abiotic stress environments (Li, 2013; Calderwood and Kopriva, 2014; Hancock and Whiteman, 2014; Yamasaki and Cohen, 2016). The buildup of intrinsic H_2S has turned out to be a frequent response of plants to environmental stresses, comprising salinity, heavy metals, water-deficit, high- and low-temperature stress, in addition to pathogen attack, which may be directly linked with the development of stress resistance in plants (Li, 2013; Calderwood and Kopriva, 2014; Hancock and Whiteman, 2014). More fascinatingly, exogenous application of H_2S, released from its donors, for instance NaHS and morpholin-4-ium 4-methoxyphenyl (morpholino) phosphinodithioate (GYY4137), has revealed considerable beneficial impacts on seed germination (Li et al., 2012; Wojtyla et al., 2016), organogenesis and development (Lin et al., 2012; Fang et al., 2014), modulation of senescence (Zhang et al., 2011), along with the development of stress tolerance against salinity (Christou et al., 2013), heavy metals (Chen et al., 2013), drought (Christou et al., 2013), heat (Li et al., 2013a,b; Li, 2015), and low temperature (Fu et al., 2013). These findings suggest that H_2S may act as a candidate signal molecule in plant cross-acclimatization.

It is notable that H_2S donors when given exogenously stimulate the endogenous H_2S concentration in plants. For example, endogenous concentration of H_2S was stimulated in *Arabidopsis* under drought conditions, most probably owing to greater expression levels of L-desulfhydrase and D-desulfhydrase enzymes (Jin et al., 2011). Likewise, H_2S donor NaHS triggered the internal H_2S levels in maize (Li et al., 2013a). Furthermore, H_2S, when given exogenously, alleviates the damaging effects of diverse abiotic stresses. It was found to be a main feature in Cd sequestration in *Populus euphratica* cells under Cd stress (Sun et al., 2013). Interaction of H_2S with ABA assisted in drought tolerance in *Arabidopsis* by regulating stomatal closure (Jin et al., 2013). Likewise, exogenous treatments of H_2S to plants, for example, *Vitis vinifera* L. activated gene expression involved in the production of secondary metabolites and several defensive compounds which increased the plant growth and abiotic resistance (Ma and Yang 2018). H_2S is predominantly active and can act together with as well as alter various other cellular signals. Hence, there may be numerous pathways of H$_2$S perception and signaling which requires to be disentangled.

Hydrogen sulfide has the capability to act together with thiol (−SH) groups which are present in peptides, for example, reduced glutathione (GSH), and with proteins that change their roles. This type of communication, modifying the cysteine thiols (−SH) into persulfide (−SSH) groups is referred to as persulfidation (Corpas, 2019). Protein persulfidation, which is an oxidative posttranslational conversion of cysteine residues, symbolizes a signaling process by H$_2$S. It is also involved in biosynthetic pathways that require sulfur transfer, for example, iron-sulfur clusters, biotin, thiamine, lipoic acid, molybdopterin, and sulfur-containing bases in RNA (Filipovic et al., 2018). These posttranslational changes in cysteine residues may function as a defensive mechanism under oxidative stress produced by different environmental conditions. Various biochemical and genetic facts have undoubtedly recognized the signaling influence of H$_2$S in cells via persulfidation, with vital results for different physiological and pathological mechanisms in animals and plants (Yuan et al., 2017; Paul and Snyder, 2018; Aroca et al., 2020). Therefore this chapter is an attempt to review cross talk of various phytohormones with H$_2$S in regulating plant growth and development under normal and adverse environmental conditions along with the mechanisms by which H$_2$S influences various plant processes. This chapter also focuses on understanding the cellular signaling cascade involving phytohormones in combination with H$_2$S.

14.2 Cross talk of phytohormones with H$_2$S

Phytohormones are vital regulators of plant growth and development as well as play a crucial role in plant responses to different abiotic stresses (Nazir et al., 2022; Khan et al., 2021a,b,c). In addition, they are the main signaling molecules and every potential signaling molecule ultimately interacts with them to produce any downstream response in plants (Jin et al., 2011; Cheng et al., 2013; Banerjee et al., 2018). Plants recognize and respond to H$_2$S, various lines of evidence indicate that H$_2$S strongly interacts with the phytohormones, for example, auxin, ABA, ETH, SA, GA, and JA (Table 14.1) throughout diverse phases of plant cycle under natural or stress environments (He et al., 2019a). This subject of research is rapidly gaining importance, given that a distinct interaction of H$_2$S with plant hormones can offer unique signaling nodes which can be genetically aimed for developing manifold stress-tolerant characters in the susceptible plant cultivars.

14.2.1 Cross talk of auxin with H$_2$S

Auxin is a main plant growth regulator which vitally controls different physiological and developmental processes, such as embryogenesis, apical dominance, phyllotaxis, initiation of main and lateral roots, tropic movements, and vascular differentiation (Wang et al., 2001; Woodward and Bartel, 2005), thereby coordinating the alterations in plant growth and morphology in response to environmental conditions (Zhao, 2010). Indole acetic acid (IAA) is the foremost recognized plant hormone; however, its biosynthetic pathway at the genetic stage is ambiguous (Fahad et al., 2015). Although IAA has broadly been recognized for its role in plant

TABLE 14.1 The effects of interaction between phytohormones and hydrogen sulfide on plants under unstressed and stressed conditions.

Plant	Phytohormone	Phytohormone concentration	H₂S form	H₂S concentration	Stressed/unstressed	Effects	References
Ipomoea batatas, Salix matsudana, and *Glycine max*	Auxin	20 μmol/L (indole-3-acetic acid, IAA)	Sodium hydrosulfide (NaHS)	0.1, 0.2, 0.4, 0.6, 0.8, and 1.0 mmol/L	Unstressed	Promotes root organogenesis	Zhang et al. (2009)
Solanum lycopersicum L.	Auxin	100 nM (naphthalene acetic acid, NAA)	NaHS	0.01, 0.1, 1, 5, 10 mM	Unstressed	Lateral root formation	Fang et al. (2014)
Arabidopsis thaliana	Auxin	10 μM (N-1-naphthylphthalamic acid, NPA)	NaHS	50–1000 μM	Unstressed	Alteration in root development	Jia et al. (2015)
Cucumis sativus L. Jinyou 35	Auxin	25, 50, 75, and 100 μM IAA	NaHS	0.5, 1.0, 1.5, 2.0, and 2.5 mM	Cold stress	Tolerance to cold stress	Zhang et al. (2021)
Solanum lycopersicum	Gibberellin	100 mg/L	NaHS	0.2 mM	Stressed	Alleviates boron toxicity	Kaya et al. (2020)
Triticum aestivum	Gibberellin	50 μM	NaHS	0.1, 1, 10, and 100 μM	Unstressed	Delays GA-triggered programmed cell death	Xie et al. (2014)
Arabidopsis	Abscisic acid	10 and 50 μM	NaHS	80 μM	Stressed	Stomatal regulation	Jin et al. (2013)
Nicotiana tabacum	Abscisic acid	25, 50, 75, 100, and 125 μM	NaHS	50 μM	Stressed	Mediates heat tolerance	Li and Jin (2016)
Triticum aestivum	Abscisic acid	–	NaHS	500 μM	Stressed	Alleviation of drought stress	Ma et al. (2016)
Arabidopsis	Abscisic acid	10 μM	NaHS	100 μM	Stressed	Stomatal regulation	Zhang et al. (2019)
Solanum lycopersicum	Ethylene	40 μL (ethephon)	NaHS	1 M solution	Unstressed	Inhibits ethylene-induced petiole abscission	Liu et al. (2020)
Vicia faba	Ethylene	0.4% (ethephon)	NaHS	0.2 mmol/L	Unstressed	Mediates ethylene-induced stomatal closure	Jing et al. (2012)

Species	Hormone	Concentration (treatment)	H$_2$S donor	H$_2$S concentration	Stress status	Effect	Reference
Solanum lycopersicum	Ethylene	1 g/L (ethephon)	NaHS	1 mmol/L	Unstressed	Inhibits fruit softening	Hu et al. (2019)
Actinidia deliciosa	Ethylene	0.4 g/L (ethephon)	NaHS	1 mM	Unstressed	Alleviates fruit ripening and senescence	Li et al. (2017)
Vigna mungo/Vigna radiata	Ethylene	25 µM (ethephon)	NaHS	10 µM	Stressed	Alleviates chromium stress	Husain et al. (2021)
Prunus persica	Ethylene	-	H$_2$S gas	20 µl/L	Unstressed	Inhibiting ethylene biosynthesis	Zhu et al. (2019)
Ocimum basilicum, Brassica oleracea	Ethylene	1, 0.1, and 0.01 µL/L	H$_2$S gas	50 and 100 µL/L	Unstressed	Inhibits leaf senescence	Al Ubeed et al. (2017)
Solanum lycopersicum	Ethylene	1.0 g/L (ethephon)	NaHS	0.90 mmol/L	Unstressed	Delayed fruit ripening	Yao et al. (2018)
Prunus persica	Ethylene	–	NaHS	0.02, 0.05, 0.1, 0.2, 0.3 mM	Stressed	Alleviates water logging stress	Xiao et al. (2020)
Zea mays	Salicylic acid	0.1, 0.3, 0.5, 0.7, and 0.9 mM	NaHS	0.5 mM	Stressed	Improved heat tolerance	Li et al. (2015)
Arabidopsis thaliana	Jasmonic acid	1, 5, and 10 µM (methyl jasmonate, MeJA)	NaHS	0.1,0.2, 0.4, 0.6, and 0.8 mM	Unstressed	Negatively regulates stomatal movement	Deng et al. (2020)
Arabidopsis thaliana	Brassinosteroid	0.1, 0.5, 1, 5, and 10 µmol/L (2,4-epibrassinolide)	Endogenous H$_2$S	–	Unstressed	Induction of stomatal closure	Ma et al. (2020)

growth and expansion; still it can regulate plant growth response to stressful environments (Eyidogan et al., 2012). IAA concentrations have been shown to increase rapidly in diverse plants treated with H_2S. In the course of lateral root development, NaHS treatment quickly raises the level of auxin and elevates the number and length of adventitious roots, indicating an intimate cross talk between H_2S and auxin (Zhang et al., 2009). Auxin generally reduces abscission of organs, and further studies have revealed that genes belonging to IAA/auxin family (IAA3 and IAA4) are frequently upregulated by H_2S (Liu et al., 2020). In sweet potato cuttings, excised *Salix matsudana* shoots and *Glycine max* seedlings, both IAA polar transport inhibitor NPA (naphthylphthalamic acid) and NO scavenger 2-(4-carboxyphenyl)-4,4,5,5-tetra-methylimidazoline-1-oxyl-3-oxide (cPTIO) perturbed the development of roots mediated by H_2S. It is reasoned that H_2S functions as upstream of IAA and NO to support root hair growth or control organ abscission (Zhang et al., 2009). Fang et al. (2014) also emphasized the synergistic functions of H_2S and auxins during lateral root initiation in NaHS-treated *Solanum lycopersicum* seedlings. Diminution in the cellular auxin concentration considerably checked the action of H_2S anabolic enzyme, that is, DES. Exogenous treatment with auxin, naphthalene acetic acid (NAA) elevated DES1 expression, DES activity and also the H_2S level. These collectively triggered the growth of lateral root system (Fang et al., 2014). Supplementation of H_2S scavenger, hypotaurine decreased the growth of lateral roots which was reversed exclusively by NaHS and not NAA. Lateral root growth coordinated by the synchronization of both H_2S and NAA was also distinguished by the upregulation of particular cell cycle genes, for instance cyclin-dependent kinases SlCDKA;1 and SlCYCA2;1, and downregulation of cyclin-dependent kinase inhibitor, SlKRP2 in the tomato plants (Fang et al., 2014). Such modulation of gene expression supports cell division and organ formation in plants (Atkins and Cross, 2018). These findings indicate existence of a feedback regulation between H_2S and auxin during plant growth, in which H_2S can upregulate the expression of IAA family genes, and IAA can also influence both expression and activity of DES1. During pathogen attack, the expression of auxin signaling F-box protein 1 (AFB1), AFB2, and AFB3 are negatively modulated by H_2S (Shi et al., 2015). Moreover, low temperature stimulates the buildup of H_2S and also activates the intrinsic IAA system. Treatment with NaHS considerably elevated the activity of flavin monooxygenase and relative transcription of FMO-like protein YUCCA2 in cucumber seedlings, which in turn elevated the concentration of endogenous IAA and enhanced cold tolerance, manifested as reduction in electrolyte leakage, increase in reactive oxygen species (ROS), gene expression, and enzyme activities linked with photosynthesis (Zhang et al., 2021). The use of IAA or elimination of H_2S had slight impact on the signaling of other molecules, but the IAA polar transport inhibitor NPA repressed H_2S-mediated cold tolerance and defense gene expression (Zhang et al., 2021). IAA takes part in H_2S -induced stress endurance in plants as a downstream signaling molecule, whereas H_2S supports auxin signal transduction by modulating the expression of auxin-related genes and the production of auxin, thus increasing the plant tolerance to unfavorable environmental circumstances. The process is reliant on the formation of an auxin gradient as a result of polar auxin transport (PAT) from the aerial to the below ground tissues (Normanly, 2010). Such PAT is restricted by high levels of H_2S, leading to modified root structures. In *Arabidopsis* roots, it has been found that H_2S altered the expression of various ACTIN-BINDING PROTEINS (ABPs) and decreased the percentage of occupancy of the F-acting bundles (Jia et al., 2015). Because the allocation of the PINFORMED (PIN) proteins accountable for regulating PAT is dependent on ABPs, it was considerably modified by H_2S

(Jia et al., 2015). Therefore the report suggested a "strongly controlled interlinked signaling network" existing between auxins, H$_2$S and ABPs. In recent times, Zhang et al. (2017) examined the impacts of NaHS application on the primary root growth of *Arabidopsis* seedlings. Elevated cellular levels of H$_2$S stimulated the synthesis of NO, triggered the MITOGEN-ACTIVATED PROTEIN KINASE 6 (*MPK6*) and caused accumulation of ROS. These severely suppressed the development of the primary root system. It was also noticed that NO was produced owing to the H$_2$S -induced generation of ROS. The triggered *MAPK 6* modulated the simultaneous synthesis of H$_2$S-induced NO (Zhang et al., 2017). Hence, the H$_2$S-mediated suppression of root growth in *Arabidopsis* due to modified PAT is controlled at the same time by interactive interactions among H$_2$S and NO. Nevertheless, optimal concentrations of H$_2$S have been found to stimulate the development of lateral roots through probable interactions with NO, CO, and IAA. IAA was also found to increase the H$_2$S biosynthesis by stimulating the expression of L-cysteine desulfhydrase (LCD) (Li et al., 2014) (Fig. 14.1).

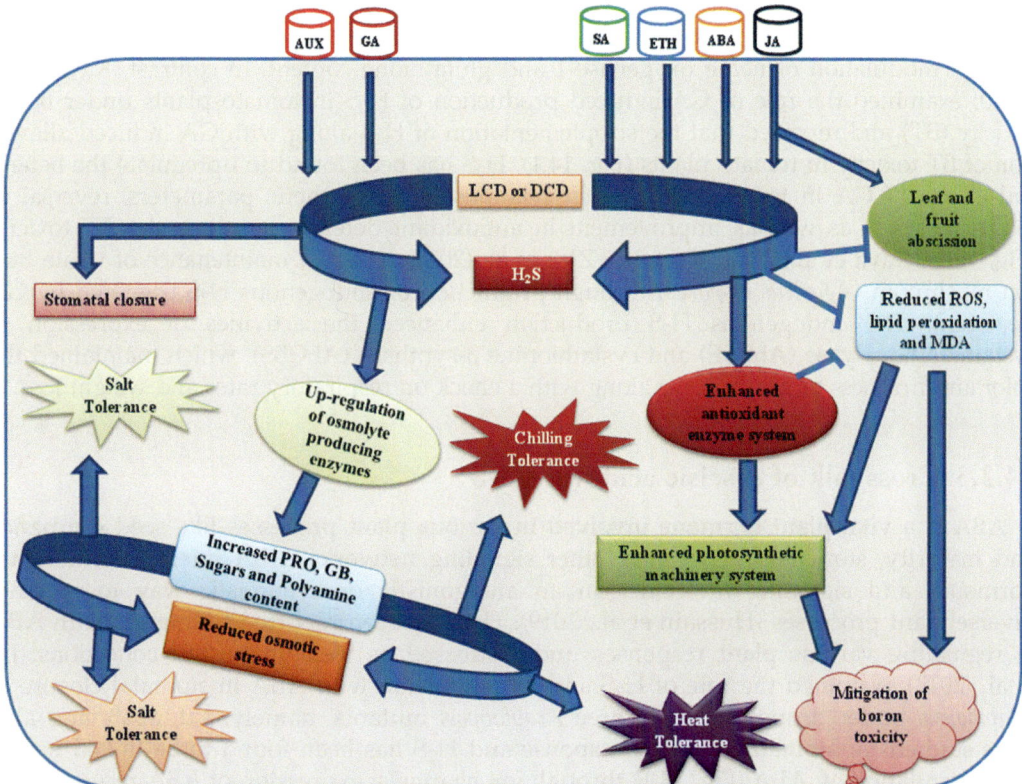

FIGURE 14.1 Signaling cascade involving phytohormones and H$_2$S (Li et al., 2021). *AUX*, Auxin; *ABA*, abscisic acid; *DCD*, D-cysteine desulfhydrase; *ETH*, ethylene; *GA*, gibberellic acid; *GB*, glycine betaine; *H$_2$S*, hydrogen sulfide; *JA*, jasmonic acid; *LCD*, L-cysteine desulfhydrase; *MDA*, malondialdehyde; *PRO*, proline; *ROS*, reactive oxygen species; *SA*, salicylic acid. Source: *Adapted from Li, Z.G., Xiang, R.H., Wang, J.Q., 2021. Hydrogen sulfide—phytohormone interaction in plants under physiological and stress conditions. J. Plant Growth Regul. 1—9.*

14.2.2 Cross talk of gibberellins with H$_2$S

GAs are tetracyclic diterpenoids, which were first identified from a plant pathogen *Gibberella fujikuroi* (*Fusarium fujikuroi*), causing "foolish-seedling disease" in rice. The infected plants produce large amounts of GA, which makes them elongated, slender, chlorotic, and partially infertile, incapable of standing straight and carrying their own weight (Hedden, 2020). GA is involved in many developmental processes in plants including enhanced seed germination, cell division and cell elongation, and induction of flowering and fruit development and maturation (Sponsel, 2016; Hedden, 2020). While performing these functions, GA can interact with several other signaling biomolecules as well as other phytohormones. H$_2$S is one of the gasotransmitters and active signaling molecule, which interacts with GAs and regulates its effects on plants (He et al., 2019b; Arif et al., 2021). GAs have been reported to modulate many plant processes including programmed cell death, which involves a steady decrease in endogenous H$_2$S production as well as activity of L-cysteine desulfhydrase (Fig. 14.1) (primary source of H$_2$S synthesis in plants) (Xie et al., 2014). The exogenous application of H$_2$S in the form of sodium hydrosulfide (NaHS) delays GA-induced programmed cell death in wheat aleurone layer cells through modulation of heme oxygenase-1 and glutathione content. In contrast, Kaya et al. (2020) examined the role of GA-induced production of H$_2$S in tomato plants under boron toxicity (BT) and reported that the supplementation of H$_2$S along with GA induced alleviation of BT toxicity in tomato plants (Fig. 14.1). H$_2$S has been found to upregulate the beneficial effects of GA in terms of leaf water relations, photosynthetic parameters, reversal of oxidative stress as well as improvement in antioxidant defense system under BT toxicity (Fig. 14.1) (Kaya et al., 2020). Recently Zhu et al. (2021) reported maintenance of white button mushroom (*Agaricus bisporus*) through production of endogenous H$_2$S triggered by GA treatment. The endogenous H$_2$S production enhanced the activities of expression of cystathionine γ-lyase (AbCSE) and cystathionine β-synthase (AbCBS), which maintained the color and firmness of mushrooms along with a check on respiratory rates and weight loss.

14.2.3 Cross talk of abscisic acid with H$_2$S

ABA is a vital plant hormone involved in various plant processes like seed dormancy and maturity, stomatal closure and other signaling networks. ABA interacts with other hormones and signaling molecules in an antagonistic or synergistic way to regulate diverse plant processes (Hussain et al., 2019). H$_2$S has been reported to interact with ABA in regulating various plant responses under stressed as well unstressed conditions. Jin et al. (2013) examined the role of H$_2$S and its interaction with ABA in stomatal closure in *Arabidopsis* under drought stress. Three *Arabidopsis* mutants, namely, lcd, aba3, and abi1 were studied for stomatal aperture response and H$_2$S has been found to be linked in stomatal regulation by ABA (Fig. 14.1) through ion channels; expression of ABA receptor candidates as well as ABA-induced H$_2$S production. Similar results were demonstrated by Scuffi et al. (2014) where L-cysteine desulfhydrase (DES1) knockout *Arabidopsis* mutant plants (des1) were used to study the participation of H$_2$S and NO in modulation of ABA-Dependent Stomatal Closure in *Arabidopsis thaliana*. The exogenous application of H$_2$S restored the effects of ABA in closing stomata in isolated epidermal strips of des1 mutants, confirming that H$_2$S is required for proper response of guard cells in stomatal closure. Furthermore, the *Arabidopsis* clade-A PROTEIN PHOSPHATASE2C mutant abscisic acid-insensitive1 (abi1−1) did not close the stomata when epidermal strips were treated with H$_2$S, suggesting that H$_2$S required a functional ABI1. Li and Jin (2016) reported accumulation of endogenous H$_2$S in ABA-treated tobacco cells and culture medium due to ABA-induced activity of L-cysteine desulfhydrase under heat stress. The endogenous buildup of H$_2$S helped in improving the heat tolerance of tobacco cells. Moreover, the exogenous application of sodium hydrosulfide (NaHS–H$_2$S donor) also partly mediated in the acquisition of heat tolerance induced by ABA (Fig. 14.1) in tobacco cell cultures. Ma et al. (2016) reported that exogenous pretreatment of H$_2$S (NaHS) to *Triticum aestivum* exhibited a differential response in roots and leaves under drought conditions (PEG + H$_2$S treatment upregulated the expression of ABA biosynthetic and reactivation genes in leaves; while in roots the expression of ABA biosynthetic and catabolism genes was boosted). The expression levels of two genes encoding ABA receptors, that is, *TaRCAR* and *TaCHLH*, were upregulated in both roots and leaves under H$_2$S + PEG treatment. Furthermore, the expression levels of *TaGLU1* and *TaGLU4*, which were involved in ABA reactivation exhibited an increase in leaves under H$_2$S + PEG treatment whereas no such upregulation was observed in roots. Zhang et al. (2020) studied the guard cells of *Arabidopsis* where H$_2$S participated in ABA-induced stomatal closure through the activity of L-cysteine desulfhydrase (DES1). The activation of DES1 by ABA led to the production of H$_2$S in guard cells and this in situ DES1/H$_2$S function allowed the stressed stomata to respond to change that used ABA as a signal. Moreover, this DES1/H$_2$S function could also help in fully complementing the response in otherwise impaired guard cell ABA phenotype of the des1 mutant. Chen et al. (2020) showed that ABA signaling is positively regulated through persulfidation of Open Stomata 1 (OST1)/SNF1-RELATED PROTEIN KINASE2.6 (SnRK2.6) by ABA-induced production of H$_2$S in the guard cells of *Arabidopsis* seedlings. The researchers identified two cysteine (Cys) sites, namely, Cys131 and Cys137, to undergo persulfidation on the surface of SnRK2.6, which are involved in the interaction with ABA response element-binding factor 2 (ABF2). When these Cys residues in SnRK2.6 were substituted with serine (S), H$_2$S-induced activity of SnRK2.6 as well as SnRK2.6–ABF2 interaction was found to be partially or completely compromised. Chen et al. (2020) revealed a novel posttranslational regulatory mechanism, which involved persulfidation of SnRK2.6 by H$_2$S for regulation of ABA signaling and ABA-induced stomatal closure.

14.2.4 Cross talk of ethylene with H$_2$S

ETH is a gaseous plant hormone, an important regulator involved in plant growth, development especially during stressed conditions (Khan et al., 2019, 2020a,b; Poór et al., 2021a). ETH production induces senescence in plants, promotes ripening of fruits and affects local or neighboring cells in stressed or infected plants. ETH has been reported to interact with other gasotransmitter signal molecules in plants, especially H$_2$S in modulating fruit ripening and senescence (Mukherjee, 2019). Jing et al. (2012) reported ETH-induced production of H$_2$S through increased activity of L-/D-cysteine desulfhydrase (Fig. 14.1) in leaves of *Vicia faba*. The exogenous application of H$_2$S caused closure of stomata (Fig. 14.1) in a concentration-dependent manner. ETH in the form of ethephon induced stomatal closure while increasing the endogenous concentration of H_2S in the leaf tissues of *V. faba*, indicating that H_2S plays a vital role in ETH signaling in stomatal closure. Al Ubeed et al. (2017) studied the effect of H_2S on the ETH-induced leaf senescence of green leafy vegetables/herbs, including pak choy (*Brassica rapa* subsp. *Chinensis*), green curly kale (*Brassica oleracea* var. *sabellica*) and sweet Italian basil (*Ocimum basilicum*). It was found that H_2S repressed the production of endogenous ETH along with reduction in chlorophyll loss, ion leakage, weight loss, and respiration. Jia et al. (2018) revealed feedback regulation of ETH biosynthesis in tomato plants by H_2S through inhibition of ACC oxidases (ACOs) activity under osmotic stress. ETH-induced H_2S negatively regulates ETH biosynthesis through persulfidation of *LeACO1* and *LeACO2* in a dose-dependent manner. Liquid chromatography—tandem mass spectrometry analysis revealed that H_2S caused persulfidation of one specific cysteine residue Cys60 of the LeACO proteins. Yao et al. (2018) reported that exogenous application of H_2S led to the reduction of ETH-induced production of hydrogen peroxide (H_2O_2), superoxide anion ($O_2\bullet^-$) and malondialdehyde in tomato fruit. Moreover, H_2S enhanced antioxidant activities, including superoxide dismutase, ascorbate peroxidase, catalase and guaiacol peroxidase. H_2S acted as antagonistic to ETH and regulated fruit ripening through enhanced expression of some antioxidant genes including *SlCuZnSOD*, *SlCAT1*, *SlAPX2*, and *SlPOD12*. Similar results were reported by Zhu et al. (2019) on exogenous application of H_2S on fruits of *Prunus persica*. Fumigation with H_2S significantly reduced the rot index, titratable acid contents and soluble solid contents in the peach fruits. Therefore, treatment with H_2S inhibited the ripening and softening of peach fruit along with a significant reduction in Aminocyclopropane-1-carboxylic acid (ACC) content as well as activities of ACC synthase and oxidase which were directly linked with the H_2S-induced reduction in ETH production. Hu et al. (2019) showed that exogenous fumigation of H_2S on tomato fruits could delay fruit color transition from green to red by inhibiting endogenous ETH biosynthesis as well as regulating the ETH signal transduction. ETH treatment was found to be significantly upregulating the expression of *SlACO1*, *SlACO3*, and *SlACO4*, which was attenuated by the application of H_2S. This suggested that H_2S is involved in inhibiting ETH synthesis at the expression level. Furthermore, H_2S treatment also downregulated the expression of other ETH biosynthesis genes like *SlACS2*, *SlACS3* as well as ETH receptor genes, such as *SlETR1*, *SlETR2*, *SlETR3*, *SlETR5*, and *SlETR6*. Therefore, exogenous application of H_2S inhibited tomato fruit softening by antagonizing the biosynthesis signal transduction of ETH. H_2S has been found to inhibit ETH-induced abscission of tomato petiole in a concentration-dependent manner (Liu et al., 2020). ETH treatment upregulated the expression of genes involved in encoding cell-wall modifying enzymes, such as *Cel5*, *TAPG2*, *TAPG4*, and *Expansin1*. However, real-time qPCR assays showed that H_2S treatment downregulated the expression of these four genes along with suppression of other ETH biosynthetic genes, such as *ACO1*, *ACO4*, *ACS6*, *ETR4*, and *ERF1*. Xiao et al. (2020) investigated the effect of exogenous H_2S on peach seedlings under waterlogged conditions. The histochemical staining and physiological and biochemical tests revealed that waterlogging stress induced ETH biosynthesis leading to increase in amount of ROS, number of cell deaths, and electrolyte permeability. Application of H_2S inhibited ETH biosynthesis leading to improved antioxidative system and significantly reduced cell death in roots of peach seedlings. Recently Husain et al. (2021) tested effect of H_2S and ETH in alleviation of hexavalent chromium [Cr(VI)] stress in *Vigna mungo* and *Vigna radiata*. The results showed that ETH can only alleviate Cr(VI) toxicity if endogenous H_2S is produced in both the crops. The presence of any H_2S inhibitor can reduce the beneficial effects of ETH in both the crops. Therefore endogenous H_2S production was necessary for alleviation of Cr (VI) toxicity in both the crop plants as it appeared to be a downstream signal of ETH.

14.2.5 Cross talk of salicylic acid with H_2S

The endogenously synthesized phenolic phytohormone SA has been reported widely in plants. The transportation of seven carbon skeletal phenolic compound occurs through phloem and plays a significant role in the alleviation of heat and drought stress as well as biotic stress in agricultural crops (Raskin,1992). It has been reported by Li et al. (2015) that the pretreatment of SA upregulated the LCD activity (Fig. 14.1) and led to the enhanced endogenous level of H_2S in maize seedlings under heat stress. The SA-mediated enhanced heat resistance occurs by the increased supplementation of NaHS. However, the increased supplementation of H_2S synthesizing inhibitors (PAG and HT) reduces the SA-mediated heat resistance. Reports suggested that there was not much significant adverse effect upon the SA biosynthetic enzymes and its endogenous level. The pretreatment with Paclobutrazol and 2-aminoindan-2-phosphonic acid (the SA biosynthetic inhibitors) did not affect NaHS-induced heat tolerance (Li et al., 2015). These results suggested that the cross talk of H_2S with SA showed the positive role in increasing the heat stress tolerance in plants (Fig. 14.1). It has also been studied that H_2S-mediated downregulation of SA phytohormone played a pivotal role in the development of chilling tolerance in *Cucumis sativus* L. seedlings via the upregulation of antioxidant machinery (Pan et al., 2020) (Fig. 14.1).

14.2.6 Cross talk of jasmonate with H_2S

Jasmonic acid (JA) is an essential phytohormone of higher plants and is endogenously synthesized from α-linolenic acid. It is considered as an important environmental signaling molecule which plays a pivotal role in the improvement of tolerance level against both biotic and abiotic stresses in plants (Fig. 14.1) (Devoto and Turner, 2003; Balbi and Devoto, 2008). It has been reported that the transcription of JA and jasmonate insensitive (JIN/MYC) plays a key role in the regulation of stomatal behavior in *Arabidopsis* (Han et al., 2018). The experimental reports revealed that the reduced level of H_2S enhanced the stomatal number which was inhibited by the application of JA. However, the supplementation of NaHS reduced the stomatal inhibition level in the *Arabidopsis* mutant *myc234* which was deficient in JA signaling. The application of H_2S reduced the transcriptome level of genes associated with stomata by decreasing the key components of stomata signaling pathway, such as STOMATAL DENSITY AND DISTRIBUTION1 (SDD1), TOO MANY MOUTHS (TMM), and SPEECHLESS (SPCH). It has been reported that the supplementation of methyl jasmonate (MeJA) reduced the stomatal density in *A. thaliana*. However, the application of a scavenger of H_2S, hypotaurin (HT) eradicated the role of MeJA-mediated decreased stomatal density in wild-type plants (Deng et al., 2020).

14.2.7 Cross talk of brassinosteroids with H_2S

Acting as a secondary signaling messenger molecule, the cross talk between H_2S and BRs plays an important role in the improvement of abiotic stress tolerance in plants (Hu et al., 2021). It has been reported by Ma et al. (2020) that the single application of epibrassinosteroid (EBR) enhanced the stomatal closure (Fig. 14.1) in *A. thaliana*. However, the treatment of H_2S biosynthetic inhibitors [hypotaurin (HT), aminooxy acetic acid (AOA), hydroxylamine (NH_2OH)] and the producer of L-/D-cysteine desulfhydrase (L-/D-CDes) ($C_3H_3KO_3$ + NH_3) significantly halt the EBR-assisted stomatal closure (Ma et al., 2020). In addition, the supplementation of EBR plays a pivotal role in the improvement of L-/D-CDes activity which in turn catalyzes the transformation of cysteine into H_2S. Reports suggested that the treatment of HT, NH_2OH, AOA, and $C_3H_3KO_3$ + NH_3 reduced the EBR-mediated enhanced activity of L-/D-CDes and H_2S content. Thus H_2S shows the synergistic approach in the EBR-mediated stomatal closure in plants (Ma et al., 2020). The cross talk of H_2S with EBR plays a pivotal role in the upregulation of stomatal movement and photosynthetic apparatus. The signaling molecule H_2S downregulates the BR-assisted signaling transduction pathway and plays a significant role in plant development (Hu et al., 2021).

14.3 Conclusions and future prospects

Plant hormones affect the plant developmental physiology and play critical roles in both normal and stressed conditions. Ongoing researches on H_2S have disclosed its several and diverse modulatory functions in biology and have generated considerable interest in this gasotransmitter. In spite of being regarded as a noxious gaseous molecule, latest researches have confirmed the growth enhancing and physiological implications of H_2S in plants. Nowadays, it is accepted that H_2S promotes seed germination, root growth, photosynthesis, stomatal movement, and plant senescence. Besides these, H_2S also modulates plant responses to stress by stimulating antioxidant defenses, increasing expression of genes encoding tolerance-related enzymes, and interacting with diverse signaling molecules. The extrinsic application of H_2S indisputably has a positive effect on different plant species, particularly those of considerable agronomic relevance under extreme environmental conditions. The treatment of exogenous H_2S, comprising a signaling mechanism, leads to an increase in several components of the antioxidant system at gene as well as protein level. H_2S is a vital factor in the resistance of cells to oxidative stress generated by an array of abiotic conditions. Nevertheless, numerous issues still remain to be explored. For instance, it has been proved that optimal concentration of H_2S causes marked impacts on plant growth and responses to stress, although diverse plants have different tolerance levels to H_2S. Hence, it is very important to check the concentration of H_2S in cells. Second, majority of the present investigations have concentrated on how exogenous application of H_2S improves plants' tolerance to stress, nevertheless the mechanism(s) by which intrinsic H_2S functions is unclear. It is also not known how external stimulus activates the accumulation of H_2S, how plants perceive H_2S signal and what are the targets and downstream cascades of H_2S signal transduction. Several reports have documented that H_2S shows cross talk with the signaling pathways of plant hormones and develops an intricate regulatory network related to plant growth and development, but the interactive mechanisms of H_2S with all the known phytohormones still remain to be clarified and should be the focus of future research.

References

Al Ubeed, H.M.S., Wills, R.B.H., Bowyer, M.C., Vuong, Q.V., Golding, J.B., 2017. Interaction of exogenous hydrogen sulfide and ethylene on senescence of green leafy vegetables. Postharvest Biol. Technol. 133, 81–87.

Arif, Y., Hayat, S., Yusuf, M., Bajguz, A., 2021. Hydrogen sulfide: a versatile gaseous molecule in plants. Plant Physiol. Biochem. 158, 372–384.

Aroca, A., Gotor, C., Romero, L.C., 2018. Hydrogen sulfide signaling in plants: emerging roles of protein persulfidation. Front. Plant Sci. 9, 1369.

Aroca, A., Gotor, C., Bassham, D.C., Romero, L.C., 2020. Hydrogen sulfide: from a toxic molecule to a key molecule of cell life. Antioxidants 9, 621.

Atkins, K.C., Cross, F., 2018. Inter-regulation of CDKA/CDK1 and the plant-specific cyclindependent kinase CDKB in control of the Chlamydomonas cell cycle. Plant Cell 30 (2), 429–446.

Balbi, V., Devoto, A., 2008. Jasmonate signalling network in Arabidopsis thaliana: crucial regulatory nodes and new physiological scenarios. New Phytol. 177 (2), 301–318.

Banerjee, A., Tripathi, D.A., Roychoudhury, A., 2018. Hydrogen sulfide trapeze: environmental stress amelioration and phytohormone crosstalk. Plant Physiol. Biochem. 132, 46–53.

Calderwood, A., Kopriva, S., 2014. Hydrogen sulfide in plants: from dissipation of excess sulfur to signalling molecule. Nitric Oxide 41, 72–78. Available from: https://doi.org/10.1016/j.niox.2014.02.005.

Chen, J., Wu, F.H., Wang, W.H., Zheng, C.J., Lin, G.H., Dong, X.J., et al., 2011. Hydrogen sulphide enhances photosynthesis through promoting chloroplast biogenesis, photosynthetic enzyme expression, and thiol redox modification in Spinacia oleracea seedlings. J. Exp. Bot. 62 (13), 4481–4493.

Chen, J., Wang, W.H., Wu, F.H., You, C.Y., Liu, T.W., Dong, X.J., et al., 2013. Hydrogen sulfide alleviates aluminum toxicity in barley seedlings. Plant Soil. 362, 301–318. Available from: https://doi.org/10.1007/s11104-012-1468-0.

Chen, S., Jia, H., Wang, X., Shi, C., Wang, X., Ma, P., et al., 2020. Hydrogen sulfide positively regulates abscisic acid signaling through persulfidation of SnRK2. 6 in guard cells. Mol. Plant 13 (5), 732–744.

Cheng, W., Zhang, L., Jiao, C.J., Su, M., Yang, T., Zhou, L.N., et al., 2013. Hydrogen sulfide alleviates hypoxia-induced root tip death in *Pisum sativum*. Plant Physiol. Biochem. 70, 278–286. Available from: https://doi.org/10.1016/j.plaphy.2013.05.042.

Christou, A., Manganaris, G.A., Papadopoulos, I., Fotopouls, V., 2013. Hydrogen sulfide induces systemic tolerance to salinity and non-ionic osmotic stress in strawberry plants through modification of reactive species biosynthesis and transcriptional regulation of multiple defence pathways. J. Exp. Bot. 64, 1953–1966. Available from: https://doi.org/10.1093/jxb/ert055.

Corpas, F.J., 2019. Hydrogen sulfide: a new warrior against abiotic stress. Trends Plant Sci. 24 (11), 983–988.

Davies, P.J., 2004. Plant hormones: biosynthesis, signal transduction, action. Kluwer Academic Publishers, Dordrecht, Netherlands.

Deng, G., Zhou, L., Wang, Y., Zhang, G., Chen, X., 2020. Hydrogen sulfide acts downstream of jasmonic acid to inhibit stomatal development in Arabidopsis. Planta 251, 656–669.

Devoto, A., Turner, J.G., 2003. Regulation of jasmonate-mediated plant responses in *Arabidopsis*. Ann. Bot. 92 (3), 329–337.

Eyidogan, F., Oz, M.T., Yucel, M., Oktem, H.A., 2012. Signal transduction of phytohormones under abiotic stresses. In: Khan, N.A., Nazar, R., Iqbal, N., Anjum, N.A. (Eds.), Phytohormones and Abiotic Stress Tolerance in Plants. Springer, Berlin, pp. 1–48.

Fahad, S., Hussain, S., Bano, A., Saud, S., Hassan, S., Shan, D., et al., 2015. Potential role of phytohormones and plant growth-promoting rhizobacteria in abiotic stresses: consequences for changing environment. Environ. Sci. Pollut. Res. 22 (7), 4907–4921. Available from: https://doi.org/10.1007/s11356-014-3754-2.

Fang, T., Cao, Z.Y., Li, J.L., Shen, W.B., Huang, L.Q., 2014. Auxin-induced hydrogen sulfide generation is involved in lateral root formation in tomato. Plant Physiol. Biochem. 76, 44–51. Available from: https://doi.org/10.1016/j.plaphy.2013.12.024.

Filipovic, M.R., Zivanovic, J., Alvarez, B., Banerjee, R., 2018. Chemical biology of H₂S signaling through persulfidation. Chem. Rev. 118, 1253–1337.

Fleet, C., Williams, M., 2011. Gibberellins. Teaching tools in plant biology: lecture notes. Plant Cell 110.

Fu, Y., Tang, J., Yao, G.F., Huang, Z.Q., Li, Y.H., Han, Z., et al., 2018. Central role of adenosine 5′-phosphosulfate reductase in the control of plant hydrogen sulfide metabolism. Front. Plant Sci. 9, 1404.

Fu, P.N., Wang, W.J., Hou, L.X., Liu, X., 2013. Hydrogen sulfide is involved in the chilling stress response in *Vitis vinifera* L. Acta Soc. Bot. Pol. 82, 295–302. Available from: https://doi.org/10.5586/asbp.2013.031.

Gomez-Roldan, V., Fermas, S., Brewer, P.B., Puech-Pagès, V., Dun, E.A., Pillot, J.P., et al., 2008. Strigolactone inhibition of shoot branching. Nature 455 (7210), 189–194.

Gotor, C., García, I., Aroca, Á., Laureano-Marín, A.M., Arenas-Alfonseca, L., Jurado-Flores, A., et al., 2019. Signaling by hydrogen sulfide and cyanide through post-translational modification. J. Exp. Bot. 70, 4251–4265.

Han, X., Hu, Y.R., Zhang, G.S., Jiang, Y.J., Chen, X.L., Yu, D.Q., 2018. Jasmonate negatively regulates stomatal development in *Arabidopsis* cotyledons. Plant Physiol. 176, 2871–2885.

Hancock, J.T., Whiteman, M., 2014. Hydrogen sulfide and cell signaling: team player or referee? Plant Physiol. Biochem. 78, 37–42. Available from: https://doi.org/10.1016/j.plaphy.2014.02.012.

Harrison, M.A., 2012. Cross-talk between phytohormone signaling pathways under both optimal and stressful environmental conditions. In: Khan, N.A., Nazar, R., Iqbal, N., Anjum, N.A. (Eds.), Phytohormones and Abiotic Stress Tolerance in Plants. Springer, Berlin, pp. 49–76.

He, H., Li, Y., He, L.F., 2019b. Role of nitric oxide and hydrogen sulfide in plant aluminum tolerance. Biometals 32, 1–9.

He, H., Garcia-Mata, C., He, L.F., 2019a. Interaction between hydrogen sulfide and hormones in plant physiological responses. Plant Growth Regul. 87 (1), 175–186.

Hedden, P., 2020. The current status of research on gibberellin biosynthesis. Plant Cell Physiol. 61 (11), 1832–1849.

Hu, D., Wei, L., Liao, W., 2021. Brassinosteroids in plants: crosstalk with small-molecule compounds. Biomolecules 11, 1800.

Hu, K.D., Zhang, X.Y., Wang, S.S., Tang, J., Yang, F., Huang, Z.Q., et al., 2019. Hydrogen sulfide inhibits fruit softening by regulating ethylene synthesis and signaling pathway in tomato (*Solanum lycopersicum*). HortScience 54 (10), 1824–1830.

Husain, T., Suhel, M., Prasad, S.M., Singh, V.P., 2021. Ethylene needs endogenous hydrogen sulfide for alleviating hexavalent chromium stress in *Vigna mungo* L. and *Vigna radiata* L. Environ. Pollut. 290, 117968.

Hussain, S., Gomes, M.M., Yano, K., Nambara, E., 2019. Interactions between abscisic acid and other hormones. Adv. Bot. Res. 92, 255–280.

Jia, H., Hu, Y., Fan, T., Li, J., 2015. Hydrogen sulfide modulates actin-dependent auxin transport via regulating ABPs results in changing of root development in Arabidopsis. Sci. Rep. 5, 8251.

Jia, H., Chen, S., Liu, D., Liesche, J., Shi, C., Wang, J., et al., 2018. Ethylene-induced hydrogen sulfide negatively regulates ethylene biosynthesis by persulfidation of ACO in tomato under osmotic stress. Front. Plant Sci. 9, 1517.

Jin, Z., Shen, J., Qiao, Z., Yang, G., Wang, R., Pei, Y., 2011. Hydrogen sulfide improves drought resistance in *Arabidopsis thaliana*. Biochem. Biophys. Res. Commun. 414, 481–486. Available from: https://doi.org/10.1016/j.bbrc.2011.09.090.

Jin, Z., Xue, S., Luo, Y., Tian, B., Fang, H., Li, H., et al., 2013. Hydrogen sulfide interacting with abscisic acid in stomatal regulation responses to drought stress in Arabidopsis. Plant Physiol. Biochem. 62, 41–46. Available from: https://doi.org/10.1016/j.plaphy.2012.10.017.

Jing, L.I.U., Hou, Z.H., Liu, G.H., Hou, L.X., Xin, L.I.U., 2012. Hydrogen sulfide may function downstream of nitric oxide in ethylene-induced stomatal closure in *Vicia faba* L. J. Integ. Agric. 11 (10), 1644–1653.

Kaya, C., Sarıoğlu, A., Ashraf, M., Alyemeni, M.N., Ahmad, P., 2020. Gibberellic acid-induced generation of hydrogen sulfide alleviates boron toxicity in tomato (*Solanum lycopersicum* L.) plants. Plant Physiol. Biochem. 153, 53–63.

Khan, M.I.R., Jahan, B., Alajmi, M.F., Rehman, M.T., Khan, N.A., 2019. Exogenously-sourced ethylene modulates defense mechanisms and promotes tolerance to zinc stress in mustard (*Brassica juncea* L.). Plants 8 (12), 540.

Khan, M.I.R., Jahan, B., AlAjmi, M.F., Rehman, M.T., Khan, N.A., 2020a. Ethephon mitigates nickel stress by modulating antioxidant system, glyoxalase system and proline metabolism in Indian mustard. Physiol. Mol. Biol. Plants 26 (6), 1201–1213.

Khan, M.I.R., Trivellini, A., Chhillar, H., Chopra, P., Ferrante, A., Khan, N.A., et al., 2020b. The significance and functions of ethylene in flooding stress tolerance in plants. Environ. Exp. Bot. 179, 104188.

Khan, M.I.R., Ashfaque, F., Chhillar, H., Irfan, M., Khan, N.A., 2021b. The intricacy of silicon, plant growth regulators and other signaling molecules for abiotic stress tolerance: an entrancing crosstalk between stress alleviators. Plant Physiol. Biochem. 162, 36−47.

Khan, M.I.R., Jahan, B., AlAjmi, M.F., Rehman, M.T., Iqbal, N., Irfan, M., et al., 2021c. Crosstalk of plant growth regulators protects photosynthetic performance from arsenic damage by modulating defense systems in rice. Ecotoxicol. Environ. Saf. 222, 112535.

Li, J., Jia, H., Wang, J., Cao, Q., Wen, Z., 2014. Hydrogen sulfide is involved in maintaining ion homeostasis via regulating plasma membrane Na^+/H^+ antiporter system in the hydrogen peroxide-dependent manner in salt-stress *Arabidopsis thaliana* root. Protoplasma 251, 899−912. Available from: https://doi.org/10.1007/s00709-013-0592-x.

Li, Z.G., Gong, M., Liu, P., 2012. Hydrogen sulfide is a mediator in H_2O_2-induced seed germination in Jatropha curcas. Acta Physiologiae Plantarum 34, 2207−2213.

Li, Z.G., 2013. Hydrogen sulfide: a multifunctional gaseous molecule in plants. Russ. J. Plant Physiol. 60, 733−740. Available from: https://doi.org/10.1134/S1021443713060058.

Li, Z.G., 2015. Synergistic effect of antioxidant system and osmolyte in hydrogen sulfide and salicylic acid crosstalk-induced heat tolerance in maize (*Zea mays* L.) seedlings. Plant Signal. Behav. 10, e1051278. Available from: https://doi.org/10.1080/15592324.2015.1051278.

Li, Z.G., Jin, J.Z., 2016. Hydrogen sulfide partly mediates abscisic acid-induced heat tolerance in tobacco (*Nicotiana tabacum* L.) suspension cultured cells. Plant Cell Tissue Organ. Cult. 125 (2), 207−214.

Li, Z.G., Ding, X.J., Du, P.F., 2013a. Hydrogen sulfide donor sodium hydrosulfide-improved heat tolerance in maize and involvement of proline. J. Plant Physiol. 170, 741−747. Available from: https://doi.org/10.1016/j.jplph.2012.12.018.

Li, T.T, Li, Z.-R., Hu, K.-D., Hu, L.-Y., Chen, X.-Y., et al., 2017. Hydrogen sulfide alleviates Kiwifruit ripening and senescence by antagonizing effect of ethylene. HortScience 52 (11), 1556−1562.

Li, Z.G., Xiang, R.H., Wang, J.Q., 2021. Hydrogen sulfide−phytohormone interaction in plants under physiological and stress conditions. J. Plant Growth Regul. 40, 2476−2484.

Li, Z.G., Xie, L.R., Li, X.J., 2015. Hydrogen sulfide acts as a downstream signal molecule in salicylic acid-induced heat tolerance in maize (*Zea mays* L.) seedlings. J. Plant Physiol. 177, 121−127.

Li, Z.G., Yang, S.Z., Long, W.B., Yang, G.X., Shen, Z.Z., 2013b. Hydrogen sulfide may be a novel downstream signal molecule in nitric oxide-induced heat tolerance of maize (*Zea mays* L.) seedlings. Plant Cell Environ. 36, 1564−1572. Available from: https://doi.org/10.1111/pce.12092.

Lin, Y.T., Li, M.Y., Cui, W.T., Lu, W., Shen, W.B., 2012. Haem oxygenase-1 is involved in hydrogen sulfide-induced cucumber adventitious root formation. J. Plant Growth Regul. 31, 519−528. Available from: https://doi.org/10.1007/s00344-012-9262-z.

Liu, D., Li, J., Li, Z., Pei, Y., 2020. Hydrogen sulfide inhibits ethylene-induced petiole abscission in tomato (*Solanum lycopersicum* L.). Hortic. Res. 7, 14.

Luo, S., Calderon-Urrea, A., Jihua, Y.U., Liao, W., Xie, J., Lv, J., et al., 2020. The role of hydrogen sulfide in plant alleviates heavy metal stress. Plant Soil. 449, 1−10.

Ma, Q., Yang, J., 2018. Transcriptome profiling and identification of functional genes involved in H_2S response in grapevine tissue cultured plantlets. Genes Genomics 40 (12), 1287−1300.

Ma, D., Ding, H., Wang, C., Qin, H., Han, Q., Hou, J., et al., 2016. Alleviation of drought stress by hydrogen sulfide is partially related to the abscisic acid signaling pathway in wheat. PLoS One 11 (9), e0163082.

Ma, Y.L., Shao, L.H., Zhang, W., Zheng, F.X., 2020. Hydrogen sulfide induced by hydrogen peroxide mediates brassinosteroid-induced stomatal closure of *Arabidopsis thaliana*. Funct. Plant Biol. 48, 195−205.

Mukherjee, S., 2019. Recent advancements in the mechanism of nitric oxide signaling associated with hydrogen sulfide and melatonin crosstalk during ethylene-induced fruit ripening in plants. Nitric Oxide 82, 25−34.

Nazir, F., Hussain, A., Fariduddin, Q., 2019a. Hydrogen peroxide modulate photosynthesis and antioxidant system in tomato (*Solanum lycopersicum* L.) plants under copper stress. Chemosphere 230, 544−558.

Nazir, F., Hussain, A., Fariduddin, Q., 2019b. Interactive role of epibrassinolide and hydrogen peroxide in regulating stomatal physiology, root morphology, photosynthetic and growth traits in *Solanum lycopersicum* L. under nickel stress. Environ. Exp. Bot. 162, 479−495.

Nazir, F., Hussain, A., Fariduddin, Q., Tanveer, A.K., 2021. Brassinosteroid and hydrogen peroxide improve photosynthetic efficiency and maintain chloroplast ultrastructure, stomatal movement, root morphology, cell viability and reduce Cu-triggered oxidative burst in tomato. Ecotoxicol. Environ. Saf. 207, 111081.

Nazir, F., Fariduddin, Q., Tanveer, A.K., 2022. Interaction between brassinosteroids and hydrogen peroxide networking signal molecules in plants. In: Brassinosteroids Signalling, Intervention with Phytohormones and Their Relationship in Plant Adaptation to Abiotic Stresses, pp. 59–79.

Normanly, J., 2010. Approaching cellular and molecular resolution of auxin biosynthesis and metabolism. Cold Spring Harb. Perspect. Biol. 2, 001594–001601.

Pan, D.Y., Fu, X., Zhang, X.W., Liu, F.J., Bi, H.G., Ai, X.Z., 2020. Hydrogen sulfide is required for salicylic acid-–induced chilling tolerance of cucumber seedlings. Protoplasma 257 (6), 1543–1557.

Paul, B.D., Snyder, S.H., 2018. Gasotransmitter hydrogen sulfide signaling in neuronal health and disease. Biochemical Pharmacol. 149, 101–109.

Poór, P., Nawaz, K., Gupta, R., Ashfaque, F., Khan, M.I.R., 2021a. Ethylene involvement in the regulation of heat stress tolerance in plants. Plant Cell Rep 1–24.

Raskin, I., 1992. Role of salicylic acid in plants. Annu. Rev. Plant Biol. 43 (1), 439–463.

Scuffi, D., Álvarez, C., Laspina, N., Gotor, C., Lamattina, L., García-Mata, C., 2014. Hydrogen sulfide generated by L-cysteine desulfhydrase acts upstream of nitric oxide to modulate abscisic acid-dependent stomatal closure. Plant Physiol. 166 (4), 2065–2076.

Shi, H., Ye, T., Han, N., Bian, H., Liu, X., Chan, Z., 2015. Hydrogen sulfide regulates abiotic stress tolerance and biotic stress resistance in *Arabidopsis*. J. Integr. Plant Biol. 57, 628–640. Available from: https://doi.org/10.1111/jipb.12302.

Skirycz, A., Inzé, D., 2010. More from less: plant growth under limited water. Curr. Opin. Biotechnol. 21, 197–203.

Sponsel, V.M., 2016. Signal achievements in gibberellin research: the second half-century. Annu. Plant. Rev. 49, 1–36.

Sun, J., Wang, R., Zhang, X., Yu, Y., Zhao, R., Li, Z., et al., 2013. Hydrogen sulfide alleviates cadmium toxicity through regulations of cadmium transport across the plasma and vacuolar membranes in *Populus euphratica* cells. Plant Physiol. Biochem. 65, 67–74. Available from: https://doi.org/10.1016/j.plaphy.2013.01.003.

Wang, Y., Mopper, S., Hasentein, K.H., 2001. Effects of salinity on endogenous ABA, IAA, JA, and SA in Iris hexagona. J. Chem. Ecol. 27, 327–342.

Went, F.W., 1935. Auxin, the plant growth-hormone. Bot. Rev. 1 (5), 162–182.

Wojtyla, L., Lechowska, K., Kubala, S., Garnczarska, M., 2016. Different modes of hydrogen peroxide action during seed germination. Front. Plant Sci. 7, 66. Available from: https://doi.org/10.3389/fpls.2016.00066.

Woodward, A.W., Bartel, B., 2005. Auxin: regulation, action, and interaction. Ann. Bot. 95, 707–735.

Xiao, Y., Wu, X., Sun, M., Peng, F., 2020. Hydrogen sulfide alleviates waterlogging-induced damage in peach seedlings via enhancing antioxidative system and inhibiting ethylene synthesis. Front. Plant Sci. 11, 696.

Xie, Y., Zhang, C., Lai, D., Sun, Y., Samma, M.K., Zhang, J., et al., 2014. Hydrogen sulfide delays GA-triggered programmed cell death in wheat aleurone layers by the modulation of glutathione homeostasis and heme oxygenase-1 expression. J. Plant Physiol. 171 (2), 53–62.

Xu, J., Zhang, S., 2015. Ethylene biosynthesis and regulation in plants. In: Wen, CK (Ed.), Ethylene in plants. Springer, Dordrecht, pp. 1–25.

Yamasaki, H., Cohen, M.F., 2016. Biological consilience of hydrogen sulfide and nitric oxide in plants: gases of primordial earth linking plant, microbial and animal physiologies. Nitric Oxide 55–56, 91–100.

Yao, G.F., Wei, Z.Z., Li, T.T., Tang, J., Huang, Z.Q., Yang, F., et al., 2018. Modulation of enhanced antioxidant activity by hydrogen sulfide antagonization of ethylene in tomato fruit ripening. J. Agric. Food Chem. 66 (40), 10380–10387.

Yuan, S., Shen, X., Kevil, C.G., 2017. Beyond a gasotransmitter: hydrogen sulfide and polysulfide in cardiovascular health and immune response. Antioxid. Redox Signal. 27, 634–653.

Zhang, L., Shi, X., Zhang, Y., Wang, J., Yang, J.J., et al., 2019. CLE9 peptide-induced stomatal closure is mediated by abscisic acid, hydrogen peroxide, and nitric oxide in Arabidopsis thaliana. Plant Cell Environ 42, 1033–1044.

Zhang, H., Tang, J., Liu, X.P., Wang, Y., Yu, W., Peng, W.Y., et al., 2009. Hydrogen sulfide promotes root organogenesis in Ipomoea batatas, *Salix matsudana* and *Glycine max*. J. Integr. Plant Biol. 51, 1086–1094.

Zhang, H., Dou, W., Jiang, C.-X., Wei, Z.-J., Liu, J., Jones, R.L., 2010. Hydrogen sulfide stimulates β-amylase activity during early stages of wheat grain germination. Plant Signal. Behav. 8, 1031–1033.

Zhang, H., Hu, S.L., Zhang, Z.J., Hu, L.Y., Jiang, C.X., Wei, Z.J., et al., 2011. Hydrogen sulfide acts as a regulator of flower senescence in plants. Postharvest Biol. Technol. 60 (3), 251–257.

Zhang, J., Zhou, M., Ge, Z., Shen, J., Zhou, C., Gotor, C., et al., 2020. Abscisic acid-triggered guard cell l-cysteine desulfhydrase function and in situ hydrogen sulfide production contributes to heme oxygenase-modulated stomatal closure. Plant Cell Environ. 43 (3), 624−636.

Zhang, T.Y., Li, F.C., Fan, C.M., Li, X., Zhang, F.F., He, J.M., 2017. Role and interrelationship of MEK1-MPK6 cascade, hydrogen peroxide and nitric oxide in darkness-induced stomatal closure. Plant Sci. 262, 190−199.

Zhang, X., Zhang, Y., Xu, C., Liu, K., Bi, H., Ai, X., 2021. H_2O_2 functions as a downstream signal of IAA to mediate H_2S-induced chilling tolerance in cucumber. Int. J. Mol. Sci. 22 (23), 12910.

Zhao, Y.D., 2010. Auxin biosynthesis and its role in plant development. Annu. Rev. Plant Biol. 61, 49−64.

Zhou, M., Zhou, H., Shen, J., Zhang, Z., Gotor, C., Romero, L.C., et al., 2021. H_2S action in plant life cycle. Plant Growth Regul. 94, 1−9.

Zhu, D., Wang, C., Liu, Y., Ding, Y., Winters, E., Li, W., et al., 2021. Gibberellic acid maintains postharvest quality of *Agaricus bisporus* mushroom by enhancing antioxidative system and hydrogen sulfide synthesis. J. Food Biochem. 45 (10), e13939.

Zhu, L., Du, H., Wang, W., Zhang, W., Shen, Y., Wan, C., et al., 2019. Synergistic effect of nitric oxide with hydrogen sulfide on inhibition of ripening and softening of peach fruits during storage. Sci. Hortic. 256, 108591.

Emerging trends in plant metabolomics and hormonomics to study abiotic stress tolerance associated with rhizospheric probiotics

Gaurav Yadav[1], Priyanka Prajapati[2], Devendra Singh[3], Sandhya Hora[4], Sneha Singh[5], Kanchan Vishwakarma[6] and Iffat Zareen Ahmad[1]

[1]Natural Product Laboratory, Department of Bioengineering, Integral University, Lucknow, Uttar Pradesh, India [2]Centre for Advanced Studies in Botany, Department of Botany, Institute of Science, Banaras Hindu University, Varanasi, Uttar Pradesh, India [3]Department of Biotechnology, Motilal Nehru National Institute of Technology Allahabad, Allahabad, Uttar Pradesh, India [4]Department of Molecular and Cellular Medicine, Institute of Liver and Biliary Sciences, New Delhi, India [5]Department of Phytochemistry, Council of Scientific and Industrial Research (CSIR)—Central Institute of Medicinal and Aromatic Plants, Lucknow, Uttar Pradesh, India [6]Department of Microbial Technology, Amity Institute of Microbial Technology, Amity University, Noida, Uttar Pradesh, India

15.1 Introduction

Soil being the essential part of the Earth comprises organic and mineral components that are required for plant growth and maintenance against variety of diseases. An alternative of protecting plants from stresses is to increase the fertility of soil, which can be achieved by the application of microbes, such as microbe-mediated carbon sequestration in soil (Vishwakarma et al., 2016). Microbes can epitomize variety of diverse products by producing different metabolites. Such metabolite product includes vaccines, vitamins, polysaccharides, inhibitors, etc., that can be employed to protect plants and also act against abiotic stresses (Mishra et al., 2018). Plant residue decomposition is facilitated by soil to retrieve the nutrients; however, the organic plant residue carbon from such decomposition is recognizable in soil after a long time. Hence, the need to maintain the soil carbon can be fulfilled by soil microorganisms that can sequester carbon into soil from atmospheric carbon dioxide.

The utilization of microorganisms has been increasing to promote the plant growth and improve health status and soil fertility. Applications of microbes has also been witnessed in alleviation of abiotic stresses in plants, that is, salinity, heavy metals, ultraviolet, drought, harmful nanomaterials, and biotic stresses, such as pests and pathogenic diseases. It is imperative that to diminish the negative effects of stresses, plant—microbe, and soil symbiosis plays a huge role (Upadhyay et al., 2019; Vishwakarma et al., 2020a). Recently a plant growth promoting rhizobacteria also known as rhizospheric probiotics has been utilized to overcome toxic effects of silver nanoparticles (AgNPs) in *Brassica juncea* (Vishwakarma et al., 2020b). These microbial strain secreted indole acetic acid (Vishwakarma et al., 2018) that proved to be beneficial for mitigating the harmful impact of AgNPs by improving redox status of plant cell, reducing oxidative stress, and increasing antioxidant enzyme activities in the cell (Vishwakarma et al., 2020b).

Endophytes are a class of microorganisms that dwell within plants and functions as plant growth promoters. They have been explicitly utilized to increase nutrients and mineral uptake, prevent growth of phytopathogens and enhance the plant's tolerance to pathogens (Vishwakarma et al., 2021). Their physiological role involves the production of hormones, such as auxins, cytokinins (CKs), and other growth-promoting molecules, such as vitamins. Plants with residing endophytes have proven pest resistance and tolerance to abiotic stresses, such as drought (Cheplick et al., 2000; Eerens et al., 1998), acidity (Lewis, 2004), heavy metals (Monnet et al., 2001), and high pH levels (Waller et al., 2005). The symbiosis between the plant roots and microbes is also facilitated by root exudates, the broad array of compounds secreted by roots, that help in establishment and regulation of the plant—microbe symbiosis (Vishwakarma et al., 2017a,b). This symbiosis ultimately influences the mineral uptake in the soil deficient in nutrients by directly chelating with the nutrients or indirectly by modifying the microbial activity in rhizosphere. For example, Rhizobium, which forms symbiosis with legumes and can survive under low nitrogen concentrations, is also involved in assisting plant growth promotion by supplying nutrients especially nitrogen, protect plant directly and indirectly from environmental stresses (Kumar et al., 2020). The advances made in microbial genome studies through techniques, such as metabolomics, proteomics, genomics, and bioinformatics, have made possible the discovery of new products, which empowers resilience against abiotic stress and also imparts beneficial metabolites to medicinal plants.

Carbohydrates and amino acids, as primary metabolites in plant stress tolerance, have been the focus of the majority research investigations aimed at enhancing abiotic stress tolerance. Carbohydrates can bring out and can be used as energy storage during photosynthesis. As N-containing compounds, amino acids are also precursors of plant metabolites related to plant defense (Szepesi, 2020). Drought-induced changes in leaf-to-root connections were explored in two different tomato genotypes in an experiments conducted by Moles et al. (2018). Another studies reported several agronomic properties of *Triticum aestivum*, *Triticale*, and *Triti pyrum* species during drought stress (Shanazari et al., 2018). The adaptation reactions of durum wheat roots to salt were examined by Annunziata et al. (2017). They discovered that the N-containing metabolites and cellular sucrose content of durum wheat roots could be remodeled. Li et al. (2018a,b) compared the physiological and metabolic responses to low temperature in two zoysia grass genotypes native to high and low latitudes. The metabolomics-based response of wheat grains to heat stress contributes to a constant filling rate (Wang et al., 2018a,b). In a study, two different barley species were compared in their reactions to extended salt stress and it was discovered that they have varied metabolite accumulation times (Ferchichi et al., 2018).

Phytohormones play a key role in plants ability to adjust to abiotic stressors by negotiating a variety of adaptive responses. They frequently change gene expression quickly by promoting or inhibiting transcriptional regulator breakdown via the ubiquitin−proteasome system. Abscisic acid (ABA) synthesis is one of the fastest responses of plants to abiotic stress, triggering ABA-inducible gene expression and causing stomatal closure thereby reducing water loss via transpiration and eventually restricting cellular growth. Numerous genes associated with ABA de novo biosynthesis and genes encoding ABA receptors and downstream signal relays have been characterized in *Arabidopsis thaliana* (Peleg and Blumwald, 2011).

Phytohormones, such as ABA, gibberellins (GAs), ethylene (ET), auxin (indole-3-acetic acid), CKs, and brassinosteroids, are involved in influencing a number of physiological processes and biological signaling in sessile plants (Raza et al., 2019).

15.2 Metabolomics: a comprehensive tool to study metabolites

Metabolites are tiny molecules that are structurally different and altered in the duration of cell metabolism. Metabolism is defined as the quantitative collection of all the low molecular weight compounds present in a cell or organism at a particular physiological or developmental state (Jorge et al., 2016). Metabolomics is a fast-emerging technology that provides a complete knowledge of biological metabolites and characterizes the mas metabolome. Metabolomics implications in plant biotechnology are enormous and plays a vital role in analyzing functional genomics. On the other side, the term metabolome refers to the study of a complete set of small molecules (metabolites) and their interactions with the biological system (Fiehn, 2002). The idea of metabolomics has arisen from systems biology that tries to measure the various facets, such as genes, metabolites, proteins, and transcripts of a living system, to generate data related to the interactions among the small molecules. It is very challenging to estimate the whole metabolites even for the *A. thaliana*, such as model systems, in which genome has been completely and extensively sequenced. According to the estimates, if the total number of compounds of a specific plant species in the metabolome will be determined, the genome size varies for various plant species (5000−25,000) (Trethewey, 2004; Oksman-Caldentey and Saito, 2005). At the same time, other approaches, such as metabolite profiling, try to quantify and identify the specific chemically related metabolites classes that normally share the chemical properties which enable the instantaneous analysis (Gates et al., 1977).

For many years, experiments related to metabolic profiling have been continuously executed. However, it has recently benefited due to the increasing interest in metabolomics for effective analysis of metabolite. High-throughput methodologies for the biomolecules' (nucleic acid microarrays, protein identification, and DNA sequencing) characterization are required before the experiments, such as proteomic, transcriptomic, and genomic, should be performed; whereas the "omics" related experiments were performed by metabolomics that displayed less dependency on any technological breakthrough. This is partially due to the higher inconsistency in the chemical properties of pools of metabolite in comparison with the proteins or nucleic acids and also since metabolites are nonbiopolymers they do not get advantage by the higher accessibility of sequenced genome data. Rather, the metabolomics experiments have been performed by using various analytical tools, mainly mass spectrometry (MS) (Dettmer et al., 2007), nuclear magnetic resonance (Eisenreich and Bacher, 2007), and infrared spectroscopy (Ellis et al., 2003). Metabolomics is used in assisting genome annotation (May et al., 2008) as well as to analyze the plant abiotic and biotic stress responses (Guy et al., 2008; Leiss et al., 2009). Phytomedicinal plants' metabolic profiling is a proven and effective tool for the optimization of plant yield and quality control (Van der Kooy et al., 2008).

This tool is also effective for reviewing the ecological and environmental interactions for exploring the unexpected bioactive compounds involved in the plant—herbivore interactions and also in discovering the fungal infections induced metabolites (Abdel-Farid et al., 2009). This metabolite fingerprints technique is also used for the purpose of identifying the compounds linked with the specific traits for quick selection in various plant breeding specific programs (Beckmann et al., 2007). Regardless of the reality that the genomic data cannot be used for the analysis of metabolite, the metabolomics approach can be effectively applied for species analysis that lacks sequenced genomes. Therefore species that are not genome enabled can be studied using metabolomics, while proteomics and transcriptomics approaches cannot be used.

15.3 Plant—microbe interactions: a metabolome insight

Decoding the interactions between plants and microbes is a highly promising aspect that helps understand the adverse as well as the beneficial effect of microbes on the plant. These interactions may be adverse, neutral, and fruitful and will directly affect the plant health, productivity, and growth (Newton et al., 2010). The interaction between plants and microbes is highly diversified because the interactions occur above the soil, below the soil, and also within the plants (Bulgarelli et al., 2013). Furthermore, the interactions with microbes may be both epiphytic and endophytic, and it may also be with the neighboring soil and environment in the vicinity of crop roots. However, majority of the microbes' interaction with plants may be pathogenic resulting in various plant infections (Strange and Scott, 2005). Both pathogenic, as well as commensal interactions require the particular signaling pathways for response or interactions, such as rice blast infection or the growth of root nodules (Riely et al., 2006).

All microbes have evolved different strategies to survive in diverse environmental situations and also to alter other microorganisms. One approach that evolved convergent in various kingdoms via strategic associations and transformative supports races with other microbes is the release of a variety of metabolites. The interaction between the microbes and the environment around the plant roots demonstrates how the secreted metabolites affect plant health, growth, and development. Since the plants secrete a significant amount of metabolites as root exudates, thus rhizosphere attracts varied microbes and prepares the ground for the battle (Raaijmakers et al., 2009).

Several metabolites secreted by the plants suppress the harmful species or attract the fruitful species that help shape the rhizosphere microbiome accumulation and thus influence the health of the plant (Pascale et al., 2020). Plants use specific chemical substances to control the development, defense, and microbial communication (Gelvin, 2012; Shah et al., 2014). Such metabolite-mediated systems in plants probably represent the transformative opportunities for several pathogens to manipulate or control such processes by secreting similar metabolites or their functional or structural analogs (Spaepen, 2015). The infection caused by *Agrobacterium tumefaciens* shows how microbes can control plant development and growth via secreted DNA. A T-DNA insertion inside the plant genomes produces plant hormones, such as CKs and auxin, and also produces the amino acid derivatives, opines that acts as a source of nitrogen for *A. tumefaciens* (Gelvin, 2012). Several microbes use a specific enzyme that alter the plant metabolites into forms which is helpful to the microbes.

Mainly a complex connection coexists among the plant and the microbes present in soil within the microscopic region (rhizosphere) that surrounds the plant roots. In this region both beneficial organisms as well as injurious microbes, having either positive or negative effects on plant survival, growth and also on its yield are found (Vacheron et al., 2013). Plant growth promoting rhizobacteria, also known as Rhizospheric probiotics are the microorganisms living in this area, plays a vital role in promoting plant health and also protects the plants from the various pathogens either indirectly by the production of several antibiotics, hydrolases, or by resistance induction or directly by nitrogen fixation, solubilization of nutrient and by the production of siderophores and phytohormones (Yadav et al., 2017) as shown in Fig. 15.1.

The interaction between the probiotics and plant are of symbiotic type means benefits and fitness cost is shared between them (Bulgarelli et al., 2013). They possess various qualities, such as promoting plant growth, bioremediation, and control disease, and also decrease the abiotic and biotic environmental stresses effect by suppressing the soil herbivores and pathogens by releasing lytic enzymes, siderophores, toxins, and antibiotics (Compant et al., 2005).

Numerous studies have revealed that the inoculating plants with probiotics exometabolites or probiotics for improving defense responses, resulted in various biochemical actions linked with the modification of plant cell wall, metabolite modification, secondary metabolites' biosynthesis, and defense genes' expression (Conrath et al., 2006).

Plant metabolomics has come up with most influential research tool for analyses of biochemical mechanisms in concern with several abiotic stresses, such as drought, extreme temperature, and salinity, with the plant growth and development (Jorge and António, 2018). In plants, numerous number of metabolites (approximately 200,000 metabolites) are

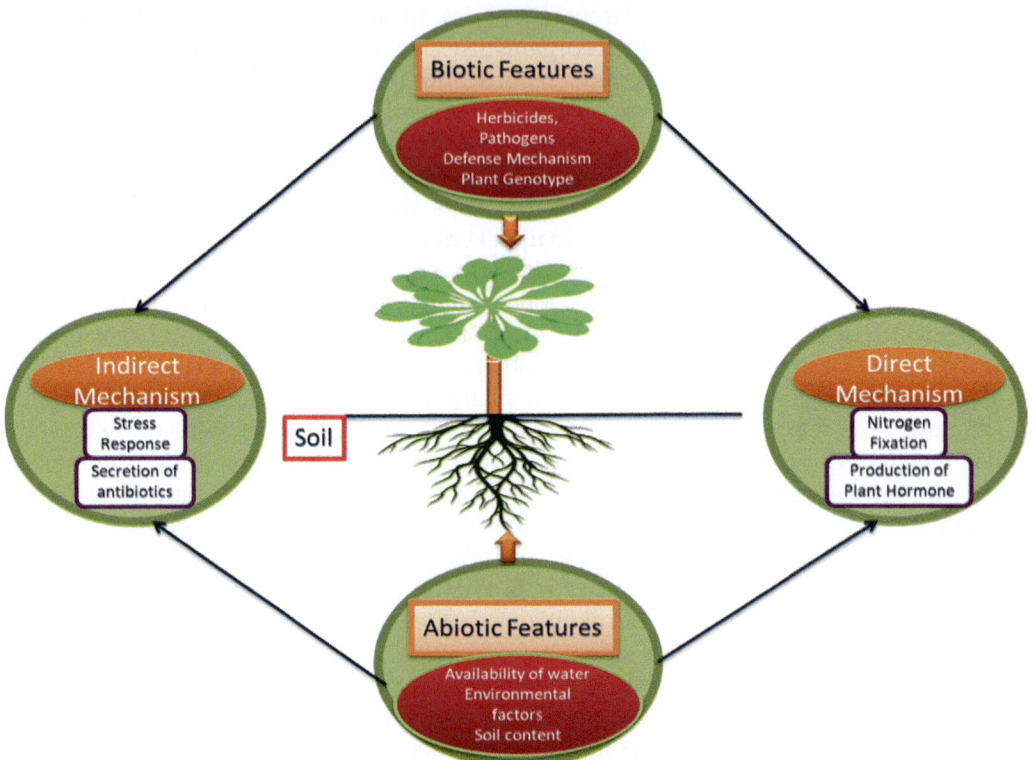

FIGURE 15.1   The biotic and abiotic features influencing rhizospheric probiotics interactions in the rhizosphere.

extracted that can vary in concentration from femtomolar to millimolar and approximately 7000–15,000 metabolites are found in one individual species. Approximately 3000–5000 were specifically found in leaves. Moreover, in a wide range of plant species, such as rice, lettuce, tobacco, potato, and wheat, metabolite profiling has already been performed. In this wide range of plants species, poplar and eucalyptus are also included.

Metabolite profiling is a difficult task due to the wide range of molecules with different structures and chemical properties. For example, it is estimated that a single accession of *Arabidopsis* contains more than 5000 metabolites, most of them yet unspecified. The most popular metabolomics techniques focus on metabolites with specific and similar chemical properties and it is known that metabolite profiling covers only a fraction of the metabolome. To achieve a comprehensive coverage of the vast range of metabolites present in the plant kingdom, several analytical techniques consisting of a separation technique couple to a detection device (usually MS) are combined (Arbona et al., 2013).

Total metabolite pool of a plant is characterized in a nonbiased way by plant metabolomics in response to its microenvironment. This metabolite pool incorporates broad range of metabolites with different physical properties from ionic inorganic mixtures to biochemically determined hydrophilic carbohydrates, natural and amino acids and a scope of hydrophobic lipid-related mixtures. Current plant metabolomics examines numerous insightful instruments because of such metabolic variety, to acquire far-reaching metabolite inclusion from a complex organic plant sample. Mass spectroscopy, such as liquid chromatography–mass spectrometry (LC–MS) and gas chromatography–mass spectrometry (GC–MS), has been most commonly used to receive comprehensive information of plant metabolome in a broad range of plant species (Jorge and António, 2018).

There are some compatible small molecular weight metabolites which act as osmoprotectants and perform defensive mechanism under abiotic stress condition in plants. These vary in their natural way and include amino acids and amines. In addition, they include carbohydrates, raffinose and polyols, and antioxidants. In these osmolytes, solubility at high levels in cell and enzyme inhibition is found in high concentrations (Krasensky and Jonak, 2012).

A few examinations have been executed to comprehend the gainful Impacts of these metabolites in plant resilience against ecological stimuli. Proline is outstanding amongst other model and the interconnection between stress resistance and proline accumulation has been all around deciphered in Bermunda grass during water stress conditions (Barnett and Naylor, 1966).

Extensive research work has been performed for other metabolites, such as γ-aminobutyric acid (GABA), betaine, glycine, trehalose, raffinose, and polyamines (such as putrescine, spermidine, and spermine); these metabolites have been proved as efficient protectors against some abiotic stresses. Under high stress levels, GABA (nonprotein amino acid) accumulates rapidly (Kaplan and Guy, 2004; Renault et al., 2010). Glycine betaine (GB) is a quaternary ammonium compound that occurs widely in the plant and synthesized from glycine and choline (Chen and Murata, 2011). Under salt, cold, and drought stress conditions, plants with natural production of GB exhibited increased the accumulation of GB. Trehalose is a nonreducing disaccharide that accumulates in higher concentrations in *Myrothamnus flabellifolius* (desiccation-tolerant plant). Trehalose is present in small amount in plants, under high-stress conditions trehalose increases moderately. Raffinose family oligosaccharides (RFOs) include stachyose, verbascose, and raffinose, which accumulate in different plant species (in leaves and desiccated seeds) under environmental stress, such as heat, cold, salinity, and drought stress. For plant's survival, metabolome adjustments play a very important role; hence, regulation of the metabolic pathways (catabolism and biosynthesis) is very critical to enhance tolerance mechanisms in plants against environmental stimuli (Arbona et al., 2013).

15.4 Abiotic stresses and its impact on plants

In a world of 7 billion people, agriculture is facing serious challenges to ensure a sufficient food supply while maintaining high productivity and quality standards. In addition, to an ever-increasing demographic demand, alterations in weather patterns due to changes in climate are affecting crop productivity globally. Unfavorable climate that results in abiotic stresses causes changes in agro-ecological conditions and indirectly affects growth and distribution of incomes, thus increasing the agricultural production demand (Fraire-Velázquez and Balderas-Hernández, 2013). An adverse environmental factor includes water scarcity (drought) (Farooq et al., 2009), nutrients deficiency (Trubat et al., 2006), high or low temperature (Hasanuzzaman et al., 2013), contamination, and ion irradiance in soil by higher levels of metals or salt.

They significantly restrain the growth and productivity of crop species world-wide (Okçu et al., 2005). For example, heat stress reduce crop yield considerably, thereby causing low resource use efficiency. Drought stress impair seed germination leading to poor stem strength, as well as reducing dry weight of roots and shoot (Zeid and Shedeed, 2006). Intergovernmental Panel on Climate Change reported that the many abiotic stress factors have been predicted to increase in future due to global climate change (http://www.ipcc.ch/). Thus a very deep understanding about abiotic stress and plant responses to various stress events is an important topic in plant research.

15.5 Plant metabolic responses to individual abiotic stresses

Many studies have been performed to analyze the metabolic responses to different abiotic stresses, such as drought, heat, or salinity. Comprehensive studies have been found in different literature on this topic. In this section, advanced applications of MS-based metabolomics approaches to examine the plant responses to abiotic stress will be described. It will emphasize on identification of stress-responsive metabolites that may contribute to plant growth and development with increased abiotic stress tolerance (Jorge and António, 2018) eventually, as shown in Fig. 15.2.

15.5.1 Metabolite responses to drought stress

Drought is a significant abiotic stress, which leads to decrease in crop yields causing morpho-physiological changes, such as depletion in shoot development (Chaves et al., 2003), decreases in transpiration and photosynthesis rates as an immediate consequence of ABA-mediated leaf stomata closure, and changes in signaling pathways and transcriptional and posttranscriptional modifications of few stress related genes (Chaves and Oliveira, 2004; Xue et al., 2008). Also, under drought conditions, plant metabolism is readjusted eventually through the accumulation of compatible solutes of osmolytes (Slama et al., 2015). These little particles gather in higher complexity in the cell without distressing cell assimilation and furthermore incorporate alcohol and solvent sugars containing saccharose, glucose, and mannitol and RFOs, such as stachyose, raffinose, and verbascose amino acids and polyamines (Slama et al., 2015). Because of this osmolyte collection, a decrease in the osmotic limit of the cell is noticed and the turgor pressure reinforced as the cell takes up water, consequently, help in balancing out layers, chemicals, and proteins or keeping up cell turgor by osmotic change. Extensive omics contemplates have been accounted for to examine plant reactions to dry season pressure (Mata et al., 2016). Gechev and his associates studied the drought tolerance of a resurrection plant *Haberlea rhodopensis* through metabolomics approaches. The utilization of gas chromatography—time of flight-mass spectroscopy (GC—TOF-MS) and liquid chromatography (LC—MS) metabolomics uncovered significant accumulation of maltose and solvent sugars sucrose; similarly, stachyose and verbascose are noticed in *H. rhodopensis* plants under drought conditions highlighting the mechanisms contributing to extreme dry conditions (Gechev et al., 2013).

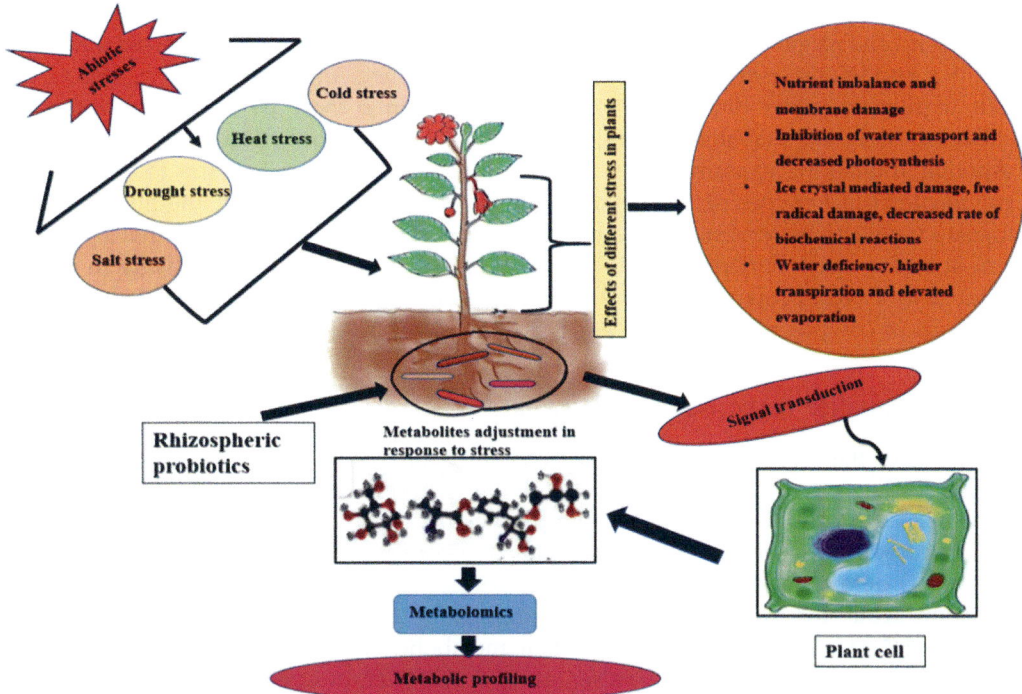

FIGURE 15.2 Abiotic stresses on plants trigger the signal transduction from rhizospheric probiotics to plant cells for a stress response through the regulation of metabolites, which can be determined through plant metabolomics.

A comparable metabolomics approach was also applied to analyze the revival of the plant in hydrated *Selaginella lepidophylla* (Yobi et al., 2013) from ultra-high performance chromatography-pair mass spectrometry (UHPLC−MS/MS) and gas chromatography (GC−MS) analysis and revealed that the assemblies of metabolites associated with the glycolytic pathway (glucose-6-phosphate, fructose-6-phosphate, and pyruvate) and tricarboxylic acid (TCA cycle) (2-oxoglutarate, fumarate, succinate, fumarate, and oxaloacetate). Similarly, the moderate impacts of long-haul dry season pressure in 21 rice cultivars (*Oryza sativa* L. sp. *indica* and *japonica*) were examined by utilizing physiological profiling examination of GC−TOF-MS metabolomics (Do et al., 2013) and were unveiled that in rice, dry spell condition prompts an accumulation of spermine along these lines leading to an organized change of polyamine absorption which is in concurrence with an osmoprotectant metabolite under dry season pressure (Do et al., 2013). The GC−TOF-MS metabolite profile in roots and leaves of two genotypes of grain (*Hordeum vulgare* L.), with heterogenous resistance to dry conditions, has recognized around 100 metabolites delicate to dry season pressure, of which the amino acid is the generally influenced class of metabolites. Moreover, few sugars accumulated were subjective in tissues of the two genotypes exposed to dryness (Chmielewska et al., 2016).

The molecular mechanisms of drought resistance have been shown through characterization of metabolic information in sunflower (*Helianthus annus* L.). Metabolomics investigation recognized 54 essential metabolites, including different amino acids, natural acids, sugar, and sugar alcohols, and uncovered that most amino acids showed low levels under dry conditions, with exemption of glycine, tyramine, malonate, proline, and GABA, which aggregated under dry season conditions. As a rule, these outcomes anticipated the putative job of these metabolites during stress feedback in sunflower (Moschen et al., 2017).

Another interesting examination was performed for osmotic alteration to dry stress in *Vigna unguiculata* (cowpea's) leaves and roots by GC—TOF-MS. From the 88 metabolites recognized, galactinol, proline, and a quercetin subordinate were the ones that responded the most to dry spell. Furthermore, these metabolites accumulated distinctively in roots as compared to leaves, proposing a more moderate system to adapt to dry season in the fragile portions of cowpea plants (Goufo et al., 2017).

15.5.2 Metabolite responses to salt stress

Soil salinity has been viewed as a worldwide issue, which altogether diminishes crop yield that influences estimated 20% of inundated land and as a result, plant development, and richness are diminished, and untimely senescence happens (Chen et al., 2015). Plant resilience to salt stress firmly relies upon the two components that are utilized by plants, these are: (1) detoxification of reactive oxygen species inside the plant cells and (2) Na^+ ion elimination from the roots or compartmentalization of these particles in the vacuoles (Munns and Tester, 2008). Plants change their metabolic status to adapt to salt stress, albeit this metabolic change extensively varies among salt-tolerant species, different basic salt-stress metabolite responses are found within the plant kingdom (Sanchez et al., 2008). Based on their salt resistance properties, plants are generally classified as halophytes (salt tolerant) and glycophytes (salt sensitive). For halophytic plants, the precollection and differential reactions of osmoprotectant metabolites contrast among plant species. Albeit in glycophytic plants, there is an expanding proof that sugar, sugar alcohols, amino acids, and TCA cycle intermediates structure the center of metabolite acclimations to salinity stress.

Curiously, a relative report utilizing both salt-tolerant and salt-sensitive Lotus species has displayed that around half of all metabolites have an equivalent reaction to salinity (Sanchez et al., 2008). Comparable situation was noticed for *A. thaliana* (salt sensitive), and its inaccessible relative *Thellungiella halophila* (salt tolerant), both accumulating solvent sugars (glucose, fructose, sucrose, and raffinose) and proline (Gong et al., 2005). Among crops, an intriguing examination explored on *H. vulgare* L. grain cultivars that varied in salt-stress tolerance for their metabolite responses to long term salt stress (Widodo et al., 2009). Rice is quite possibly the most delicate cereal harvests, in any case, GC—TOF-MS investigation uncovered lower levels of TCA cycle intermediates and other organic acids in the underlying foundations of more-tolerant rice cultivars than in more sensitive rice cultivars (Zuther et al., 2007). A cutting-edge metabolomics approach, in particular GC and LC combined with a triple quadrupole mass spectrometer, was applied for the quantitative profiling of a wide range of metabolites from two gram varieties *Cicer arietinum* L. with reverse responses to salt pressure. The LC—MS metabolite profiling method permitted to quantitatively assess 28 biogenic amines and amino acids despite the fact that GC—MS approach allowed the quantitatively break down of 48 essential metabolites, varying from sugars and sugar phosphates to natural acids. Likewise, this methodology demonstrated that the metabolic contrasts between the two different varieties rely upon the metabolites engaged with carbon assimilation, the TCA cycle just as amino acid assimilation (Dias et al., 2015).

Actinorhizal plants are a collection of enduring dicotyledonous angiosperms. These plants are profoundly resistant to outrageous conditions and have monetary significance (creation of wood and subordinates). The model actinorhizal plant *Casuaurina glauca* can withstand undeniable degrees of salinity. Jorge and his partners have developed a metabolomics study to explore the effect of salt stress in *C. glauca* nodulated and nonnodulated plants exposed to various degrees of salinity (Jorge et al., 2017). GC−TOF-MS metabolic profiling information uncovered significant metabolite level uniqueness in amino acid assimilation in both plant groups which incorporates proline, sugars and ornithine.

15.5.3 Metabolite responses to heat stress

Heat stress is characterized by an increase in optimum temperature range (normally 10−15°C), for an enough timeframe, to make an irreversible harm to plant development and growth. When plants are exposed to heat stress they exhibit increase in lipid membrane fluidity, denaturation of proteins and inactivation of key compounds in several organelles, such as mitochondria and chloroplasts. Metabolomics applied on plants exposed to temperature stress has revealed the accumulation of osmolytes, especially dissolvable sugars, proline, and glycine-betaine (Wahid and Close, 2007). Du and his colleagues applied a GC−MS metabolite profiling to characterize metabolites related with differential heat resilience between two grass species, in particular C3 Kentucky twang and C4 Bermudagrass (Du et al., 2011). In both grass species, 36 temperature stress-sensitive metabolites were distinguished ranging from natural amino acids to sugars and sugar alcohols. It was observed that Bermudagrass accumulated more metabolites than Kentucky twang. Among the diverse aggregated metabolites, this investigation portrayed seven sugars (xylose, maltose, sucrose, fructose, galactose, floridoside, and melibiose) a sugar alcohol (inositol), six natural acids (methyl malonic acid, malic acid, citrus extract, threonic acid, galacturonic acid, and isocitric acid), and nine amino acids (methionine, lysine, glycine, isoleucine, GABA, threonine, valine, alanine, and asparagine) (Du et al., 2011). Li et al. (2016) utilized GC−MS to explore whether higher levels of GABA could improve heat resilience in *Agrostis stolonifera* L. (Bentgrass) under cool season. Upon exogenous supplementation of GABA, metabolite profiling uncovered an amassing in the degree of five natural acids (oxalic acid, malic acid, succinic acid, aconitic acid, and threonic acid), five sugars (maltose, galactose, glucose, fructose, and sucrose), six amino acids (aspartic corrosive, alanine, threonine, serine, valine, and glutamic corrosive) and two sugar alcohols (myo-inositol and mannitol). This investigation showed that the GABA instigated heat resilience in crawling bentgrass.

15.6 Plant hormonomics

Plant associations with diverse microbial communities are the outcome of millions of years of coevolution and have assisted plant migration and ability to adapt to land. Plant hormones play a significant role in the plant–microbiota communication process. Hormones are chemical messengers that play an endocrine (activity remote from biosynthesis site) or paracrine (function in cells adjacent to biosynthesis site) role in cellular and physiological processes. Hormones are responsible in host factors (such as nutrition and signaling pathways) along with microbial adaptation to their hosts. Hormones can contribute to microbial diversity in the endosphere and distinct root compartments by influencing plant defense and development, or in the rhizosphere through direct or indirect interactions between bacteria and excreted hormones (Eichmann et al., 2021).

The rhizosphere and indeed the mammalian gut probiotics have some functional similarities; both are places where nutrients are absorbed and contribute to plant/animal health and development. The rhizosphere's microbial community composition improve the plant's functional capacities (especially against abiotic stress).

Plant hormones play an important role in microbial sensing and subsequent plant signaling. Plant hormones are beginning to play a role in transforming root microbiome composition, either directly or indirectly, to sustain plant development in the face of abiotic stress (Carvalhais et al., 2017). Plant hormones can be released into the rhizosphere, where they may have a significant influence on plant-interacting microorganisms and the root microbiome as a whole. The existence of most plant-derived hormones in root exudates has been proven (Faure et al., 2009). On the other hand, hormones may have indirect effects on root exudates, which are utilized to interact with the microbial population (Carvalhais et al., 2015).

15.6.1 Strigolactones—class of hormones specific to plant synergies

Strigolactones (SLs) are a class of plant hormones that have a role in root system design, pathogenic weed propagation, and plant–microbe communication, among many other things (Aliche et al., 2020). Because of the role of SLs in the establishment of plant–microbe symbioses, numerous studies have employed amplicon sequencing for microbiome profiling to investigate the influence of SLs on the microbiota (Liu et al., 2020). SLs' signaling is thought to have a role in creating rhizosphere probiotic communities, as per research. Members of *Fusarium solani*, *Rhizobiaceae*, predatory *Bdellovibrio*, and *Shinella* were more prevalent in the rhizospheres of soybean plants overexpressing SLs' biosynthesis and signaling genes (Liu et al., 2020). SLs' biosynthesis and signaling-deficient rice (*O. sativa*) mutants, on the other hand, demonstrated a drop in various good bacteria groups in the mutant rhizospheres, as well as a decrease in the pathogens *Olpidium brassicae* (Nasir et al., 2019). When taken as a whole, SL has a demonstrable impact on the form of rhizosphere communities, although without a clear positive or negative effect.

Most terrestrial plants experience nutrient inadequacy, such as a lack of phosphorus (P) or nitrogen (N) in the soil (Marzec, 2016). Inorganic phosphate (Pi) is the only accessible form of P for plants in soil. The roots are the primary site for Pi uptake which leads to significant modifications below ground, such as enhanced root growth and decreased shoot/root ratio, suppression of shoot branching, restricted primary root growth, and considerable lateral root and root hair development. The release and production of SLs are promoted by low levels of phosphorus present in the soil (Sun et al., 2014).

The potential of SLs to interact with nitrogen-fixing bacteria remains unaffected by SL exudation caused due to the lack of nitrogen. Some of the legumes, such as red clover, alfalfa and crimson clover, respond to P deficiency by raising SL secretions while most nonlegumes react to both N and P deficiencies (Peláez-Vico et al., 2016). A decrease in shoot P levels is the actual stimuli for SL induction under nutrient deprivation as N deficiency affects shoot phosphorus content in some species. The effects of N insufficiency on SL production depend on plant species, nutrient, extent of nutritional stress, and pathways of macronutrient absorption. Therefore SLs' nutritional deprivation activates the SL-mediated stress adaptive responses (Sun et al., 2014).

To interact and communicate with plants and other microorganisms, probiotics regulates phytohormones and modulate hormone signaling. Plant-associated microorganisms create a variety of hormones and hormone-like compounds, as well as enzymes that affect hormone levels in the rhizosphere. Rhizospheric probiotic colonization is linked to alterations in plant metabolism, signaling, and hormone balance (Tsukanova et al., 2017). Different strains of rhizospheric probiotics can manufacture phytohormones, metabolize them, or influence plant hormone production and signaling.

Rhizospheric probiotics can improve plant nutrient availability by assisting nitrogen fixation (Kuan et al., 2016), phosphate absorption (Mehta et al., 2015), siderophore generation (Zhou et al., 2016), mycorrhizal symbiosis development, and root architecture regulation (Navarro-Ródenas et al., 2016). Plant pathogen resistance (Sharifi and Ryu, 2016), pathogen growth (Prasannakumar et al., 2015), and the inhibitory effects of abiotic stresses, such as drought, can all be activated by rhizospheric probiotics (Lim and Kim, 2013). Since hormones control plant growth and development, the effects of rhizospheric probiotic colonization are linked to changes in hormone levels, localization, and signaling (Verbon and Liberman, 2016). It is important to note that many physiological processes in plants are controlled by dynamic interaction between many hormones rather than the quantity of a single hormone (Naseem et al., 2015). Numerous findings have investigated the subtle influence of plant growth-promoting rhizobacteria (PGPR) on the expression of plant genes, including those involved in phytohormone signaling, metabolism, and degradation (Lara-Chavez et al., 2015).

15.6.1.1 *Auxin*

Auxin is a phytohormone that is essential for plant growth and development. It is necessary for the continuation of the cell cycle (Demeulenaere and Beeckman, 2014). Despite the fact that auxin investigation has a significant finding, new information on its metabolism, reception, and transport, as well as its role in plants, is constantly being discovered (Niu et al., 2015). It is also empirical that rhizospheric probiotics can impact the level and localization of auxin, as well as the direction of auxin transport in the plant, particularly in the majority of physiological activities (Ahmed and Hasnain, 2014). How do bacteria alter auxin homeostasis in plants? First of all, auxin can be synthesized directly. There is a lot of evidence that distinct rhizospheric strains produce auxin in culture (Ahmed and Hasnain, 2014). This research implies that auxin production is the fundamental source of some rhizospheric strains which stimulates action on host plants in these circumstances.

15.6.1.2 Ethylene

ET a plant origin hormone is involved in a variety of physiological mechanisms, such as root apical meristem function and root hair growth (Vandenbussche and Van Der Straeten, 2012), development of the "triple response" in elongated seedlings (hypocotyl swelling, root shortening, and exaggeration in the roundness of the apical hook), fruit ripening, leaf senescence, stomata formation, gravitropism (Vandenbussche and Van Der Straeten, 2012), and response to abiotic stress (Verma et al., 2016).

Depending on the plant species and hormone levels, ET can either restrict or promote plant growth (Vandenbussche and Van Der Straeten, 2012). Rhizospheric probiotics can alter plant ET regulation by modulating the expression of the ET production enzymes ACC-synthase and ACC-oxidase genes. For example, in *A. thaliana* (Verbon and Liberman, 2016) and *Panicum virgatum, Burkholderia phytofirmans* PsJN promotes the expression of *ACS* and *ACO* genes (Poupin et al., 2016).

Consequently, various strains might affect the concentration of ET in the plant, either increasing or lowering the gene responsible for the production. It is critical to know the molecular pathways by which the strain alters the ET levels in the plant under various applications.

15.6.1.3 Cytokinins

CKs are a category of N6-substituted adenine derivatives that constitute the most common type of plant hormone. CKs have a role in a variety of plant functions. They are required for cell differentiation, in particular (Schaller et al., 2014). They are also involved in the development of abiotic stress tolerance in plants (O'Brien and Benková, 2013). Plant probiotics have been shown to effect the CK levels too. Many strains can produce CK (Glick, 2012), indicating that various bacteria use distinct ways to manipulate plant CK levels.

Bacterial CKs play a critical function in abiotic stress management. For example, the ability to synthesize CK is required by PGPR *Pseudomonas fluorescens* G20-18 to enhance the resistance of *A. thaliana* plants against infection with *Pseudomonas syringae*, as the mutant *P. fluorescens* G20-18, which is incapable of CK synthesis, has no effect (Großkinsky et al., 2016).

15.6.1.4 Gibberellins

Plant probiotics, such as other hormones, also regulates the quantity of endogenous GA. Various probiotics strains can even generate GAs in their metabolic processes. The quantity of endogenous GAs in the roots of host plants rises after inoculation with GA-producing PGPR strains *Bacillus cereus* MJ-1 (Joo et al., 2005) and *Promicromonospora* sp. SE188 (Kang et al., 2014). In mutant rice plants with impaired GA synthesis pathway, the GA-producing strain *Leifsonia soli* SE134 and *Enterococcus faecium* LKE12 stimulate root development. This demonstrates that these strains use bacterial GAs to compensate for the lack of plant GAs (Kang et al., 2014). Furthermore, probiotics can promote the plant endogenous GA production. When plants are inoculated with GA-producing strains of *Promicromonospora* sp. SE188 and *Bacillus amyloliquefaciens* RWL-1, the level of GAs in the plant increases, including those that are lacking from these culture medium

(Kang et al., 2014). Eventually, this method of plant growth promotion using rhizospheric probiotics increases the GA levels in the plants.

15.6.1.5 Abscisic acid

ABA is a plant hormone that activates genes involved in stress tolerance in response to abiotic stimuli (drought, cold, salt stress, and soil pollution) (Sah et al., 2016). Cellular ABA content rises when plants are infected with ABA-producing bacteria, such as *Bacillus licheniformis* Rt4M10, *P. fluorescens* Rt6M10, and *Azospirillum brasilense* Sp 245, and the plant becomes more drought tolerant (Cohen et al., 2015).

PGPR can produce a variety of phytohormones (Dodd et al., 2010). These phytohormones can also be an antagonistic, meaning that they play competing functions in plant life. During colonization, it is unknown how the balance of hormones generated by rhizospheric probiotics is maintained. As a result, predicting the impact of a particular strain on a plant is challenging. This is believed to be dependent on both the plant type and the surrounding environment. Furthermore, the presence of a minimal levels of endogenous ABA in plants may be essential to measure the stimulatory response associated with the strain. *Bacillus megaterium*, for example, promotes the development of *Solánum lycopérsicum* wild type but inhibits the growth of ABA-deficient mutants (Porcel et al., 2014). As a result, rhizospheric probiotics can influence the amount of ABA in the plant, impacting its development and resilience to abiotic stress.

The ability of rhizosheric bacteria to impact plant hormonal state through bacterial hormone synthesis or metabolism is thought to be a key method by which they stimulate plant growth and productivity. Inoculating these bacteria into the plant rhizosphere may cause positive or negative outcomes depending on the hormonal composition as well prevailing environmental conditions (Kudoyarova et al., 2019).

By synthesizing growth-stimulating hormones (Shi et al., 2017) and metabolizing growth-inhibitory hormones, microorganisms can have a direct impact on plant development (Glick, 2014). Experiments indicates that inoculation with PGPR produces critical changes in plant expression of hormone-mediated genes and have emphasized the relevance of hormones in mediating such plant/microbe interactions (Lim and Kim, 2013). Microbial-related regulation of hormones, as well as changes in plant endogenous hormone levels (Dodd et al., 2010) are triggered by volatile compounds generated by microbes and might cause changes in plant hormonal state (Zhang et al., 2007). Plant hormones, such as auxins (Spaepen and Vanderleyden, 2011), have an effect on microbial gene expression after being added to culture medium.

Hormones generated by rhizospheric probiotics are closely correlated to plant growth promotion, whereas other benefits (such as their ability to increase mineral nutrition and plant resistance to pathogens and abiotic stress) are evaluated apart from microbial effects on the plant hormonal system. Hormone-induced activation of plant antioxidant systems, such as catalase, ascorbate peroxidase (Zavaleta-Mancera et al., 2007), and Cu−Zn−superoxide dismutase (Tyburski et al., 2009) enzymes, safeguards plants from oxidative stress, which is linked with the majority of harmful environmental variables. As a result, the ability of rhizospheric bacteria to alter plant hormonal state is likely implicated in most known processes of rhizosphere microbial community growth promotion. Auxins, CKs, ABA, and ET have all been explored in relation to their impact on plant hormonal state. However, there methods of action may help to draw broad inferences about additional plant hormones mediated by rhizospheric bacteria. Furthermore, because these bacteria can synthesis ABA (Shahzad et al., 2017), plant absorption of the microbially generated hormone should affect ABA-mediated activities.

15.6.2 Plant hormone profiling through hormonomics

Targeted and nontargeted metabolomic-based approach, such as phytometabolomics, sensomics, lipidomics, and hormonomics, can aid in the discovery of metabolic pathways, signal transmissions and metabolites with unique bioactivities that can be up- or down-regulated by stress conditions at the same time. For all components investigated, an ideal technique should offer both a qualitative analysis and an exact quantitative data. Due to the low quantities of phytohormones (in comparison to primary and secondary metabolites) and wide ranges of chemical stability, it also necessitates proper sample preparation and good instrumental performance (in terms of both robustness and sensitivity).

Šimura et al. (2018) described a technique for LCMS/Ms-based plant hormone profiling that is quick, sensitive, and concurrent. Bioactive forms of hormones, their precursors, and metabolites are among the analytes used to get quantitative snapshots of the physiological condition of examined tissues. The method was used to describe phytohormone profiles in salt-stressed roots of *A. thaliana* seedlings to test the practical applicability of this hormonomics methodology. The seedlings' hormonal responses to salt stress, one of the primary variables in crop production, were revealed using statistical data analysis and cross-validation with transcriptome data (Munns and Tester, 2008).

15.6.3 Profiling of phytohormone-related compounds by ultrafast high performance liquid chromatography—mass spectroscopy

Many of the molecules exhibit identical core structures, including isomers with similar MS fragmentation patterns, which makes profiling plant hormones complex (e.g., *cis*- and *trans*-zeatin, topolin isomers, brassinolide and 24-epibrassinolide, and castasterone and 24-epicastasterone). The above-mentioned isomeric compounds had improved peak morphologies and peak-to-peak resolution when separated using reverse phase UHPLC with charged-surface hybrid and ET-bridged hybrid polymer (Floková et al., 2014). The mobile phase composition and the employment of various mobile phase additives had a significant impact on separation, peak shape, and analyteionisation. Cao et al. (2016) discovered that raising the concentration of formic acid from 0.05 to 0.2% decreased the separation of CKs. However, utilizing a charged-surface hybrid column and isocratic elution with 0.01% formic acid in both mobile phase solutions, the best baseline separation of CK isomers may indeed be accomplished.

Positive electrospray ionization-based mass spectroscopy can resolve the metabolites of CKs, auxins, GAs, and certain ABA precursors and their amino acid conjugates (Tarkowská et al., 2016).

15.6.4 Method validation

The determined endogenous levels of *Arabidopsis* seedling extracts should be adjusted from the quantities of nonlabelled standards added to verify the UHPLC—MS technique (Šimura et al., 2018). Finally, the predicted concentrations of each analyte are compared to the known quantities given to samples to determine method accuracy, which ranges from 5.2 to 8.68% bias on average. The relative standard deviation of analyte concentrations is used to calculate method precision, which varies from 1.06 to 7.8%. Intraday and interday precisions for endogenous phytohormones measured in *Arabidopsis* seedlings varied from 0.6 to 10.5% and 1.3 to 17.3%, respectively.

15.6.5 Phytohormone quantification in plants under salinity stress

Šimura et al. (2018) examined hormone-related transcript and metabolite levels in samples of root and shoot tissues of stressed *Arabidopsis* plants and controls to test the applicability of the targeted metabolomics technique. Plant stress chemicals, such as ABA, have been demonstrated to improve salt tolerance (Ryu and Cho, 2015). Increases in ABA, its oxidation products phaseic acid and dihydrophaseic acid, as well as upregulation of biosynthesis and oxidation genes, were linked to salt stress in roots. Other stress hormone-related metabolite profiles were shown to have a similar relationship with the varied responses to salt stress in roots via plant hormonomics.

Under salinity stress, auxin and CK-related metabolites showed a more dynamic balance that could not be easily linked to biosynthesis and metabolism. Recent developments in analytical technologies have made it possible to identify more hormone metabolites (precursors, catabolites, and conjugates) in a single sample, allowing researchers to learn more about the hormone metabolome's overall pattern (Novák et al., 2017).

Although plant hormones, such as auxins, may be quantified using LC—MS, their distribution patterns in plant organs, tissues, and cells are still unknown (Novák et al., 2017). The measurement of exact levels of single members of the phytohormone family has been the subject of a wide range of targeted approaches (Novák et al., 2017).

In the future, analyzing a mix of derivatized and nonderivatized phytohormone classes in one approach might be a very important and helpful step in hormonomics (Dawid and Hille, 2018). In the future, strategies that combination of biological and molecular structural characterizations of marker metabolites with metabolomics techniques, such as phytometabolomics or sensomics approaches, will be effective in producing high-quality phytoalexin-enriched functional meals.

15.7 Conclusion

Despite the beneficial role of rhizospheric probiotics for improved tolerance to abiotic stresses, the extensively acknowledged protection value of plants metabolites has showcased their potential. Among omics science, metabolomics is an emerging application that is primarily concerned with high throughput polaroid of a metabolome at a given point and under specific physiological conditions. Metabolomics is of special relevance to understand the plants response to environmental signal. Because metabolites are the final product of cellular regulatory processes their studies on plant are potentially a more direct indicator of functional phenotype (Marzec, 2016). A future prospect of plant metabolomics is to enhance the agronomic and medicinal crop improvement for advanced agriculture.

Hormones have developed into go-betweens for microorganisms and plants to create conditions for the formation of niches at the rhizosphere and to allow targeted host adaptation for endosphere colonization as a type of shared chemical language. While the relevance of hormones in the plant—microbe interactions is widely understood, the go-between role of plant and microbe-derived hormones in root microbiome assembly is still an area to be explored.

Recent breakthroughs in omics-based technologies, such as meta-transcriptomics, meta-proteomics, and metabolomics, in combination with amplicon sequencing, can aid in the cataloging of hormonal processes in plant/root holobionts. The discovery of molecules influenced by hormone signaling with the ability to change the microbiome will be aided by metabolomics analysis of root exudates, as well as a better understanding of the functional molar ranges of exuded plant-derived and microbe-derived hormones.

Acknowledgments

The authors are thankful to Integral University Lucknow and Amity University Uttar Pradesh for providing necessary facilities for the completion of this chapter.

References

Abdel-Farid, I.B., Jahangir, M., van den Hondel, C.A., Kim, H.K., Choi, Y.H., Verpoorte, R., 2009. Fungal infection-induced metabolites in *Brassica rapa*. Plant. Sci. 176 (5), 608—615.

Ahmed, A., Hasnain, S., 2014. Auxins as one of the factors of plant growth improvement by plant growth promoting rhizobacteria. Pol. J. Microbiol. 63 (3), 261.

Aliche, E.B., Screpanti, C., De Mesmaeker, A., Munnik, T., Bouwmeester, H.J., 2020. Science and application of strigolactones. N. Phytol. 227 (4), 1001—1011.

Annunziata, M.G., Ciarmiello, L.F., Woodrow, P., Maximova, E., Fuggi, A., Carillo, P., 2017. Durum wheat roots adapt to salinity remodeling the cellular content of nitrogen metabolites and sucrose. Front. Plant Sci. 7, 2035.

Arbona, V., Manzi, M., Ollas, C.D., Gómez-Cadenas, A., 2013. Metabolomics as a tool to investigate abiotic stress tolerance in plants. Int. J. Mol. Sci. 14 (3), 4885—4911.

Barnett, N.M., Naylor, A.W., 1966. Amino acid and protein metabolism in Bermuda grass during water stress. Plant Physiol. 41 (7), 1222—1230.

Beckmann, M., Enot, D.P., Overy, D.P., Draper, J., 2007. Representation, comparison, and interpretation of metabolome fingerprint data for total composition analysis and quality trait investigation in potato cultivars. J. Agric. Food Chem. 55 (9), 3444—3451.

Bulgarelli, D., Schlaeppi, K., Spaepen, S., Van Themaat, E.V., Schulze-Lefert, P., 2013. Structure and functions of the bacterial microbiota of plants. Annu. Rev. Plant Biol. 64, 807—838.

Cao, Z.Y., Sun, L.H., Mou, R.X., Zhang, L.P., Lin, X.Y., Zhu, Z.W., et al., 2016. Profiling of phytohormones and their major metabolites in rice using binary solid-phase extraction and liquid chromatography-triple quadrupole mass spectrometry. J. Chromatogr. A 1451, 67—74.

Carvalhais, L.C., Dennis, P.G., Badri, D.V., Kidd, B.N., Vivanco, J.M., Schenk, P.M., 2015. Linking jasmonic acid signaling, root exudates, and rhizosphere microbiomes. Mol. Plant Microbe Interact. 28 (9), 1049—1058.

Carvalhais, L.C., Schenk, P.M., Dennis, P.G., 2017. Jasmonic acid signalling and the plant holobiont. Curr. Opin. Microbiol. 37, 42—47.

Chaves, M.M., Oliveira, M.M., 2004. Mechanisms underlying plant resilience to water deficits: prospects for water-saving agriculture. J. Exp. Bot. 55 (407), 2365–2384.

Chaves, M.M., Maroco, J.P., Pereira, J.S., 2003. Understanding plant responses to drought-from genes to the whole plant. Funct. Plant Biol. 30 (3), 239–264.

Chen, T.H., Murata, N., 2011. Glycinebetaine protects plants against abiotic stress: mechanisms and biotechnological applications. Plant Cell Environ. 34 (1), 1–20.

Chen, T.W., Kahlen, K., Stützel, H., 2015. Disentangling the contributions of osmotic and ionic effects of salinity on stomatal, mesophyll, biochemical and light limitations to photosynthesis. Plant Cell Environ. 38 (8), 1528–1542.

Cheplick, G.P., Perera, A., Koulouris, K., 2000. Effect of drought on the growth of *Lolium perenne* genotypes with and without fungal endophytes. Funct. Ecol. 14, 657–667.

Chmielewska, K., Rodziewicz, P., Swarcewicz, B., Sawikowska, A., Krajewski, P., Marczak, Ł., et al., 2016. Analysis of drought-induced proteomic and metabolomic changes in barley (*Hordeum vulgare* L.) leaves and roots unravels some aspects of biochemical mechanisms involved in drought tolerance. Front. Plant Sci. 7, 1108.

Cohen, A.C., Bottini, R., Pontin, M., Berli, F.J., Moreno, D., Boccanlandro, H., et al., 2015. *Azospirillum brasilense* ameliorates the response of *Arabidopsis thaliana* to drought mainly via enhancement of ABA levels. Physiol. Plant. 153 (1), 79–90.

Compant, S., Duffy, B., Nowak, J., Clément, C., Barka, E.A., 2005. Use of plant growth-promoting bacteria for biocontrol of plant diseases: principles, mechanisms of action, and future prospects. Appl. Environ. Microbiol. 71 (9), 4951–4959.

Conrath, U., Beckers, G.J., Flors, V., García-Agustín, P., Jakab, G., Mauch, F., et al., 2006. Priming: getting ready for battle. Mol. Plant Microbe Interact. 19 (10), 1062–1071.

Dawid, C., Hille, K., 2018. Functional metabolomics—a useful tool to characterize stress-induced metabolome alterations opening new avenues towards tailoring food crop quality. Agronomy. 8 (8), 138.

Demeulenaere, M.J., Beeckman, T., 2014. The interplay between auxin and the cell cycle during plant development. Auxin and Its Role in Plant Development. *Springer*, Vienna, pp. 119–141.

Dettmer, K., Aronov, P.A., Hammock, B.D., 2007. Mass spectrometry-based metabolomics. Mass Spectrom. Rev. 26 (1), 51–78.

Dias, D.A., Hill, C.B., Jayasinghe, N.S., Atieno, J., Sutton, T., Roessner, U., 2015. Quantitative profiling of polar primary metabolites of two chickpea cultivars with contrasting responses to salinity. J. Chromatogr. B 1000, 1–3.

Do, P.T., Degenkolbe, T., Erban, A., Heyer, A.G., Kopka, J., Köhl, K.I., et al., 2013. Dissecting rice polyamine metabolism under controlled long-term drought stress. PLoS One 8 (4), e60325.

Dodd, I.C., Zinovkina, N.Y., Safronova, V.I., Belimov, A.A., 2010. Rhizobacterial mediation of plant hormone status. Ann. Appl. Biol. 157 (3), 361–379.

Du, H., Wang, Z., Yu, W., Liu, Y., Huang, B., 2011. Differential metabolic responses of perennial grass *Cynodontransvaalensis* × *Cynodondactylon* (C4) and *PoaPratensis* (C3) to heat stress. Physiol. Plant. 141 (3), 251–264.

Eerens, J.P.J., Lucas, R.J., Easton, S., White, J.G.H., 1998. Influence of the ryegrass endophyte (*Neotyphodiumlolii*) in a cool-moist environment II. Sheep production. N. Z. J. Agric. Res. 41, 191–199.

Eichmann, R., Richards, L., Schäfer, P., 2021. Hormones as go-betweens in plant microbiome assembly. Plant J. 105 (2), 518–541.

Eisenreich, W., Bacher, A., 2007. Advances of high-resolution NMR techniques in the structural and metabolic analysis of plant biochemistry. Phytochemistry. 68 (22–24), 2799–2815.

Ellis, D.I., Harrigan, G.G., Goodacre, R., 2003. Metabolic fingerprinting with Fourier transform infrared spectroscopy. In: Harrigan, G.G., Goodacre, R. (Eds.), Metabolic Profiling: Its Role in Biomarker Discovery and Gene Function Analysis. *Springer*, Boston, MA, pp. 111–124.

Farooq, M., Wahid, A., Kobayashi, N.S., Fujita, D.B., Basra, S.M., 2009. Plant drought stress: effects, mechanisms and management. Sustain. Agric. 153–188.

Faure, D., Vereecke, D., Leveau, J.H., 2009. Molecular communication in the rhizosphere. Plant Soil 321 (1), 279–303.

Ferchichi, S., Hessini, K., Dell'Aversana, E., D'Amelia, L., Woodrow, P., Ciarmiello, L.F., et al., 2018. *Hordeumvulgare* and *Hordeummaritimum* respond to extended salinity stress displaying different temporal accumulation pattern of metabolites. Funct. Plant Biol. 45, 10961109.

Fiehn, O., 2002. Metabolomics—the link between genotypes and phenotypes. Funct. Genomics 155—171.

Floková, K., Tarkowská, D., Miersch, O., Strnad, M., Wasternack, C., Novák, O., 2014. UHPLC—MS/MS based target profiling of stress-induced phytohormones. Phytochemistry. 105, 147—157.

Fraire-Velázquez, S., Balderas-Hernández, V.E., 2013. Abiotic stress in plants and metabolic responses. Abiotic Stress-Plant Responses and Applications in Agriculture. InTech, Rijeka, Croatia, pp. 25—48.

Gates, S.C., Young, N.D., Holland, J.F., Sweeley, C.C., 1977. Automated multicomponent analysis of biological mixtures by gas chromatography-mass spectrometry. Adv. Mass Spectrom. Biochem. Med. 2, 171—181.

Gechev, T.S., Benina, M., Obata, T., Tohge, T., Sujeeth, N., Minkov, I., et al., 2013. Molecular mechanisms of desiccation tolerance in the resurrection glacial relic *Haberlea rhodopensis*. Cell. Mol. Life Sci. 70 (4), 689—709.

Gelvin, S.B., 2012. Traversing the cell: *Agrobacterium* T-DNA's journey to the host genome. Front. Plant Sci. 3, 52.

Glick, B.R., 2012. Plant growth-promoting bacteria: mechanisms and applications. Scientifica. 2012.

Glick, B.R., 2014. Bacteria with *ACC deaminase* can promote plant growth and help to feed the world. Microbiol. Res. 169 (1), 30—39.

Gong, Q., Li, P., Ma, S., InduRupassara, S., Bohnert, H.J., 2005. Salinity stress adaptation competence in the extremophile *Thellungiella halophila* in comparison with its relative *Arabidopsis thaliana*. Plant J. 44 (5), 826—839.

Goufo, P., Moutinho-Pereira, J.M., Jorge, T.F., Correia, C.M., Oliveira, M.R., Rosa, E.A., et al., 2017. Cowpea (*Vigna unguiculata* L. Walp.) metabolomics: osmoprotection as a physiological strategy for drought stress resistance and improved yield. Front. Plant Sci. 8, 586.

Großkinsky, D.K., Tafner, R., Moreno, M.V., Stenglein, S.A., García de Salamone, I.E., Nelson, L.M., et al., 2016. Cytokinin production by *Pseudomonas fluorescens* G20-18 determines biocontrol activity against *Pseudomonas syringae* in *Arabidopsis*. Sci. Rep. 6 (1), 1.

Guy, C., Kaplan, F., Kopka, J., Selbig, J., Hincha, D.K., 2008. Metabolomics of temperature stress. Physiologiaplantarum 132 (2), 220—235.

Hasanuzzaman, M., Nahar, K., Alam, M., Roychowdhury, R., Fujita, M., 2013. Physiological, biochemical, and molecular mechanisms of heat stress tolerance in plants. Int. J. Mol. Sci. 14 (5), 9643—9684.

Joo, G.J., Kim, Y.M., Kim, J.T., Rhee, I.K., Kim, J.H., Lee, I.J., 2005. Gibberellins-producing rhizobacteria increase endogenous gibberellins content and promote growth of red peppers. J. Microbiol. 43 (6), 510—515.

Jorge, T.F., António, C., 2018. Plant metabolomics in a changing world: metabolite responses to abiotic stress combinations. Plant, Abiotic Stress and Responses to Climate Change. pp. 111—132.

Jorge, T.F., Rodrigues, J.A., Caldana, C., Schmidt, R., van Dongen, J.T., Thomas-Oates, J., et al., 2016. Mass spectrometry-based plant metabolomics: metabolite responses to abiotic stress. Mass Spectrom. Rev. 35 (5), 620—649.

Jorge, T.F., Duro, N., Da Costa, M., Florian, A., Ramalho, J.C., Ribeiro-Barros, A.I., et al., 2017. GC-TOF-MS analysis reveals salt stress-responsive primary metabolites in *Casuarina glauca* tissues. Metabolomics. 13 (8), 1—3.

Kang, S.M., Khan, A.L., You, Y.H., Kim, J.G., Kamran, M., Lee, I.J., 2014. Gibberellin production by newly isolated strain *Leifsonia soli* SE134 and its potential to promote plant growth. J. Microbiol. Biotechnol. 24 (1), 106—112.

Kaplan, F., Guy, C.L., 2004. β-Amylase induction and the protective role of maltose during temperature shock. Plant. Physiol. 135 (3), 1674—1684.

Krasensky, J., Jonak, C., 2012. Drought, salt, and temperature stress-induced metabolic rearrangements and regulatory networks. J. Exp. Bot. 63 (4), 1593—1608.

Kuan, K.B., Othman, R., Abdul Rahim, K., Shamsuddin, Z.H., 2016. Plant growth-promoting rhizobacteria inoculation to enhance vegetative growth, nitrogen fixation and nitrogen remobilisation of maize under greenhouse conditions. PLoS One 11 (3), e0152478.

Kudoyarova, G., Arkhipova, T., Korshunova, T., Bakaeva, M., Loginov, O., Dodd, I.C., 2019. Phytohormone mediation of interactions between plants and non-symbiotic growth promoting bacteria under edaphic stresses. Front. Plant Sci. 1368.

Kumar, N., Srivastava, P., Vishwakarma, K., Kumar, R., Kuppala, H., Maheshwari, S.K., et al., 2020. The rhizobium—plant symbiosis: state of the art. Plant Microbe Symbiosis 1—20.

Lara-Chavez, A., Lowman, S., Kim, S., Tang, Y., Zhang, J., Udvardi, M., et al., 2015. Global gene expression profiling of two switchgrass cultivars following inoculation with *Burkholderia phytofirmans* strain PsJN. J. Exp. Bot. 66 (14), 4337—4350.

Leiss, K.A., Choi, Y.H., Abdel-Farid, I.B., Verpoorte, R., Klinkhamer, P.G., 2009. NMR metabolomics of thrips (*Frankliniellaoccidentalis*) resistance in Senecio hybrids. J. Chem. Ecol. 35 (2), 219—229.

Lewis, G.C., 2004. Effects of biotic and abiotic stress on the growth of three genotypes of *Loliumperenne* with and without infection by the fungal endophyte *Neotyphodium lolii*. Ann. Appl. Biol. 144, 53−63.

Li, Z., Yu, J., Peng, Y., Huang, B., 2016. Metabolic pathways regulated by γ-aminobutyric acid (GABA) contributing to heat tolerance in creeping bentgrass (*Agrostis stolonifera*). Sci. Rep. 6 (1), 1−6.

Li, S., Yang, Y., Zhang, Q., Liu, N., Xu, Q., Hu, L., 2018a. Differential physiological and metabolic response to low temperature in two zoysiagrass genotypes native to high and low latitude. PLoS One 13 (6), e0198885.

Li, Z., Peng, Y., Huang, B., 2018b. Alteration of transcripts of stress-protective genes and transcriptional factors by γ-aminobutyric acid (GABA) associated with improved heat and drought tolerance in creeping bentgrass (*Agrostis stolonifera*). Int. J. Mol. Sci. 19 (6), 1623.

Lim, J.H., Kim, S.D., 2013. Induction of drought stress resistance by multi-functional PGPR *Bacillus licheniformis* K11 in pepper. Plant Pathol. J. 29 (2), 201.

Liu, F., Rice, J.H., Lopes, V., Grewal, P., Lebeis, S.L., Hewezi, T., et al., 2020. Overexpression of strigolactone-associated genes exerts fine-tuning selection on soybean rhizosphere bacterial and fungal microbiome. Phytobiomes J. 4 (3), 239−251.

Marzec, M., 2016. Strigolactones as part of the plant defence system. Trends Plant Sci. 21 (11), 900−903.

Mata, A.T., Jorge, T.F., Pires, M.V., Antonio, C., 2016. Drought stress tolerance in plants: insights from metabolomics. In: Hossain, M., Wani, S., Bhattacharjee, S., Burritt, D., Tran, L.S. (Eds.), Drought Stress Tolerance in Plants, Vol 2. *Springer*, Cham, pp. 187−216.

May, P., Wienkoop, S., Kempa, S., Usadel, B., Christian, N., Rupprecht, J., et al., 2008. Metabolomics-and proteomics-assisted genome annotation and analysis of the draft metabolic network of *Chlamydomonasreinhardtii*. Genetics 179 (1), 157−166.

Mehta, P., Walia, A., Kulshrestha, S., Chauhan, A., Shirkot, C.K., 2015. Efficiency of plant growth-promoting P-solubilizing *Bacillus circulans* CB7 for enhancement of tomato growth under net house conditions. J. Basic Microbiol. 55 (1), 33−44.

Mishra, M., Vishwakarma, K., Singh, J., Jain, S., Kumar, V., Tripathi, D.K., et al., 2018. Exploring the multifaceted role of microbes in pharmacology. Microbial Biotechnology. *Springer*, Singapore, pp. 319−329.

Moles, T.M., Mariotti, L., De Pedro, L.F., Guglielminetti, L., Picciarelli, P., Scartazza, A., 2018. Drought induced changes of leaf-to-root relationships in two tomato genotypes. Plant. Physiol. Biochem. 128, 24−31.

Monnet, F., Vaillant, N., Hitmi, A., Coudret, A., Sallanon, H., 2001. Endophytic *Neotyphodium lolii* induced tolerance to Zn stress in *Loliumperenne*. Physiol. Plant. 113, 557−563.

Moschen, S., Di Rienzo, J.A., Higgins, J., Tohge, T., Watanabe, M., González, S., et al., 2017. Integration of transcriptomic and metabolic data reveals hub transcription factors involved in drought stress response in sunflower (*Helianthus annuus* L.). Plant Mol. Biol. 94 (4), 549−564.

Munns, R., Tester, M., 2008. Mechanisms of salinity tolerance. Annu. Rev. Plant Biol. 59, 651−681.

Naseem, M., Kaltdorf, M., Dandekar, T., 2015. The nexus between growth and defence signalling: auxin and cytokinin modulate plant immune response pathways. J. Exp. Bot. 66 (16), 4885−4896.

Nasir, F., Shi, S., Tian, L., Chang, C., Ma, L., Li, X., et al., 2019. Strigolactones shape the rhizomicrobiome in rice (*Oryza sativa*). Plant Sci. 286, 118−133.

Navarro-Ródenas, A., Berná, L.M., Lozano-Carrillo, C., Andrino, A., Morte, A., 2016. Beneficial native bacteria improve survival and mycorrhization of desert truffle mycorrhizal plants in nursery conditions. Mycorrhiza. 26 (7), 769−779.

Newton, A.C., Fitt, B.D., Atkins, S.D., Walters, D.R., Daniell, T.J., 2010. Pathogenesis, parasitism and mutualism in the trophic space of microbe−plant interactions. Trends Microbiol. 18 (8), 365−373.

Niu, Y., Jin, G., Li, X., Tang, C., Zhang, Y., Liang, Y., et al., 2015. Phosphorus and magnesium interactively modulate the elongation and directional growth of primary roots in *Arabidopsis thaliana* (L.) Heynh. J. Exp. Bot. 66 (13), 3841−3854.

Novák, O., Napier, R., Ljung, K., 2017. Zooming in on plant hormone analysis: tissue-and cell-specific approaches. Annu. Rev. Plant. Biol. 68, 323−348.

O'Brien, J.A., Benková, E., 2013. Cytokinin cross-talking during biotic and abiotic stress responses. Front. Plant Sci. 4, 451.

Okçu, G., Kaya, M.D., Atak, M., 2005. Effects of salt and drought stresses on germination and seedling growth of pea (*Pisumsativum* L.). Turk. J. Agric. For. 29 (4), 237−242.

Oksman-Caldentey, K.M., Saito, K., 2005. Integrating genomics and metabolomics for engineering plant metabolic pathways. Curr. Opin. Biotechnol. 16 (2), 174−179.

Pascale, A., Proietti, S., Pantelides, I.S., Stringlis, I.A., 2020. Modulation of the root microbiome by plant molecules: the basis for targeted disease suppression and plant growth promotion. Front. Plant. Sci. 10, 1741.

Peláez-Vico, M.A., Bernabéu-Roda, L., Kohlen, W., Soto, M.J., López-Ráez, J.A., 2016. Strigolactones in the Rhizobium-legume symbiosis: stimulatory effect on bacterial surface motility and down-regulation of their levels in nodulated plants. Plant. Sci. 245, 119−127.

Peleg, Z., Blumwald, E., 2011. Hormone balance and abiotic stress tolerance in crop plants. Curr. Opin. Plant Biol. 14 (3), 290−295.

Porcel, R., Zamarreño, Á.M., García-Mina, J.M., Aroca, R., 2014. Involvement of plant endogenous ABA in *Bacillus megaterium* PGPR activity in tomato plants. BMC Plant Biol. 14 (1), 1−2.

Poupin, M.J., Greve, M., Carmona, V., Pinedo, I., 2016. A complex molecular interplay of auxin and ethylene signaling pathways is involved in Arabidopsis growth promotion by *Burkholderia phytofirmans* PsJN. Front. Plant Sci. 7, 492.

Prasannakumar, S.P., Gowtham, H.G., Hariprasad, P., Shivaprasad, K., Niranjana, S.R., 2015. *Delftia tsuruhatensis* WGR−UOM−BT1, a novel rhizobacterium with PGPR properties from *Rauwolfia serpentina* (L.) Benth. exKurz also suppresses fungal phytopathogens by producing a new antibiotic—AMTM. Lett. Appl. Microbiol. 61 (5), 460−468.

Raaijmakers, J.M., Paulitz, T.C., Steinberg, C., Alabouvette, C., Moënne-Loccoz, Y., 2009. The rhizosphere: a playground and battlefield for soilborne pathogens and beneficial microorganisms. Plant Soil 321 (1), 341−361.

Raza, A., Mehmood, S.S., Tabassum, J., Batool, R., 2019. Targeting plant hormones to develop abiotic stress resistance in wheat. Wheat Production in Changing Environments. Springer, Singapore, pp. 557−577.

Renault, H., Roussel, V., El Amrani, A., Arzel, M., Renault, D., Bouchereau, A., et al., 2010. The *Arabidopsis* pop2-1 mutant reveals the involvement of *GABA transaminase* in salt stress tolerance. BMC Plant Biol. 10 (1), 1−6.

Riely, B.K., Mun, J.H., Ané, J.M., 2006. Unravelling the molecular basis for symbiotic signal transduction in legumes. Mol. Plant Pathol. 7 (3), 197−207.

Ryu, H., Cho, Y.G., 2015. Plant hormones in salt stress tolerance. J. Plant Biol. 58 (3), 147−155.

Sah, S.K., Reddy, K.R., Li, J., 2016. Abscisic acid and abiotic stress tolerance in crop plants. Front. Plant Sci. 7, 571.

Sanchez, D.H., Siahpoosh, M.R., Roessner, U., Udvardi, M., Kopka, J., 2008. Plant metabolomics reveals conserved and divergent metabolic responses to salinity. Physiologiaplantarum 132 (2), 209−219.

Schaller, G.E., Street, I.H., Kieber, J.J., 2014. Cytokinin and the cell cycle. Curr. Opin. Plant Biol. 21, 7−15.

Shah, J., Chaturvedi, R., Chowdhury, Z., Venables, B., Petros, R.A., 2014. Signaling by small metabolites in systemic acquired resistance. Plant J. 79 (4), 645−658.

Shahzad, R., Khan, A.L., Bilal, S., Waqas, M., Kang, S.M., Lee, I.J., 2017. Inoculation of abscisic acid-producing endophytic bacteria enhances salinity stress tolerance in *Oryzasativa*. Environ. Exp. Bot. 136, 68−77.

Shanazari, M., Golkar, P., MirmohammadyMaibody, A.M., 2018. Effects of drought stress on some agronomic and bio-physiological traits of Triticum aestivum, Triticale, and Triti pyrum genotypes. Arch. Agron. Soil. Sci. 64 (14), 2005.

Sharifi, R., Ryu, C.M., 2016. Are bacterial volatile compounds poisonous odors to a fungal pathogen *Botrytis cinerea*, alarm signals to *Arabidopsis* seedlings for eliciting induced resistance, or both? Front. Microbiol. 7, 196.

Shi, T.Q., Peng, H., Zeng, S.Y., Ji, R.Y., Shi, K., Huang, H., et al., 2017. Microbial production of plant hormones: opportunities and challenges. Bioengineered. 8 (2), 124−128.

Šimura, J., Antoniadi, I., Široká, J., Tarkowská, D.E., Strnad, M., Ljung, K., et al., 2018. Plant hormonomics: multiple phytohormone profiling by targeted metabolomics. Plant Physiol. 177 (2), 476−489.

Slama, I., Abdelly, C., Bouchereau, A., Flowers, T., Savouré, A., 2015. Diversity, distribution and roles of osmoprotective compounds accumulated in halophytes under abiotic stress. Ann. Bot. 115 (3), 433−447.

Spaepen, S., 2015. Plant hormones produced by microbes. In: Lugtenberg, B. (Ed.), Principles of Plant-Microbe Interactions. *Springer*, Cham, pp. 247−256.

Spaepen, S., Vanderleyden, J., 2011. Auxin and plant-microbe interactions. Cold Spring Harb. Perspect. Biol. 3 (4), a001438.

Strange, R.N., Scott, P.R., 2005. Plant disease: a threat to global food security. Annu. Rev. Phytopathol. 43, 83−116.

Sun, H., Tao, J., Liu, S., Huang, S., Chen, S., Xie, X., et al., 2014. Strigolactones are involved in phosphate-and nitrate-deficiency-induced root development and auxin transport in rice. J. Exp. Bot. 65 (22), 6735−6746.

Szepesi, Á., 2020. Role of metabolites in abiotic stress tolerance. Plant Life Under Changing Environment. Academic Press, pp. 755–774.

Tarkowská, D., Novák, O., Oklestkova, J., Strnad, M., 2016. The determination of 22 natural brassinosteroids in a minute sample of plant tissue by UHPLC–ESI–MS/MS. Anal. Bioanal. Chem. 408 (24), 6799–6812.

Trethewey, R.N., 2004. Metabolite profiling as an aid to metabolic engineering in plants. Curr. Opin. Plant Biol. 7 (2), 196–201.

Trubat, R., Cortina, J., Vilagrosa, A., 2006. Plant morphology and root hydraulics are altered by nutrient deficiency in *Pistacialentiscus* (L.). Trees 20 (3), 334.

Tsukanova, K.A., Meyer, J.J., Bibikova, T.N., 2017. Effect of plant growth-promoting Rhizobacteria on plant hormone homeostasis. South Afr. J. Bot. 113, 91–102.

Tyburski, J., Dunajska, K., Mazurek, P., Piotrowska, B., Tretyn, A., 2009. Exogenous auxin regulates H_2O_2 metabolism in roots of tomato (*Lycopersiconesculentum* Mill.) seedlings affecting the expression and activity of CuZn-superoxide dismutase, catalase, and peroxidase. Acta Physiol. Plant. 31 (2), 249–260.

Upadhyay, N., Vishwakarma, K., Singh, J., Verma, R.K., Prakash, V., Jain, S., et al., 2019. Plant-microbe-soil interactions for reclamation of degraded soils: potential and challenges. Phyto and Rhizo Remediation. *Springer*, Singapore, pp. 147–173.

Vacheron, J., Desbrosses, G., Bouffaud, M.L., Touraine, B., Moënne-Loccoz, Y., Muller, D., et al., 2013. Plant growth-promoting rhizobacteria and root system functioning. Front. Plant Sci. 4, 356.

Van der Kooy, F., Verpoorte, R., Meyer, J.M., 2008. Metabolomic quality control of claimed anti-malarial *Artemisia afra* herbal remedy and *A. afra* and *A. annua* plant extracts. South Afr. J. Bot. 74 (2), 186–189.

Vandenbussche, F., Van Der Straeten, D., 2012. The role of ethylene in plant growth and development. Annu. Plant Rev. Plant Horm. Ethyl. 44, 222.

Verbon, E.H., Liberman, L.M., 2016. Beneficial microbes affect endogenous mechanisms controlling root development. Trends Plant Sci. 21 (3), 218–229.

Verma, V., Ravindran, P., Kumar, P.P., 2016. Plant hormone-mediated regulation of stress responses. BMC Plant Biol. 16 (1), 1.

Vishwakarma, K., Sharma, S., Kumar, N., Upadhyay, N., Devi, S., Tiwari, A., 2016. In: Singh, D., Singh, H., Prabha, R. (eds.), Contribution of microbial inoculants to soil carbon sequestration and sustainable agriculture. Microbialinoculants in Sustainable Agricultural Productivity. Springer, New Delhi, pp. 101–113.

Vishwakarma, K., Sharma, S., Kumar, V., Upadhyay, N., Kumar, N., Mishra, R., et al., 2017a. Current scenario of root exudate–mediated plant-microbe interaction and promotion of plant growth. In: Kumar, V., Kumar, M., Sharma, S., Prasad, R. (Eds.), Probiotics in Agroecosystem. *Springer*, Singapore, pp. 349–369.

Vishwakarma, K., Mishra, M., Jain, S., Singh, J., Upadhyay, N., Verma, R.K., et al., 2017b. Exploring the role of plant-microbe interactions in improving soil structure and function through root exudation: a key to sustainable agriculture. In: Singh, D., Singh, H., Prabha, R. (Eds.), Plant-Microbe Interactions in Agro-Ecological Perspectives. *Springer*, Singapore, pp. 467–487.

Vishwakarma, K., Kumar, V., Tripathi, D.K., Sharma, S., 2018. Characterization of rhizobacterial isolates from *Brassica juncea* for multitrait plant growth promotion and their viability studies on carriers. Environ. Sustain. 1 (3), 253–265.

Vishwakarma, K., Kumar, N., Shandilya, C., Mohapatra, S., Bhayana, S., Varma, A., 2020a. Revisiting plant–microbe interactions and microbial consortia application for enhancing sustainable agriculture: a review. Front. Microbiol. 11, 3195.

Vishwakarma, K., Singh, V.P., Prasad, S.M., Chauhan, D.K., Tripathi, D.K., Sharma, S., 2020b. Silicon and plant growth promoting rhizobacteria differentially regulate AgNP-induced toxicity in *Brassica juncea*: Implication of nitric oxide. J. Hazard. Mater. 390, 121806.

Vishwakarma, K., Kumar, N., Shandilya, C., Varma, A., 2021. Unravelling the role of endophytes in micronutrient uptake and enhanced crop productivity. Symbiotic Soil Microorganisms. *Springer*, Cham, pp. 63–85.

Wahid, A., Close, T.J., 2007. Expression of dehydrins under heat stress and their relationship with water relations of sugarcane leaves. Biol. Plant. 51 (1), 104–109.

Waller, F., Achatz, B., Baltruschat, H., Fodor, J., Becker, K., Fischer, M., et al., 2005. The endophytic fungus *Piriformosporaindica* reprograms barley to salt-stress tolerance, disease resistance, and higher yield. Proc. Natl. Acad. Sci. U. S. A. 102 (38), 13386–13391.

Wang, X., Hou, L., Lu, Y., Wu, B., Gong, X., Liu, M., et al., 2018a. Metabolic adaptation of wheat grains contributes to a stable filling rate under heat stress. J. Exp. Bot.

Wang, Q., Zeng, S., Wu, X., Lei, H., Wang, Y., Tang, H., 2018b. Interspecies developmental differences in metabonomic phenotypes of *Lyciumruthenicum* and *L. barbarum* fruits. J. Proteome Res. 17 (9), 3223–3236.

Widodo, Patterson, J.H., Newbigin, E., Tester, M., Bacic, A., Roessner, U., 2009. Metabolic responses to salt stress of barley (*HordeumVulgare* L.) cultivars, Sahara and clipper, which differ in salinity tolerance. J. Exp. Bot. 60, 4089–4103.

Xue, G.P., McIntyre, C.L., Glassop, D., Shorter, R., 2008. Use of expression analysis to dissect alterations in carbohydrate metabolism in wheat leaves during drought stress. Plant Mol. Biol. 67 (3), 197–214.

Yadav, G., Vishwakarma, K., Sharma, S., Kumar, V., Upadhyay, N., Kumar, N., et al., 2017. Emerging significance of rhizospheric probiotics and its impact on plant health: current perspective towards sustainable agriculture. In: Kumar, V., Kumar, M., Sharma, S., Prasad, R. (Eds.), Probiotics and Plant Health. *Springer*, Singapore, pp. 233–251.

Yobi, A., Wone, B.W., Xu, W., Alexander, D.C., Guo, L., Ryals, J.A., et al., 2013. Metabolomic profiling in *Selaginella lepidophylla* at various hydration states provides new insights into the mechanistic basis of desiccation tolerance. Mol. Plant 6 (2), 369–385.

Zavaleta-Mancera, H.A., López-Delgado, H., Loza-Tavera, H., Mora-Herrera, M., Trevilla-Garcia, C., Vargas-Suárez, M., et al., 2007. Cytokinin promotes catalase and ascorbate peroxidase activities and preserves the chloroplast integrity during dark-senescence. J. Plant Physiol. 164 (12), 1572–1582.

Zeid, I.M., Shedeed, Z.A., 2006. Response of alfalfa to putrescine treatment under drought stress. Biol. Plant. 50 (4), 635–640.

Zhang, H., Kim, M.S., Krishnamachari, V., Payton, P., Sun, Y., Grimson, M., et al., 2007. Rhizobacterial volatile emissions regulate auxin homeostasis and cell expansion in Arabidopsis. Planta. 226 (4), 839–851.

Zhou, C., Guo, J., Zhu, L., Xiao, X., Xie, Y., Zhu, J., et al., 2016. *Paenibacilluspolymyxa* BFKC01 enhances plant iron absorption via improved root systems and activated iron acquisition mechanisms. Plant. Physiol. Biochem. 105, 162–173.

Zuther, E., Koehl, K., Kopka, J., 2007. Comparative metabolome analysis of the salt response in breeding cultivars of rice. In: Jenks, M.A., Hasegawa, P.M., Jain, S.M. (Eds.), Advances in Molecular Breeding toward Drought and Salt Tolerant Crops. Springer, Dordrecht, pp. 285–315.

16

The main fungal pathogens and defense-related hormonal signaling in crops

Nadeem Iqbal, Zalán Czékus, Attila Ördög and Péter Poór

Department of Plant Biology, Faculty of Science and Informatics, University of Szeged, Szeged, Hungary

16.1 Introduction

Plants in natural environments are continuously facing a wide variety of different pathogens, such as viruses, bacteria, fungi, nematodes, and insects. These kinds of pathogens deploy different strategies to infect and cause diseases, resulting in reduced yield and plant survival. These invaders possess specific characteristics to make a parasitic relationship with host plants. Upon the attack and infection caused by microbial pathogens, plants induce defensive responses to arrest foreign invaders (Glazebrook, 2005; Panstruga et al., 2009). The immune responses in plants activated against pathogens depend on their ability to identify pathogens and/or pathogen/microbe-associated molecular patterns using membrane protein receptor-like kinases known as pattern-recognition receptors stimulating pattern-triggered immunity (PTI). To achieve a successful attack, invaders can produce virulence molecules known as effector proteins, which supress PTI. In addition, plants have developed resistance (R) genes encoding cytoplasmic receptors to identify specific pathogenic effectors and stimulate effector-triggered immunity (ETI) along with hypersensitive response (HR) leading to programmed cell death (PCD) to prevent the infection of pathogens (Dodds and Rathjen, 2010; Erb et al., 2012).

Several fungal species interact with plants in either a beneficial or harmful manner making fungal symbiosis or disease development, respectively (Sanders, 2011; Dean et al., 2012). Both symbiotic and pathogenic fungi penetrate their hyphae into the host to obtain nutrients without breaking the plasma membrane. The fungal cell wall (chitin) is recognized via specific plant receptors, resulting in the activation of basal immunity. Both mutualistic and pathogenic interactions are regulated by chitin perception (Gust et al., 2012). Fungi have developed different types of effectors and metabolites to either hinder plant defense or promote their invasion in plant tissues under favorable conditions (Kamoun, 2007; Chanclud and Morel, 2016). In addition to the production of effectors, fungi can produce different hormones resembling phytohormones to control plant development and induce signaling events in plants (De Vleesschauwer et al., 2013; Pozo et al., 2015). Interestingly, plant pathogens secrete different effector proteins to interfere with plant defense strategies and assist in pathogen colonization (Dangl et al., 2013; Lo Presti et al., 2015).

Mycotoxins are secondary metabolites produced by various fungi that can affect the growth and development of plants and can induce PCD in cells (Ismaiel and Papenbrock, 2015). Toxigenic fungi can be characterized by the followings based on Miller (1995): plant pathogens, such as *Fusarium graminearum* and *Alternaria alternata*; fungi that grow and accumulate toxins on stressed or senescent plants, such as *Fusarium moniliforme*; fungi that initially colonize plants but increase growing after harvesting, such as *Aspergillus flavus*; fungi that are in the soil or decaying plant material that occur on developing kernels in the field or proliferate later during the storage, such as *Penicillium verrucosum* and *Aspergillus ochraceus* (Miller, 1995). The accumulation of mycotoxins induces various molecular, biochemical, and physiological responses of plants mediated by the defense-related phytohormones (Iqbal et al., 2021).

Plants have evolved highly developed mechanisms to recognize pathogens and to utilize particular defensive responses to protect themselves against pathogens or mycotoxins. The signaling network of plant hormones is activated the downstream of ETI and PTI and plays a vital role in plant immunity against pathogens (Pieterse et al., 2009; Katagiri and Tsuda, 2010). Moreover, PTI leads to various cellular responses, such as generation of reactive oxygen species (ROS), changes in cytosolic ionic flux, activation of mitogen-activated protein kinase (MAPK) or calcium-dependent protein kinase cascades, synthesis of various defense-related hormones, and strengthening physical barriers (Macho and Zipfel, 2014; Couto and Zipfel, 2016). The most important phytohormones are salicylic acid (SA), jasmonate (JA), ethylene (ET), and abscisic acid (ABA), which can act as key defensive players in the modulation of signaling networks needed for the establishment of resistance against different pathogens (Robert-Seilaniantz et al., 2011; Pieterse et al., 2012; Verma et al., 2016). Several scientific reports have documented the involvement of each phytohormone at various organs and developmental stages and their crucial role under stress conditions depending on their concentration (Ku et al., 2018; Cortleven et al., 2019; Yu et al., 2019).

These plant hormones can work both synergistically and antagonistically depending upon the plant growth stage and their concentration (Yu et al., 2020). They have been reported to defend the plant against various plant pathogens in a various way (Koornneef and Pieterse, 2008; Ghorbel and Brini, 2021). SA signaling provokes both local and systemic resistance against pathogens (biotrophic) that feed and reproduce on living host cells. On the other side, JA and ET are needed mostly for resistance against necrotrophic pathogens and herbivore insects, which extract nutrients from host cells using specific feeding structures (Spoel et al., 2007; Bari and Jones, 2009; Panstruga et al., 2009). All of them play an essential role in the induction of induced systemic resistance (ISR), respectively. In addition, other phytohormones, such as auxins (indole-3-acetic acid, IAA), brassinosteroids (BRs), cytokinins (CKs), and melatonin, play an indispensable role in the regulation of defense responses against pathogens and are also involved in the regulation of plant growth and development (Robert-Seilaniantz et al., 2011; Pieterse et al., 2012; Arnao and Hernandez-Ruiz, 2014; Torres-Vera et al., 2014). Plant fitness can eventually be reduced as the maintenance of resistance is costly without pathogenic attacks (Yang et al., 2015). The crosstalk among these phytohormones assists the plants to utilize particular defense strategies for specific pathogens resulting in the induction of signaling networks to optimize plant responses against these invaders or toxins (Takatsuji and Jiang, 2014). This kind of defense strategy induced in plants helps to use resources effectively and efficiently.

16.2 Salicylic acid

SA can induce both local and systemic acquired resistance (SAR) in several plant species under pathogenic attacks of biotrophs and hemibiotrophs and enhances resistance against numerous other pathogens (De Vos et al., 2005; Glazebrook, 2005). In dicot plants, that is, tobacco and *Arabidopsis*, SA production is associated with the expression of pathogenesis-related proteins (PR). Nevertheless, in the case of monocots, the role of SA has not been fully elucidated under biotic stress (Qi et al., 2012). In contrast, five fungal species, such as *Aspergillus*, *Glomerella*, *Rhodotorula*, *Trichosporon*, and *Trichoderma*, have been documented to metabolize SA by converting SA into catechol and/or gentisate which are further catabolized into 3-oxoadipate and metabolized in the tricarboxylic acid cycle (Dodge and Wackett, 2005). Another study was conducted to determine the role of SA in wheat spikelets infected by *F. graminearum* and where increased expressions of *PR1* and *PR4* and a significant decline in the JA marker gene, *Pdf1.2* level were found. In the case of *Arabidopsis*, *PR1*, *PR4*, and *Pdf1.2* levels were high under *F. graminearum* infection depicting the involvement of SA and JA (Makandar et al., 2010). Therefore the different expressions of *Pdf1.2* upon *F. graminearum* infection and *PR1* expression in response to SA application suggest the existence of respective and specific defense signaling networks in wheat and *Arabidopsis* (Qi et al., 2012).

SA, as a secondary metabolite, is responsible for the activation of defense responses in plant—pathogen interactions (Vlot et al., 2009). Interestingly, plants have well-distinguished SA biosynthesis pathways categorized into isochorismate pathway and phenylalanine pathway. Both pathways are mediated by two specific enzymes: isochorismate synthase (ICS) and phenylalanine ammonia lyase (PAL), respectively. The presence of these pathways is needed for both local and systemic acquired resistance and PCD as well (Wildermuth et al., 2001). Therefore these findings suggest the vital role of SA in the induction of SAR (Netea et al., 2011). The methylated form of SA, the gaseous methyl salicylate, plays a crucial role in the signal transduction of SAR in plants for the prevention of pathogens (viruses, bacteria, and fungi) invasion (Park et al., 2007; Qi et al., 2012; Yang et al., 2015). The important role of SA in fungal disease control was revealed by limiting *F. graminearum* germination and growth, which is responsible for *Fusarium* head blight in wheat (Qi et al., 2012). SA application induced gene expressions of *PR1, PR3, PR5,* and

PR9 and activated resistance against bacterial invasions, as confirmed by higher endoge-nous levels of SA (Yang et al., 2015).

Intriguingly, SA can also induce resistance response via foliar treatment against *Fusarium* wilt in bean, as evidenced by enhanced levels of free and conjugated SA levels and elevated contents of PAL and peroxidases activities (Xue et al., 2013). Exogenous application of SA in tomato resulted in the induction of gene expressions of PR2 and PR3 proteins, enhanced PAL activity and antioxidants, and ta reduction in potato viral infec-tions (Falcioni et al., 2013). Moreover, plant disease resistance is affected by overexpressing of SA biosynthesis genes or genetic mutations related to SA (Verberne et al., 2000; Wildermuth et al., 2001). In addition, a higher SA accumulation was reported when two bacterial SA biosynthesis genes were overexpressed in tobacco and also resulted in upregulation of PR proteins for enhanced plant resistance against pathogenic infections (Yang et al., 2015). Mutant tobacco plants with low or absence of SA biosynthesis ability are unable to induce SAR. However, exogenous SA treatment to SA biosynthesis mutants can induce SAR and elevate their disease resistance in response to pathogenic attacks (Yang et al., 2015). Furthermore, *Lolium perenne* was treated with SA and its analog ben-zothiadiazole exogenously and higher contents of *PR* genes and callose deposition were observed resulting in the reduction of gray leaf spot disease (Rahman et al., 2014).

However, a direct and efficient way for the SA-elicited defense for fungi is to avoid SA formation. Various filamentous plant pathogens deploy the strategy, such as *Ustilago may-dis* (an agent of corn smut), releasing a chorismate mutase Cmu1 which forms prephenate from chorismate (Djamei et al., 2011). In a scientific study when plants were infected with *cmu1* deletion mutants, the results depicted higher levels of SA accumulation and showed higher immunity against pathogens as compared to wild type. So, the *Cmu1* is a cyto-plasmic effector that interferes with SA biosynthesis (Djamei et al., 2011). Secreted choris-mate mutases are also found in some necrotrophic fungi, such as *Sclerotinia sclerotiorum*, illustrating that reducing SA levels could be a common approach for suppressing the host defense system (Kabbage et al., 2013; Derbyshire et al., 2017). Nevertheless, *Cmu1* has been reported for interaction with maize kiwellin *ZmKWL1* in the apoplast. *ZmKWL1* is a pro-tein linked with the defense system, which can inhibit the chorismate mutase activity of *Cmu1* (Han et al., 2019). Another alternative way has been documented to lower SA con-tent exploited by *Phytophthora sojae* and *Verticillium dahliae* due to the secretion of isochor-ismatases *PsIsc1* and *VdIsc1*, respectively (Liu et al., 2014). *PsIsc1* is functional inside plants as well. Thus isochorismate is converted by isochorismatases into 2,3-dihydro-2,3-dihy-droxybenzoate and pyruvate, hindering isochorismate from being used in SA biosynthesis. Furthermore, silenced *PsIsc1* in *P. sojae* or inactivated *VdIsc1* in *V. dahliae* enhanced SA accumulation in infected plants and induced SA marker gene *PR1* (Liu et al., 2014). In addition, some fungal species can produce salicylate hydroxylases for SA degradation into catechol explaining the lower SA levels in infected plant tissues (Han and Kahmann, 2019).

SA-mediated signaling is modulated by two transcriptional factors, such as calmodulin-binding protein 60g (CBP60g) and SAR deficient1 (SARD1), via binding to the promoter region of gene ICS1 participating in the SA biosynthesis (Wang et al., 2011; Sun et al., 2015, 2018). For instance, *Arabidopsis* double mutants (*cbp60g/sard1*) were found to be more susceptible to *Verticillium dahliae* infection confirming the crucial role of cbp60g/sard1 in enhancing defense against *V. dahliae* (Qin et al., 2018). Surprisingly, the lack of VdSCP41 in *V. dahliae* decreased virulence while *Arabidopsis* plants with VdSCP41 showed reduced expression of ICS1 resulting in disease symptoms after *V. dahliae* infection (Qin et al., 2018). In addition, the nonexpressor of PR genes1 (NPR1) is considered a master regulator of SA-mediated signaling and defense responses (Wang et al., 2006). SA accumulation in response to pathogenic attacks causes phosphorylation of NPR1 followed by a monomerization and subsequently, its translation in the nucleus to stimulate expression of PR genes. However, NPR1 oligomers are localized in an inactive form in the cytosol in uninfected plants (Lee et al., 2015). Moreover, NPR1 is responsible for the modulation of PR gene expression via interacting with TGA transcriptional factors (Fu and Dong, 2013). Thus PR1 gene expression has been broadly deployed as a SA marker because of its abrupt induction and response during pathogenic attacks to promote SA-mediated plant defense responses (Dong, 2004; Han and Kahmann, 2019). It was showed that PR1 can bind to sterols to hinder the growth of *Phytophthora brassicae*, but no effect was observed in the case of other fungal species, such as *Aspergillus niger* and *Botrytis cinerea* (Gamir et al., 2017). However, a connection between growth inhibition of *P. brassicae* and sterol-binding activity exists due to alleviation of PR-1 protein P14c (from tomato) upon addition of cholesterol (Gamir et al., 2017). In another study, the cerato-platanin-like SsCP1 secreted by *S. sclerotiorum* was found to be involved in virulence via interaction with apoplastic PR1 in *Arabidopsis*, resulting in necrosis like cell death with high doses and activation of the SA-mediated pathway (Yang et al., 2018).

Furthermore, two other effectors, ToxA and Tox3, are secreted by the necrotrophic *Parastagonospora nodorum* pathogen in wheat targeting some isoforms of PR1 (Lu et al., 2014; Breen et al., 2017). PR1−5 increased the necrosis induction of purified ToxA in wheat leaves (Lu et al., 2014). Interestingly, during SA-mediated response signaling, papain-like cysteine proteases (PLCPs) are released in the apoplast and are involved in plant defense response against pathogenic attacks (van der Linde et al., 2012; Misas-Villamil et al., 2016). SA treatment to maize induced the secretion of PLCPs like CP1, CP2, and XCP2 (van der Linde et al., 2012). Moreover, SA increases PLCP release in tomato (Kovács et al., 2016), which is considered to be part of SA-triggered defense response. Furthermore, an immune signaling peptide 1 (Zip1) in *Zea mays* has been reported to be secreted from its propeptide precursor due to SA-evoked PLCPs, which could stimulate SA-mediated signaling downstream (Ziemann et al., 2018).

As it was mentioned earlier, plant responses to pathogenic attacks mediated by SA are dependent upon the developmental stage. For example, SA application induced resistance in mature rice plants against *Magnaporthe grisea*, but no response was observable in the case of young plants. Intriguingly, lesions of the hypersensitive reaction were formed as a result of elevated levels of SA (Iwai et al., 2007). In addition to pathogens, environmental factors, such as the intensity and quality of light or temperature, influenced plant resistance by modulating SA accumulation. For instance, cucumber plants showed elevated resistance to *Sphaerotheca fuliginea* under red light exposure, and interestingly, this resistance was linked with SA-dependent signaling (Wang et al., 2010).

SA stimulates plant resistance not only against biotrophs but also against root-knot nematodes (Branch et al., 2004; Mostafanezhad et al., 2014). Nematodes are sedentary endoparasites feeding on the plant cells throughout their life cycle (Gheysen and

Mitchum, 2011). They cause devastating yield loss in agriculture accounting for 5%−12% annual crop loss globally. Furthermore, SA showed contribution to the resistance against *Meloidogyne incognita* nematodes in tomato plants and increased the glutathione levels (Meher et al., 2011). Moreover, SA triggered SAR response against nematodes in tomato roots upon foliar application (Molinari et al., 2014). Several studies reported that activation of SA pathway is capable of the induction of defense response in plants against biotrophic pathogens (Murray et al., 2007; Fabro et al., 2008). Nevertheless, some studies demonstrated the conflict with the current model. For example, Nováková and co-workers (2014) revealed that SA induced resistance against the necrotrophic pathogen *S. sclerotiorum* which was reported earlier with ET and JA signaling pathways. Thus SA accumulation can be an essential marker for plant disease resistance under both biotrophic and necrotrophic pathogenic attacks (Nováková et al., 2014).

Many scientific reports unveiled that sphingolipids are involved in the induction and regulation of plant hormones, particularly, in SA accumulation mediating the SA-signaling pathway (Asai et al., 2000; Iqbal et al., 2021). The high levels of SA result in the deposition of ROS and RNS causing the oxidation of lipids and proteins and activating HR-like cell death in infected plant tissues to prevent pathogens invasion (Loake and Grant, 2007; Ding and Ding, 2020). Conversely, SA triggers SAR to strengthen plant immunity to repressing pathogen attacks. Thus ROS metabolism, the antioxidant synthesis, and activation of antioxidant enzymes are the main defense responses regulated by the SA pathway. Furthermore, SA has the capability to induce the expression and accumulation of PR proteins (Klessig et al., 2018; Zhang and Li, 2019; Ding and Ding, 2020). The enhanced expressions of *PR1*, *PR2*, and *PR5* have been reported against fumonisin B1 (FB1) in *Arabidopsis* leaves in a mycotoxin dose-dependent manner (Stone et al., 2000). FB1 exposure caused SA accumulation in maize seedlings as well (De La Torre-Hernandez et al., 2010). However, NahG transgenic *Arabidopsis* lines degrading SA showed the higher survival rate of protoplasts after FB1 treatment suggesting the positive role of SA in PCD (Asai et al., 2000). It was found that FB1-mediated ROS generation increased PAL activity and SA contents in *Arabidopsis* (Xing et al., 2013). Furthermore, the interaction between SA signaling and sphingolipid metabolism is depending on other phytohormones, Ca^{2+} concentration, ROS changes, and MAPK6 (Sanchez-Rangel et al., 2015).

Numerous studies explained that pathogens and insects secrete effector proteins that intervene in the plant defense processes leading to disease development (Brooks et al., 2005; Nomura et al., 2005; El Oirdi et al., 2011). The bacterial virulence factor coronatine produced by virulent *Pseudomonas syringae* is structurally and functionally similar to methyl jasmonate (MeJA) and reduces SA-dependent immune defenses. Thus hormone-regulated defense signaling is used by pathogens to develop vulnerability in the host (El Oirdi et al., 2011). On the other hand, the fungal pathogen *S. sclerotiorum* degrades SA actively and thus reduces the levels of SA disturbing growth or oxalate production (Penn and Daniel, 2013). *Hyaloperonospora arabidopsidis* results in downy mildew in *Arabidopsis* and is responsible for the suppression of SA-mediated SAR (Asai et al., 2014). Intriguingly, mollusks release some substances similar to phytohormones in their mucous containing a significant amount of SA and induce the expression of several SA-related genes in the leaves of *Arabidopsis* (Kästner et al., 2014). Based on these observations it can be concluded that SA plays a crucial role in the activation of defense mechanisms in plants and the application of SA could be utilized as a suitable strategy to recover the yield reduction induced by biotic stress in the agriculture sector.

16.3 Jasmonic acid

JA along with its metabolites including the gaseous MeJA are derivatives of lipid compounds stemming from linolenic acid in the plasma membrane. These compounds serve as signal molecules reconciliating plant growth and development and inducing signal transduction under biotic and abiotic stresses (Schaller and Stintzi, 2009). Several scientific reports demonstrated that JA stimulates the plant defense systems in response to pathogens, chewing herbivores, and phloem-feeding insects (Glazebrook, 2005). JA also plays a role in plant—pathogen interactions during the formation of SAR. For instance, the exogenous application of JA to rice plants resulted in enhanced resistance against the necrotrophic pathogen *Rhizoctonia solani* via inducing the phenylpropanoid pathway (Taheri and Tarighi, 2010). Mainly, the coronatine insensitive 1 (COI1) F-box protein is responsible for the mediation of JA responses. Interestingly, *coi1* mutants showed more (enhanced) resistance to bacterial infections and higher SA levels (Kloek et al., 2001). In contrast, *coi1* plants showed vulnerability to *Alternaria brassicicola* and *Botrytis cinerea*. However, JA-dependent and coi1-independent defense responses were reported in *Arabidopsis thaliana* against *S. sclerotiorum* (Stotz et al., 2011). In addition, AUXIN RESPONSE FACTOR 2 (ARF2) serves as a negative regulator of defense responses against *S. sclerotiorum*. Moreover, different environmental factors can modulate plant defense responses, such as light quality, which can affect plant resistance through regulating JA signaling pathways. Similarly, low red/far-red ratios resulted in a decrease in resistance in *Arabidopsis* against *B. cinerea* and also influenced JA responses and SA-independent responses (Cerrudo et al., 2012).

Furthermore, it has been reported that JA is crucial for plant defense responses against nematodes (Bhattarai et al., 2007). The exogenous foliar application of MeJA to tomato plants enhanced the expression of genes related to JA biosynthesis upon *M. incognita* infection (Fujimoto et al., 2011). Transcriptional analysis revealed the significance of the JA signaling pathway, which was needed for tomato vulnerability to nematodes (Bhattarai et al., 2008). Numerous studies elucidated that other hormones could suppress nematodes via crosswalk with the JA pathway. It was documented that BRs suppressed JA biosynthesis and signaling pathways in response to nematode attack in rice and a crosstalk between BR and JA pathway was found (Nahar et al., 2013). The exogenous treatment of MeJA to rice shoots induced a strong systemic defense in the roots preventing nematodes infection, thus enhancing plant resistance. On the other hand, ET stimulates JA signaling and its biosynthesis, suggesting the pivotal role of the JA pathway systemically against nematodes (Bhattarai et al., 2007). However, various environmental factors, such as the availability of CO_2, play an essential role in determining the severity of the infection via stimulating the JA pathway (Sun et al., 2010; 2011). Furthermore, in spite of inducing resistance against necrotrophic pathogens in several plant species, JA has the ability to stimulate plant resistance against biotrophic pathogens (Duan et al., 2014; Ross et al., 2014).

Exogenous MeJA also induced the expressions of *PRs* and elevated the powdery mildew resistance of vulnerable wheat varieties revealing that JA plays a crucial role in defense against wheat infection caused by powdery mildew. The manipulation of the JA pathway might improve resistance against powdery mildew in wheat (Duan et al., 2014). Intriguingly, exogenous JA treatment can enhance resistance to bacterial blight and blast caused by *Xanthomonas oryzae* and *Magnaporthe oryzae*, respectively, in rice (Yamada et al., 2012; Riemann et al., 2013). However, rice plants increased expression of pathogen-responsive *WRKY30* gene with enhanced resistance against fungal pathogens *R. solani* and blast fungus *M. grisea* with higher endogenous JA accumulation and JA biosynthesis-related genes (Peng et al., 2012). Some necrotrophic pathogens use the SA signaling pathway that functions in an antagonistic manner with the JA pathway resulting in the reduction of plant resistance against pathogens (Xin and He, 2013). For example, necrotrophic pathogen *B. cinerea* can produce an elicitor that acts antagonistically with JA signaling pathway assisting the fungus to develop the disease in tomato (El Oirdi et al., 2011). Similarly, necrotrophic pathogens *Alternaria solani* and *B. cinerea* might manipulate SA signaling pathway via NPR1, where NPR1 needs a transcriptional factor TGA1 to reduce expression of two JA-dependent defense genes, PROTEINASE INHIBITORS I and II for the development of disease symptoms (Rahman et al., 2012). These phenomena focus on the strategy manipulated by pathogens to defend and overcome the plant's defensive system and giving them an advantage to develop the disease within the host. However, JA plays a vital role in plant immunity and coordinates plant defense responses against pathogens via different mechanisms. Exogenous JA treatment or manipulated genes encoding for JA provides a huge opportunity to utilize them upon biotic stress in agricultural crop production at a large scale.

It has been reported that the JA pathway is involved in the survival of plants under adverse environmental conditions including pathogenic attacks as well (Glazebrook, 2005). Recently, JA-induced defense responses in plants against biotrophic or hemibiotrophic pathogens have been revealed to contribute resistance in plants (Riemann et al., 2013; Lemarie et al., 2015). For instance, the JA-induced signaling pathway confirmed its role in rice against rice blast caused by *M. oryzae* (Riemann et al., 2013). This hemibiotrophic fungus releases hydroxylated JA molecule (12OH-JA) during early infection to overcome JA-mediated defense response. In this case, the fungus secretes an enzyme monooxygenase Abm that converts JA to 12OH-JA. Furthermore, *M. oryzae* utilizes Abm for the conversion of JA to 12OH-JA to prevent the induction of JA-mediated defense responses (Patkar et al., 2015). Actually, an *abm* mutant of *M. oryzae* could not show blast symptoms on rice plants due to higher accumulation of MeJA in infected parts, which evoked strong defense responses in host. This finding suggests that 12OH-JA blocks JA-stimulated defense and acts as a metabolite effector (Patkar et al., 2015). Conversely, a hemibiotrophic fungus *F. oxysporum* causes root wilt but produces JAs (Cole et al., 2014). However, *Arabidopsiscoi1* mutants with defective JA signaling showed higher resistance against *F. oxysporum* as compared to wild types demonstrating that *F. oxysporum* needs active COI1-elicited JA defense pathway to develop disease symptoms (Thatcher et al., 2009). Interestingly, it was found that endogenous JA biosynthesis is essential for COI1-induced JA signaling during *F. oxysporum* infection (Cole et al., 2014). In addition, it has been reported that virulence-promoting secreted in xylem (SIX) effector Fo5176-SIX4 stimulates JA signaling

(Thatcher et al., 2012). The necrotrophic grapevine pathogen *Lasiodiplodia mediterranea* can also stimulate JA signaling due to the production of JA ester lasiojasmonate A (LasA) (Chini et al., 2018). LasA is converted into JA-Ile acting as a strong inducer of cell death and an activator of the JA signaling pathway. Thus LasA is considered as a metabolite effector in the final infection stage that stimulates JA-evoked cell death (Chini et al., 2018).

16.4 Ethylene

ET is an unsaturated hydrocarbon gas molecule and acts as a crucial regulator of PTI in plants against pathogens (Anver and Tsuda, 2015; Kim et al., 2014). Generally, ET is synthesized from *S*-adenosylmethionine (SAM), which is generated by the transformation of amino acid methionine. SAM is further converted into 1-aminocyclopropane-1-carboxylic acid (ACC) in the presence of ACC synthase (ACS). Thereafter, the ACC oxidase enzyme converts ACC into ET. ET is one of the main regulators of plant immunity (Wang et al., 2002). Plant and pathogen interactions determine the disease resistance modulated by ET (Broekaert et al., 2006; Groen and Whiteman, 2014). Based on the microarray analysis, genes encoding for ET biosynthesis and response in tomato were significantly altered during *Clavibacter michiganensis* infection. In addition, *Never ripe* (*Nr*) tomato plants (impaired ET perception) and transgenic plants (reduced ET biosynthesis) exhibited delayed disease symptoms (wilting) as compared to wild-type plants infected with *C. michiganensi*. These findings demonstrate that ET plays a vital role in regulating tomato vulnerability upon pathogen attack (Balaji et al., 2008). Particularly, mutants with impaired ET signaling or biosynthesis did not compromise *M. incognita*-elicited resistance, although showed vulnerability to potato aphids. So, ET is responsible for the regulation of defense responses and takes part in aphid vulnerability in tomato (Mantelin et al., 2009).

ET also has an important role in the development of disease resistance in wheat plants. Enhanced levels of ET and the induction of its biosynthetic genes expression were found in both resistant and susceptible plants after aphid infection. Moreover, higher ET emission was observed in blast-resistant rice cultivars as compared to wild-type plants indicating the significance of ET biosynthesis in the development of resistance in response to blast fungus. Furthermore, the exogenous treatment with ethephon (an ET precursor) can enhance plant defense response against blast. In addition, more lesions were expanded due to the inhibition of ET biosynthesis (Iwai et al., 2006). Mutant rice overexpressing the *ACS2* gene has increased ET production and showed enhanced resistance to both necrotrophic and hemibiotrophic pathogens, including *M. oryzae* and *R. solani* (Helliwell et al., 2013). OsEDR1 stimulates ET biosynthesis and, nonetheless, results in the reduction of SA and JA accumulation. In addition, it inhibits SA- and JA-related defense responses, thus negatively affecting the development of plant resistance against *X. oryzae* (agent of bacterial blight) (Shen et al., 2011). ET and JA signaling function antagonistically and in parallel to suppress SA signaling in rice. In contrast, this kind of interaction was not observed in *Arabidopsis* in which ET cooperates with JA to inhibit the SA signaling pathway (Mantelin et al., 2013).

ET has the potential to modulate resistance not only locally but the SAR as well. ET, as being a gas molecule can easily overcome the vascular barriers (Frost et al., 2007;

Robert-Seilaniantz et al., 2011; Pieterse et al., 2012), which is particularly essential in tall plants because volatile signaling can reach far away but vascular signals reach only a few centimeters in hours. ET perception is needed to initiate SAR upon TMV infection that consequently activates SA accumulation and SAR development in uninfected leaves (Verberne et al., 2003). The foreign invaders might use ET signaling to infect the host either via the production of ET or ET-stimulating effectors, which target ET signaling to weaken plant immunity (Wi et al., 2012). For instance, *Ralstonia solanacearum* (pathogen)-produced ET regulates the gene expression associated with the host (Valls et al., 2006). Surprisingly, *B. cinerea* has the ability to produce ET in tomato and *in vitro* (Cristescu et al., 2002). Nonetheless, low levels of ET are adequate to activate plant defense responses during infection (Chague et al., 2006). In addition, the ET receptor (ETR1) is needed for pathogenesis by *F. oxysporum*. The *etr1* mutants of *Arabidopsis* exhibited significantly fewer wilt symptoms as compared to control plants. Moreover, reduced vascular growth and enhanced expression of SA-responsive genes were also reported (Pantelides et al., 2013). All these findings illustrate that ET is required for the activation of plant defense responses and can be utilized for the protection of crops against disease development.

ET plays also a crucial role in fruit ripening and senescence (Bleecker and Kende, 2000). ET-insensitive *Arabidopsis* and soybean plants showed more enhanced susceptibility to pathogenic attacks. On the contrary, the activation of ET signaling induces the establishment of plant resistance against pathogens indicating the significant role of ET in plant defense responses (Berrocal-Lobo et al., 2002; Yang et al., 2017a). ET production is highly dependent on the activity of ACS (Li et al., 2011; Helliwell et al., 2013). The effector PsAvh238 of *P. sojae* is upregulated at the time of infection and is important for pathogenicity (Yang et al., 2017b). A recent report revealed that PsAvh238 interacted with soybean Type2 ACSs (GmACSs), suppressed ET biosynthesis, and assisted in infection (Yang et al., 2018). Interestingly, the inhibition of ET synthesis or signaling and silencing of *GmACSs* enhanced virulence of *P. sojae*, while overexpressed *GmACSs* in the leaves of tobacco increased the resistance (Yang et al., 2018). Hence, a *PsAvh238* mutant exhibited no ability to inhibit ET signaling and displayed decreased virulence (Yang et al., 2018). In contrast, *Cochliobolus miyabeanus* (a necrotrophic fungus and an agent of brown spot of rice) needs ET signaling for virulence (Van Bockhaven et al., 2015). Thus the exogenous treatment of ethephon resulted in disease development and ET-insensitive rice plants showed more resistance to *C. miyabeanus* (De Vleesschauwer et al., 2010). Moreover, *C. miyabeanus* induced ET production which established the source of ET in infected tissues. Blocking of the fungal ET synthesis compromised the colonization of *C. miyabeanus* in rice leaves suggesting that *C. miyabeanus* exploits ET to promote virulence, which acts as an effector (Van Bockhaven et al., 2015).

ET plays an essential part along with sphingolipid metabolism to enhance immunity against pathogens. ET promotes PCD through ROS accumulation in a dose-dependent manner and functions antagonistically with SA (Trobacher, 2009; Poor et al., 2013). Using ET receptor mutant *etr1—1 Arabidopsis* plants it was revealed that FB1-elicited PCD is dependent on ET signaling in defense responses (Asai et al., 2000). A rapid and enhanced PCD and chlorophyll degradation were reported in ET receptor mutant *etr1—1*; however, the same trends were not found in other ET receptor mutants (Plett et al., 2009). Intriguingly, it was also documented that FB1 exposure resulted in increased gene expressions of ET response factors *ERF1* and *ERF102* (Mase et al., 2013). In addition, exogenous application of ACC resulted in a significant reduction of FB1-elicited PCD in *Arabidopsis* (Wu et al., 2015). Furthermore, different alterations in sphingolipid compositions were reported in ET mutants impaired in ET biosynthesis or signaling. For example, the *ctr1−1* mutants with active ET signaling exhibited a fewer number of ceramides and hydroxyceramides than wild-type plants. In addition, other mutants, such as *etr1−1* and *ein2*, with defective ET signaling showed higher sensitivity to FB1 treatment, indicating the pivotal role of ET in rescuing FB1-elicited cell death (Wu et al., 2015; Huby et al., 2019).

16.5 Abscisic acid

ABA is an essential plant hormone that plays vital roles in plant growth and development, such as in the regulation of seed germination, seed dormancy, and growth of seedlings or in defense reaction of plants by inducing stomatal closure. In addition, ABA is known as a stress hormone due to its crucial role in abiotic stress responses, such as high salinity, high temperature, and drought (Wasilewska et al., 2008; Gordon et al., 2016). Recent studies have elucidated the involvement of ABA in the responses against pathogenic attacks. One of the most significant findings was the identification of families posing soluble ABA receptors (Ma et al., 2009; Park et al., 2009). These are small steroidogenic acute regulatory-related lipid transfer (START) domain proteins containing a part of a bigger Betv1 superfamily. In *Arabidopsis*, around thirteen ABA receptor homologs have been identified and characterized and named as Regulatory Component of ABA Receptors (RCARs) (Nishimura, et al., 2009; Santiago et al., 2009a). Interestingly, various expansive families of ABA receptors have been identified in dicots including citrus, grapes, cucumber, soybean, and strawberry (Chai et al., 2012; Li et al., 2011; Jia et al., 2011; Boneh et al., 2012, Romero et al., 2012; Bai et al., 2013).

Furthermore, the roles of these receptors have widely been associated with ABA sensitivity, stress perception, and ripening. Moreover, ABA receptor families have been identified in monocots, such as rice and barley, and found to be regulators of development and abiotic stress responses (Kim et al., 2012; Seiler et al., 2014; Tian et al., 2015). In addition to ABA's role under abiotic stresses, ABA has also been reported recently to participate in the regulation of defense responses to several plant diseases. However, ABA can modulate defense responses in both ways, positively or negatively depending on the pathogen. More pieces of evidence are required to assess how ABA is working in the case of fungal pathogens infection (Mauch-Mani and Mauch, 2008; Asselbergh et al., 2008; Cao et al., 2011). While the overexpression of *AtRCAR3* exhibited resistance against *P. syringae* via the regulation of stomatal aperture in *Arabidopsis* (Lim et al., 2014), ABA increased the susceptibility of *Arabidopsis* to *F. oxysporum* (Anderson et al., 2004; Trusov et al., 2009). However, ABA knocked-down plants exhibited elevated resistance to *Plectosphaerella cucumerina* (Sanchez-Vallet et al., 2012). Surprisingly, regarding cereal crops, a spray of ABA on rice plants enhanced the severity of infection by blast fungus *M. grisea* (Koga et al., 2004). In addition, exogenous treatment of ABA has promoted disease symptoms of *M. oryzae* in barley (Ulferts et al., 2015).

Moreover, a few recent studies documented that the treatment of wheat heads with ABA enhanced its susceptibility to *F. graminearum* resulting in *Fusarium* head blight (FHB) (Buhrow et al., 2016; Qi et al., 2016). FHB is a fungal disease of shriveled or discolored grains contaminated with one of the *Fusarium* mycotoxins, the deoxynivalonol (DON) (McCormick, 2003; Gordon et al., 2016). Some reports revealed that exogenous ABA treatment was not associated with the regulation of pathogen growth or sporulation, gene expression of trichothecenes, DON accumulation, or JA/SA accumulation. However, ABA was linked with increased expression of the pathogens infecting genes including hydrolases and cytoskeletal reorganization genes (Buhrow et al., 2016). Concurrently, *F. graminearum* showed the potential to produce ABA itself suggesting a link between ABA and the pathogen virulence (Qi et al., 2016). In addition, ABA was found to act antagonistically with other defense-related phytohormones, such as SA, JA, and ET (Cao et al., 2011; Qi et al., 2016).

The ABA along with other phytohormones, such as ET, initiates signaling pathways and enhances senescence (Woo et al., 2019). Interestingly, the overexpression of *CmWRKY15* in *Chrysanthemum* caused the inhibition of *NCED3A/NCED3* expressions involved in ABA biosynthesis. In addition, the expressions of *ARF4/ARI4/ABI5/RAB18* were downregulated upon inhibition of ABA biosynthesis consequently leading to the inhibition of stomatal closure. Therefore the reduced ABA concentration led to stomatal opening in mutant plants facilitating the invasion of *A. alternata*. Moreover, the ABA content and the *PP2C* expression were significantly enhanced after the infection with *A. alternata* in both wild-type and mutant plants (Feng et al., 2016; Jia et al., 2021). Similarly, the ABA signaling pathway induced by *NaMPK4* is needed for wild tobacco resistance to *A. alternata* via stomatal closure responses, and ABA levels were enhanced after infection (Sun et al., 2014). A scientific study revealed that ABA contents were significantly higher in susceptible apple plants than those of resistant ones after 24 h of inoculation with *A. alternata* (Chen et al., 2012) and reduced ABA concentrations were observed to induce the growth of *A. alternata* (Kepczynska and Kepczynski, 2005). Exogenous ABA treatment induced the expression of senescence-associated genes (*SAGs*) and caused leaf abscission, respectively (Lim et al., 2007). Several biotic and abiotic stressors increased endogenous ABA contents and stimulated the ABA signaling pathway leading to senescence of plants (Liang et al., 2014; Yang et al., 2014; Mao et al., 2017).

The role of ABA has been explored under different kinds of biotic and abiotic stresses (Lim et al., 2015; Niu et al., 2018; Chen et al., 2020). ABA biosynthetic pathways are activated at various growth stages, such as leaf senescence and cotyledons formation. Furthermore, ABA biosynthesis is induced under unfavorable environmental conditions, such as low temperature, dehydration, or enhanced sodium content in the soil (Shu et al., 2016; Zhao et al., 2016). Recent reports have illustrated that cytosolic ABA level is regulated in guard cells via the activation of biosynthesis, transport across the membrane, and catabolism, respectively (Chen et al., 2020). However, the ABA contents in plants are controlled by the difference of the rate at which ABA synthesizes and its converts into phasic acid (PA) (Buckley, 2019). It is a vital hormone for the modulation of plant development and plant responses to both biotic and abiotic stresses (Cutler et al., 2010).

ABA also plays a negative role in the plant resistance development against biotrophic pathogens, such as *F. graminearum*, *H. arabidopsidis*, *Golovinomyces cichoracearum*, and *M. oryzae* (Buhrow et al., 2016; Xiao et al., 2017). On the other side, ABA signaling pathway is needed for the induction of plant resistance against many necrotrophic pathogens, such as *P. cucumerina*, *Pythium irregulare*, *A. brassicicola*, and *C. miyabeanus* (Fan et al., 2009; De Vleesschauwer et al., 2010). Intriguingly, some fungal species, such as *M. oryzae*, *U. maydis*, and *B. cinerea*, are capable to produce ABA (Bruce et al., 2011; Spence et al., 2015). A mutant of *M. oryzae* (*aba4*) deprived of one of ABA synthesis genes showed reduced ABA production and more severe inhibition in appressoria formation as compared to wild type. Moreover, the exogenous ABA treatment restored this defect (Spence et al., 2015). However, this *aba4* mutant failed to infect rice and also exhibited growth defects and morphological disorders which could not be compensated by exogenous ABA applications (Spence et al., 2015). Therefore it is impossible to isolate the endogenous functions from ABA from functions as a pathogenicity promoter in this process.

16.6 The interplay among salicylic acid, jasmonate, ethylene, and abscisic acid in the defense responses of crops

The hormonal signaling pathways of plants are widely interconnected leading to both synergistic and antagonistic functions (Pieterse et al., 2012; Naseem et al., 2015; Berens et al., 2017). For instance, antagonistic relationship between SA and JA while synergism between JA and ET are exhibited in defense signaling (Spoel and Dong, 2008; Zhu et al., 2011; Han and Kahmann, 2019). In addition, growth-promoting and defense-related phytohormones rely on crosstalk for better and balanced regulation (Huot et al., 2014; Berens et al., 2017). Tomato leaves treated with exopolysaccharide (EPS) showed increased susceptibility to *B. cinerea* and higher SA levels, however, reduced gene expressions related to JA, that is, *Protease Inhibitor I* and *II* (El Oirdi et al., 2011). The induction of *NPR1* expression was also observable after *B. cinerea* infection while knockdown of *NPR1* increased JA-marker gene expressions. These findings indicate that EPS acts as an effector, which is involved in the activation of SA signaling and inhibition of the JA signaling pathway through NPR1 (El Oirdi et al., 2011; Han and Kahmann, 2019).

In general, ET along with JA function against necrotrophic pathogens to induce defense responses via activating transcriptional factor ERF1 (Derksen et al., 2013). For example, the development of ISR needs also JA and ET pathways and functional NPR1, which is the regulatory gene of SAR (Dempsey and Klessig, 2012). Furthermore, ET exhibited interaction with SA signaling pathways. For instance, ET stimulates the *PR1* gene expression in *Arabidopsis* while SA increased expressions of genes encoding for ET biosynthesis suggesting the positive interaction of these hormones (De Vos et al., 2006). Nevertheless, ET and SA also showed an antagonistic interaction. ET diminished *PR1* expression at the 7th day of the postinoculation with *Verticillium longisporum* (Johansson et al., 2006). The ET-insensitive mutant *ein3* was beneficial against bacterial pathogens while in the case of *ein3eil1* (double mutants) accumulated SA in high contents indicating a negative role of ET signaling on SA biosynthesis (Chen et al., 2009).

SA-induced resistance and ET/JA-induced defense responses were found to be effective against several pathogens. Interestingly, these three plant hormones interact among themselves both synergistically and antagonistically under pathogenic attacks. *S. sclerotiorum* is a necrotrophic pathogen that induces *ACS2* gene for ET biosynthesis, *PR1*and*LOX3* genes, as well as JA and SA accumulation in *Arabidopsis* (Nováková et al., 2014). When necrotrophic fungal species, such as *A. alternata*, infected tomato plants, JA, SA, and ET hormones were found to actively participate in defense signaling against its AAL toxin (Jia et al., 2013). While, in spite of synergistic interplay, antagonistic interactions were also reported among phytohormones against invading pathogens. Analysis of wheat genome revealed different gene expressions in wheat upon *Fusarium* head blight disease. The gene expression results illustrated that the resistance pathways are modulated by different hormonal signaling pathways, such as JA and ET, while SAR-inducing SA was negligible (Li and Yen, 2008). JA signaling pathways are essential in decreasing vulnerability while SA-mediated defense assists plants to develop resistance against AAL. These findings exhibit that JA plays a vital role in regulating sensitivity to AAL that partly depends upon ET perception and biosynthesis, while the SA pathway promotes defense responses against AAL and however antagonizes ET response (Jia et al., 2013).

Interestingly, JA is synthesized rapidly under pathogenic attacks and plays a crucial part in the regulation of PCD changing the status of ROS metabolism. However, JA and SA showed antagonistic interaction with each other (Glazebrook, 2005; Coll et al., 2011). It was found that overexpression of sphingosine-1-phosphate lyase 1 (*OsSPL1*) in tobacco reduced the expression level of *PR1* genes encoding for SA pathway and resulted in higher susceptibility to *P. syringae* while on the other side, the expression of JA-dependent *PDF1.2* was elevated confirming that both SA and JA function antagonistically in defense signaling (Zhang et al., 2014). The connection between SA–JA pathways in sphingolipids metabolism was shown in sphingosine kinase 1 (*SPHK1*) overexpressing plants that displayed SA accumulation and ROS production upon FB1 treatments leading to PCD. Conversely, knocked-down *SPHK1* expression resulted in higher JA levels under FB1 exposure. Therefore the *SPHK1* gene modulates the balance between long-chain bases (LCBs) and LCB phosphates upon FB1 application and also serves as a signaling molecule for both JA and SA pathways under FB1 exposure (Qin et al., 2017). Moreover, FB1-mediated cell death is dependent on both JA and SA, demonstrated by the higher survival rate of JA biosynthetic mutant *jar1−1* protoplasts than wild-type under FB1 treatment (Asai et al., 2000). Furthermore, FB1 applications not only enhanced expression of SA-related gene (*PR1*) but also of *PDF1.2* in *Arabidopsis* (Stone et al., 2000). Thereafter, it was revealed that JA signaling pathway is needed by FB1-mediated PCD, which is further controlled by two ligases, RING DOMAIN LIGASE3 (RGLG3) and RGLG4 by repressing SA pathway in *Arabidopsis* (Zhang et al., 2015). Concurrently, the defense responses against FB1are also dependent on the interactions between ET and other plant hormones. Intriguingly, PR1 expression induced byFB1 treatment is dependent on *PDF1.2* transcript levels but not on *etr1−1* and *etr2−1* indicating that ET regulates defense genes via both ET and JA signaling pathways (Plett et al., 2009).

Other phytohormones including ABA, auxin, BRs, CKs, and melatonin also contribute to the regulation of plant–pathogen interactions but, are also involved in the suppression of plant defense against biotic stress via interactions with JA, SA, and ET signaling pathways. ABA has been found to play role not only under abiotic stress but also under plant–pathogen interactions acting as a signaling molecule (Adie et al., 2007; Asselbergh et al., 2008; Chen and Yu, 2014). Enhanced endogenous levels of ABA or exogenous ABA treatment induced susceptibility to pathogens via hormonal interactions (Achuo et al., 2006; Truman et al., 2006;

Fan et al., 2009; Jiang et al., 2010). ABA accumulated when tomato was infected with *P. syringae*, on the other hand, exogenous treatment of ABA caused susceptibility to *X. oryzae* and *M. oryzae* by repressing SA-mediated signaling (Xu et al., 2013).

Likewise, ABA application suppresses the SAR induction by the inhibition of the SA signaling pathway, which was further found to be independent of JA/ET signaling illustrating an antagonistic interaction between SA-mediated SAR and ABA-mediated abiotic stress response (Yasuda et al., 2008). Furthermore, these findings focus on the negative effects of plant hormones on plant defense pathways. ABA application compromised rice resistance toward nematodes (*Hirschmanniella oryzae*) and showed antagonistic interaction with SA/ET/JA defense signaling systems (Nahar et al., 2012). However, ABA also promoted resistance against plant pathogens (Flors et al., 2008). For instance, ABA treatment increased resistance against the pathogenic necrotroph *C. miyabeanus* in rice mesophyll cells by the suppression of ET signaling (Vleesschauwer et al., 2010). ABA serves as an important regulator of defensive gene expressions for plant defense responses to prevent infection via influencing JA biosynthesis and defense responses in *Arabidopsis*. Furthermore, ABA mutants are more vulnerable to pathogenic attacks (Adie et al., 2007). These results revealed that ABA plays an essential and prominent role under biotic stress by the regulation of various plant—pathogen interactions, which could be elucidated further by the examination of the interplay between SA and JA signaling pathways under biotic stress (Fan et al., 2009).

16.7 Conclusions

Conclusively, plant hormones especially SA, JA, ABA, and ET regulate a complex network of signaling pathways against pathogen attacks. The interplay of these hormones assists plants in the fine and efficient allocation of energy with reduced fitness costs. These studies have elucidated that the SA signaling pathway is mostly responsible for the suppression of biotrophic fungal pathogens while JA and ET function together to prevent necrotrophic pathogen attacks. Moreover, ET and JA are commonly found in antagonistic interactions with SA signaling pathway except for few exceptions to arrest foreign intruders. In addition to these primary hormones, other phytohormones, such as ABA, also play part in regulating defense responses and signaling. Several studies have reported the defense mechanisms and immune responses against these pathogens under controlled conditions but for large-scale application, a lot of research is still required. Then, more research work could help to understand the complex interactions of these phytohormones and their mechanisms under both biotic and abiotic stress conditions to develop an effective and efficient disease management strategy.

Acknowledgments

We apologize to those colleagues whose work was not reviewed here. This work was supported by the grants from the National Research, Development and Innovation Office of Hungary—NKFIH (Grant no. NKFIH FK 1324871 and 138867) and by the UNKP-21—5 New National Excellence Program of the Ministry of Human Capacities. P.P. and A.Ö. were supported by the János Bolyai Research Scholarship of the Hungarian Academy of Sciences.

Conflicts of interest

The authors declare that the research was conducted in the absence of any commercial or financial relationships that could be construed as a potential conflict of interest.

Author contributions

Conceptualization P.P.; writing—original draft preparation, N.I.; writing—review and editing, N.I., Z.C., A.Ö., and P.P.

References

Achuo, E., Prinsen, E., Höfte, M., 2006. Influence of drought, salt stress and abscisic acid on the resistance of tomato to *Botrytis cinerea* and *Oidium neolycopersici*. Plant Pathol 55 (2), 178–186.

Adie, B.A., Pérez-Pérez, J., Pérez-Pérez, M.M., Godoy, M., Sánchez-Serrano, J.J., Schmelz, E.A., et al., 2007. ABA is an essential signal for plant resistance to pathogens affecting JA biosynthesis and the activation of defenses in Arabidopsis. Plant Cell 19 (5), 1665–1681.

Anderson, J.P., Badruzsaufari, E., Schenk, P.M., Manners, J.M., Desmond, O.J., Ehlert, C., et al., 2004. Antagonistic interaction between abscisic acid and jasmonate-ethylene signaling pathways modulates defense gene expression and disease resistance in Arabidopsis. Plant Cell 16, 3460–3479.

Anver, S., Tsuda, K., 2015. Ethylene and plant immunity. Ethylene In Plants. Springer, Dordrecht, pp. 205–221.

Arnao, M.B., Hernandez-Ruiz, J., 2014. Melatonin: plant growth regulator and/or biostimulator during stress? Trends Plant Sci. 19 (12), 789–797.

Asai, S., Rallapalli, G., Piquerez, S.J., Caillaud, M.C., Furzer, O.J., Ishaque, N., et al., 2014. Expression profiling during arabidopsis/downy mildew interaction reveals a highly-expressed effector that attenuates responses to salicylic acid. PLoS Pathog. 10 (10), e1004443.

Asai, T., Stone, J.M., Heard, J.E., Kovtun, Y., Yorgey, P., Sheen, J., et al., 2000. Fumonisin B1–induced cell death in Arabidopsis protoplasts requires jasmonate-, ethylene-, and salicylate-dependent signaling pathways. Plant Cell 12, 1823–1835.

Asselbergh, B., De Vleesschauwer, D., Hofte, M., 2008. Global switches and fine-tuning-ABA modulates plant pathogen defense. Mol. Plant Microbe Interact. 21, 709–719.

Bai, G., Yang, D.H., Zhao, Y., Ha, S., Yang, F., Ma, J., et al., 2013. Interactions between soybean ABA receptors and type 2C protein phosphatases. Plant Mol. Biol. 83, 651–664.

Balaji, V., Mayrose, M., Sherf, O., Jacob-Hirsch, J., Eichenlaub, R., Iraki, N., et al., 2008. Tomato transcriptional changes in response to *Clavibacter michiganensis* subsp. michiganensis reveal a role for ethylene in disease development. Plant Physiol. 146 (4), 1797–1809.

Bari, R., Jones, J.D., 2009. Role of plant hormones in plant defence responses. Plant Mol. Biol. 69 (4), 473–488.

Berens, M.L., Berry, H.M., Mine, A., Argueso, C.T., Tsuda, K., 2017. Evolution of hormone signaling networks in plant defense. Annu. Rev. Phytopathol. 55, 401–425.

Berrocal-Lobo, M., Molina, A., Solano, R., 2002. Constitutive expression of ethylene-response-factor1 in Arabidopsis confers resistance to several necrotrophic fungi. Plant J. 29, 23–32.

Bhattarai, K.K., Li, Q., Liu, Y., Dinesh-Kumar, S.P., Kaloshian, I., 2007. The MI-1-mediated pest resistance requires Hsp90 and Sgt1. Plant Physiol. 144 (1), 312–323.

Bhattarai, K.K., Xie, Q.G., Mantelin, S., Bishnoi, U., Girke, T., Navarre, D.A., et al., 2008. Tomato susceptibility to root-knot nematodes requires an intact jasmonic acid signaling pathway. Mol. Plant Microbe Interact. 21 (9), 1205–1214.

Bleecker, A.B., Kende, H., 2000. Ethylene: a gaseous signal molecule in plants. Annu. Rev. Cell Dev. Biol. 16, 1–18.

Boneh, U., Biton, I., Zheng, C., Schwartz, A., Ben-Ari, G., 2012. Characterization of potential ABA receptors in Vitis vinifera. Plant Cell Rep. 31, 311–321.

Branch, C., Hwang, C.F., Navarre, D.A., Williamson, V.M., 2004. Salicylic acid is part of the Mi-1-mediated defense response to rootknot nematode in tomato. Mol. Plant Microbe Interact. 17 (4), 351–356.

Breen, S., Williams, S.J., Outram, M., Kobe, B., Solomon, P.S., 2017. Emerging insights into the functions of pathogenesis-related protein 1. Trends Plant Sci. 22, 871–879.

Broekaert, W.F., Delaure, S.L., De Bolle, M.F., Cammue, B.P., 2006. The role of ethylene in host-pathogen interactions. Annu. Rev. Phytopathol. 44, 393–416.

Brooks, D.M., Bender, C.L., Kunkel, B.N., 2005. The Pseudomonas syringae phytotoxin coronatine promotes virulence by overcoming salicylic acid-dependent defences in *Arabidopsis thaliana*. Mol. Plant Pathol. 6 (6), 629–639.

Bruce, S.A., Saville, B.J., Emery, R.N., 2011. *Ustilago maydis* produces cytokinins and abscisic acid for potential regulation of tumor formation in maize. J. Plant Growth Regul. 30, 51–63.

Buckley, T.N., 2019. How do stomata respond to water status? New Phytol. 224 (1), 21–36.

Buhrow, L.M., Cram, D., Tulpan, D., Foroud, N.A., Loewen, M.C., 2016. Exogenous abscisic acid and gibberellic acid elicit opposing effects on *Fusarium graminearum* infection in wheat. Phytopathology 106, 986–996.

Cao, F.Y., Yoshioka, K., Desveaux, D., 2011. The roles of ABA in plant-pathogen interactions. J. Plant Res. 124, 489–499.

Cerrudo, I., Keller, M.M., Cargnel, M.D., Demkura, P.V., de Wit, M., Patitucci, M.S., et al., 2012. Low red/far-red ratios reduce Arabidopsis resistance to *Botrytis cinerea*and jasmonate responses via a COI1-JAZ10-dependent, salicylic acid-independent mechanism. Plant Physiol. 158 (4), 2042–2052.

Chague, V., Danit, L.V., Siewers, V., Gronover, C.S., Tudzynski, P., Tudzynski, B., et al., 2006. Ethylene sensing and gene activation in *Botrytis cinerea*: a missing link in ethylene regulation of fungus-plant interactions? Mol. Plant Microbe Interact. 19 (1), 33–42.

Chai, Y.M., Jia, H.F., Li, C.L., Dong, Q.H., Shen, Y.Y., 2012. FaPYR1 is involved in strawberry fruit ripening. J. Exp. Bot. 62, 5079–5089.

Chanclud, E., Morel, J.B., 2016. Plant hormones: a fungal point of view. Mol. Plant Pathol. 17 (8), 1289–1297.

Chen, H.M., Xue, L., Chintamanani, S., Germain, H., Lin, H.Q., Cui, H.T., et al., 2009. Ethylene Insensitive3 and Ethylene Insensitive3-Like1 Repress Salicylic Acid Induction Deficient2 expression to negatively regulate plant innate immunity in Arabidopsis. Plant Cell 21 (8), 2527–2540.

Chen, K., Li, G.J., Bressan, R.A., Song, C.P., Zhu, J.K., Zhao, Y., 2020. Abscisic acid dynamics, signaling, and functions in plants. J. Integr. Plant Biol. 62 (1), 25–54.

Chen, L., Yu, D., 2014. ABA regulation of plant response to biotic stresses. Abscisic Acid: Metabolism, Transport and Signaling. Springer, pp. 409–429.

Chen, Y., Zhang, C., Cong, P., 2012. Dynamics of growth regulators during infection of apple leaves by *Alternaria alternata* apple pathotype. Australas. Plant Pathol. 41 (3), 247–253.

Chini, A., Cimmino, A., Masi, M., Reveglia, P., Nocera, P., Solano, R., et al., 2018. The fungal phytotoxin lasiojasmonate A activates the plant jasmonic acid pathway. J. Exp. Bot. 69, 3095–3102.

Cole, S.J., Yoon, A.J., Faull, K.F., Diener, A.C., 2014. Host perception of jasmonates promotes infection by *Fusarium oxysporum* formae speciales that produce isoleucine- and leucine-conjugated jasmonates. Mol. Plant Pathol. 15, 589–600.

Coll, N.S., Epple, P., Dangl, J.L., 2011. Programmed cell death in the plant immune system. Cell Death Differ. 18, 1247–1256.

Cortleven, A., Leuendorf, J.E., Frank, M., Pezzetta, D., Bolt, S., Schmülling, T., 2019. Cytokinin action in response to abiotic and biotic stresses in plants. Plant Cell Environ. 42 (3), 998–1018.

Couto, D., Zipfel, C., 2016. Regulation of pattern recognition receptor signalling in plants. Nat. Rev. Immunol. 16, 537–552.

Cristescu, S.M., De Martinis, D., Hekkert, S.T., Parker, D.H., Harren, F.J.M., 2002. Ethylene production by *Botrytis cinerea* in vitro and in tomatoes. Appl. Environ. Microbiol. 68 (11), 5342–5350.

Cutler, S.R., Rodriguez, P.L., Finkelstein, R.R., Abrams, S.R., 2010. Abscisic acid: emergence of a core signaling network. Annu. Rev. Plant Biol. 61, 651–679.

Dangl, J.L., Horvath, D.M., Staskawicz, B.J., 2013. Pivoting the plant immune system from dissection to deployment. Science 341, 746–751.

De La Torre-Hernandez, M.E., Rivas-San Vicente, M., Greaves-Fernandez, N., Cruz-Ortega, R., Plasencia, J., 2010. Fumonisin B1 induces nuclease activation and salicylic acid accumulation through long-chain sphingoid base build-up in germinating maize. Physiol. Mol. Plant Pathol. 74, 337–345.

De Vleesschauwer, D., Gheysen, G., Hofte, M., 2013. Hormone defense networking in rice: tales from a different world. Trends Plant Sci. 18, 555–565.

De Vleesschauwer, D., Yang, Y., Cruz, C.V., Hofte, M., 2010. Abscisic acid-induced resistance against the brown spot pathogen *Cochliobolus miyabeanus* in rice involves MAP kinase-mediated repression of ethylene signaling. Plant Physiol. 152, 2036–2052.

De Vos, M., Van Oosten, V.R., Van Poecke, R.M.P., Van Pelt, J.A., Pozo, M.J., Mueller, M.J., et al., 2005. Signal signature and transcriptome changes of Arabidopsis during pathogen and insect attack. Mol. Plant Microbe Interact. 18, 923–937.

De Vos, M., Van Zaanen, W., Koornneef, A., Korzelius, J.P., Dicke, M., Van Loon, L.C., et al., 2006. Herbivore-induced resistance against microbial pathogens in Arabidopsis. Plant Physiol. 142 (1), 352–363.

Dean, R., Van Kan, J.A.L., Pretorius, Z.A., Hammond-Kosack, K.E., DiPietro, A., Spanu, P.D., et al., 2012. The Top 10 fungal pathogens in molecular plant pathology. Mol. Plant Pathol. 13, 414–430.

Dempsey, D.A., Klessig, D.F., 2012. SOS - too many signals for systemic acquired resistance? Trends Plant Sci. 17 (9), 538–545.

Derbyshire, M., Denton-Giles, M., Hegedus, D., Seifbarghy, S., Rollins, J., van Kan, J., et al., 2017. The complete genome sequence of the phytopathogenic fungus *Sclerotinia sclerotiorum* reveals insights into the genome architecture of broad host range pathogens. Genome Biol. Evol. 9, 593–618.

Derksen, H., Rampitsch, C., Daayf, F., 2013. Signaling cross-talk in plant disease resistance. Plant Sci. 207, 79–87.

Ding, P., Ding, Y., 2020. Stories of salicylic acid: a plant defense hormone. Trends Plant Sci. 25, 549–565.

Djamei, A., Schipper, K., Rabe, F., Ghosh, A., Vincon, V., Kahnt, J., et al., 2011. Metabolic priming by a secreted fungal effector. Nature 478, 395–398.

Dodds, P.N., Rathjen, J.P., 2010. Plant immunity: towards an integrated view of plant-pathogen interactions. Nat. Rev. Genet. 11 (8), 539–548.

Dodge, A.G., Wackett, L.P., 2005. Metabolism of bismuth subsalicylate and intracellular accumulation of bismuth by *Fusarium sp.* strain BI. Appl. Environ. Microbiol. 71, 876–882.

Dong, X., 2004. NPR1, all things considered. Curr. Opin. Plant Biol. 7, 547–552.

Duan, Z., Lv, G., Shen, C., Li, Q., Qin, Z., Niu, J., 2014. The role of jasmonic acid signalling in wheat (*Triticum aestivum L.*) powdery mildew resistance reaction. Eur. J. Plant Pathol. 140 (1), 169–183.

El Oirdi, M., El Rahman, T.A., Rigano, L., El Hadrami, A., Rodriguez, M.C., Daayf, F., et al., 2011. Botrytis cinerea manipulates the antagonistic effects between immune pathways to promote disease development in tomato. Plant Cell 23, 2405–2421.

Erb, M., Meldau, S., Howe, G.A., 2012. Role of phytohormones in insect specific plant reactions. Trends Plant Sci. 17 (5), 250–259.

Fabro, G., Di Rienzo, J.A., Voigt, C.A., Savchenko, T., Dehesh, K., Somerville, S., et al., 2008. Genome-wide expression profiling Arabidopsis at the stage of *Golovinomyces cichoracearum* haustorium formation. Plant Physiol. 146 (3), 1421–1439.

Falcioni, T., Ferrio, J.P., del Cueto, A.I., Giné, J., Achón, M.Á., Medina, V., 2013. Effect of salicylic acid treatment on tomato plant physiology and tolerance to potato virus X infection. Eur. J. Plant Pathol. 138 (2), 331–345.

Fan, J., Hill, L., Crooks, C., Doerner, P., Lamb, C., 2009. Abscisic acid has a key role in modulating diverse plant-pathogen interactions. Plant Physiol. 150, 1750–1761.

Feng, G., Xu, Q., Wang, Z., Zhuoma, Q., 2016. AINTEGUMENTA negatively regulates age-dependent leaf senescence downstream of AUXIN RESPONSE FACTOR 2 in *Arabidopsis thaliana*. Plant Biotechnol. 33 (2), 71–76.

Flors, V., Ton, J., van Doorn, R., Jakab, G., Garcia-Agustin, P., Mauch-Mani, B., 2008. Interplay between JA, SA and ABA signalling during basal and induced resistance against Pseudomonas syringaeand *Alternaria brassicicola*. Plant J. 54 (1), 81–92.

Frost, C.J., Appel, M., Carlson, J.E., De Moraes, C.M., Mescher, M.C., Schultz, J.C., 2007. Within-plant signalling via volatiles overcomes vascular constraints on systemic signalling and primes responses against herbivores. Ecol. Lett. 10 (6), 490–498.

Fu, Z.Q., Dong, X., 2013. Systemic acquired resistance: turning local infection into global defense. Annu. Rev. Plant Biol. 64, 839–863.

Fujimoto, T., Tomitaka, Y., Abe, H., Tsuda, S., Futai, K., Mizukubo, T., 2011. Expression profile of jasmonic acid-induced genes and the induced resistance against the root-knot nematode (*Meloidogyne incognita*) in tomato plants (*Solanum lycopersicum*) after foliar treatment with methyl jasmonate. J. Plant Physiol. 168 (10), 1084–1097.

Gamir, J., Darwiche, R., Van'tHof, P., Choudhary, V., Stumpe, M., Schneiter, R., et al., 2017. The sterol-binding activity of pathogenesis-related protein 1 reveals the mode of action of an antimicrobial protein. Plant J. 89, 502—509.

Gheysen, G., Mitchum, M.G., 2011. How nematodes manipulate plant development pathways for infection. Curr. Opin. Plant Biol. 14 (4), 415—421.

Ghorbel, M., Brini, F., 2021. Hormone mediated cell signaling in plants under changing environmental stress. Plant Gene 28, 100335.

Glazebrook, J., 2005. Contrasting mechanisms of defense against biotrophic and necrotrophic pathogens. Annu. Rev. Phytopathol. 43, 205—227.

Gordon, C.S., Rajagopalan, N., Risseeuw, E.P., Surpin, M., Ball, F.J., Barber, C.J., et al., 2016. Characterization of Triticum aestivum abscisic acid receptors and a possible role for these in mediating Fusairum head blight susceptibility in wheat. PLoS One 11 (10), e0164996.

Groen, S.C., Whiteman, N.K., 2014. The evolution of ethylene signaling in plant chemical ecology. J. Chem. Ecol. 40 (7), 700—716.

Gust, A.A., Willmann, R., Desaki, Y., Grabherr, H.M., Nurnberger, T.€, 2012. Plant LysM proteins: modules mediating symbiosis and immunity. Trends Plant Sci. 17, 495—502.

Han, X., Kahmann, R., 2019. Manipulation of phytohormone pathways by effectors of filamentous plant pathogens. Front. Plant Sci. 10, 822.

Han, X., Altegoer, F., Steinchen, W., Binnebesel, L., Schuhmacher, J., Glatter, T., et al., 2019. A kiwellin disarms the metabolic activity of a secreted fungal virulence factor. Nature. 565, 650—653.

Helliwell, E.E., Wang, Q., Yang, Y., 2013. Transgenic rice with inducible ethylene production exhibits broad spectrum disease resistance to the fungal pathogens *Magnaporthe oryzae* and *Rhizoctonia solani*. Plant Biotechnol. J. 11 (1), 33—42.

Huby, E., Napier, J.A., Baillieul, F., Michaelson, L.V., Dhondt-Cordelier, S., 2019. Sphingolipids: towards an integrated view of metabolism during the plant stress response. New Phytol. 225, 659—670.

Huot, B., Yao, J., Montgomery, B.L., He, S.Y., 2014. Growth-defense tradeoffs in plants: a balancing act to optimize fitness. Mol. Plant 7, 1267—1287.

Iqbal, N., Czékus, Z., Poór, P., Ördög, A., 2021. Plant defence mechanisms against mycotoxin fumonisin B1. Chem. Biol. Interact. 109494.

Ismaiel, A.A., Papenbrock, J., 2015. Mycotoxins: producing fungi and mechanisms of phytotoxicity. Agriculture 5 (3), 492—537.

Iwai, T., Miyasaka, A., Seo, S., Ohashi, Y., 2006. Contribution of ethylene biosynthesis for resistance to blast fungus infection in young rice plants. Plant Physiol. 142 (3), 1202—1215.

Iwai, T., Seo, S., Mitsuhara, I., Ohashi, Y., 2007. Probenazole-induced accumulation of salicylic acid confers resistance to *Magnaporthe grisea* in adult rice plants. Plant Cell Physiol. 48 (7), 915—924.

Jia, H.F., Chai, Y.M., Li, C.L., Lu, D., Luo, J.J., Qin, L., et al., 2011. Abscisic acid plays an important role in the regulation of strawberry fruit ripening. Plant Physiol. 157, 188—199.

Jia, B., Li, G., Yang, X., Liu, L., Heng, W., Liu, P., 2021. Physiological and transcriptional responses of *Alternaria alternata* induced abnormal leaf senescence in Pyrus pyrifolia. Sci. Hortic. 277, 109786.

Jia, C., Zhang, L., Liu, L., Wang, J., Li, C., Wang, Q., 2013. Multiple phytohormone signalling pathways modulate susceptibility of tomato plants to *Alternaria alternata* f. sp. *lycopersici*. J. Exp. Bot. 64 (2), 637—650.

Jiang, C.-J., Shimono, M., Sugano, S., Kojima, M., Yazawa, K., Yoshida, R., et al., 2010. Abscisic acid interacts antagonistically with salicylic acid signaling pathway in rice-*Magnaporthe grisea* interaction. Mol. Plant Microbe Interact. 23 (6), 791—798.

Johansson, A., Staal, J., Dixelius, C., 2006. Early responses in the Arabidopsis-*verticillium longisporum* pathosystem are dependent on NDR1, JA- and ET-associated signals via cytosolic NPR1 and RFO1. Mol. Plant Microbe Interact. 19 (9), 958—969.

Kabbage, M., Williams, B., Dickman, M.B., 2013. Cell death control: the interplay of apoptosis and autophagy in the pathogenicity of *Sclerotinia sclerotiorum*. PLoS Pathog. 9, e1003287.

Kamoun, S., 2007. Groovy times: filamentous pathogen effectors revealed. Curr. Opin. Plant Biol. 10, 358—365.

Kästner, J., von Knorre, D., Himanshu, H., Erb, M., Baldwin, I.T., Meldau, S., 2014. Salicylic acid, a plant defense hormone, is specifically secreted by a molluscan herbivore. PLoS One 9 (1), e86500.

Katagiri, F., Tsuda, K., 2010. Understanding the plant immune system. Mol. Plant Microbe Interact. 23 (12), 1531—1536.

Kepczynska, E., Kepczynski, J., 2005. Inhibitory effect of methyl jasmonate on development of phytopathogen *Alternaria alternata* [Fr.] Keissl. and its reversal by ethephon and ACC. Acta Physiol. Plant 4 (27).

Kim, H., Hwang, H., Hong, J.W., Lee, Y.N., Ahn, I.P., Yoon, I.S., et al., 2012. A rice orthologue of the ABA receptor, OsPYL/RCAR5, is a positive regulator of the ABA signal transduction pathway in seed germination and early seedling growth. J. Exp. Bot. 63, 1013–1024.

Kim, Y., Tsuda, K., Igarashi, D., Hillmer, R.A., Sakakibara, H., Myers, C.L., et al., 2014. Mechanisms underlying robustness and tunability in a plant immune signaling network. Cell Host Microbe 15 (1), 84–94.

Klessig, D.F., Choi, H.W., Dempsey, D.M.A., 2018. Systemic acquired resistance and salicylic acid: past, present, and future. Mol. Plant Microbe Interact. 31, 871–888.

Kloek, A.P., Verbsky, M.L., Sharma, S.B., Schoelz, J.E., Vogel, J., Klessig, D.F., et al., 2001. Resistance to *Pseudomonas syringae*conferred by an *Arabidopsis thaliana* coronatine-insensitive (coi1) mutation occurs through two distinct mechanisms. Plant J. 26 (5), 509–522.

Koga, H., Dohi, K., Mori, M., 2004. Abscisic acid and low temperatures suppress the whole plant-specific resistance reaction of rice plants to the infection of *Magnaporthe grisea*. Physiol. Mol. Plant Pathol. 65, 3–9.

Koornneef, A., Pieterse, C.M.J., 2008. Cross talk in defense signalling. Plant Physiol. 146, 839–844.

Kovács, J., Poór, P., Szepesi, Á., Tari, I., 2016. Salicylic acid induced cysteine protease activity during programmed cell death in tomato plants. Acta Biol. Hung. 67 (2), 148–158.

Ku, Y.S., Sintaha, M., Cheung, M.Y., Lam, H.M., 2018. Plant hormone signaling crosstalks between biotic and abiotic stress responses. Int. J. Mol. Sci. 19 (10), 3206.

Lee, H.J., Park, Y.J., Seo, P.J., Kim, J.H., Sim, H.J., Kim, S.G., et al., 2015. Systemic immunity requires SnRK2.8-mediated nuclear import of NPR1 in Arabidopsis. Plant Cell 27, 3425–3438.

Lemarie, S., Robert-Seilaniantz, A., Lariagon, C., Lemoine, J., Marnet, N., Jubault, M., et al., 2015. Both the jasmonic acid and the salicylic acid pathways contribute to resistance to the biotrophic clubroot agent *Plasmodiophora brassicae* in Arabidopsis. Plant Cell Physiol. 56, 2158–2168.

Li, C., Jia, H., Chai, Y., Shen, Y., 2011. Abscisic acid perception and signaling transduction in strawberry: a model for non-climacteric fruit ripening. Plant Signal. Behav. 6, 1950–1953.

Li, G., Yen, Y., 2008. Jasmonate and ethylene signaling pathway may mediate Fusarium head blight resistance in wheat. Crop. Sci. 48 (5), 1888–1896.

Liang, C., Wang, Y., Zhu, Y., Tang, J., Hu, B., Liu, L., et al., 2014. OsNAP connects abscisic acid and leaf senescence by fine-tuning abscisic acid biosynthesis and directly targeting senescence-associated genes in rice. Proc. Natl. Acad. Sci. U. S. A 111 (27), 10013–10018.

Lim, C.W., Luan, S., Lee, S.C., 2014. A prominent role for RCAR3-mediated ABA signaling in response to *Pseudomonas syringae pv.* tomato DC3000 infection in Arabidopsis. Plant Cell Physiol. 55, 1691–1703.

Lim, C.W., Baek, W., Jung, J., Kim, J.H., Lee, S.C., 2015. Function of ABA in stomatal defense against biotic and drought stresses. Int. J. Mol. Sci. 16 (7), 15251–15270.

Lim, P.O., Kim, H.J., Gil Nam, H., 2007. Leaf senescence. Annu. Rev. Plant Biol. 58, 115–136.

Liu, T., Song, T., Zhang, X., Yuan, H., Su, L., Li, W., et al., 2014. Unconventionally secreted effectors of two filamentous pathogens target plant salicylate biosynthesis. Nat. Commun. 5, 4686.

Lo Presti, L., Lanver, D., Schweizer, G., Tanaka, S., Liang, L., Tollot, M., et al., 2015. Fungal effectors and plant susceptibility. Annu. Rev. Plant Biol. 66, 513–545.

Loake, G., Grant, M., 2007. Salicylic acid in plant defence—the players and protagonists. Curr. Opin. Plant Biol. 10, 466–472.

Lu, S., Faris, J.D., Sherwood, R., Friesen, T.L., Edwards, M.C., 2014. A dimeric PR-1-type pathogenesis-related protein interacts with ToxA and potentially mediates ToxA-induced necrosis in sensitive wheat. Mol. Plant Pathol. 15, 650–663.

Ma, Y., Szostkiewicz, I., Korte, A., Moes, D., Yang, Y., Christmann, A., et al., 2009. Regulators of PP2C phosphatase activity function as abscisic acid sensors. Science. 324, 1064–1068.

Macho, A.P., Zipfel, C., 2014. Plant PRRs and the activation of innate immune signaling. Mol. Cell 54, 263–272.

Makandar, R., Nalam, V., Chaturvedi, R., Jeannotte, R., Sparks, A.A., Shah, J., 2010. Involvement of interaction between salicylic acid and jasmonate signaling pathways in Arabidopsis interaction with *Fusarium graminearum*. Mol. Plant Microbe Interact. 23, 861–870.

Mantelin, S., Bhattarai, K.K., Jhaveri, T.Z., Kaloshian, I., 2013. Mi-1-mediated resistance to *Meloidogyne incognita* in tomato may not rely on ethylene but hormone perception through ETR3 participates in limiting nematode infection in a susceptible host. PLoS One 8 (5), e63281.

Mantelin, S., Bhattarai, K.K., Kaloshian, I., 2009. Ethylene contributes to potato aphid susceptibility in a compatible tomato host. New Phytol. 183 (2), 444–456.

Mao, C., Lu, S., Lv, B., Zhang, B., Shen, J., He, J., et al., 2017. A rice NAC transcription factor promotes leaf senescence via ABA biosynthesis. Plant Physiol. 174 (3), 1747–1763.

Mase, K., Ishihama, N., Mori, H., Takahashi, H., Kaminaka, H., Kodama, M., et al., 2013. Ethylene-responsive AP2/ERF transcription factor MACD1 participates in phytotoxin-triggered programmed cell death. Mol. Plant Microbe Interact. 26, 868–879.

Mauch-Mani, B., Mauch, F., 2008. The role of abscisic acid in plant-pathogen interactions. Curr. Opin. Plant Biol. 8, 409–414.

McCormick, S., 2003. The role of DON in pathogenicity. In: Leonard, K.J., Bushnell, W.R. (Eds.), Fusarium Head Blight of Wheat and Barley. American Phytopathological Society, St. Paul, MN, USA, pp. 165–183.

Meher, H.C., Gajbhiye, V.T., Singh, G., 2011. Salicylic acid-induced glutathione status in tomato crop and resistance to root-knot nematode, *Meloidogyne incognita* (Kofoid & White) Chitwood. J. Xenobiot. 1 (1), 22–28.

Miller, J.D., 1995. Fungi and mycotoxins in grain: implications for stored product research. J. Stored Products Res. 31 (1), 1–16.

Misas-Villamil, J.C., van der Hoorn, R.A., Doehlemann, G., 2016. Papain-like cysteine proteases as hubs in plant immunity. New Phytol. 212, 902–907.

Molinari, S., Fanelli, E., Leonetti, P., 2014. Expression of tomato salicylic acid (SA)-responsive pathogenesis-related genes in Mi-1 mediated and SA- induced resistance to root-knot nematodes. Mol. Plant Pathol. 15 (3), 255–264.

Mostafanezhad, H., Sahebani, N., Zarghani, S.N., 2014. Control of rootknot nematode (*Meloidogyne javanica*) with combination of *Arthrobotrys oligospora* and salicylic acid and study of some plant defense responses. Biocontrol Sci. Technol. 24 (2), 203–215.

Murray, S.L., Ingle, R.A., Petersen, L.N., Denby, K.J., 2007. Basal resistance against *Pseudomonas syringae* in Arabidopsis involves WRKY53 and a protein with homology to a nematode resistance protein. Mol. Plant Microbe Interact. 20 (11), 1431–1438.

Nahar, K., Kyndt, T., Hause, B., Höfte, M., Gheysen, G., 2013. Brassinosteroids suppress rice defense against root-knot nematodes through antagonism with the jasmonate pathway. Mol. Plant Microbe Interact. 26 (1), 106–115.

Nahar, K., Kyndt, T., Nzogela, Y.B., Gheysen, G., 2012. Abscisic acid interacts antagonistically with classical defense pathways in rice migratory nematode interaction. New Phytol. 196 (3), 901–913.

Naseem, M., Kaltdorf, M., Dandekar, T., 2015. The nexus between growth and defence signalling: auxin and cytokinin modulate plant immune response pathways. J. Exp. Bot. 66, 4885–4896.

Netea, M.G., Quintin, J., van Der Meer, J.W., 2011. Trained immunity: a memory for innate host defense. Cell Host Microbe 9 (5), 355–361.

Nishimura, N., Hitomi, K., Arvai, A.S., Rambo, R.P., Hitomi, C., Cutler, S.R., et al., 2009. Structural mechanism of abscisic acid binding and signaling by dimeric PYR1. Science. 326, 1373–1379.

Niu, M., Huang, Y., Sun, S., Sun, J., Cao, H., Shabala, S., et al., 2018. Root respiratory burst oxidase homologue-dependent H2O2 production confers salt tolerance on a grafted cucumber by controlling Na + exclusion and stomatal closure. J. Exp. Bot. 69 (14), 3465–3476.

Nomura, K., Melotto, M., He, S.-Y., 2005. Suppression of host defense in compatible plant *Pseudomonas syringae* interactions. Curr. Opin. Plant Biol. 8 (4), 361–368.

Nováková, M., Šašek, V., Dobrev, P.I., Valentová, O., Burketová, L., 2014. Plant hormones in defense response of *Brassica napus* to *Sclerotinia sclerotiorum* reassessing the role of salicylic acid in the interaction with a necrotroph. Plant Physiol. Biochem 80, 308–317.

Panstruga, R., Parker, J.E., Schulze-Lefert, P., 2009. SnapShot: plant immune response pathways. Cell 136 (5), 978.

Pantelides, I.S., Tjamos, S.E., Pappa, S., Kargakis, M., Paplomatas, E.J., 2013. The ethylene receptor ETR1 is required for *Fusarium oxysporum* pathogenicity. Plant Pathol. 62 (6), 1302–1309.

Park, S.Y., Fung, P., Nishimura, N., Jensen, D.R., Fujii, H., Zhao, Y., et al., 2009. Abscisic acid inhibits type 2C protein phosphatases via the PYR/PYL family of START proteins. Science. 324, 1068–1071.

Park, S.W., Kaimoyo, E., Kumar, D., Mosher, S., Klessig, D.F., 2007. Methyl salicylate is a critical mobile signal for plant systemic acquired resistance. Science 318 (5847), 113–136.

Patkar, R.N., Benke, P.I., Qu, Z., Chen, Y.Y., Yang, F., Swarup, S., et al., 2015. A fungal monooxygenase-derived jasmonate attenuates host innate immunity. Nat. Chem. Biol. 11, 733–740.

Peng, X., Hu, Y., Tang, X., Zhou, P., Deng, X., Wang, H., et al., 2012. Constitutive expression of rice WRKY30 gene increases the endogenous jasmonic acid accumulation, PR gene expression and resistance to fungal pathogens in rice. Planta 236 (5), 1485–1498.

Penn, C.D., Daniel, S.L., 2013. Salicylate degradation by the fungal plant pathogen *Sclerotinia sclerotiorum*. Curr. Microbiol. 67 (2), 218–225.

Pieterse, C.M., Van der Does, D., Zamioudis, C., Leon-Reyes, A., Van Wees, S.C., 2012. Hormonal modulation of plant immunity. Annu. Rev. Cell Dev. Biol. 28, 489–521.

Pieterse, C.M., Leon-Reyes, A., Van der Ent, S., Van Wees, S.C., 2009. Networking by small-molecule hormones in plant immunity. Nat. Chem. Biol. 5 (5), 308–316.

Plett, J.M., Cvetkovska, M., Makenson, P., Xing, T., Regan, S., 2009. Arabidopsis ethylene receptors have different roles in Fumonisin B1-induced cell death. Physiol. Mol. Plant Pathol. 74 (1), 18–26.

Poor, P., Kovacs, J., Szopko, D., Tari, I., 2013. Ethylene signaling in salt stress- and salicylic acid-induced programmed cell death in tomato suspension cells. Protoplasma 250, 273–284.

Pozo, M.J., López-Ráez, J.A., Azcón-Aguilar, C., García-Garrido, J.M., 2015. Phytohormones as integrators of environmental signals in the regulation of mycorrhizal symbioses. New Phytol. 205, 1431–1436.

Qi, P.-F., Balcerzak, M., Rocheleau, H., Leung, W., Wei, Y.-M., Zheng, Y.-L., et al., 2016. Jasmonic acid and abscisic acid play important roles in host-pathogen interaction between *Fusarium graminearum* and wheat during the early infection stages of Fusarium head blight. Physiol. Mol. Plant Pathol. 93, 39–48.

Qi, P.F., Johnston, A., Balcerzak, M., Rocheleau, H., Harris, L.J., Long, X.Y., et al., 2012. Effect of salicylic acid on *Fusarium graminearum*, the major causal agent of fusarium head blight in wheat. Fungal Biol. 116 (3), 413–426.

Qin, J., Wang, K., Sun, L., Xing, H., Wang, S., Li, L., et al., 2018. The plant-specific transcription factors CBP60g and SARD1 are targeted by a Verticillium secretory protein VdSCP41 to modulate immunity. eLife 7, e34902.

Qin, X., Zhang, R.X., Ge, S., Zhou, T., Liang, Y.K., 2017. Sphingosine kinase AtSPHK1 functions in fumonisin B1-triggered cell death in Arabidopsis. Plant Physiol. Biochem. 119, 70–80.

Rahman, A., Kuldau, G.A., Uddin, W., 2014. Induction of salicylic acidmediated defense response in perennial ryegrass against infection by *Magnaporthe oryzae*. Phytopathology 104 (6), 614–623.

Rahman, T.A., Oirdi, M.E., Gonzalez-Lamothe, R., Bouarab, K., 2012. Necrotrophic pathogens use the salicylic acid signaling pathway to promote disease development in tomato. Mol. Plant Microbe Interact. 25 (12), 1584–1593.

Riemann, M., Haga, K., Shimizu, T., Okada, K., Ando, S., Mochizuki, S., et al., 2013. Identification of rice allene oxide cyclase mutants and the function of jasmonate for defence against *Magnaporthe oryzae*. Plant J. 74, 226–238.

Robert-Seilaniantz, A., Grant, M., Jones, J.D., 2011. Hormone crosstalk in plant disease and defense: more than just jasmonate-salicylate antagonism. Annu. Rev. Phytopathol. 49, 317–343.

Romero, P., Lafuente, M.T., Rodrigo, M.J., 2012. The citrus ABA signalosome: identification and transcriptional regulation during sweet orange fruit ripening and leaf dehydration. J. Exp. Bot. 63, 4931–4945.

Ross, A., Yamada, K., Hiruma, K., Yamashita-Yamada, M., Lu, X.L., Takano, Y., et al., 2014. The Arabidopsis PEPR pathway couples local and systemic plant immunity. EMBO J. 33 (1), 62–75.

Sanchez-Rangel, D., Rivas-San Vicente, M., de la Torre-Hernandez, M.E., Najera-Martinez, M., Plasencia, J., 2015. Deciphering the link between salicylic acid signaling and sphingolipid metabolism. Front. Plant Sci. 6, 125.

Sanchez-Vallet, A., Lopez, G., Ramos, B., Delgado-Cerezo, M., Riviere, M.P., Lorente, F., et al., 2012. Disruption of abscisic acid signaling constitutively activates Arabidopsis resistance to the necrotrophic fungus *Plectosphaerella cucumerina*. Plant Physiol. 160, 2109–2124.

Sanders, I.R., 2011. Mycorrhizal symbioses: how to be seen as a good fungus. Curr. Biol. 21, R550–R552.

Santiago, J., Dupeux, F., Round, A., Antoni, R., Park, S.Y., Jamin, M., et al., 2009a. The abscisic acid receptor PYR1 in complexwith abscisic acid. Nature. 462, 665–668.

Schaller, A., Stintzi, A., 2009. Enzymes in jasmonate biosynthesisstructure, function, regulation. Phytochemistry 70 (13), 1532—1538.

Seiler, C., Harshavardhan, V.T., Reddy, P.S., Hensel, G., Kumlehn, J., Eschen-Lippold, L., et al., 2014. Abscisic acid signaling responses and impact assimilation efficiency in barley under terinal drought stress. Plant Physiol. 164, 1677—1696.

Shen, X., Liu, H., Yuan, B., Li, X., Xu, C., Wang, S., 2011. OsEDR1 negatively regulates rice bacterial resistance via activation of ethylene biosynthesis. Plant Cell Environ. 34 (2), 179—191.

Shu, K., Chen, Q., Wu, Y., Liu, R., Zhang, H., Wang, S., et al., 2016. ABSCISIC ACID-INSENSITIVE 4 negatively regulates flowering through directly promoting Arabidopsis FLOWERING LOCUS C transcription. J. Exp. Bot. 67 (1), 195—205.

Spence, C.A., Lakshmanan, V., Donofrio, N., Bais, H.P., 2015. Crucial roles of abscisic acid biogenesis in virulence of rice blast fungus *Magnaporthe oryzae*. Front. Plant Sci. 6, 1082.

Spoel, S.H., Dong, X., 2008. Making sense of hormone crosstalk during plant immune responses. Cell Host Microbe 3, 348—351.

Spoel, S.H., Johnson, J.S., Dong, X., 2007. Regulation of tradeoffs between plant defenses against pathogens with different lifestyles. Proc. Natl. Acad. Sci. U. S. A. 104 (47), 18842—18847.

Stone, J.M., Heard, J.E., Asai, T., Ausubel, F.M., 2000. Simulation of fungal-mediated cell death by fumonisin B1 and selection of fumonisin B1—resistant (fbr) Arabidopsis mutants. Plant Cell 12, 1811—1822.

Stotz, H.U., Jikumaru, Y., Shimada, Y., Sasaki, E., Stingl, N., Mueller, M.J., et al., 2011. Jasmonate-dependent and COI1-independent defense responses against *Sclerotinia sclerotiorum* in *Arabidopsis thaliana*: auxin is part of COI1-independent defense signaling. Plant Cell Physiol. 52 (11), 1941—1956.

Sun, H., Hu, X., Ma, J., Hettenhausen, C., Wang, L., Sun, G., et al., 2014. Requirement of ABA signalling-mediated stomatal closure for resistance of wild tobacco to *Alternaria alternata*. Plant Pathol. 63 (5), 1070—1077.

Sun, T., Busta, L., Zhang, Q., Ding, P., Jetter, R., Zhang, Y., 2018. TGACG-binding factor 1 (TGA1) and TGA4 regulate salicylic acid and pipecolic acid biosynthesis by modulating the expression of systemic acquired resistance deficient 1 (SARD1) and calmodulin-binding protein 60g (CBP60g). New Phytol. 217, 344—354.

Sun, T., Zhang, Y., Li, Y., Zhang, Q., Ding, Y., Zhang, Y., 2015. ChIP-seq reveals broad roles of SARD1 and CBP60g in regulating plant immunity. Nat. Commun. 6, 10159.

Sun, Y.C., Cao, H.F., Yin, J., Kang, L., Ge, F., 2010. Elevated CO_2 changes the interactions between nematode and tomato genotypes differing in the JA pathway. Plant Cell Environ. 33 (5), 729—739.

Sun, Y.C., Yin, J., Cao, H.F., Li, C.Y., Kang, L., Ge, F., 2011. Elevated CO_2 influences nematode-induced defense responses of tomato genotypes differing in the JA pathway. PLoS One 6 (5), e19751.

Taheri, P., Tarighi, S., 2010. Riboflavin induces resistance in rice against *Rhizoctonia solani* via jasmonate-mediated priming of phenylpropanoid pathway. J. Plant Physiol. 167 (3), 201—208.

Takatsuji, H., Jiang, C.J., 2014. Plant Hormone Crosstalks under Biotic Stresses. Springer, Dordrecht, pp. 323—350.

Thatcher, L.F., Gardiner, D.M., Kazan, K., Manners, J.M., 2012. A highly conserved effector in *Fusarium oxysporum* is required for full virulence on Arabidopsis. Mol. Plant Microbe Interact. 25, 180—190.

Thatcher, L.F., Manners, J.M., Kazan, K., 2009. *Fusarium oxysporum* hijacks COI1-mediated jasmonate signaling to promote disease development in Arabidopsis. Plant J. 58, 927—939.

Tian, X., Wang, Z., Li, X., Lv, T., Liu, H., Wang, L., et al., 2015. Characterization and functional analysis of pyrabactin resistance-like abscisic acid receptor family in rice. Rice 8, 28.

Torres-Vera, R., García, J.M., Pozo, M.J., López-Ráez, J.A., 2014. Do strigolactones contribute to plant defence? Mol. Plant Pathol. 15 (2), 211—216.

Trobacher, C.P., 2009. Ethylene and programmed cell death in plants. Botany 87, 757—769.

Truman, W., de Zabala, M.T., Grant, M., 2006. Type III effectors orchestrate a complex interplay between transcriptional networks to modify basal defence responses during pathogenesis and resistance. Plant J. 46 (1), 14—33.

Trusov, Y., Sewelam, N., Rookes, J.E., Kunkel, M., Nowak, E., Schenk, P.M., et al., 2009. Heterotrimeric G-proteins-mediated resistance to necrotrophic pathogens includes mechanisms independent of salicylic acid-jasmonic acid/ethylene- and abscisic acid-mediated defense signaling. Plant J. 58, 69—81.

Ulferts, S., Delventhal, R., Splivallo, R., Karlovsky, P., Schaffrath, U., 2015. Abscisic acid negatively interferes withbasal defense of barley against *Magnaporthe oryzae*. BMC Plant Biol.

Valls, M., Genin, S., Boucher, C., 2006. Integrated regulation of the type III secretion system and other virulence determinants in *Ralstonia solanacearum*. PLoS Pathog. 2 (8), 798–807.

Van Bockhaven, J., Spichal, L., Novak, O., Strnad, M., Asano, T., Kikuchi, S., et al., 2015. Silicon induces resistance to the brown spot fungus *Cochliobolus miyabeanus* by preventing the pathogen from hijacking the rice ethylene pathway. New Phytol. 206, 761–773.

van der Linde, K., Hemetsberger, C., Kastner, C., Kaschani, F., van der Hoorn, R.A., Kumlehn, J., et al., 2012. A maize cystatin suppresses host immunity by inhibiting apoplastic cysteine proteases. Plant Cell 24, 1285–1300.

Verberne, M.C., Hoekstra, J., Bol, J.F., Linthorst, H.J., 2003. Signaling of systemic acquired resistance in tobacco depends on ethylene perception. Plant J. 35 (1), 27–32.

Verberne, M.C., Verpoorte, R., Bol, J.F., Mercado-Blanco, J., Linthorst, H.J., 2000. Overproduction of salicylic acid in plants by bacterial transgenes enhances pathogen resistance. Nat. Biotechnol. 18 (7), 779–783.

Verma, V., Ravindran, P., Kumar, P.P., 2016. Plant hormone-mediated regulation of stress responses. BMC Plant Biol. 16 (1), 1–10.

Vleesschauwer, D., Yang, Y., Cruz, C., Hofte, M., 2010. Abscisic acid-induced resistance against the brown spot pathogen *Cochliobolus miyabeanusin* rice involves MAP kinase-mediated repression of ethylene signaling. Plant Physiol. 152, 2036–2052.

Vlot, A.C., Dempsey, D.M.A., Klessig, D.F., 2009. Salicylic acid, a multifaceted hormone to combat disease. Annu. Rev. Phytopathol. 47, 177–206.

Wang, D., Amornsiripanitch, N., Dong, X., 2006. A genomic approach to identify regulatory nodes in the transcriptional network of systemic acquired resistance in plants. PLoS Pathog. 2, e123.

Wang, H., Jiang, Y.P., Yu, H.J., Xia, X.J., Shi, K., Zhou, Y.H., et al., 2010. Light quality affects incidence of powdery mildew, expression of defence-related genes and associated metabolism in cucumber plants. Eur. J. Plant Pathol. 127 (1), 125–135.

Wang, K.L.-C., Li, H., Ecker, J.R., 2002. Ethylene biosynthesis and signaling networks. Plant Cell 14 (Suppl), S131–S151.

Wang, L., Tsuda, K., Truman, W., Sato, M., Nguyen le, V., Katagiri, F., et al., 2011. CBP60g and SARD1 play partially redundant critical roles in salicylic acid signaling. Plant J. 67, 1029–1041.

Wasilewska, A., Vlad, F., Sirichandra, C., Redko, Y., Jammes, F., Valon, C., et al., 2008. An update on abscisic acid signaling in plants and more. Mol. Plant 1, 198–217.

Wi, S.J., Ji, N.R., Park, K.Y., 2012. Synergistic biosynthesis of biphasic ethylene and reactive oxygen species in response to hemibiotrophic *Phytophthora parasitica* in tobacco plants. Plant Physiol. 159 (1), 251–265.

Wildermuth, M.C., Dewdney, J., Wu, G., Ausubel, F.M., 2001. Isochorismate synthase is required to synthesize salicylic acid for plant defence. Nature 414 (6863), 562–565.

Woo, H.R., Kim, H.J., Lim, P.O., Nam, H.G., 2019. Leaf senescence: systems and dynamics aspects. Annu. Rev. Plant Biol. 70, 347–376.

Wu, J.X., Wu, J.L., Yin, J., Zheng, P., Yao, N., 2015. Ethylene modulates sphingolipid synthesis in Arabidopsis. Front. Plant Sci. 6, 1122.

Xiao, X., Cheng, X., Yin, K., Li, H., Qiu, J.L., 2017. Abscisic acid negatively regulates post-penetration resistance of Arabidopsis to the biotrophic powdery mildew fungus. Sci. China Life Sci 60, 891–901.

Xin, X.F., He, S.Y., 2013. *Pseudomonas syringae* pv. tomato DC3000: amodel pathogen for probing disease susceptibility and hormone signaling in plants. Annu. Rev. Phytopathol. 51, 473–498.

Xing, F., Li, Z., Sun, A., Xing, D., 2013. Reactive oxygen species promote chloroplast dysfunction and salicylic acid accumulation in fumonisin B1-induced cell death. FEBS Lett. 587, 2164–2172.

Xu, J., Audenaert, K., Hofte, M., De Vleesschauwer, D., 2013. Abscisic acid promotes susceptibility to the rice leaf blight pathogen *Xanthomonas oryzae* pv oryzae by suppressing salicylic acid-mediated defenses. PLoS One 8 (6), e67413.

Xue, R.F., Wu, J., Wang, L.F., Blair, M.W., Wang, X.M., De Ge, W., et al., 2013. Salicylic acid enhances resistance to *Fusarium oxysporum* f. sp. phaseoli in Common Beans (*Phaseolus vulgaris* L.). J. Plant Growth Regul. 33, 470–476.

Yamada, S., Kano, A., Tamaoki, D., Miyamoto, A., Shishido, H., Miyoshi, S., et al., 2012. Involvement of OsJAZ8 in jasmonate-induced resistance to bacterial blight in rice. Plant Cell Physiol. 53 (12), 2060–2072.

Yang, B., Wang, Q., Jing, M., Guo, B., Wu, J., Wang, H., et al., 2017b. Distinct regions of the Phytophthora essential effector Avh238 determine its function in cell death activation and plant immunity suppression. New Phytol. 214, 361–375.

Yang, C., Li, W., Cao, J., Meng, F., Yu, Y., Huang, J., et al., 2017a. Activation of ethylene signaling pathways enhances disease resistance by regulating ROS and phytoalexin production in rice. Plant J. 89, 338–353.

Yang, G., Tang, L., Gong, Y., Xie, J., Fu, Y., Jiang, D., et al., 2018. A cerato-platanin protein SsCP1 targets plant PR1 and contributes to virulence of *Sclerotinia sclerotiorum*. New Phytol. 217, 739–755.

Yang, J., Worley, E., Udvardi, M., 2014. A NAP-AAO3 regulatory module promotes chlorophyll degradation via ABA biosynthesis in Arabidopsis leaves. Plant Cell 26 (12), 4862–4874.

Yang, Y.X., Ahammed, G., Wu, C., Fan, S.Y., Zhou, Y.H., 2015. Crosstalk among jasmonate, salicylate and ethylene signaling pathways in plant disease and immune responses. Curr. Protein Peptide Sci. 16 (5), 450–461.

Yasuda, M., Ishikawa, A., Jikumaru, Y., Seki, M., Umezawa, T., Asami, T., et al., 2008. Antagonistic interaction between systemic acquired resistance and the abscisic acid-mediated abiotic stress response in Arabidopsis. Plant Cell 20 (6), 1678–1692.

Yu, S.G., Kim, J.H., Cho, N.H., Oh, T.R., Kim, W.T., 2020. Arabidopsis RING E3 ubiquitin ligase JUL1 participates in ABA-mediated microtubule depolymerization, stomatal closure, and tolerance response to drought stress. Plant J. 103 (2), 824–842.

Yu, Z., Zhang, D., Xu, Y., Jin, S., Zhang, L., Zhang, S., et al., 2019. CEPR2 phosphorylates and accelerates the degradation of PYR/PYLs in Arabidopsis. J. Exp. Bot. 70 (19), 5457–5469.

Zhang, H., Jin, X., Huang, L., Hong, Y., Zhang, Y., Ouyang, Z., et al., 2014. Molecular characterization of rice sphingosine-1-phosphate lyase gene OsSPL1 and functional analysis of its role in disease resistance response. Plant Cell Rep. 33, 1745–1756.

Zhang, X., Wu, Q., Cui, S., Ren, J., Qian, W., Yang, Y., et al., 2015. Hijacking of the jasmonate pathway by the mycotoxin fumonisin B1 (FB1) to initiate programmed cell death in Arabidopsis is modulated by RGLG3 and RGLG4. J. Exp. Bot. 66, 2709–2721.

Zhang, Y., Li, X., 2019. Salicylic acid: biosynthesis, perception, and contributions to plant immunity. Curr. Opin. Plant Biol. 50, 29–36.

Zhao, Y., Chan, Z., Gao, J., Xing, L., Cao, M., Yu, C., et al., 2016. ABA receptor PYL9 promotes drought resistance and leaf senescence. Proc. Natl Acad. Sci. U. S. A. 113 (7), 1949–1954.

Zhu, Z., An, F., Feng, Y., Li, P., Xue, L., Mu, A., et al., 2011. Derepression of ethylene-stabilized transcription factors (EIN3/EIL1) mediates jasmonate and ethylene signaling synergy in Arabidopsis. Proc. Natl. Acad. Sci. U. S. A. 108, 12539–12544.

Ziemann, S., van der Linde, K., Lahrmann, U., Acar, B., Kaschani, F., Colby, T., et al., 2018. An apoplastic peptide activates salicylic acid signalling in maize. Nat. Plants 4, 172–180.

Index

CPI Antony Rowe
Eastbourne, UK
April 22, 2024